NUMERICAL METHODS IN ENGINEERING

Theories with MATLAB, Fortran, C and Pascal Programs

NUMERICAL METHODS IN ENGINEERING

Theories with MATLAB, Fortran, C and Pascal Programs

P. Dechaumphai
N. Wansophark

Alpha Science International Ltd.
Oxford, U.K.

Numerical Methods in Engineering
Theories with MATLAB, Fortran, C and Pascal Programs
502 pgs. | 214 figs. | 63 tbls.

P. Dechaumphai
N. Wansophark
Mechanical Engineering Department
Chulalongkorn University
Payathai Road, Pathumwan
Bangkok 10330, Thailand

ALPHA SCIENCE INTERNATIONAL LTD.
7200 The Quorum, Oxford Business Park North
Garsington Road, Oxford OX4 2JZ, U.K.

www.alphasci.com

Printed from the camera-ready copy provided by the Authors.

ISBN 978-1-84265-649-5

Printed in India

Preface

The book, *Numerical Methods in Engineering: Theories with MATLAB, Fortran, C and Pascal Programs*, is written in a clear, easy-to-understand manner on theories and the use of the numerical methods. Topics and materials of the methods in this book were taught at George Washington University, NASA Langley Research Center campus while the first author was a NASA aerospace engineer. Such materials were also taught at Old Dominion University, Norfolk, Virginia, and have been currently taught at Chulalongkorn University. By teaching and performing research on the numerical methods for the past 30 years, the materials in this book have been improved and updated continuously. The main objective of this book is to present the numerical methods in their simplest form so that engineers and scientists can understand them easily and quickly.

The book contains 9 chapters which are essential in the study of the numerical methods. The materials in these chapters are suitable to be used in both the undergraduate and graduate levels. The first chapter introduces the methods and the need to study them for solving practical engineering problems today. The chapter also explains different types of numerical errors and the use of hardware and software. The second chapter explains several methods for finding roots from a single nonlinear equation. The methods are extended to find roots from a set of nonlinear equations. Popular methods for finding roots by solving a set of linear simultaneous equations are presented in chapter 4. These methods are classified into two groups of the direct and iterative techniques. Their detailed computational procedures including advantages and disadvantages are presented. Interpolation and extrapolation methods for finding an appropriate function to represent a set of data are presented in chapter 4. Chapter 5 explains several least-squares regression methods to provide a function that best fit a set of data. Many types of functions to best fit sets of linear and nonlinear data are presented. Numerical integration and differentiation methods are explained in chapter 7. Basic and popular integration methods that are employed in commercial software for analyzing practical engineering problems are explained. Chapter 7 presents several methods for solving the ordinary differential equations. The methods can be used to analyze the first- and higher-order ordinary differential equations. Methods for solving partial differential equations are presented in chapter 8. The finite difference methods for analyzing the elliptic, parabolic and hyperbolic differential equations are explained in details. These methods are simple and suitable for problems with regular geometry. For problems that have complex geometry, the finite element method is preferred. The finite element method is introduced in chapter 9. The chapter explains details of the method for solving one- and two-dimensional problems. For all the methods presented in these chapters, listings of the corresponding computer programs are provided. These computer programs are written in Matlab, Fortran, C and Pascal so that readers can select the preferred computer language. The programs can be modified to solve other types of problems that may arise in other courses and research work.

The first author would like to thank his former Professor, Dr. Earl A. Thornton, and his supervisor, Dr. Allan R. Wieting of the Aerothermal Loads Branch at NASA Langley Research Center. He expresses his appreciation to the students at NASA Langley Research Center, Old Dominion University and Chulalongkorn University who took his courses on the numerical methods and helped him to improve the presentation of materials in this book. The authors wish to thank Mr. Sascha J. Mehra, the Director and the staff of Alpha Science International Ltd. for their advice and cooperation. Finally, the authors would like to thank their wives Mrs. Yupa Dechaumphai and Patcharin Wansophark for the understanding and support in writing this book.

P. Dechaumphai

N.Wansophark

Contents

Chapter

1

First Step to Numerical Methods

1.1 Introduction

Solving problems in sciences and engineering today requires knowledge in numerical methods. The methods are based on the use of mathematics, computational procedures, computer software and hardware for analyzing practical problems that normally have complex geometry. For examples, the finite volume method may be used to determine the flow behavior surrounding a moving vehicle. The computed pressure can be used in the modification of the vehicle body in order to reduce the drag force. The finite element method may be used to analyze the strength of the vehicle body structure to reduce its damage during a collision. The method can be also applied to analyze the temperature and associated thermal stress that occur on the automobile engine during running. Understanding such phenomena from the solutions by using the numerical methods helps designers to significantly reduce the time and cost for designing new products.

Designing a vehicle body structure with maximum strength or developing a new engine with reduced thermal stress was not possible in the past by using classical mathematics for exact solutions. The exact solutions can not be obtained because both of the geometries and boundary conditions of these problems are normally complex. The numerical methods, such as the finite element, finite volume and finite difference methods, are being used to obtain approximate solutions. These methods play a very important role in engineering analysis and design today. Figure 1.1 shows a finite difference mesh for an analysis of flow field surrounding a fighter jet. For problems that have complex geometries, the finite element method is often applied to obtain solutions. Figure 1.2 shows a finite element mesh of a vehicle body structure during its collision.

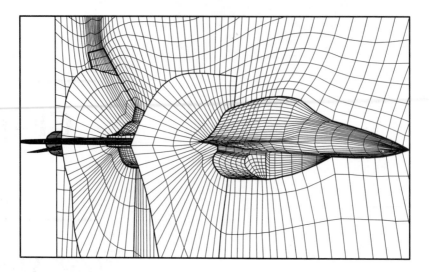

Figure 1.1 A finite difference mesh for an analysis of flow field surrounding a fighter jet.

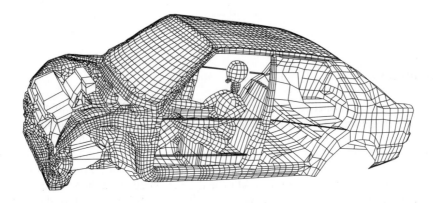

Figure 1.2 A finite element mesh of a vehicle body structure during collision.

These numerical methods significantly help designers and analysts to understand the problem behaviors in order to improve their designs. However, designers and analysts must understand the numerical methods prior to using them. These numerical methods are not difficult to understand and currently being taught in most of the sciences and engineering schools.

1.2 What are the Numerical Methods?

The numerical methods are techniques for obtaining approximate solutions. The methods consist of computational procedures that can be followed in the form of patterns. Generally, the procedures consist of arithmetic operations. These operations are simply the addition, subtraction, multiplication and division. A computational procedure can be performed by using a calculator for a small problem. The same procedure needs help from a computer program to provide solutions for a larger problem. The numerical methods are thus similar to the recipes in a cookbook. Users can follow the recipes or procedures to generate solutions. At present, many commercial software have been developed and used widely. These software contain numerical methods with some specific procedures to provide approximate solutions. Thus, it is very important for users to understand the computational procedures that are built inside the software prior to using them.

With the explanation above, it is clear that the concepts of the numerical methods are not new. Many numerical methods have been developed long time ago. They have not been used effectively, however, due to the lack of digital computers. At present, high efficient personal computers are available at low cost. The methods are thus popular and widely used by students and designers for obtaining solutions that are not possible in the past by using classical mathematics.

However, there are many practical applications today that can not be solved effectively by the numerical methods and current computers. For an example, analysis of flow field surrounding an aerospace vehicle that travels many times faster than the sound speed. The flow field is very complex and requires a computational mesh with a large number of grid points. Such problem is still a challenging problem today because it needs a very large computer memory as well as a substantial computational time.

1.3 Need for Studying Numerical Methods

In this section, an example is presented to demonstrate advantages of using a numerical method as compared to the classical mathematics method for solving a typical problem. In addition to the presentation of the solution procedure, the example highlights the importance in understanding the physical meaning of the problem. Understanding physical meanings before solving a problem is important because a proper numerical method can be selected to provide a solution with high accuracy. Understanding physical meaning of a problem is also very helpful, especially when analyzing more complex problems. Such understanding is required mainly because:

(a) there is no single numerical method that can provide solution to every problem,

(b) error of the solution always occurs from a numerical method, and

(c) there is no single numerical method that is the best for all problems.

With the above statements, it is very important to understand the computational procedures of the numerical methods clearly. Understand the computational procedure can lead to a more accurate solution. In addition, it also provides confidence for the user on the obtained solution, especially when solving a complex problem.

In order to demonstrate that the numerical methods are not difficult to understand as compared to the classical mathematics, a following example is studied.

Example 1.1   Apply the Newton's second law to develop a governing differential equation for approximately determining the space shuttle velocity during its descending as shown in Fig. 1.3. Derive the exact solution and develop a numerical procedure for obtaining an approximate solution of the velocity. Compare the approximate velocity solution with the exact solution.

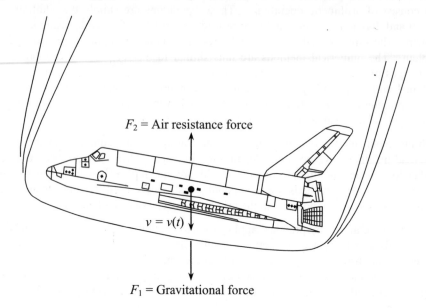

F_2 = Air resistance force

$v = v(t)$

F_1 = Gravitational force

Figure 1.3 Forces on space shuttle during descending.

Solution From the Newton's second law,

$$F = ma \qquad (1.1)$$

Where F represents the net force which is

$$F = F_1 - F_2 \qquad (1.2)$$

In Eq. (1.2), F_1 is the gravitational force defined by

$$F_1 = mg \qquad (1.3)$$

where m is the mass of the shuttle and g is the gravitational acceleration constant. If the air resistance force F_2 is assumed to vary linearly with the velocity v, then

$$F_2 = cv \qquad (1.4)$$

where c represents the drag coefficient which depends on the shuttle geometry. By substituting Eqs. (1.3) and (1.4) into Eq. (1.2), Eq. (1.1) becomes

$$mg - cv = ma$$

Because acceleration is the rate of change of the velocity, then

$$mg - cv = m\frac{dv}{dt}$$

The above equation is a linear ordinary differential equation that can be written as,

$$\frac{dv}{dt} + \frac{c}{m}v = g \tag{1.5}$$

where the unknown is the velocity v which depends on time t.

The velocity v can be determined by solving the differential Eq. (1.5) using

(a) a mathematical method for an exact solution, or

(b) a numerical method for an approximate solution.

The exact solution can be derived by using the method of separation of variables. Equation (1.5) is rewritten as

$$\frac{dv}{dt} = g - \frac{c}{m}v$$

i.e., separate the independent variable t and the dependent variable v so that they are on opposite side of the equation. Integration is then performed on both sides

$$\int \frac{dv}{g - \frac{c}{m}v} = \int dt \tag{1.6}$$

to yield

$$-\frac{m}{c}\ln\left(g - \frac{c}{m}v\right) = t + A \tag{1.7}$$

where A is the integrating constant that can be determined from the given initial condition. For example, if the velocity $v = 0$ at time $t = 0$, then Eq. (1.7) gives

$$A = -\frac{m}{c}\ln g \tag{1.8}$$

By substituting A from Eq. (1.8) into Eq. (1.7),

$$-\frac{m}{c}\ln\left(g - \frac{c}{m}v\right) = t - \frac{m}{c}\ln g$$

$$\frac{m}{c}\ln\left(g - \frac{c}{m}v\right) - \frac{m}{c}\ln g = -t$$

$$\ln\left(1 - \frac{c}{mg}v\right) = -\frac{c}{m}t$$

$$1 - \frac{c}{mg}v = e^{-\frac{c}{m}t}$$

$$\frac{c}{mg}v = 1 - e^{-\frac{c}{m}t}$$

$$v = \frac{mg}{c}\left(1 - e^{-\frac{c}{m}t}\right) \tag{1.9}$$

The exact velocity solution as shown in Eq. (1.9) indicates that the velocity is zero at time $t = 0$ as given by the initial condition. The velocity then increases with time and reaches a constant value as the time approaches infinity,

$$v(t \to \infty) \;=\; \frac{mg}{c} \qquad\qquad (1.10)$$

At such condition, the air resistance force F_2 and the gravitational force F_1 are equal.

By assigning the following data:

mass of space shuttle m =   90,000   kg

drag coefficient c =   450   kg/sec   (1.11)

gravitational acceleration constant g =   9.8   m/sec^2

then, the expression for the shuttle velocity in Eq. (1.9) is,

$$v(t) \;=\; 1,960\left(1 - e^{-0.005t}\right) \qquad\qquad (1.12)$$

The values of the velocity at every 30 seconds are shown in Table 1.1.

Table 1.1 Exact shuttle velocities at every 30 seconds according to Eq. (1.12).

Time t, sec	Velocity v, m/sec
0	0
30	273
60	508
90	710
120	884
150	1,034
180	1,163
⋮	⋮
∞	1,960

In stead of finding the exact solution from the differential equation, an approximate solution can be derived. The rate of change of the velocity dv/dt in Eq. (1.5) can be approximated by considering the plot of the velocity versus time in Fig. 1.4. The rate of change of the velocity dv/dt, which is the slope of the velocity v with respect to time t at point A, may be approximated by

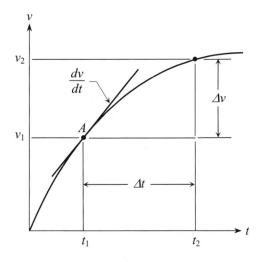

Figure 1.4 Plot of the velocity with time.

$$\frac{\Delta v}{\Delta t} = \frac{v_2 - v_1}{t_2 - t_1}$$

(1.13)

Such approximation $\Delta v/\Delta t$ becomes dv/dt as $\Delta t \to 0$. This means the approximation of Eq. (1.13) is accurate if the time step Δt is small. Using such approximation, Eq. (1.5) may be written as

$$\frac{v_2 - v_1}{\Delta t} + \frac{c}{m}v_1 = g$$

The above equation can be written in a more general form for $i = 1, 2, 3, \ldots$ as

$$\frac{v_{i+1} - v_i}{\Delta t} + \frac{c}{m}v_i = g$$

(1.14)

Or,

$$v_{i+1} = v_i + \Delta t \left(g - \frac{c}{m}v_i \right)$$

(1.15)

With the values of the shuttle mass m, the drag coefficient c and the gravitational acceleration constant g as shown in Eq. (1.11), Eq. (1.15) becomes,

$$v_{i+1} = v_i + \Delta t \left(9.8 - 0.005v_i \right)$$

(1.16)

Equation (1.16) suggests that if the time step Δt and the velocity at step i are known, the velocity at step $i+1$ can be determined directly. By using the time step $\Delta t = 30$ s and the initial velocity of zero, table 1.2 shows the approximate solution of Eq. (1.16) as compared to the exact solution of Eq. (1.12).

Table 1.2 Comparative exact and approximate velocity solutions by using the time step $\Delta t = 30$ seconds.

i	$i+1$	t, sec	*v*, m/sec Approximate solution Eq. (1.16)	Exact solution Eq. (1.12)
0	1	30	294	273
1	2	60	544	508
2	3	90	756	710
3	4	120	937	884
4	5	150	1,090	1,034
5	6	180	1,221	1,163
⋮	⋮	⋮	⋮	⋮
49	50	1,500	1,959	1,959

Both of the exact and approximate solutions are compared by the plot as shown in Fig. 1.5. With the time step $\Delta t = 30$ s, the approximate solution is in good agreement with the exact solution. From the derivation of the exact solution, the computational procedure for generating the approximate solution and the comparison of both the solutions in Fig. 1.5, the following details are observed:

(a) the approximate solution in Eq. (1.15) obtained from the numerical method can be derived easily as compared to the derivation of the exact solution,

(b) the approximate solution in Eq. (1.15) can be computed easily by developing a short computer program,

(c) the computer program can be executed by using a small time step to produce a more accurate solution,

(d) If the air resistance force F_2 does not vary linearly with the velocity, e.g., if it varies with the velocity in the form,

$$F_2 = cv^4 \tag{1.17}$$

then, the governing differential equation becomes nonlinear, i.e.,

$$\frac{dv}{dt} + \frac{c}{m}v^4 = g \tag{1.18}$$

In this case, the exact solution is difficult to find. However, the approximate solution can be determined by using the same procedure as explained earlier.

From the above reasons, the numerical methods are popular and being used by scientists and engineers for solving problems. Practical problems can be analyzed effectively if users have some backgrounds on both the computer hardware and software as explained in the following section.

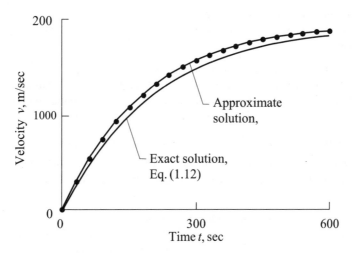

Figure 1.5 Comparison between the exact and approximate solutions for
the shuttle velocity with respect to time.

1.4 Computer Hardware and Software

Studying a numerical method requires two basic ingredients: (1) the understanding of the theory and computational procedure of the method, and (2) the experience in using computers on both the hardware and software. The following chapters on this book explain theories of the numerical methods and their computational procedures. The computational procedures are translated into computer programs in different languages of Fortran, MATLAB, Pascal and C. These computer programs can be executed on different types of computers. Students are encouraged to study the theories and their computational procedures prior to using the computer programs.

During the past decades, computer hardware has been improved significantly. A large size problem can be analyzed quickly and conveniently on a personal computer today. In the past, such a problem must be solved on a mainframe computer or a workstation. A mainframe computer was operated in an air-conditioned room with controlled temperature. The computer also required an operator for maintenance and service during its operation. At present, personal computers or notebooks are widely used to analyze many practical problems. A large size problem may be solved by connecting the personal computers together in parallel. For a very large size problem, a supercomputer is normally used. These supercomputers are expensive and employed only by some big companies or government agencies.

It should be kept in mind that, like a calculator, these computers are used to perform basic operations of addition, subtraction, multiplication and division. But they can perform such operations at a very high efficiency, especially if they were instructed by computer software. A software for performing computational procedure of a numerical method may be written by using different computer languages. Popular languages are such as Fortran, Pascal, C, Java and Basic. Nowadays, many software packages that contain different numerical methods have been developed and used widely. Examples of these software packages are MATLAB, Mathematica and Maple. Because these software packages are handy and easy to use, they are accepted by students to solve small size problems or employ for their projects. For the large size problems that occur in engineering applications, special software are employed. These special software, sometimes known as Computer-Aided Engineering (CAE) software, are very expensive. The software are being used for analysis and design of new products in automotive, electronics, medical and aerospace industry. They were developed by experience programmers who understand both the theories and computational procedures very well.

As a student who needs to learn the numerical methods, a simple computer language should be selected. The most important aspect is that the methods and their computational procedures must be understood clearly prior to developing any computer program. Such understanding will provide basis for solving more complex problems that occur in practical applications.

Few key ingredients are needed during developing a computer program:

(a) *Understanding computer commands.* There are few computer commands that are frequently used during developing or running a computer program. These commands are such as copying, deleting, editing, reading files, etc.

(b) *Knowing how to edit a file.* Once a computer program is created as a file, editing and correcting it are needed before it can be used. File editing is quite simple today. Many computer systems allow users to edit the files conveniently on the monitor screen.

(c) *Understanding computer languages.* Understanding a computer language clearly is a must for developing a good computer program. The developed program should be simple and provides high computational efficiency at the same time.

To help readers on computer programming, this book contains a number of computer programs that correspond to the computational procedures of the presented numerical methods. The programs are written in Fortran, MATLAB, Pascal and C languages. An example of a computer program is shown in the example below.

Example 1.2 The approximate solution of the space shuttle velocity as shown in Eq. (1.16) and Table 1.2 can be obtained by using the computer programs written in Fortran and MATLAB in Fig. 1.6.

Fortran

```
PROGRAM  SHUTTLE
T  = 0.
DT = 30.
V  = 0.
DO 10  I=1,50
V = V + DT*(9.8 - .005*V)
T = T + DT
WRITE(6,100)  T, V
100 FORMAT(2F12.0)
10 CONTINUE
STOP
END
```

MATLAB

```
% Program Shuttle
T  = 0.;
DT = 30.;
V  = 0.;
for I = 1:50
V = V + DT*(9.8 - 0.005*V);
T = T + DT;
fprintf('T = %8.2f   ', T);
fprintf('V = %8.2f\n  ', V);
end
```

Figure 1.6 Computer programs for determining the shuttle velocity according to Eq. (1.16).

Example 1.3 The sine and cosine functions can be written in the form of infinite series as

$$\sin x = x - \frac{x^3}{3!} + \frac{x^5}{5!} - \frac{x^7}{7!} + \cdots \qquad (1.19)$$

$$\cos x = 1 - \frac{x^2}{2!} + \frac{x^4}{4!} - \frac{x^6}{6!} + \cdots \qquad (1.20)$$

where *x* is the angle in radians. Develop a computer program to compute values of both functions from 0 to 180 degrees with an increment of every 10 degrees.

The computer programs in Fortran and MATLAB for determing both sine and cosine functions are shown in Fig. 1.7. Solutions of the sine and cosine functions obtained from these programs are presented in Table 1.3.

Fortran

```
        PROGRAM  SINCOS
C.....PROGRAM FOR COMPUTING SIN AND COSINE
C.....FUNCTIONS FROM 0 TO 180 DEGREES
C.....WITH INCREMENT AT EVERY 10 DEGREES
        PI = 4.*ATAN(1.)
        DEG = 0.
        DEL = 10.
        WRITE(6,100)
100 FORMAT(/, 5X, 'DEGREES',
     *        10X, 'SIN', 12X, 'COS', /)
        DO 10  IDEG=1,19
        X = PI*DEG/180.
        SUMS   = X
        SUMC   = 1.
        TERMS  = X
        TERMC  = 1.
        SIGN  =-1.
        DO 20  N=1,100
        MS = 2*N + 1
        MC = 2*N
        TERMS = TERMS*X*X/(MS*(MS-1))
        TERMC = TERMC*X*X/(MC*(MC-1))
        SUMS   = SUMS + SIGN*TERMS
        SUMC   = SUMC + SIGN*TERMC
        SIGN   = -SIGN
20  CONTINUE
        WRITE(6,200)  DEG, SUMS, SUMC
200 FORMAT(F10.0, 2F16.6)
        DEG = DEG + DEL
10  CONTINUE
        STOP
        END
```

MATLAB

```
%  Program sin & cos
%  Program for computing sin and cosine
%  function for angles from 0 to 180 degrees
%  with increment at every 10 degrees
deg = 0.;
del = 10.;
fprintf ('   Degrees         SIN           COS \n');
for ideg = 1:19
    x     = pi*deg/180.;
    sums  = x;
    sumc  = 1.;
    terms = x;
    termc = 1.;
    sign  = -1.;
    for n = 1:100
        ms    = 2*n + 1;
        mc    = 2*n;
        terms = terms*x*x/(ms*(ms-1));
        termc = termc*x*x/(mc*(mc-1));
        sums  = sums + sign*terms;
        sumc  = sumc + sign*termc;
        sign  = -sign;
    end
    fprintf('%10.0f %16.6f %16.6f\n', deg, sums, sumc);
    deg = deg + del;
end
```

Figure 1.7 Computer programs for determining sine and cosine functions.

Table 1.3 Values of sine and cosine functions computed from the infinite series by using the computer programs in Fig. 1.7.

DEGREES	SIN	COS
0.	.000000	1.000000
10.	.173648	.984808
20.	.342020	.939693
30.	.500000	.866025
40.	.642788	.766044
50.	.766044	.642788
60.	.866025	.500000
70.	.939693	.342020
80.	.984808	.173648
90.	.000000	.000000
100.	.984808	−.173648
110.	.939693	−.342020
120.	.866025	−.500000
130.	.766044	−.642788
140.	.642788	−.766044
150.	.500000	−.866025
160.	.342020	−.939693
170.	.173648	−.984808
180.	.000000	−1.000000

1.5 Errors

Example 1.1 shows a computational procedure for determining the space shuttle velocity by using a simple numerical method. The numerical method requires less effort to provide a solution as compared to the use of classical mathematics. However, the approximated solution obtained from the numerical method has error which occurs from the use of a large time step. The error can be reduced by using a smaller time step, but the problem needs more computational time. For practical problems with a large number of unknowns, using a small time step may be prohibited because too large computational time is required. A proper time step, thus, must be decided prior to analyzing a large size problem.

In addition to the error that occurs from the use of time step as explained above, there are other types of errors that may arise from different sources. These errors are:

(a) *Modeling error.* Analysis of a problem normally starts from the discretization of the problem domain into small chunks of elements. These elements are connected at grid points where the unknowns are located and determined. The error is introduced because the continuum model is transformed into a discrete model. A large error occurs if the discrete model contains only few elements. The discrete model using small elements will produce a solution with less error. However, the model with small elements contains a large number of grid points and unknowns. High computer memory and computational time are required thus for the solution.

(b) *Propagation of error.* An error may propagate from one solution to another. As in the example for determination of the shuttle velocity, the error that occurs from the computed solution at 30 seconds can propagate to produce an additional error in the computed solution at 60 seconds. Or in the example of the car collision as shown in Fig. 1.2, the error that occurs in the deformed structure solution at an early time can propagate to produce more error of the structure solution at the later time. The propagation of error must be realized, especially when analyzing a practical problem containing a large number of unknowns.

(c) *Error from data.* Uncertain data produces error in the solution. As in Ex. 1.1, the space shuttle mass may not be exactly 90,000 kg and thus the computed solution is altered from the actual situation. Or in the example of the car collision in Ex. 1.2, the actual material stiffness may be different from that used in the computation. However, in many cases, actual data may not be available. Analysts must make judgment and be very careful prior to selecting proper data in order to obtain accurate and reasonable solutions.

(d) *Blunder error.* Careless programmers can produce serious error in the computed solution. Such error may come from typing data incorrectly, writing a computer program without checking it thoroughly, etc. The error may come from incorrect statements in computer programs. A new computer program must be inspected or debugged comprehensively to assure that it will not produce such error.

(e) *Truncation error.* The truncation error occurs when some terms are omitted or excluded from the equations during computation. For example, higher-order terms are omitted in the computation of the infinite series for the sine and cosine functions in Ex. 1.3. Truncation error always occurs in the computation of infinite series that are arisen from exact solutions of academic type problems.

(f) *Round-off error.* The round-off error arises from the use of computers that have limited capability in storing values. For example, the value of π that consists of 25 significant figures is

$$\pi = 3.141592653589793238462643 \tag{1.21}$$

Such the value of π with 25 significant figures can be stored on any computer today. Few decades ago, a typical computer may store only 10 significant figures, so that the value 3.141592653 is used for π in the computation.

The number of significant figures is used to indicate the accuracy of solution obtained from a numerical method. If the value of π is given by 3.14159, it has six significant figures. Frequently, most of the values used in numerical methods are in the floating point format. The π value of 3.14159 is written in the floating point format that has six significant figures as 0.314159×10^1. Thus, the following numbers of $0.0001278, 0.001278$ and 0.01278 have the same number of floating points of four. These numbers can be written in the form of the floating point format as 0.1278×10^{-3}, 0.1278×10^{-2} and 0.1278×10^{-1}, respectively.

Numbers are stored using the binary system in computers. The binary system is different from the decimal or base-10 system. As an example, the value of 107 in the decimal system is determined from

$$1 \times 10^2 + 0 \times 10^1 + 7 \times 10^0 \;=\; 107$$

The same value is represented by 1101011 in the binary system, which is determined from

$$1 \times 2^6 + 1 \times 2^5 + 0 \times 2^4 + 1 \times 2^3 + 0 \times 2^2 + 1 \times 2^1 + 1 \times 2^0 \;=\; 107$$

The value of 1101011 in the binary system consists only the numbers 1 and 0 which are called bits. A total of 32 bits may be needed to represent a single value or a word. The value of 107 may be stored in the computer under the binary system as

$$00000000000000000000000001101011$$

In the above set of 32 bits, the first bit is used to indicate the positive or negative value of the number (0 = positive, 1 = negative). The following 31 bits are used to represent the number that can go up to 2,147,483,647 (or $2^{31} - 1$). This means a typical 32 bit computer can store integer values ranging from $-2,147,483,647$ to $+2,147,483,647$. For a value that is beyond such range, it is stored in the floating point format. The 32 bits are divided into four parts. The first two parts represent the sign of the number and exponent, while the last two parts are used for storing the magnitudes of the exponent and mantissa. Thus, a typical word can store a floating point value in the range of -10^{38} to 10^{38}. Understanding how values are stored in the computers help users to be aware of the very small or large numbers during analyzing a problem.

As explained earlier, numerical error can arise from different sources. If an exact solution of the problem is known, the true error E_t can be determined. In the example of the space shuttle velocity, the true error is determined from

$$E_t \;=\; v_e - v_a \tag{1.22}$$

where v_e and v_a are the exact and approximate solutions, respectively. The true percentage error ε_t can also be computed as

$$\varepsilon_t \;=\; \frac{v_e - v_a}{v_e} \times 100\% \tag{1.23}$$

For example, the exact and approximate shuttle velocities at 30 seconds in Table 1.2 are 273 and 294 m/sec, respectively. Then, the true percentage error is

$$\varepsilon_t \;=\; \frac{273 - 294}{273} \times 100\% \;=\; -7.69\% \tag{1.24}$$

Because exact solutions are not available in practical problems, the approximate error is normally used to measure the solution error. For the example of the space shuttle velocity, the approximate error may be defined by the difference between the two approximate solutions. In this case, the approximate percentage error ε_a is determined from

$$\varepsilon_a = \frac{v_{new} - v_{old}}{v_{new}} \times 100\% \tag{1.25}$$

where v_{new} and v_{old} are the two approximate velocities. For example, the two approximate velocities at 30 seconds obtained from using the time steps Δt of 30 and 10 seconds are 294 and 280 m/sec, respectively. Thus, the approximate percentage error is

$$\varepsilon_a = \frac{280 - 294}{280} \times 100\% = -5.00\% \tag{1.26}$$

1.6 Closure

The chapter presents an overview of the numerical methods for analyzing science and engineering problems. The chapter started from showing benefits of the methods for solving a variety of practical problems. Solutions obtained from the methods help analysts to understand behaviors that occur on the problems. Understanding the behaviors of the problems can lead to the improved designs. The chapter explained the key ingredients of the numerical methods that consist of the understanding of basic theories and the use of computer software and hardware. Several examples were presented to demonstrate advantages of the numerical methods for obtaining solutions as compared to the classical mathematics. In solving a simple problem, an approximate solution can be obtained easily by using a numerical method as compared to the finding of an exact solution from classical mathematics. For a more general problem, the exact solution is not available and the numerical method may be the only way to obtain a useful solution.

Different types of computers and computer languages widely used in the numerical methods for analyzing problems are explained. Popular languages normally employed in academic institutes and research organizations are Fortran, Pascal, C, Java and Basic. Many computer software that include packages of numerical methods, such as MATLAB and Mathematica, are highlighted. The users, however, should understand the theories of the numerical methods behind these software prior to using them. For the cases where the software are not available and the users must develop computer programs by themselves, few computer commands must be understood. Simple computer programs using several languages are presented in this book to help readers in developing their own programs.

Exercises

1. Use Ex. 1.1 of the shuttle velocity determination to study the improved solution accuracy that will be obtained by reducing the time step. Compare the true errors by using the time steps of 10, 5 and 1 seconds. Tabulate the computed solutions and their errors between the times of 0 and 1,500 seconds with the increment of every 30 seconds.

2. Use the computer program in Ex. 1.3 to determine the computed solution accuracy of the sine and cosine functions by employing 3, 5 and 10 terms in the series. Determine the true and approximate errors by using 8 significant figures.

3. Develop a computer program to determine the exponential function which can be written in infinite series form

$$e^x = 1 + x + \frac{x^2}{2!} + \frac{x^3}{3!} + \frac{x^4}{4!} + \ldots$$

Then, use the program to determine the solutions of the function for $x = 0.1, 0.5, 1, 5$ and 50 with 8 significant figures. Explain the difficulties encountered and ways to improve the computed solutions.

4. Explain and develop a computer program to find roots of the equation,

$$ax^2 + bx + c = 0$$

where a, b and c are constants. The developed computer program should avoid problems that may occur from arbitrary values of a, b and c.

5. Develop computer programs to proof the following relations:

(a) $1 + 2 + 3 + \ldots + n = \dfrac{n(n+1)}{2}$

(b) $1^2 + 2^2 + 3^2 + \ldots + n^2 = \dfrac{n(n+1)(2n+1)}{6}$

(c) $1^3 + 2^3 + 3^3 + \ldots + n^3 = \dfrac{n^2(n+1)^2}{4}$

(d) $1^4 + 2^4 + 3^4 + \ldots + n^4 = \dfrac{n(n+1)(2n+1)(3n^2+3n-1)}{30}$

In each case, use $n = 10$, 50 and 100. Then, compute the true solution errors from the use of different values of n.

6.   Develop computer programs to proof the following equalities:

(a) $1 - \dfrac{1}{2} + \dfrac{1}{3} - \dfrac{1}{4} + \dfrac{1}{5} - \cdots \qquad = \quad \ln 2$

(b) $1 - \dfrac{1}{3} + \dfrac{1}{5} - \dfrac{1}{7} + \dfrac{1}{9} - \cdots \qquad = \quad \dfrac{\pi}{4}$

(c) $\dfrac{1}{1^2} + \dfrac{1}{2^2} + \dfrac{1}{3^2} + \dfrac{1}{4^2} + \dfrac{1}{5^2} + \cdots \quad = \quad \dfrac{\pi^2}{6}$

(d) $\dfrac{1}{1^2} - \dfrac{1}{2^2} + \dfrac{1}{3^2} - \dfrac{1}{4^2} + \dfrac{1}{5^2} - \cdots \quad = \quad \dfrac{\pi^2}{12}$

(e) $\dfrac{1}{1^4} - \dfrac{1}{2^4} + \dfrac{1}{3^4} - \dfrac{1}{4^4} + \dfrac{1}{5^4} - \cdots \quad = \quad \dfrac{7\pi^4}{720}$

In each case, use the number of terms on the left-hand side of the equations as many as possible. Then, determine the true errors with 10 significant figures.

7.   If $|x| < 1$, then the relation,

$$\frac{1}{1-x} \quad = \quad 1 + x + x^2 + x^3 + \cdots$$

is valid. Proof the validity of the relation by developing a computer program by using $x = 0.2$. Determine the true errors that occur from using 10, 50 and 100 terms on the right-hand side of the equation.

8.   Develop a computer program to show that

$$\sum_{j=0}^{n} x^j \quad = \quad \frac{1 - x^{n+1}}{1 - x}$$

by using $x = 0.01, 0.1, 0.5, 0.9$ and 0.99, respectively. Give comments on the accuracy of the computed solutions. Use the highest value of n that can be done on the computer.

9.   Develop a computer program to proof that

$$\frac{1}{2} \ln\left(\frac{1+x}{1-x}\right) \quad = \quad x + \frac{x^3}{3} + \frac{x^5}{5} + \frac{x^7}{7} + \cdots$$

for $-1 < x < 1$. Use the number of terms on the right-hand side of equation as many as possible so that the computed solutions with 8 significant figures do not alter. Test the program by using $x = -0.5$ and 0.5.

10.   Develop a computer program for determining the value of π from

$$\sum_{n=1}^{\infty} \frac{1}{(2n-1)^2} \quad = \quad \frac{\pi^2}{8}$$

Compare the computed solution with the value of π in Eq. (1.21). Explain the difficulties encountered and suggest ways to improve accuracy of the computed solution.

11. Develop a computer program to show that

$$\sum_{n=1}^{\infty} \frac{1}{4n^2-1} = \frac{1}{2}$$

Then, determine the true percentage errors if $n = 10, 50$ and 100. Show the computed solutions using 8 significant figures. Explain the difficulties encountered and suggest ways for improving the solutions.

12. Develop a computer program to show that

$$x^2 = \frac{c^2}{3} + \frac{4c^2}{\pi^2} \sum_{n=1}^{\infty} \frac{(-1)^n}{n^2} \cos\frac{n\pi x}{c}$$

where $0 \le x \le c$. If $c=4$, employ the developed program to verify the relation for $x=0.1, 1$ and 3 by using at least 8 significant figures in the computation.

13. Develop a computer program to determine the error function that is expressed in the form of infinite series as

$$erf(x) = \frac{2}{\sqrt{\pi}} \sum_{n=0}^{\infty} \frac{(-1)^n x^{2n+1}}{n!(2n+1)}$$

Determine the function for $x = 0.5, 1, 5$ and 10 by using 8 significant figures.

14. Develop computer programs to determine the Bessel functions of the first kind of order zero and one. These functions are expressed in the forms of infinite series as

$$J_0(x) = 1 - \frac{x^2}{2^2} + \frac{x^4}{2^2 \cdot 4^2} - \frac{x^6}{2^2 \cdot 4^2 \cdot 6^2} + \cdots$$

$$J_1(x) = \frac{x}{2} - \frac{x^3}{2^2 \cdot 4} + \frac{x^5}{2^2 \cdot 4^2 \cdot 6} - \frac{x^7}{2^2 \cdot 4^2 \cdot 6^2 \cdot 8} + \cdots$$

Determine the two functions for $x = 1$ and 5 by using 5 significant figures. Compare the computed solutions with those tabulated in mathematical handbooks.

15. In an examination of a class that contains 40 students, the scores vary randomly between 0 and 100. Develop a computer program to reorder the scores of the students from 100 down to 0 with the corresponding student identities.

16. A wall with the thickness of π in x-direction has an initial temperature of zero. The wall is subjected to a uniform internal heat generation so that the transient temperature distribution with respect to time t is given by

$$T(x,t) = x + \frac{8}{\pi} \sum_{n=1}^{\infty} \frac{(-1)^n}{(2n-1)^2} \exp\left[-\frac{(2n-1)^2 t}{4}\right] \sin\frac{(2n-1)x}{2}$$

Develop a computer program to determine the transient temperature distribution. Plot the computed distributions at different times and show the distribution through the thickness of the wall when the time approaches infinity.

17.   Buckling analysis of a vertical column due to its own weight leads to the need to determine the function

$$f(x) = 1 + \sum_{m=1}^{\infty} C_m x^{2m}$$

where $m = 1$; $C_1 = -\dfrac{3}{8}$

and $m \geq 2$; $C_m = -\dfrac{3 C_{m-1}}{4m(3m-1)}$

Develop a computer program to determine the function for $0 \leq x \leq 2.0$ with the increment of x at every 0.2. Print the computed solutions by using 8 significant figures.

18.   A solid sphere with radius of c has a uniform initial temperature of T_0 at time $t = 0$. If the outer surface temperature is changed abruptly to zero, the transient temperature distribution that varies with the radius r and time t is

$$T(r,t) = \frac{2T_0 c}{\pi} \sum_{n=1}^{\infty} (-1)^{n+1} \exp\left(-\frac{n^2 \pi^2 t}{c^2}\right) \frac{1}{nr} \sin \frac{n\pi r}{c}$$

Develop a computer program to compute and plot the radial temperature distributions at different times by using the value $c = 5$ and $T_0 = 100$.

Chapter 2

Root of Equations

2.1 Introduction

In the process for solving many engineering and scientific problems, roots of equations are needed. If an equation is represented by a function $f(x)$, then the root x is such that it makes the value of the function to be zero. For example, the roots of the second-order polynomial,

$$f(x) = ax^2 + bx + c = 0 \tag{2.1}$$

where a, b and c are constants, are determined from the formula,

$$x = \frac{-b \pm \sqrt{b^2 - 4ac}}{2a} \tag{2.2}$$

Some polynomials are in higher-order form, such as

$$f(x) = 2x^4 - 7x^3 + 4x^2 + 7x - 6 = 0 \tag{2.3}$$

The factorization technique may be used to rewrite the polynomial into the form

$$(x-1)(x+1)(x-2)(2x-3) = 0 \tag{2.4}$$

so that the roots of the equation are obtained easily.

For general problems in engineering and scientific applications, the function $f(x)$ may not be in the form of polynomials as shown in Eq. (2.1) or (2.3). For examples, in the design for a proper cross-sectional area of a stop-sign pole, vibration of the pole due to the pressure load from the wind must be analyzed. In the analysis process, roots of the transcendental function in the form

$$f(x) = \cosh x \cos x + 1 = 0 \tag{2.5}$$

are required. Or in the analysis process for determining the shock wave angle generated from a 20 degree wedge, the root of the transcendental function

$$f(x) = 2\cot x \left[\frac{9\sin^2 x - 1}{9(1.4 + \cos 2x) + 2} \right] - \tan\frac{\pi}{9} = 0 \tag{2.6}$$

is needed. Roots of the transcendental equation,

$$f(x) = \sinh\frac{4}{9x} - \frac{5}{9x} = 0 \tag{2.7}$$

are required in the analysis process for determining tension in a cable that is suspended between two poles. In the buckling analysis of a vertical column due to its own weight, the root of the function in the form of an infinite series

$$f(x) = \sum_{n=1}^{\infty} \frac{3}{8} x^{2n} - 1 = 0 \tag{2.8}$$

is required. Or, in the example of the space shuttle in section 1.1, the velocity v of the vehicle during gliding into atmosphere is determined from

$$v = \frac{mg}{c}\left(1 - e^{-\frac{c}{m}t}\right) \tag{2.9}$$

where m is mass of the shuttle (90,000 kg), g is gravitational acceleration constant (9.81 m/sec^2), t is time in second, and c is the coefficient of drag force in kg/sec. At time $t = 500$ second and the shuttle velocity is 5 times speed of sound (about 1,650 m/sec) and if the coefficient of drag force c is needed, Eq. (2.9) leads to a transcendental equation in the form

$$1,650 = \frac{(90,000)(9.81)}{c}\left(1 - e^{-\frac{c}{90,000}(500)}\right) \tag{2.10}$$

The examples above are few problems that arise in engineering problems. These examples need to find values of x which are the roots of

$$f(x) = 0 \tag{2.11}$$

Popular methods to find such values are presented in this chapter. These methods are: (1) the graphical method, (2) the bisection method, (3) the false-position method, (4) the one-point iteration method, (5) the Newton-Raphson method, and (6) the secant method. All methods consist of simple computational procedures that can be understood easily. Moreover, these procedures can be used for developing computer programs directly.

2.2 Graphical Method

The simplest method for finding the root of function $f(x) = 0$ is the graphical method. The idea of this method is to determine values of the function at different x by plotting. The example below helps understanding the method procedure clearly.

Example 2.1 Find the root $\bar{x}$ of the function,

$$f(x) = e^{-x/4}(2-x)-1 = 0 \qquad (2.12)$$

by using the graphical method. Plot a graph to display behavior of the function.

Solution Values of the function $f(x)$ in Eq. (2.12) can be calculated and plotted as shown in Fig. 2.1

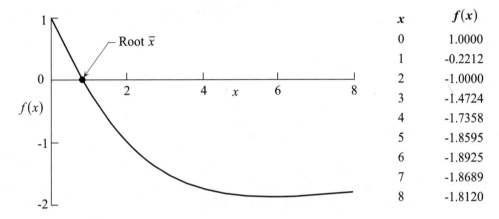

x	$f(x)$
0	1.0000
1	-0.2212
2	-1.0000
3	-1.4724
4	-1.7358
5	-1.8595
6	-1.8925
7	-1.8689
8	-1.8120

Figure 2.1 Distribution of $f(x)$ in Eq. (2.12) and its values at different x locations.

By considering Fig. 2.1, the function $f(x)$ is zero when it intersects the x-axis in the interval of $0.75 < x < 0.80$. After recalculating and plotting the function within this interval, the graph and its values are obtained as shown in Fig. 2.2.

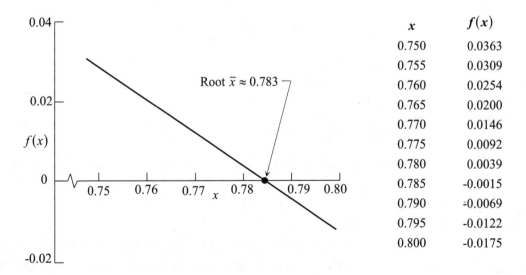

x	$f(x)$
0.750	0.0363
0.755	0.0309
0.760	0.0254
0.765	0.0200
0.770	0.0146
0.775	0.0092
0.780	0.0039
0.785	-0.0015
0.790	=0.0069
0.795	-0.0122
0.800	-0.0175

Figure 2.2 Distribution of $f(x)$ in Eq. (2.12) and its values at finer x locations.

From Fig. 2.2, a more accurate value of the root $\bar{x}$ about 0.783 is obtained. It is an approximate value and may not be suitable if a solution with high accuracy is needed. However, the example shows that the graphical method is simple especially if a computer program for generating values of the function is available. Many commercial software today allow users to input a function, so that the software can generate values of the function and plot its variation directly on the monitor screen.

Even though the graphical method is easy but it is time consuming if a solution with high accuracy is needed. Other methods that can provide higher solution accuracy are presented in the following sections.

2.3 Bisection Method

The main idea of the bisection method is based on the fact that the values of the function have different signs when x are less and greater than the root $\bar{x}$ as shown in Fig. 2.2. From example 2.1, the approximate value of the root $\bar{x}$ is 0.783. When x is less than the root $\bar{x}$, the sign of the function is positive. The sign of the function becomes negative when x is greater than $\bar{x}$. In general, the sign of a function may change from a positive to negative value across the x-axis such as the graph in Fig. 2.2. The sign of a function may also change from a negative to positive quantity as shown in Fig. 2.3.

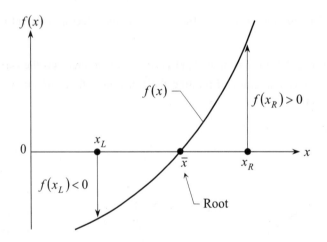

Figure 2.3 Bisection method for finding the root of $f(x) = 0$.

Figure 2.3 shows that the function $f(x)$ changes from a negative value at $x = x_L$ (subscript L denotes Left side of $\bar{x}$) to a positive value at $x = x_R$ (subscript R denotes Right side of $\bar{x}$). The figure indicates that, as the sign of the values $f(x_L)$ and $f(x_R)$ are different, the root $\bar{x}$ of the function must be between x_L and x_R.

The idea for finding root of the equation $f(x) = 0$ by this method is to reduce the interval from x_L to x_R by a half and then properly select the sub-interval where the sign change occurs. This sub-interval, that contains the root $\bar{x}$, is used as a new interval for the next calculation. The computational procedure of the method are as follows.

<u>Step 1</u> Find the mean value x_M from given x_L and x_R,

$$x_M = \frac{x_L + x_R}{2} \tag{2.13}$$

Then determine the value $f(x_M)$ of the function at point x_M. Such value can be a positive (case A) or negative (case B) quantity as shown in Fig. 2.4.

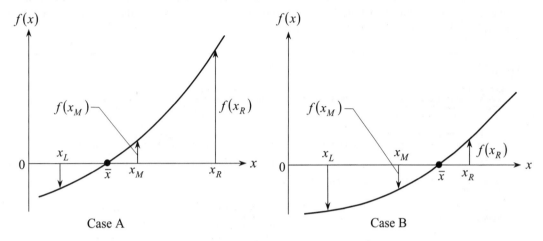

Case A   Case B

Figure 2.4 Value of $f(x_M)$ which can be a positive or negative quantity.

<u>Step 2</u> Multiply $f(x_M)$ by $f(x_R)$,

If $f(x_M)\cdot f(x_R)>0$ the result is case A where the root $\bar{x}$ is in the interval $x_L < \bar{x} < x_M$.

If $f(x_M)\cdot f(x_R)<0$ the result is case B, where the root $\bar{x}$ is in the interval $x_M < \bar{x} < x_R$.

<u>Step 3</u> Reassign the value of x_L or x_R to reduce the size of current interval.

If the result is case A, reassign value of x_R to x_M.

If the result is case B, reassign value of x_L to x_M.

<u>Step 4</u> Check for convergence of the computed solution by using a criterion such as,

$$|f(x_M)| < \varepsilon \tag{2.14}$$

where ε is the acceptable error or tolerance. Another form of the convergence criterion is

$$\left| \frac{x_M^{new} - x_M^{old}}{x_M^{new}} \right| \times 100\% < \varepsilon_S \tag{2.15}$$

where ε_S is the stopping tolerance, for example 0.05%. If the computed result reaches the convergence criterion as specified in Eq. (2.14) or (2.15), the computation stops. If not, the computation is repeated by going back to step 1.

Example 2.2   Develop a computer program by using the four steps of the bisection method as explained above to find the root of Eq. (2.12). Use $x_L = 0$ and $x_R = 2$ with the stopping tolerance ε_S = 0.001% in the form of Eq. (2.15).

A computer program to find the root of Eq. (2.12) by using the bisection method is shown in Fig. 2.5. The computed solutions at different iterations are presented in Table 2.1. The listing of computer program as shown in Fig. 2.5 can be modified to find roots of other equations. User can simply replace the function statement in the program by the function of a new problem.

Fortran

```
      PROGRAM  BISECT
C.....PROGRAM FOR COMPUTING ROOT OF NONLINEAR
C.....EQUATION USING THE BISECTION METHOD
C.....DEFINE THE FOLLOWING GIVEN VALUES:
C.....    XL = LEFT VALUE OF X
C.....    XR = RIGHT VALUE OF X
C.....    ES = STOPPING CRITERION TOLERANCE (%)
      XL = 0.
      XR = 2.
      ES = 0.001
C.....CHECK WHETHER THE ROOT IS IN GIVEN RANGE:
      FXL = FUNC(XL)
      FXR = FUNC(XR)
      AA  = FXL*FXR
      IF(AA.GE.0.)  THEN
          WRITE(6,10)
          STOP
      ENDIF
   10 FORMAT(/, ' ROOT IS NOT IN THE GIVEN RANGE')
      WRITE(6,20)
   20 FORMAT(/, 3X, 'ITERATION NO.', 9X, 'X', /)
      DO 100  ITER=1,500
      XM  = (XL+XR)/2.
      FXM = FUNC(XM)
      FXR = FUNC(XR)
      AA  = FXM*FXR
        IF(AA.GT.0.)  THEN
C.....CASE  A:   XL < ROOT < XM
          XR = XM
        ELSE
C.....CASE  B:   XM < ROOT < XR
          XL = XM
        ENDIF
C.....CHECK FOR TOLERANCE:
      XN = (XL+XR)/2.
      WRITE(6,50)  ITER, XN
   50 FORMAT(1X, I8, 8X, E14.6)
      TOL = ABS((XN-XM)*100./XN)
      IF(TOL.LT.ES)  GO TO 200
  100 CONTINUE
      WRITE(6,110)
  110 FORMAT(/, ' ROOT CAN NOT BE REACHED FOR',
     *          ' THE GIVEN CONDITIONS'         )
      GO TO 300
  200 WRITE(6,210)  XN
  210 FORMAT(/, ' THE ROOT IS ', E14.6)
  300 CONTINUE
      STOP
      END
C-------------------------------------------------
      FUNCTION FUNC(X)
      FUNC = EXP(-X/4.)*(2.-X) - 1.
      RETURN
      END
```

MATLAB

```
% PROGRAM BISECT
% PROGRAM FOR COMPUTING ROOT OF NONLINEAR
% EQUATION USING THE BISECTION METHOD
% DEFINE THE FOLLOWING GIVEN VALUES:
%         XL = LEFT VALUE OF X
%         XR = RIGHT VALUE OF X
%         ES = STOPPING CRITERION TOLERANCE (%)
%-----------------------------------------------------
func = inline('exp(-X/4.0)*(2.-X) - 1.', 'X');
%-----------------------------------------------------
XL  = 0.;
XR  = 2.;
ES  = 0.001;
%   Check whether the root is in given range:
FXL = func(XL);
FXR = func(XR);
AA  = FXL*FXR;
if AA >= 0.
    disp(' ROOT IS NOT IN THE GIVEN RANGE');
    break
end
fprintf('\n   ITERATION NO.        X \n');
for iter = 1:500
    XM  = (XL+XR)/2.;
    FXM = func(XM);
    FXR = func(XR);
    AA  = FXM*FXR;
    if AA > 0.
%  CASE  A:   XL < ROOT < XM
        XR = XM;
    else
%  CASE  B:   XM < ROOT < XR
        XL = XM;
    end
%  CHECK FOR TOLERANCE:
    XN = (XL+XR)/2.;
    fprintf(' %8d        %14.6e\n', iter, XN);
    tol = abs((XN-XM)*100./XN);
    if tol < ES
        fprintf('\n   THE ROOT IS %14.6e\n', XN)
        break
    end
end
if tol > ES
    fprintf(' ROOT CAN NOT BE REACHED FOR\n');
    fprintf('   THE GIVEN CONDITIONS');
    break
end
```

Figure 2.5 Computer program for finding the root of Eq. (2.12)
by the bisection method.

Table 2.1 Solution convergence to the root of Eq. (2.12) by the bisection method.

Iteration number	Root x
1	0.500000
2	0.750000
3	0.875000
4	0.812500
5	0.781250
6	0.796875
7	0.789063
8	0.785156
9	0.783203
10	0.784180
11	0.783691
12	0.783447
13	0.783569
14	0.783630
15	0.783600
16	0.783585
17	0.783592

In order to find a root of an equation by using the bisection method, user must know approximate position of root $\bar{x}$. The method needs a starting interval between x_L and x_R that contains root $\bar{x}$. The method keeps reducing the interval by a half until a converged solution is obtained. Because a starting interval between x_L and x_R is needed prior to the calculation, the method is sometimes called the bracketing method. The idea of the bisection method also leads to a more efficient method as presented in the next section.

2.4 False-position Method

The idea of the false-position method for finding the root of an equation is similar to that of the bisection method. Instead of bisecting the interval, the false-position method locates the root of equation by drawing a straight line to join the values of $f(x_L)$ and $f(x_R)$. The location x_1 that occurs from intersecting the straight line and the x-axis is the new estimated root of the equation as shown in Fig. 2.6. The figure gives the relation,

$$\tan \theta = \tan \beta$$

$$\frac{f(x_R)}{x_R - x_1} = \frac{f(x_L)}{x_L - x_1} \tag{2.16}$$

$$x_L \, f(x_R) - x_1 \, f(x_R) = x_R \, f(x_L) - x_1 \, f(x_L)$$

Thus, the location x_1 can be determined from

$$x_1 = \frac{x_L \, f(x_R) - x_R \, f(x_L)}{f(x_R) - f(x_L)} \qquad (2.17)$$

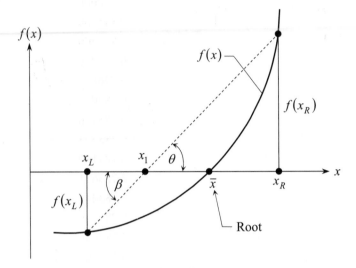

Figure 2.6 False-position method for finding the root of $f(x) = 0$.

The procedure of the false-position method starts from specifying the values of x_L and x_R. The location x_1 is then determined by using Eq. (2.17). Either the value of x_L or x_R is updated by a new appropriate value so that the current interval is reduced. Detailed steps of the false-position method are as follows.

<u>Step 1</u>   From the given locations x_L and x_R, determine the corresponding values of the function, $f(x_L)$ and $f(x_R)$. Then compute the location x_1 by using Eq. (2.17) and calculate the value of the function $f(x_1)$ at this location. The computed value of the function can either be positive or negative as shown in Fig. 2.7.

<u>Step 2</u> Multiply $f(x_1)$ by $f(x_R)$.

      If $f(x_1) \cdot f(x_R) < 0$ the result is case A where the root $\bar{x}$
                                                 is in the interval $x_1 < \bar{x} < x_R$

      If $f(x_1) \cdot f(x_R) > 0$ the result is case B where the root $\bar{x}$
                                                 is in the interval $x_L < \bar{x} < x_1$

<u>Step 3</u> Reassign the value of x_L or x_R according to the result from step 2.

      If the result is in case A, reassign value of x_L to x_1.

      If the result is in case B, reassign value of x_R to x_1.

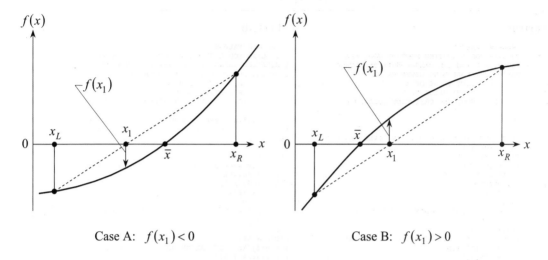

Case A: $f(x_1) < 0$ Case B: $f(x_1) > 0$

Figure 2.7 Value of $f(x_1)$ which can be positive or negative.

Step 4 Check for convergence of the computed solution by using a criterion as shown in Eq. (2.14) or (2.15). If the specified convergence criterion is met, stop the computation. If not, the computation is repeated by going back to step 1.

Example 2.3 Develop a computer program for finding the root of Eq. (2.12) by using the false-position method. Use the initial values $x_L = 0$ and $x_R = 2$ with the stopping tolerance ε_S in Eq. (2.15) as 0.001%.

The developed computer program is shown in Fig. 2.8. The computed solutions during the iteration process are shown in Table 2.2.

Table 2.2 Solution convergence to the root of Eq. (2.12) by the false-position method.

Iteration number	Root x
1	1.000000
2	0.818867
3	0.789254
4	0.784501
5	0.783741
6	0.783619
7	0.783600
8	0.783597

Fortran

```
      PROGRAM FALPOS
C.....PROGRAM FOR COMPUTING ROOT OF NONLINEAR
C.....EQUATION USING THE FALSE-POSITION METHOD
C.....DEFINE THE FOLLOWING GIVEN VALUES:
C.....    XL = LEFT VALUE OF X
C.....    XR = RIGHT VALUE OF X
C.....    ES = STOPPING CRITERION TOLERANCE (%)
      XL = 0.
      XR = 2.
      ES = 0.001
C.....CHECK WHETHER THE ROOT IS IN GIVEN RANGE:
      FXL = FUNC(XL)
      FXR = FUNC(XR)
      AA  = FXL*FXR
      IF(AA.GE.0.)  THEN
          WRITE(6,10)
          STOP
      ENDIF
   10 FORMAT(/, ' ROOT IS NOT IN THE GIVEN RANGE')
      X1OLD = XL
      WRITE(6,20)
   20 FORMAT(/, 3X, 'ITERATION NO.', 9X, 'X', /)
      DO 100  ITER=1,500
      FXL = FUNC(XL)
      FXR = FUNC(XR)
      X1  = (XL*FXR - XR*FXL)/(FXR - FXL)
      FX1 = FUNC(X1)
      AA  = FX1*FXR
         IF(AA.LT.0.)  THEN
C.....CASE  A:   X1 < ROOT < XR
           XL = X1
         ELSE
C.....CASE  B:   XL < ROOT < X1
           XR = X1
         ENDIF
C.....CHECK FOR TOLERANCE:
      WRITE(6,50)   ITER, X1
   50 FORMAT(1X, I8, 8X, E14.6)
      TOL = ABS((X1-X1OLD)*100./X1)
      IF(TOL.LT.ES)  GO TO 200
      X1OLD = X1
  100 CONTINUE
      WRITE(6,110)
  110 FORMAT(/, ' ROOT CAN NOT BE REACHED FOR',
     *          ' THE GIVEN CONDITIONS'          )
      GO TO 300
  200 WRITE(6,210)  X1
  210 FORMAT(/, 3X, 'THE ROOT IS ', E14.6)
  300 CONTINUE
      STOP
      END
C-------------------------------------------------
      FUNCTION FUNC(X)
      FUNC = EXP(-X/4.)*(2.-X) - 1.
      RETURN
      END
```

MATLAB

```
% PROGRAM FALPOS
% PROGRAM FOR COMPUTING ROOT OF NONLINEAR
% EQUATION USING THE FALSE-POSITION METHOD
% DEFINE THE FOLLOWING GIVEN VALUES:
%       XL = LEFT VALUE OF X
%       XR = RIGHT VALUE OF X
%       ES = STOPPING CRITERION TOLERANCE (%)
%--------------------------------------------------
func = inline('exp(-X/4.0)*(2.-X) - 1.', 'X');
%--------------------------------------------------
XL = 0.;
XR = 2.;
ES = 0.001;
%   Check whether the root is in given range:
FXL = func(XL);
FXR = func(XR);
AA  = FXL*FXR;
if AA >= 0.
   disp(' ROOT IS NOT IN THE GIVEN RANGE');
   break
end
X1OLD = XL;
fprintf('\n  ITERATION NO.       X \n');
for iter = 1:500
    FXL = func(XL);
    FXR = func(XR);
    X1  = (XL*FXR - XR*FXL)/(FXR - FXL);
    FX1 = func(X1);
    AA  = FX1*FXR;
    if AA < 0.
% CASE  A:   X1 < ROOT < XR
        XL = X1;
    else
% CASE  B:   XL < ROOT < X1
        XR = X1;
    end
% CHECK FOR TOLERANCE:
      fprintf(' %8d        %14.6e\n', iter, X1);
      tol = abs((X1-X1OLD)*100./X1);
      if tol < ES
         fprintf('\n   THE ROOT IS  %14.6e\n', X1)
         break
      end
      X1OLD = X1;
end
if tol > ES
   fprintf('   ROOT CAN NOT BE REACHED FOR\n');
   fprintf('   THE GIVEN CONDITIONS');
   break
end
```

Figure 2.8 Computer program for finding the root of Eq. (2.12)
by the false-position method.

By comparing convergence rates between the bisection and false-position methods as shown in Table 2.1 and 2.2, it is apparent that the false-position method converges faster than the bisection method. A schematic diagram for the solution convergence of the false-position method is shown in Fig. 2.8 (a).

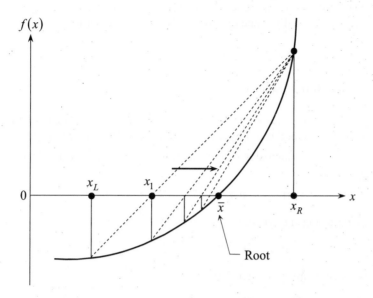

Figure 2.8(a) Schematic diagram for solution convergence of the false-position method.

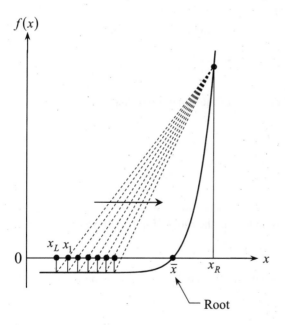

Figure 2.8(b) Slow convergence rate by the false-position method.

Figure 2.8 (a) shows that the specified positions of the initial locations x_L and x_R affect the convergence rate. If the initial location x_L in Fig. 2.8(a) moves to the left-hand side which is farther from the root $\bar{x}$, the number of iterations will increase. For some special functions such as that shown in Fig. 2.8(b), a large number of iterations is needed with low convergence rate. Both the bisection and false-position methods need an initial interval that contains the root of equation. Knowing a proper interval is the main advantage of the bracketing method because the iteration process always leads to a root of the equation. Before using these methods, user should have a priori knowledge of the proper interval that contains the root of equation.

For a practical problem, the location of the root $\bar{x}$ is not known a priori. It will be very convenient for the computation if only one initial value of x is used for finding the root of equation. Such method is sometimes called the open method. Different types of the open method are presented in the following sections.

2.5 One-point Iteration Method

The one-point iteration method is one of the simplest open methods for finding the root of equation. The idea of the method starts by arranging the give function $f(x) = 0$ such that the single variable x is placed on the left-hand side of the equation. For example, if the function $f(x)$ is given by

$$f(x) \quad = \quad \sinh 4x + 7x^2 - x + 3 \quad = \quad 0 \tag{2.18}$$

Equation (2.18) can be arranged to yield,

$$x \quad = \quad \sinh 4x + 7x^2 + 3 \tag{2.19}$$

Then, Eq. (2.19) is rewritten in an iterative form as,

$$x_{i+1} \quad = \quad \sinh 4x_i + 7x_i^2 + 3 \tag{2.20}$$

Eq. (2.20) suggests that the new estimated value of root x_{i+1} is determined from the terms on the right-hand side of equation that use the old value of x_i.

If the given function $f(x)$ does not contain the single variable x that can be separated to the left-hand side of the equation, such as

$$f(x) \quad = \quad \cos x \quad = \quad 0 \tag{2.21}$$

In this case, the variable x can be added on both sides of the equation, so that

$$x \quad = \quad \cos x + x \tag{2.22}$$

Then, Eq. (2.22) is written in an iterative form as,

$$x_{i+1} \quad = \quad \cos x_i + x_i \tag{2.23}$$

Example 2.4 Use the one-point iteration method to find the root of Eq. (2.12) which is

$$f(x) \quad = \quad e^{-x/4}(2-x) - 1 \quad = \quad 0$$

The equation above can be rewritten such that the single variable x is placed on the left-hand side of the equation as,

$$x \quad = \quad 2 - e^{x/4} \tag{2.24}$$

Equation (2.24) is then written in an iterative form as,

$$x_{i+1} \quad = \quad 2 - e^{x_i/4} \tag{2.25}$$

Equation (2.25) is used to develop a computer program as shown in Fig. 2.9. The computational procedure starts from an initial guess x as zero with the stopping tolerance of $\varepsilon_S = 0.001\%$. The computed solutions during the iteration process are shown in Table 2.3.

Fortran

```
      PROGRAM  ONEPT
      XOLD = 0.
      ES = .001
      WRITE(6,10)
   10 FORMAT(/, 3X, 'ITERATION NO.', 9X, 'X', /)
      DO 20  I=1,100
      XNEW = 2. - EXP(XOLD/4.)
      WRITE(6,100)  I, XNEW
  100 FORMAT(1X, I8, 8X, E14.6)
      TOL = ABS((XNEW-XOLD)*100./XNEW)
      IF(TOL.LT.ES)  GO TO 30
      XOLD = XNEW
   20 CONTINUE
   30 WRITE(6,200)  XNEW
  200 FORMAT(/, 3X, 'THE ROOT IS ', E14.6)
      STOP
      END
```

MATLAB

```
% PROGRAM ONEPT
XOLD = 0.;
ES   = 0.001;
fprintf('\n   ITERATION NO.          X\n');
for i = 1:100
    XNEW = 2. - exp(XOLD/4.0);
    fprintf(' %8d       %14.6e\n',i,XNEW);
    tol  = abs((XNEW-XOLD)*100./XNEW);
    if tol < ES
        fprintf('\n   THE ROOT IS  %14.6e\n', XNEW);
        break
    end
    XOLD = XNEW;
end
```

Figure 2.9 Computer program for finding the root of Eq. (2.12)
by the one-point iteration method.

Table 2.3 Solution convergence to the root of Eq. (2.12) by the one-point iteration method.

Iteration number	Root x
1	1.000000
2	0.715975
3	0.803987
4	0.777379
5	0.785485
6	0.783021
7	0.783771
8	0.783543
9	0.783612
10	0.783591
11	0.783597

The concept for solution convergence of the one-point iteration method can be explained graphically. The function $f(x) = 0$ in Example 2.4 can be separated into two equations; the first equation is $F(x) = x$ and the second equation is $G(x) = 2 - e^{x/4}$. The intersection point can be found by equating the two equations,

$$F(x) \quad = \quad G(x)$$

or

$$x = 2 - e^{x/4}$$

It can be seen that the equation obtained from the above procedure is identical to Eq. (2.24). Thus, it may be concluded that the concept of finding the root of equation by using the one-point iteration method is to find the intersection point between two equations, herein are $F(x) = x$ and $G(x) = 2 - e^{x/4}$ as shown in Fig. 2.10.

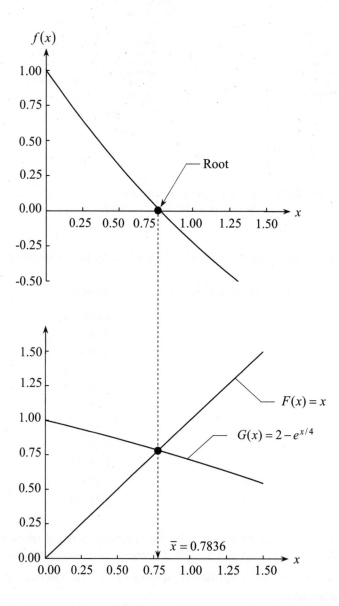

Figure 2.10 Root of equation by finding the intersection point of two functions.

Figure 2.11 shows the convergence behavior of the solution by the one-point iteration method. The initial guess value of x starts from zero (point ⓪ in Fig. 2.11). Then, this value is used to calculate the expression on right-hand side of Eq. 2.25 which is point ① on the graph. Because the value on the left-hand side must be equal to the value on the right hand side, i.e., $F(x) = G(x)$, so the new estimated value of root is now moved to point ② which is placed on function $F(x) = x$. At this point, the value of the new estimated root is x_1. After that, the point x_1 is used to calculate the expression on the right-hand side of Eq. (2.25) again that makes the solution converges to point ③ on the graph. When the functions $F(x)$ is set to be equal to $G(x)$, the new estimated value of root moves to point ④ with the value x_2. The process is repeated until the solution is converged as a spiral shape to the intersection point between the function $F(x)$ and $G(x)$ which is the root of the equation $f(x) = 0$. Another form for solution convergence of the one-point iteration method is the stair step pattern as shown by an example in Fig. 2.12. Figure 2.13 shows two examples that the one-point iteration method may lead to diverged solutions if the initial guess values are not provided properly.

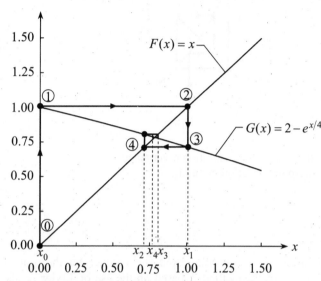

Figure 2.11 Convergence behavior of the solution by the one-point iteration method.

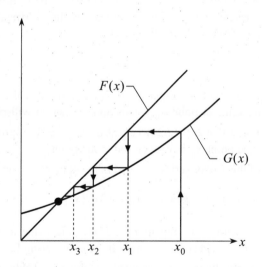

Figure 2.12 Convergence behavior of the solution by the one-point iteration method in the stair step pattern.

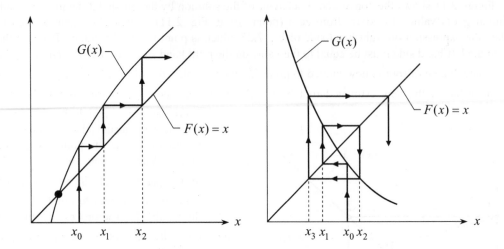

Figure 2.13 Schematic diagrams for solution divergence
of the one-point iteration method.

2.6 Newton-Raphson Method

One of the most popular open methods that can provide rapid convergence for the root of equation is the Newton-Raphson method. Because the method is based on the use of Taylor series, thus, the concept of the series is explained herein first.

The Taylor series can be used to find the value of a function at point x from the values of the function and its derivatives at a nearby point x_0. Even though the Taylor series contains an infinite number of terms but its physical meaning can be interpreted without difficulty. For example, if the Taylor series is approximated by using only the first term, i.e.,

$$f(x) \cong f(x_0) \tag{2.26}$$

it is called the zero-order approximation. Equation (2.26) implies that the value of the function at point x is approximated as its value at point x_0. This approximation will be true if the considering function is constant. It is noted that Eq. (2.26) can also be used to approximate the value of the function at point x if the selected point x_0 is very closed to the point x.

If the first two terms of the Taylor series are used in the approximation,

$$f(x) \cong f(x_0) + (x - x_0) f'(x_0) \tag{2.27}$$

it is called the first-order approximation. Equation (2.27) is linear due to the first derivative term, $f'(x_0)$. Thus, the equation will provide exact solution if the considering function is linear.

Similarly, if the considering function varies quadratically, the Taylor series with the first three terms as shown in Eq. (2.28) will give the exact solution.

$$f(x) \cong f(x_0) + (x-x_0)f'(x_0) + \frac{(x-x_0)^2}{2!}f''(x_0) \tag{2.28}$$

In general, the functions that occur in engineering and scientific applications are complex and can not be represented by the polynomials with finite terms. To represent such complex function, the Taylor series with infinite terms must be used as,

$$f(x) = f(x_0) + (x-x_0)f'(x_0) + \frac{(x-x_0)^2}{2!}f''(x_0) + \ldots + \frac{(x-x_0)^n}{n!}f^{(n)}(x_0) + \ldots \tag{2.29}$$

Representation of the Taylor series for a function is illustrated in the following example.

Example 2.5 Use the Taylor series to determine values of the function,

$$f(x) = \sin x \tag{2.30}$$

at point $x = \pi/6$ (or $x = 30°$) by using the value of the function and its derivatives at point $x_0 = \pi/12$ (or $x = 15°$). Determine the solutions by using the Taylor series with zero-order to sixth-order approximation.

It is noted that $\sin(\pi/6) = 0.5$ and

$$x - x_0 = \frac{\pi}{6} - \frac{\pi}{12} = \frac{\pi}{12}$$

If the zero-order approximation of the Taylor series with only one term is used as shown in Eq. (2.26), the approximate value of $\sin(\pi/6)$ is

$$f\left(\frac{\pi}{6}\right) \cong \sin\left(\frac{\pi}{12}\right) = 0.2588190$$

For the first-order approximation of the Taylor series with two terms as shown in Eq. (2.27), the approximate value of $\sin(\pi/6)$ is

$$f\left(\frac{\pi}{6}\right) \cong \sin\left(\frac{\pi}{12}\right) + \frac{\pi}{12}\cos\left(\frac{\pi}{12}\right) = 0.5116978$$

Similarly, for the third-order approximation of the Taylor series with three terms as shown in Eq. (2.28), the approximated value of $\sin(\pi/6)$ is

$$f\left(\frac{\pi}{6}\right) \cong \sin\left(\frac{\pi}{12}\right) + \frac{\pi}{12}\cos\left(\frac{\pi}{12}\right) - \frac{1}{2}\left(\frac{\pi}{12}\right)^2 \sin\left(\frac{\pi}{12}\right) = 0.5028282$$

The approximate values of $\sin(\pi/6)$ from the zero-order to sixth-order approximation are shown in Table 2.4. The table shows that the solution converges to the exact value which is 0.5 when the order of approximation is increased.

Table 2.4 Approximate values of $\sin(\pi/6)$ by Taylor series of order zero to six.

order n	$f^{(n)}(x)$	$f(\pi/6)$
0	$\sin x$	0.2588190
1	$\cos x$	0.5116978
2	$-\sin x$	0.5028282
3	$-\cos x$	0.4999396
4	$\sin x$	0.4999902
5	$\cos x$	0.5000001
6	$-\sin x$	0.5000000

The use of the Taylor series is the basis of the Newton-Raphson method for finding the root of equation $f(x)=0$. The idea of the method is to use the first two terms as shown in Eq. (2.27) for approximating the function $f(x)$ in the iteration process,

$$f(x) \;=\; f(x_0)+(x-x_0)\,f'(x_0) \;=\; 0 \tag{2.31}$$

or

$$(x-x_0)\,f'(x_0) \;=\; -f(x_0)$$

$$x-x_0 \;=\; -\frac{f(x_0)}{f'(x_0)} \tag{2.32}$$

The physical meaning of Eq. (2.32) is systematically explained by Fig. 2.14.

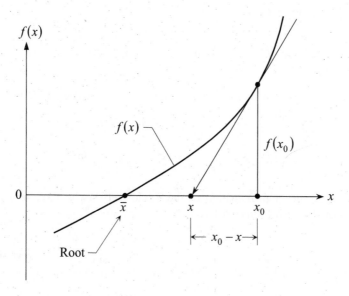

Figure 2.14 Finding the new approximate value x from the initial value x_0 by Newton-Raphson method.

The procedure of Newton-Raphson method starts from an initial value x_0 as shown in Fig. 2.14. Then, the values of function $f(x)$ and its first derivative $f'(x)$ at point x_0 are determined. These computed values are substituted into Eq. (2.32) to obtain the new value of x as shown in Fig. 2.14. The process is repeated until the new value of x converges to the root $\bar{x}$ as shown in Fig. 2.15.

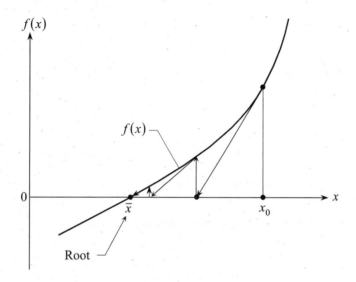

Figure 2.15 Convergence behavior of Newton-Raphson method.

Figure 2.15 highlights a rapid convergence of the solution by using the Newton-Raphson method. The method is, thus, very popular among the methods for finding the root of an equation. However, the method may provide a solution that diverges from the real root. A diverged solution may be caused by the behavior of the function and the initial guess value x_0, etc. Figure 2.16 (a-b) shows the Newton-Raphson method that can provide a converged or diverged solution depending on the initial guess value x_0.

From the above explanation, the method uses an old value to determine a new value of x through Eq. (2.32) which was approximated from the Taylor series. If Δx is the difference between the old and new values of x, Eq. (2.32) can be written as,

$$\Delta x \;=\; x - x_0 \;=\; -\frac{f(x_0)}{f'(x_0)} \tag{2.33}$$

The equation above can be used in developing a computer program with the following computational procedure.

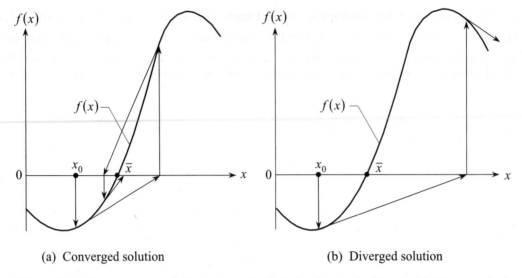

(a) Converged solution   (b) Diverged solution

Figure 2.16 Convergence and divergence solution behaviors by the Newton-Raphson method.

<u>Step 1</u>   Determine the value of the function and its first derivative at the old point x. Then, determine the increment Δx from

$$\Delta x_{k+1} = -\frac{f(x_k)}{f'(x_k)} \tag{2.34}$$

where the subscripts k and $k+1$ represent the iteration numbers k and $k+1$, respectively.

<u>Step 2</u>   Determine the new value x from

$$x_{k+1} = x_k + \Delta x_{k+1} \tag{2.35}$$

<u>Step 3</u>   Check for the convergence of the solution by using one of the following convergence criteria,

(a) $|\Delta x_{k+1}| < \varepsilon_1 \tag{2.36}$

where ε_1 is the absolute error, or

(b) $\left| \frac{\Delta x_{k+1}}{x_{k+1}} \right| < \varepsilon_2 \tag{2.37}$

where ε_2 is the relative error, or

(c) $\left| \frac{\Delta x_{k+1}}{x_{k+1}} \right| \times 100\% < \varepsilon_3 \tag{2.38}$

where ε_3 is the percentage relative error. If the computed solution does not meet the specified convergence criterion, the process is repeated by going back to step 1.

Example 2.6 Develop a computer program to find the root of Eq. (2.12) by using the Newton-Raphson method. Use the initial guess $x_0 = 3$ and the percentage relative in the form of Eq. (2.38) as $\varepsilon = 0.001\%$.

Before developing a computer program, the first derivative of the function is required. The given function in Eq. (2.12) is

$$f(x) = e^{-x/4}(2-x)-1$$

then
$$f'(x) = e^{-x/4}(-1) + \left(-\frac{1}{4}\right)e^{-x/4}(2-x) - 0$$

$$= e^{-x/4}\left(-\frac{3}{2}+\frac{x}{4}\right)$$

These equations are used in the development of a computer program as shown in Fig. 2.17. The computed solutions of x at different iterations are shown in Table 2.5.

Fortran

```
      PROGRAM  NEWRAP
C.....PROGRAM FOR COMPUTING ROOT OF NONLINEAR
C.....EQUATION USING THE NEWTON-RAPHSON METHOD
C.....DEFINE THE FOLLOWING GIVEN VALUES:
C.....     X0 = INITIAL GUESS VALUE OF X
C.....     ES = STOPPING CRITERION TOLERANCE (%)
      X0 = 3.
      ES = 0.001
      WRITE(6,20)
   20 FORMAT(/, 3X, 'ITERATION NO.', 9X, 'X', /)
      X = X0
      DO 100  ITER=1,500
      F = FUNC(X)
      DF = DERIV(X)
      DX = -F/DF
      X = X + DX
      WRITE(6,50)  ITER, X
   50 FORMAT(1X, I8, 8X, E14.6)
      TOL = ABS(DX*100./X)
      IF(TOL.LT.ES)  GO TO 200
  100 CONTINUE
      WRITE(6,110)
  110 FORMAT(/, ' ROOT CAN NOT BE REACHED FOR',
     *          ' THE GIVEN CONDITION'      )
      GO TO 300
  200 WRITE(6,210)  X
  210 FORMAT(/, 3X, 'THE ROOT IS ', E14.6)
  300 CONTINUE
      STOP
      END
C-----------------------------------------------
      FUNCTION FUNC(X)
      FUNC = EXP(-X/4.)*(2.-X) - 1.
      RETURN
      END
C-----------------------------------------------
      FUNCTION DERIV(X)
      DERIV = -EXP(-X/4.)*(1.5-X/4.)
      RETURN
      END
```

MATLAB

```
% PROGRAM NEWRAP
% PROGRAM FOR COMPUTING ROOT OF NONLINEAR
% EQUATION USING THE NEWTON-RAPHSON METHOD
% DEFINE THE FOLLOWING GIVEN VALUES:
%       X0 = INITIAL GUESS VALUE OF X
%       ES = STOPPING CRITERION TOLERANCE (%)
%-------------------------------------------------
func = inline('exp(-X/4.0)*(2.-X) - 1.', 'X');
%-------------------------------------------------
deriv = inline('-exp(-X/4.0)*(1.5-X/4.0)', 'X');
%-------------------------------------------------
X0 = 3.;
ES = 0.001;
fprintf('\n   ITERATION NO.        X \n');
X = X0;
for iter = 1:500
    F  = func(X);
    DF = deriv(X);
    DX = -F/DF;
    X  = X + DX;
    fprintf('  %8d        %14.6e\n', iter, X);
    tol = abs(DX*100./X);
    if tol < ES
        fprintf('\n   THE ROOT IS %14.6e\n', X)
        break
    end
end
if tol > ES
    fprintf('  ROOT CAN NOT BE REACHED FOR\n');
    fprintf('  THE GIVEN CONDITION');
    break
end
```

Figure 2.17 Computer program for finding the root of Eq. (2.12)
by using the Newton-Raphson method.

Table 2.5 Solution convergence of the function in Eq. (2.12) by the Newton-Raphson method.

Iteration number	Root x
1	-1.156000
2	0.189438
3	0.714043
4	0.782542
5	0.783596
6	0.783596

2.7 Secant Method

The key idea of the secant method is the similar to the Newton-Raphson method except the first derivative of the function is obtained approximately. The method is useful when the given function is complex such that its exact derivative cannot be derived easily. The first derivative of the function is estimated from the values of the function at the two points x_0 and x_1 as shown in Fig. 2.18.

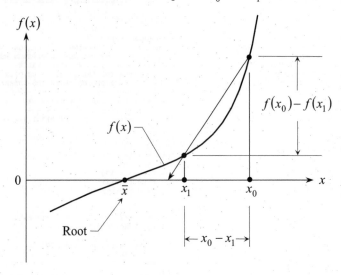

Figure 2.18 Finding the root of equation by the secant method.

$$f'(x_1) = \frac{f(x_0) - f(x_1)}{x_0 - x_1} \qquad (2.39)$$

The approximate value in Eq. (2.39) is used to determine Δx through Eq. (2.33) of the Newton-Raphson method,

$$\Delta x = -\frac{f(x_1)}{f'(x_1)} = -\frac{f(x_1) \cdot (x_0 - x_1)}{f(x_0) - f(x_1)} \qquad (2.40)$$

The computed Δx is then used for determining the new value of x. All other computational steps are identical to the Newton-Raphson method as shown in Eqs. (2.34) – (2.38).

Equation (2.39) and Fig. 2.18 show that the secant method requires two initial values of x to estimate the first derivative by using Eq. (2.39). The following example shows the results from the secant method for finding the root of Eq. (2.12).

Example 2.7 Develop a computer program for finding the root of Eq. (2.12) by using the secant method with the initial values $x_0 = 3$ and $x_1 = 2$. Use the convergence criterion $\varepsilon = 0.001\%$ as shown in Eq. (2.38).

The corresponding computer program is shown in Fig. 2.19 while the computed solutions during the iteration process are shown in Table 2.6.

Fortran

```
      PROGRAM  SECANT
C.....PROGRAM FOR COMPUTING ROOT OF NONLINEAR
C.....EQUATION USING THE SECANT METHOD
C.....DEFINE THE FOLLOWING GIVEN VALUES:
C.....    X0 = FIRST  VALUE OF INITIAL GUESS OF X
C.....    X1 = SECOND VALUE OF INITIAL GUESS OF X
C.....    ES = STOPPING CRITERION TOLERANCE (%)
      X0 = 3.
      X1 = 2.
      ES = 0.001
      WRITE(6,20)
 20   FORMAT(/, 3X, 'ITERATION NO.', 9X, 'X', /)
      DO 100  ITER=1,500
      F0 = FUNC(X0)
      F1 = FUNC(X1)
      DF = (F0-F1)/(X0-X1)
      DX = -F1/DF
      X0 = X1
      X1 = X1 + DX
      WRITE(6,50)  ITER, X1
 50   FORMAT(1X, I8, 8X, E14.6)
      TOL = ABS(DX*100./X1)
      IF(TOL.LT.ES)  GO TO 200
 100  CONTINUE
      WRITE(6,110)
 110  FORMAT(/, ' ROOT CAN NOT BE REACHED FOR',
     *          ' THE GIVEN CONDITIONS'        )
      GO TO 300
 200  WRITE(6,210)  X1
 210  FORMAT(/, 3X, 'THE ROOT IS ', E14.6)
 300  CONTINUE
      STOP
      END
C-------------------------------------------------
      FUNCTION FUNC(X)
      FUNC = EXP(-X/4.)*(2.-X) - 1.
      RETURN
      END
```

MATLAB

```
% PROGRAM SECANT;
% PROGRAM FOR COMPUTING ROOT OF NONLINEAR
% EQUATION USING THE SECANT METHOD
% DEFINE THE FOLLOWING GIVEN VALUES :
%     X0 = FIRST  VALUE OF INITIAL GUESS OF X
%     X1 = SECOND VALUE OF INITIAL GUESS OF X
%     ES = STOPPING CRITERION TOLERANCE (%)
%-------------------------------------------------
func = inline('exp(-X/4.0)*(2.-X) - 1.', 'X');
%-------------------------------------------------
X0 = 3.;
X1 = 2.;
ES = 0.001;
fprintf('\n  ITERATION NO.          X \n');
for iter = 1:500
    F0 = func(X0);
    F1 = func(X1);
    DF = (F0-F1)/(X0-X1);
    DX = -F1/DF;
    X0 = X1;
    X1 = X1 + DX;
    fprintf(' %8d          %14.6e\n', iter, X1);
    tol = abs(DX*100./X1);
    if tol < ES
        fprintf('\n  THE ROOT IS  %14.6e\n', X1)
        break
    end
end
if tol > ES
    fprintf('  ROOT CAN NOT BE REACHED FOR\n');
    fprintf('  THE GIVEN CONDITIONS');
    break
end
```

Figure 2.19 Computer program for finding the root of Eq. (2.12) by the secant method.

Table 2.6 Solution convergence to the root of Eq. (2.12) by the secant method.

Iteration number	Root x
1	-0.111700
2	1.028830
3	0.829093
4	0.781058
5	0.783622
6	0.783596

2.8 MATLAB Functions for Finding Root of Equation

Built-in functions in MATLAB for finding root of equation are explained in this section. The equation is input into MATLAB in the form of a function. MATLAB then creates the function by using the command `inline` in the form,

```
f = inline('expression','variable')
```

where   `f`   is the name of the function
        `expression`   is the expression of the function
        `variable`   contains the variables in the function

For example, the function $f(x) = x^2 - 5$ can be created by using the command

```
>> f = inline('x^2-5','x')
```

MATLAB then responds as

```
f =

     Inline function:
     f(x) = x^2-5
```

To find the value of the function, e.g., $x = 7$,

```
>> f(7)
ans =
     44
```

If the function contains many variables, the same `inline` command can also be used. As an example,

```
>> g = inline('2+5*x+3*y','x','y')
g =
     Inline function:
     g(x,y) = 2+5*x+3*y
```

Root of function is determined by using the `fzero` command. The structure of the command is,

```
y = fzero(function,x0)
```

where   `y`   is the root value
        `function`   is the function name
        `x0`   is the initial guess value

If the bracketing method is employed in the determination of the root, the two initial guess values `x0` and `x1` are specified by using the command,

$$y = \text{fzero(function, [x0 x1])}$$

Finding the root of an equation can be done easily by using the `fzero` command with an initial guess value. For example,

```
>> fzero(f,3)
ans =
      2.2361
```

If a negative value of the roots is needed, a different value for the initial guess may be input,

```
>> fzero(f,-3)
ans =
     -2.2361
```

Root of the equation may also be found by using two initial guess values. In this case, the command is

```
>> fzero(f,[-3 1])
ans =
     -2.2361
```

If the root is not contained within the two given values, MATLAB will display the error message as follows,

```
>> fzero(f,[3 5])
??? Error using ==> fzero
The function values at the interval endpoints must differ in sign.
```

Example 2.8   Use the `fzero` command to find the root of Eq. (2.12) with the initial guess value of $x = 3$.

Before the root of equation is determined, the function in Eq. (2.12) must be input into MATLAB by using the `inline` command. Then, the function `fzero` is used to find the root of the equation as follows,

```
>> f = inline('exp(-x/4).*(2-x)-1','x');
>> y = fzero(f,3)
y =
      0.7836
```

It is noted that the `inline` command can be combined together with the `fzero` command to find the root of equation,

```
>> fzero(inline('exp(-x/4).*(2-x)-1','x'),3)
ans =
      0.7836
```

If the root of the equation is imaginary, the `fzero` command should not be used. Another popular command for finding the imaginary roots of a polynomial function is the `roots` command as,

$$y = \text{roots(coef)}$$

where   y   is the vector of the roots

coef   is the vector of the polynomial coefficients

Example 2.9   Find the roots of the equation which is in the form of polynomials,

$$f(x) \;=\; x^4 - 9x^3 - 2x^2 + 120x - 130 \;=\; 0$$

The roots command is first used to create a vector that contains the coefficients of the polynomials as,

```
>> a = [1 -9 -2 120 -130];
```

The roots command is then employed to find the roots of the equation,

```
>> x = roots(a)

x =

      7.3995
     -3.6001
      3.9721
      1.2286
```

Example 2.10 Find the roots of the equation which is in the form of polynomials,

$$f(x) \;=\; x^4 - x^3 - 2.75x^2 + 5.25x - 2.5 \;=\; 0$$

The roots command can be used to find the real and imaginary roots of the equation as follows,

```
>> b = [1 -1 -2.75 5.25 -2.5];
>> x = roots(b)

x =

     -2.0000
      1.0000 + 0.5000i
      1.0000 - 0.5000i
      1.0000
```

2.9 Roots of System of Non-linear Equations

The topics presented earlier are for finding root of a single equation that is in the form of $f(x) = 0$. In practice, analyses of engineering and scientific problems often lead to a system of non-linear equations. Fundamentals of the methods explained earlier can be used to find for the roots of such system of non-linear equations.

The system of non-linear equations, that consists of n equations with n unknowns of $x_1, x_2, \ldots, x_n$, can be written in a general form as,

$$\left.\begin{array}{rcl} f_1(x_1, x_2, \ldots, x_n) &=& 0 \\ f_2(x_1, x_2, \ldots, x_n) &=& 0 \\ \vdots & & \vdots \\ f_n(x_1, x_2, \ldots, x_n) &=& 0 \end{array}\right\} \; n \; equations \qquad (2.41)$$

or in the matrix form,

$$[A]\{X\} = \{B\} \qquad (2.42)$$
$$(n \times n) \ (n \times 1) \qquad (n \times 1)$$

where $[A]=[A(x_i)]$ is the square matrix which contains the unknown variables x_i, $i=1,2,...,n$; $\{X\}$ is the column vector of the unknowns x_i, and $\{B\}$ is the vector that consists of known values. Examples of the matrices in Eq. (2.42) are shown in Example 2.11 and 2.12.

The methods for solving the system of non-linear Eq. (2.41) presented herein are: (1) the direct iteration method, and (2) the Newton-Raphson iteration method.

2.9.1 Direct iteration method

The main idea of direct iteration method for solving the system of non-linear equations is similar to the one-point iteration method for finding the root of a single equation as explained in section 2.5. The method starts from rewriting the functions in Eq. (2.41) in the form for performing iteration as,

$$x_1^{k+1} = g_1\left(x_1^k, x_2^k,..., x_n^k\right)$$
$$x_2^{k+1} = g_2\left(x_1^k, x_2^k,..., x_n^k\right)$$
$$\vdots \qquad \vdots \qquad (2.43)$$
$$x_n^{k+1} = g_n\left(x_1^k, x_2^k,..., x_n^k\right)$$

where k is the iteration number. The computational procedure of the method consists of the following steps.

Step 1   Provide the initial guess x_i^k, $i=1,2,...,n$

Step 2   Compute x_i^{k+1}, $i=1,2,...,n$ by using Eq. (2.43)

Step 3   Check for solution convergence of x_i by using the criterion such as,

$$\left|\frac{x_i^{k+1} - x_i^k}{x_i^{k+1}}\right| \times 100\% < \varepsilon \qquad (2.44)$$

If the specified convergence criterion is not met, the process is repeated by going back to step 2.

Example 2.11 Use the direct iteration method for solving the system of non-linear equations in the matrix form below,

$$\begin{bmatrix} -1 & 1 \\ 1 & x_2 \end{bmatrix} \begin{Bmatrix} x_1 \\ x_2 \end{Bmatrix} = \begin{Bmatrix} 1 \\ 5 \end{Bmatrix} \qquad (2.45)$$

The two equations in Eq. (2.45) are first expressed as,

$$-x_1 + x_2 = 1 \qquad (2.46a)$$

$$x_1 + x_2^2 = 5 \qquad (2.46b)$$

Equations (2.46a-b) are then written in the iteration form,

$$x_1^{k+1} = x_2^k - 1 \tag{2.47a}$$

$$x_2^{k+1} = \sqrt{5 - x_1^k} \tag{2.47b}$$

By providing the initial guess of $x_1 = x_2 = 0$, the new values of x_1 and x_2 are determined from Eqs. (2.47a-b). The computed solutions of x_1 and x_2 at each iteration are shown in Table 2.7.

Table 2.7 Convergence of solutions to the roots of Eq. (2.45) by the direct iteration method.

Iteration number	x_1	x_2
0	0.000	0.000
1	-1.000	2.236
2	1.236	2.449
3	1.449	1.940
4	0.940	1.884
⋮	⋮	⋮
12	1.000	2.000

2.9.2 Newton-Raphson Iteration Method

The basic idea of the Newton-Raphson method for solving a set of non-linear equations in the form of Eq. (2.41) is similar to that for finding the root of a single equation as explained in section 2.6. The only difference is the use of the Taylor series for n variables. For a typical equation i, the Taylor series for n variables is given by

$$f_i(x_1 + \Delta x_1,\ x_2 + \Delta x_2,\ \dots,\ x_n + \Delta x_n)$$

$$= f_i(x_1, x_2, \dots, x_n) + \sum_{j=1}^{n} \frac{\partial f_i}{\partial x_j}(x_1, x_2, \dots, x_n)\Delta x_j + \dots \tag{2.48}$$

where x_i, $i = 1, 2, \dots, n$ are the values used for determining the new values of $x_i + \Delta x_i$. If only the first two terms on the right-hand side of Eq. (2.48) are used in the approximation, the resulting equation will be in the similar form as shown in Eq. (2.31), i.e.,

$$0 = f_i(x_1, x_2, \dots, x_n) + \sum_{j=1}^{n} \frac{\partial f_i}{\partial x_j}(x_1, x_2, \dots, x_n)\Delta x_j \tag{2.49}$$

or,

$$\sum_{j=1}^{n} \frac{\partial f_i}{\partial x_j}\Delta x_j = -f_i \tag{2.50}$$

For example, if a system of non-linear equations consists of only 3 equations, then $n = 3$ and $i, j = 1, 2, 3$. Equation (2.50) can be written explicitly as,

$$
\begin{bmatrix}
\dfrac{\partial f_1}{\partial x_1} & \dfrac{\partial f_1}{\partial x_2} & \dfrac{\partial f_1}{\partial x_3} \\[2mm]
\dfrac{\partial f_2}{\partial x_1} & \dfrac{\partial f_2}{\partial x_2} & \dfrac{\partial f_2}{\partial x_3} \\[2mm]
\dfrac{\partial f_3}{\partial x_1} & \dfrac{\partial f_3}{\partial x_2} & \dfrac{\partial f_3}{\partial x_3}
\end{bmatrix}
\underbrace{\begin{Bmatrix} \Delta x_1 \\ \Delta x_2 \\ \Delta x_3 \end{Bmatrix}}_{\{\Delta x\}}
= -\underbrace{\begin{Bmatrix} f_1 \\ f_2 \\ f_3 \end{Bmatrix}}_{\{f\}}
\tag{2.51}
$$
$$
\underbrace{\phantom{\begin{bmatrix} \dfrac{\partial f_1}{\partial x_1} \end{bmatrix}}}_{[J]}
$$

For a set of n non-linear equations, the Newton-Raphson iteration method leads to,

$$
\underset{(n\times n)\ (n\times1)}{[J]\{\Delta x\}} = \underset{(n\times1)}{-\{f\}}
\tag{2.52a}
$$

where $[J]$ is called the Jacobian matrix for which its coefficients are determined from

$$
J_{ij} = \frac{\partial f_i}{\partial x_j}
\tag{2.52b}
$$

The vector $\{\Delta x\}$ contains increments of the solutions. The vector $\{f\}$ on the right-hand side of the equation consists of the function f_i, $i = 1, 2, \ldots, n$ which are evaluated at x_i. This vector $\{f\}$ is sometimes called the residual vector because it becomes zero as all x_i converge to the correct solutions.

From the above explanation, the computational procedures of the Newton-Raphson iteration method for finding the roots of non-linear equations are as follows.

<u>Step 1</u>   Solve the system of equations,

$$
[J]^k \{\Delta x\}^{k+1} = -\{f\}^k
\tag{2.53}
$$

where the superscript k denotes the iteration number.

<u>Step 2</u>   Compute the new solution values from

$$
\{x\}^{k+1} = \{x\}^k + \{\Delta x\}^{k+1}
\tag{2.54}
$$

<u>Step 3</u>   Check for solution convergence by using the criterion as shown in Eq. (2.36) – (2.38). If the computed solutions are not converged within the specified criterion, the process is repeated by going back to step 1.

Example 2.12 Use the Newton-Raphson iteration method to solve the system of non-linear equations given in matrix form,

$$
\begin{bmatrix}
1 & 2 & 1 & 4 \\
x_1 & 2x_1 & 0 & x_4^2 \\
x_1^2 & 0 & x_3 & 1 \\
0 & 3 & 0 & x_3
\end{bmatrix}
\begin{Bmatrix} x_1 \\ x_2 \\ x_3 \\ x_4 \end{Bmatrix}
=
\begin{Bmatrix} 20.700 \\ 15.880 \\ 21.218 \\ 21.100 \end{Bmatrix}
\tag{2.55}
$$

<u>Solution</u> From Eq. (2.55), the four functions are,

$$
\begin{aligned}
f_1 &= x_1 + 2x_2 + x_3 + 4x_4 - 20.700 \\
f_2 &= x_1^2 + 2x_1 x_2 + x_4^3 - 15.880 \\
f_3 &= x_1^3 + x_3^2 + x_4 - 21.218 \\
f_4 &= 3x_2 + x_3 x_4 - 21.100
\end{aligned}
\tag{2.56}
$$

It is noted that if the correct solutions of x_1, x_2, x_3 and x_4 are substituted into the Eq.(2.56), the four functions of f_i, $i = 1, 2, 3, 4$ are all zero.

The size of the Jacobian matrix for this problem is (4×4) and its coefficients are determined from Eq. (2.52b) as,

$$
[J] = \begin{bmatrix}
1 & 2 & 1 & 4 \\
2x_1 + 2x_2 & 2x_1 & 0 & 3x_4^2 \\
3x_1^2 & 0 & 2x_3 & 1 \\
0 & 3 & x_4 & x_3
\end{bmatrix}
\tag{2.57}
$$

If the initial guess values are $x_1 = x_2 = x_3 = x_4 = 1$, then Eq. (2.53) becomes,

$$
\begin{bmatrix}
1 & 2 & 1 & 4 \\
4 & 2 & 0 & 3 \\
3 & 0 & 2 & 1 \\
0 & 3 & 1 & 1
\end{bmatrix}
\begin{Bmatrix}
\Delta x_1 \\
\Delta x_2 \\
\Delta x_3 \\
\Delta x_4
\end{Bmatrix}
= -
\begin{Bmatrix}
-12.700 \\
-11.880 \\
-18.218 \\
-17.100
\end{Bmatrix}
\tag{2.58}
$$

The procedure for solving a system of Eq. (2.58) will be explained in details in chapter 3. The solutions for the system of equations above are,

$$
\begin{Bmatrix}
\Delta x_1 \\
\Delta x_2 \\
\Delta x_3 \\
\Delta x_4
\end{Bmatrix}
=
\begin{Bmatrix}
1.75037 \\
3.67630 \\
6.89579 \\
-0.82469
\end{Bmatrix}
\tag{2.59}
$$

Thus, the new values of the roots are computed by using Eq. (2.54),

$$
\begin{Bmatrix}
x_1 \\
x_2 \\
x_3 \\
x_4
\end{Bmatrix}
=
\begin{Bmatrix}
1. \\
1. \\
1. \\
1.
\end{Bmatrix}
+
\begin{Bmatrix}
1.75037 \\
3.67630 \\
6.89579 \\
-0.82469
\end{Bmatrix}
=
\begin{Bmatrix}
2.75037 \\
4.67630 \\
7.89579 \\
0.17531
\end{Bmatrix}
\tag{2.60}
$$

Convergence of the solutions is then checked by using the specified criterion. If the convergence criterion is not met, the procedure is repeated by determining the Jacobian matrix in Eq. (2.57) with the new computed values from Eq. (2.60). Table 2.8 shows the computed solutions with 6 significant figures that are obtained different iterations. The final solutions can be verified by substituting them back into the functions in Eq. (2.56) so that all functions must be zero.

Table 2.7 Converged solutions to the roots of the non-linear equations in Example 2.12.

Iteration number	x_1	x_2	x_3	x_4
0	1.00000	1.00000	1.00000	1.00000
1	2.75037	4.67630	7.89579	0.17531
2	1.34485	5.29712	5.94935	0.70289
3	1.47750	3.84372	4.34185	1.79830
.	.	.	.	.
.	.	.	.	.
.	.	.	.	.
.	.	.	.	.
8	1.20000	5.60000	4.30000	1.00000

2.10 Closure

In this chapter, the methods for finding roots of a single equation and a set of non-linear equations were presented. By understanding the procedure of these methods together with ability to develop the corresponding computer programs, the roots of the equation can be obtained easily. Finding the root of a single equation is an important topic that occurs in the analysis of many engineering problems. Understanding the methods can help analysts to determine and verify the computed solutions.

Some of the methods used for finding solution of a single equation were extended to solve for solutions of a set of non-linear equations. The set of non-linear equations often occurs in the analysis of practical applications. Some popular methods, such as the Newton-Raphson iteration method, have been used and built inside the commercial software today. Software users need to understand the procedures for solving these non-linear equations to assure the accuracy of the computed solutions.

Exercises

1. In the vibration analysis of a stop sign pole, the root of the following transcendental function is required,

$$\cosh x \, \cos x + 1 \; = \; 0$$

Find the first three positive roots by using: (a) the graphical method, (b) the bisection method, (c) the false-position method and (d) the Newton-Raphson method. Use appropriate initial guess values in order to obtain the solution with 6 significant figures.

2. Employ the bisection method to determine the value of $\sqrt[4]{13}$ by solving the equation,

$$x^4 - 13 \; = \; 0$$

Use the initial left and right values of 1.5 and 2.0. Perform iteration until the solution with six significant figures does not alter.

3.   Employ the false-position method to determine the value of $1/43$ by solving the following equation

$$\frac{1}{x} - 43 \; = \; 0$$

Use the initial left and right values of 0.02 and 0.03. Determine the solution by using the convergence criterion with the tolerance of 0.000001%.

4.   Apply the Taylor series to determine the value of the function

$$f(x) \; = \; \ln x$$

at $x=4$ by using the value of the function and its derivatives at $x = 2$. Determine values of the function from the Taylor series of order 0 to 6 and compute the errors for each case.

5.   Apply the Taylor series and develop a computer program to determine the value of the function

$$f(x) \; = \; e^{-x}$$

at $x = 2$ by using the value of function and its derivatives at $x = 1$. Determine values of the function from the Taylor series of order 0 to 6 and compute the errors for each case.

6.   In the determination for the oblique shock angle β generated from a flow of Mach M over an inclined plan that makes angle θ with the horizontal axis, the root of the equation,

$$\tan \theta \; - \; 2 \cot \beta \left[\frac{M^2 \sin^2 \beta - 1}{M^2 (\gamma + \cos 2\beta) + 2} \right] \; = \; 0$$

is needed. In the above equation, γ is the specific heat ratio for air which is 1.4. The flow is at Mach $M = 3$ and $\theta = \pi/9$. Determine the angle β of the oblique shock by using: (a) the bisection method with the initial left and right values of 0 and $\pi/4$, and (b) the secant method with the initial left and right values of $\pi/6$ and $\pi/4$. The computed solution of the angle β must have accuracy up to 6 significant figures.

7.   Find the roots of the following equations by using the Newton-Raphson method. The numbers in parentheses are the initial guess values for each case. The computed solutions must have their accuracy up to 4 significant figures.

(a)	$x^2 - 2x - 3$	$= 0$	(0.5)
(b)	$xe^{-x^2} - \cos x$	$= 0$	(2.0)
(c)	$10 \ln x - x$	$= 0$	(1.0)
(d)	$e^x - \left(1 + x + x^2/2\right)$	$= 0$	(1.5)
(e)	$x^3 - 100$	$= 0$	(2.0)
(f)	$e^x - \sin(\pi x/3)$	$= 0$	(-2.8)

8. Find the value of $\sqrt{7}$ with its accuracy up to 8 significant figures by using the Newton-Raphson method. It is noted that the solution can be obtained from the equation.

$$x^2 - 7 = 0$$

Use the initial guess of $x = 2$.

9. Use the Newton-Raphson method to determine the eigenvalue that occurs in the buckling analysis of a vertical bar due to its own weight. The eigenvalue β is the root of the equation,

$$1 + \sum_{m=1}^{\infty} C_m \beta^{2m} = 0$$

where $\qquad m = 1 \qquad\qquad C_1 = -\dfrac{3}{8}$

and $\qquad m \geq 2 \qquad\qquad C_m = -\dfrac{3\,C_{m-1}}{4m\,(3m-1)}$

Use the initial guess of zero and the computed solution must have their accuracy up to 6 significant figures.

10. In the vibration analysis of a beam that has a mass attached at the middle of it, the root of the equation,

$$2 - x(\tan x - \tanh x) = 0$$

is needed. Use: (a) the bisection method, (b) the Newton-Raphson method and (c) function `fzero` of MATLAB for finding the root. The computed solutions must have their accuracy up to 6 significant figures. Hint: the root of this equation is between 0.5 and 1.5.

11. A cable with the length $\ell = 180$ m and weight $w = 36$ N/m hangs between two poles. The distance between the two poles is $L = 165$ m. The tension H in the cable is determined from

$$\sinh \frac{wL}{2H} = \frac{w\ell}{2H}$$

Use: (a) the Newton-Raphson method, (b) the secant method and (c) function `fzero` of MATLAB to find the tension H in the cable. The computed solutions must have their accuracy up to 6 significant figures.

12. Find the drag coefficient of the space shuttle as shown in Eq. (2.10) by using: (a) the bisection method, (b) the false-position method and (c) the Newton-Raphson method. The computed solutions must have their accuracy up to 6 significant figures.

13. Show that the equation,

$$x^4 - 9x^3 - 2x^2 + 120x - 130 = 0$$

has four roots. Then, use any method explained in this chapter to find those roots with their accuracy up to 8 significant figures.

14.   Derive related equations and explain computational procedures for determining the value of $\sqrt[3]{C}$ by using the Newton-Raphson method. The value C is a positive real number. Then develop a computer program to determine the value C by using the stopping criterion of $\varepsilon = 0.000001\%$ as shown in Eq. (2.38).

15.   Use the Newton-Raphson method to find the root of the equation,

$$\cos x = x e^x$$

with the initial guess $x = 1$. Perform iteration until the solution with six significant figures does not alter.

16.   Show that the nth root of any constant C can be obtained by solving the equation

$$x^n - C = 0$$

Also show that the equation for finding the root of the above equation by the Newton-Raphson method is in form

$$x_{k+1} = \frac{1}{n}\left[(n-1)x_k + \frac{C}{x_k^{n-1}}\right]$$

where the subscript k is the iteration number.

17.   Use the computer program of the secant method to find the root of the equation in Example 2.7 with initial guess values $x_0 = 3$ and $x_1 = 1$. Discuss the convergence behavior as compared to the solution in Example 2.7. Plot graph to compare the convergence rates of the two solutions.

18.   Energy loss due to friction of flow in a pipe is required for selecting an appropriate size of the water pump. If the pipe has diameter D and length L, the energy loss is determined from,

$$h_l = f \frac{L}{D} \frac{\bar{V}^2}{2}$$

where h_l is energy loss, $\bar{V}$ is the average flow velocity and f is the friction factor. If the inside surface of pipe is not smooth, the friction factor is determined from,

$$\frac{1}{f^{0.5}} = -2\log\left(\frac{\varepsilon}{3.7} + \frac{2.51}{Re\, f^{0.5}}\right)$$

where ε is the roughness value of the surface and Re is the Reynold number. Use any built-in function of MATLAB to find the value of friction factor when $\varepsilon = 0.01$ and $Re = 100,000$. Compare the computed solution with the solution obtained from any method explained in this chapter.

19.   A cable was hung between the two points at the same level. The distance between the two points is 250 m as shown in Fig. P2.19.

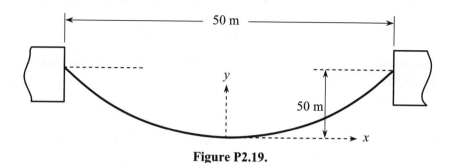

Figure P2.19.

The cable deflects by its own weight in a parabolic shape and the maximum deflection occurs at middle of the cable. The defection curve for the cable is in the form,

$$y = \frac{T_0}{w}\left(\cosh\frac{wx}{T_0} - 1\right)$$

where y is the deflection distance, w is the cable weight per unit length, T_0 is the cable tension at the position of maximum deflection, and x is the coordinate in the horizontal direction. If $w = 120$ N/m at the position $x = 125$ m and the deflection $y = 50$ m, use MATLAB to find the value of the tension T_0. Compare the computed solution with the solution obtained from the one-point iteration method.

20. A beam as shown in Fig. P2.20 is subjected to the load which increases along the beam length. The equation for the beam deflection curve is,

$$y = -\frac{wx}{360EIL}\left(3x^4 - 10L^2x^2 + 7L^4\right)$$

where y is the deflection of beam, x is the distance along the beam axis, w is the load along beam, E is the Young's modulus, I is the moment of inertia of the beam cross-section and L is the beam length. Employ MATLAB to find the location of the maximum deflection that occurs along the beam and its value. Use the following data of $L = 450$ cm, $E = 50,000$ kN/cm^2, $I = 30,000$ cm^4 and $w = 1.75$ kN/cm in the analysis. It is noted that the maximum deflection occurs at the location of $dy/dx = 0$.

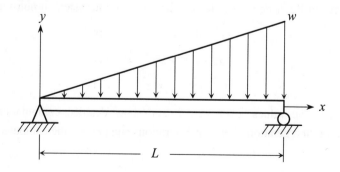

Figure P2.20.

21.   Use the `roots` command in MATLAB to find all roots of the polynomial equation,

$$x^5 - 5x^4 - 35x^3 + 125x^2 + 194x - 280 \ = \ 0$$

Compare the solutions with the results obtained from any method in this chapter. Then, explain advantages and disadvantages of each method.

22.   Use the `roots` command in MATLAB to find all roots of the polynomial equation,

$$x^5 - 3.5x^4 + 2.75x^3 + 2.125x^2 - 3.875x + 1.25 \ = \ 0$$

Compare the solutions with the results obtained from any method in this chapter.

23.   Develop a computer program by using the direct iteration method for solving the system of non-linear equations,

$$x^3 - 6y \ = \ 2$$

$$5x + 2y^2 \ = \ 12$$

Then, compare the computed solutions with the exact solutions.

24.   Use the Newton-Raphson method to find the roots for the system of non-linear equations with 3 significant figures,

$$x^2y - xy^2 + 6 \ = \ 0$$

$$y^2 + 4x^2 - 3xy - 7 \ = \ 0$$

25.   Use the Newton-Raphson method to find the roots for the system of non-linear equations

$$xy \ = \ 1 \qquad \text{and} \qquad y \ = \ \cos h \, x$$

Show the detailed Jacobian matrix and use the initial values $x = y = 1$ to compute the solutions. Perform 5 iterations and compare the computed results with the exact solutions.

26.   Use the Newton-Raphson method to find the roots for the system of non-linear equations,

$$3x^2 + 4y - 5yz \ = \ -19$$

$$-xz + 2y^2 + z^2 \ = \ 14$$

$$xy^2 - 3yz + x^2z \ = \ -11$$

Employ the initial values $x = y = z = 2$ and show detailed computation with 4 significant figures for the first iteration. Perform 5 iterations and present the computed solutions in a table.

27. Develop a computer program by using the Newton-Raphson method for solving the system of non-linear equations,

$$5x_1x_3 - 2x_1x_2 + 4x_3^2 - x_2x_4 = 9.75$$

$$6x_1 + 3x_2 + x_3 - x_4 = 5.50$$

$$2x_1^2 + x_2x_3 - 5x_3 + x_1x_4^2 = -3.50$$

$$-3x_1x_4 - 2x_2^2 + 6x_3x_4 + x_3x_4 = 16.00$$

Use the initial guess values $x_1 = x_2 = x_3 = x_4 = 1$ and perform the computation until the solutions converge to 6 significant figures of accuracy.

Chapter
3

System of Linear Equations

3.1 Introduction

In chapter 2, several techniques for finding root of a single equation often occurs during the analysis of engineering problems were studied. For practical problems, many numerical techniques such as the finite difference and finite element techniques are used for obtaining solutions that describe their physical behaviors. In the analysis process, these methods lead to a set of simultaneous algebraic equations for which their roots representing the solutions of the problems are required. For examples, the finite difference technique was used to determine the flow phenomena surrounding a fighter jet in Fig. 1.1, while the finite element technique was employed to predict the structure deformation of passenger car during its collision in Fig. 1.2. During the analyses of these two problems, both techniques generate the large sets of algebraic simultaneous equations. The unknowns of the problems which are the roots for the sets of the algebraic equations must be solved. Understanding the methods for solving such system of equations together with the ability for developing computer programs are thus very important in the process for solving practical engineering problems.

In this chapter, various methods for solving the system of equations will be explained. Advantages and disadvantages of each method will be described. To illustrate the computational procedure for each method clearly, a small size of a system of equations will be used as an example. However, the computational procedures can be applied to solve a large size of the system of equations.

To demonstrate that a set of equations is arisen during the analysis of a typical problem, the following example is studied.

Example 3.1 A square metal plate, with the thickness t and thermal conductivity k, is subjected to a surface heating q as shown in Fig. 3.1. The origin of the x-y coordinates is at the plate center as shown in the figure. Zero temperature is specified as the boundary conditions along the four edges of the plate. With the specified surface heating and boundary conditions, high temperature occurs at the plate center with the temperature distribution $T(x, y)$ as shown in the figure.

The governing equation that represents the conservation of energy for determining the temperature distribution over the plate is

$$\frac{\partial^2 T}{\partial x^2} + \frac{\partial^2 T}{\partial y^2} = -\frac{q}{kt} \tag{3.1}$$

If the size of the plate is 4×4 unit and $q/kt = 400$, the exact temperature distribution is

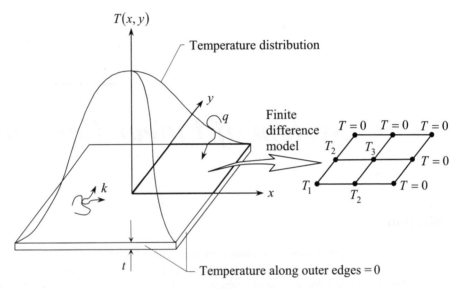

Figure 3.1 Temperature distribution on the square plate subjected to a surface heating and the finite difference model.

$$T(x, y) = 200\left(4 - x^2\right) - \frac{25{,}600}{\pi^3} \sum_{n=0}^{\infty} \frac{(-1)^n \cos\left((2n+1)\pi x/4\right) \cosh\left((2n+1)\pi y/4\right)}{(2n+1)^3 \cosh\left((2n+1)\pi/2\right)} \tag{3.2}$$

Approximate solution of the plate temperature can be obtained by using the finite difference technique. Details for the procedure of the finite difference technique will be explained in chapter 8. The technique starts from dividing the computational domain into a mesh with rectangular shape as shown in Fig. 3.1. Due to the symmetry of the temperature solution, only the upper-right quarter of the plate can be used for modeling in the analysis. If this quarter is divided into 2×2 intervals, the unknowns are the temperatures T_1, T_2 and T_3 at the grid points as shown in the figure. The finite difference technique leads to the set of equations in the form

$$\begin{bmatrix} 4 & -4 & 0 \\ -1 & 4 & -2 \\ 0 & -2 & 4 \end{bmatrix} \begin{Bmatrix} T_1 \\ T_2 \\ T_3 \end{Bmatrix} = \begin{Bmatrix} 400 \\ 400 \\ 400 \end{Bmatrix} \tag{3.3}$$

The computed solutions at these grid points are $T_1 = 450$, $T_2 = 350$ and $T_3 = 275$. It should be noted that the computed temperatures at the grid points contain errors from the model discretization. Higher solution accuracy can be obtained by modeling the plate with a finer mesh. Such the finer mesh, will lead to a set of more equations. For example, if the upper-right quarter of the plate is divided into 10×10 intervals, there will be a total 55 unknowns which are solved from 55 equations. Thus, in general, a system of equations can be written in the matrix form as,

$$\underset{(n \times n)\,(n \times 1)}{[A]\{X\}} = \underset{(n \times 1)}{\{B\}} \tag{3.4}$$

where n is the number of equations, $[A]$ is a square matrix with the size of $(n \times n)$, the vector $\{X\}$ is a column matrix containing n unknowns and the vector $\{B\}$ is a column matrix with the known values. Definitions of matrices, their properties and manipulations are explained in appendix A.

In this chapter, several numerical methods for solving the system of equations in the form of Eq. (3.4) are explained. These methods are: (1) the Cramer's rule, (2) the Gauss elimination method, (3) the Gauss-Jordan method, (4) the matrix inversion method, (5) the LU decomposition method, (6) the Cholesky decomposition method, (7) the Jacobi iteration method, (8) the Gauss-Seidel iteration method, (9) the successive over-relaxation method, and (10) the conjugate gradient method.

3.2 Cramer's Rule

Cramer's rule is the method suitable for solving a system of few equations. The method solves for the solutions by finding determinants of the matrices. The unknown x_i in Eq. (3.4) is determined from

$$x_i = \frac{\det [A]_i}{\det [A]} \tag{3.5}$$

where $\det [A]$ is the determinant of matrix $[A]$, and $\det [A]_i$ is the determinant of same matrix $[A]$ but its column i is replaced by the vector $\{B\}$. The use of Cramer's rule is illustrated by the following examples.

Example 3.2   Use Cramer's rule to solve the following system of equations,

$$\begin{bmatrix} 2 & 1 \\ 1 & -1 \end{bmatrix} \begin{Bmatrix} x_1 \\ x_2 \end{Bmatrix} = \begin{Bmatrix} 4 \\ -1 \end{Bmatrix} \tag{3.6}$$

Here, the matrix $[A]$ and vector $\{B\}$ are

$$[A] = \begin{bmatrix} 2 & 1 \\ 1 & -1 \end{bmatrix} \quad ; \quad \{B\} = \begin{Bmatrix} 4 \\ -1 \end{Bmatrix}$$

By using the Cramer's rule as shown in Eq. (3.5), the unknowns x_1 and x_2 can be determined as follows

$$x_1 = \frac{\det[A]_1}{\det[A]} = \frac{\begin{vmatrix} 4 & 1 \\ -1 & -1 \end{vmatrix}}{\begin{vmatrix} 2 & 1 \\ 1 & -1 \end{vmatrix}} = \frac{-4+1}{-2-1} = \frac{-3}{-3} = 1$$

$$x_2 = \frac{\det[A]_2}{\det[A]} = \frac{\begin{vmatrix} 2 & 4 \\ 1 & -1 \end{vmatrix}}{\begin{vmatrix} 2 & 1 \\ 1 & -1 \end{vmatrix}} = \frac{-2-4}{-2-1} = \frac{-6}{-3} = 2$$

Example 3.3   Use the Cramer's rule to solve Eq. (3.3),

$$\begin{bmatrix} 4 & -4 & 0 \\ -1 & 4 & -2 \\ 0 & -2 & 4 \end{bmatrix} \begin{Bmatrix} x_1 \\ x_2 \\ x_3 \end{Bmatrix} = \begin{Bmatrix} 400 \\ 400 \\ 400 \end{Bmatrix}$$

For this example, the matrix $[A]$ and vector $\{B\}$ are

$$[A] = \begin{bmatrix} 4 & -4 & 0 \\ -1 & 4 & -2 \\ 0 & -2 & 4 \end{bmatrix} \quad ; \quad \{B\} = \begin{Bmatrix} 400 \\ 400 \\ 400 \end{Bmatrix}$$

By using the Cramer's rule as shown in Eq. (3.5), the three unknowns are determined from

$$x_1 = \frac{\det[A]_1}{\det[A]} = \frac{\begin{vmatrix} 400 & -4 & 0 \\ 400 & 4 & -2 \\ 400 & -2 & 4 \end{vmatrix}}{\begin{vmatrix} 4 & -4 & 0 \\ -1 & 4 & -2 \\ 0 & -2 & 4 \end{vmatrix}}$$

$$= \frac{400(16-4) + 4(1{,}600+800) + 0}{4(16-4) + 4(-4-0) + 0} = \frac{14{,}400}{32} = 450$$

$$x_2 = \frac{\det[A]_2}{\det[A]} = \frac{\begin{vmatrix} 4 & 400 & 0 \\ -1 & 400 & -2 \\ 0 & 400 & 4 \end{vmatrix}}{\begin{vmatrix} 4 & -4 & 0 \\ -1 & 4 & -2 \\ 0 & -2 & 4 \end{vmatrix}}$$

$$= \frac{4(1{,}600+800) - 400(-4-0) + 0}{4(16-4) + 4(-4-0) + 0} = \frac{11{,}200}{32} = 350$$

$$x_3 = \frac{\det[A]_3}{\det[A]} = \frac{\begin{vmatrix} 4 & -4 & 400 \\ -1 & 4 & 400 \\ 0 & -2 & 400 \end{vmatrix}}{\begin{vmatrix} 4 & -4 & 0 \\ -1 & 4 & -2 \\ 0 & -2 & 4 \end{vmatrix}}$$

$$= \frac{4(1,600+800) + 4(-400-0) + 400(2-0)}{4(16-4) + 4(-4-0) + 0} = \frac{8,800}{32} = 275$$

The two examples above show that solutions of the system of equations can be obtained conveniently by using the Cramer's rule. However, it should be noted that the Cramer's rule is not used for solving a large system of equations. This is mainly because the method requires a large number of operations as compared to other methods. The number of operations required by the Cramer's rule to solve a set of n equations is $(n-1) \cdot (n+1)!$. If a set of equations contains only 10 equations, the method needs 360 million operations. Furthermore, about 10^{157} operations will be needed for solving a set of 100 equations. The method was thus never been used for solving practical problems that consist of a large number of equations. In the next topic, the Gauss elimination method will be presented. The method uses only $(4n^3 + 9n^2 - 7n)/6$ operations for solving a system of n equations. The method, thus, requires only about 700,000 operations to solve a set of 100 equations.

3.3 Gauss Elimination Method

The Gauss elimination method is one of the most popular methods for solving systems of equations generated from the computational procedures in scientific and engineering analyses. The Gauss elimination method for solving a set of equations is divided into two main steps as follows.

(a) *Forward elimination.* For example, if the system of equations consists of 3 equations as,

$$\begin{bmatrix} a_{11} & a_{12} & a_{13} \\ a_{21} & a_{22} & a_{23} \\ a_{31} & a_{32} & a_{33} \end{bmatrix} \begin{Bmatrix} x_1 \\ x_2 \\ x_3 \end{Bmatrix} = \begin{Bmatrix} b_1 \\ b_2 \\ b_3 \end{Bmatrix} \tag{3.7}$$

The first step of the method is to modify the matrix $[A]$ on the left-hand side of the equation so that all coefficients beneath its diagonal line are zero,

$$\begin{bmatrix} a_{11} & a_{12} & a_{13} \\ 0 & a'_{22} & a'_{23} \\ 0 & 0 & a''_{33} \end{bmatrix} \begin{Bmatrix} x_1 \\ x_2 \\ x_3 \end{Bmatrix} = \begin{Bmatrix} b_1 \\ b'_2 \\ b''_3 \end{Bmatrix} \tag{3.8}$$

where the coefficients with prime symbols have different values from those shown in the original Eq. (3.7).

(b) *Back substitution.* With the equations in the form of Eq. (3.8) obtained from the first step, the unknowns x_i can be determined by using back substitution from the last to the first equation as follows,

$$x_3 = b_3''/a_{33}''$$

$$x_2 = (b_2' - a_{23}' x_3)/a_{22}'$$ (3.9)

$$x_1 = (b_1 - a_{12} x_2 - a_{13} x_3)/a_{11}$$

To understand the procedure of the Gauss elimination method more clearly, the set of three equations solved by the Cramer's rule in example 3.3 is repeated as presented below.

Example 3.4 Use the Gauss elimination method to solve the system of equations as shown in Eq. (3.3)

$$\begin{bmatrix} 4 & -4 & 0 \\ -1 & 4 & -2 \\ 0 & -2 & 4 \end{bmatrix} \begin{Bmatrix} x_1 \\ x_2 \\ x_3 \end{Bmatrix} = \begin{Bmatrix} 400 \\ 400 \\ 400 \end{Bmatrix}$$ (3.10)

The above system of three equations can be written explicitly as

$$4x_1 - 4x_2 \qquad = \quad 400$$ (3.11a)

$$-x_1 + 4x_2 - 2x_3 \quad = \quad 400$$ (3.11b)

$$-2x_2 + 4x_3 \quad = \quad 400$$ (3.11c)

(a) *Forward elimination.* The method starts from dividing Eq. (3.11a) by the coefficient of x_1

$$x_1 - x_2 \qquad = \quad 100$$

Then, multiplying it with the coefficient of x_1 from Eq. (3.11b)

$$-x_1 + x_2 \qquad = \quad -100$$

The equation is subtracted from Eq. (3.11b) to obtain

$$3x_2 - 2x_3 \quad = \quad 500$$ (3.11b')

By applying the same process to Eq. (3.11c),

$$-2x_2 + 4x_3 \quad = \quad 400$$ (3.11c')

After one loop of forward elimination, the system of equations becomes

$$4x_1 - 4x_2 \qquad = \quad 400$$ (3.11a)

$$3x_2 - 2x_3 \quad = \quad 500$$ (3.11b')

$$-2x_2 + 4x_3 \quad = \quad 400$$ (3.11c')

The next loop of forward elimination starts from Eq. (3.11b') by dividing it with the coefficient of x_2

$$x_2 - \frac{2}{3}x_3 \quad = \quad \frac{500}{3}$$

and then multiplying it by the coefficient of x_2 from Eq. (3.11c')

$$-2x_2 + \frac{4}{3}x_3 = -\frac{1,000}{3}$$

By subtracting the equation above from Eq. (3.11c') to obtain

$$\frac{8}{3}x_3 = \frac{2,200}{3} \qquad\qquad (3.11\text{c}'')$$

After the second loop of forward elimination, the system of equations now becomes

$$4x_1 - 4x_2 \quad = \quad 400 \qquad\qquad (3.11\text{a})$$

$$3x_2 - 2x_3 \quad = \quad 500 \qquad\qquad (3.11\text{b}')$$

$$\frac{8}{3}x_3 \quad = \quad \frac{2,200}{3} \qquad\qquad (3.11\text{c}'')$$

If the above equations are written in matrix form, the coefficients beneath the diagonal line of the square matrix on the left-hand side of the system equations are all zero as needed.

(b) *Back substitution.* By starting from the last equation, the solution x_3 can be determined. The computed solution x_3 is used to determine the values of x_2 and then x_1, respectively,

$$x_3 \quad = \quad (2,200/3)(3/8) \quad = \quad 275$$

$$x_2 \quad = \quad (500 + 2(275))/3 \quad = \quad 350$$

$$x_1 \quad = \quad (400 + 4(350))/4 \quad = \quad 450$$

These values of x_1, x_2 and x_3 are the solutions of Eq. (3.10).

Example 3.4 demonstrates that the Gauss elimination method has a systematic procedure and can be used to develop a computer program directly. The procedure as shown in example 3.4 can be generalized for solving a set of n equations as follows.

For a system of n equations,

$$a_{11}x_1 + a_{12}x_2 + a_{13}x_3 + \ldots + a_{1n}x_n \quad = \quad b_1 \qquad\qquad (3.12\text{a})$$

$$a_{21}x_1 + a_{22}x_2 + a_{23}x_3 + \ldots + a_{2n}x_n \quad = \quad b_2 \qquad\qquad (3.12\text{b})$$

$$a_{31}x_1 + a_{32}x_2 + a_{33}x_3 + \ldots + a_{3n}x_n \quad = \quad b_3 \qquad\qquad (3.12\text{c})$$

$$\vdots \qquad \vdots \qquad \vdots \qquad\quad \vdots \qquad\qquad \vdots$$

$$a_{n1}x_1 + a_{n2}x_2 + a_{n3}x_3 + \ldots + a_{nn}x_n \quad = \quad b_n \qquad\qquad (3.12\text{n})$$

the two main steps of the forward elimination and back substitution are still applied to solve for the solutions.

Forward elimination The procedure starts from dividing the first equation (Eq. (3.12a)) by the coefficient of x_1

$$x_1 + \frac{a_{12}}{a_{11}}x_2 + \frac{a_{13}}{a_{11}}x_3 + \ldots + \frac{a_{1n}}{a_{11}}x_n = \frac{b_1}{a_{11}}$$

By multiplying by the coefficient of x_1 from the second equation (Eq. (3.12b)),

$$a_{21}x_1 + a_{21}\frac{a_{12}}{a_{11}}x_2 + a_{21}\frac{a_{13}}{a_{11}}x_3 + \ldots + a_{21}\frac{a_{1n}}{a_{11}}x_n = a_{21}\frac{b_1}{a_{11}}$$

and subtracting it from Eq. (3.12b) to eliminate the term associated with x_1. Such manipulations yield,

$$\underbrace{\left(a_{22} - a_{21}\frac{a_{12}}{a_{11}}\right)}_{a'_{22}}x_2 + \underbrace{\left(a_{23} - a_{21}\frac{a_{13}}{a_{11}}\right)}_{a'_{23}}x_3 + \ldots + \underbrace{\left(a_{2n} - a_{21}\frac{a_{1n}}{a_{11}}\right)}_{a'_{2n}}x_n = \underbrace{b_2 - a_{21}\frac{b_1}{a_{11}}}_{b'_2}$$

or,

$$a'_{22}x_2 + a'_{23}x_3 + \ldots + a'_{2n}x_n = b'_2 \tag{3.12b'}$$

The process is repeated from Eq. (3.12c) to Eq. (3.12n) to obtain the system of equations in the form,

$$a_{11}x_1 + a_{12}x_2 + a_{13}x_3 + \ldots + a_{1n}x_n = b_1 \tag{3.13a}$$
$$a'_{22}x_2 + a'_{23}x_3 + \ldots + a'_{2n}x_n = b'_2 \tag{3.13b}$$
$$a'_{32}x_2 + a'_{33}x_3 + \ldots + a'_{3n}x_n = b'_3 \tag{3.13c}$$
$$\vdots \qquad \vdots \qquad \qquad \vdots \qquad \qquad \vdots$$
$$a'_{n2}x_2 + a'_{n3}x_3 + \ldots + a'_{nn}x_n = b'_n \tag{3.13n}$$

After the first loop of forward elimination, the terms associated with x_1 from the second equation onward are eliminated.

For the second loop of forward elimination, the terms associated with x_2 from Eqs. (3.13c) through (3.13n) are eliminated. The procedure starts from dividing Eq. (3.13b) by a'_{22}, multiplying it by the coefficient a'_{32} from Eq. (3.13c) and subtracting the result from Eq. (3.13c). These operations lead to a new system of equations in the form,

$$a_{11}x_1 + a_{12}x_2 + a_{13}x_3 + \ldots + a_{1n}x_n = b_1 \tag{3.14a}$$
$$a'_{22}x_2 + a'_{23}x_3 + \ldots + a'_{2n}x_n = b'_2 \tag{3.14b}$$
$$a''_{33}x_3 + \ldots + a''_{3n}x_n = b''_3 \tag{3.14c}$$
$$\vdots \qquad \vdots \qquad \vdots$$
$$a''_{n3}x_3 + \ldots + a''_{nn}x_n = b''_n \tag{3.14n}$$

The same procedure is repeated until the loop number $(n-1)$ is reached and the system of equations finally becomes,

$$a_{11}x_1 + a_{12}x_2 + a_{13}x_3 + \ldots + a_{1n}x_n = b_1 \tag{3.15a}$$

$$a'_{22}x_2 + a'_{23}x_3 + \ldots + a'_{2n}x_n = b'_2 \tag{3.15b}$$

$$a''_{33}x_3 + \ldots + a''_{3n}x_n = b''_3 \tag{3.15c}$$

$$\vdots \qquad \vdots$$

$$a_{nn}^{(n-1)}x_n = b_n^{(n-1)} \tag{3.15n}$$

The prime symbols and the values in the superscripts represent the number of the computational loop needed for the forward elimination process.

Back substitution From Eq. (3.15), the unknown x_n can be determined directly from the last equation, Eq. (3.15n), as

$$x_n = \frac{b_n^{(n-1)}}{a_{nn}^{(n-1)}} \tag{3.16a}$$

Then, the unknowns $x_{n-1}, x_{n-2}, \ldots, x_2, x_1$ can be obtained by using back substitution as follows. The computed value of x_n is substituted into the $(n-1)^{th}$ equation to solve for the unknown x_{n-1}. Then, the computed values of x_n and x_{n-1} are substituted into the $(n-2)^{th}$ equation to solve for the unknown x_{n-2}. The same procedure is repeated to determine the remaining values of the unknowns x_i. Determination of the unknowns x_i, $i = n-1, \ldots, 1$ can be obtained using the expression,

$$x_i = \frac{b_i^{(i-1)} - \sum\limits_{j=i+1}^{n} a_{ij}^{(i-1)}x_j}{a_{ii}^{(i-1)}} \tag{3.16b}$$

where subscript i represents the ith equation in Eq. (3.15).

Procedures of the Gauss elimination method for solving the system of equations as explained from Eq. (3.12) to (3.16) are used to develop a computer program as shown in Fig. 3.2. Table 3.1 shows the input data generated from example 3.4 together with the computed solutions from the program.

Table 3.1 Example of input data from Eq. (3.10) for using with the program in Fig. 3.2 and the computed solutions by the Gauss elimination method.

Input data from Eq. (3.10)

```
3
4.   -4.    0.    400.
-1    4.   -2.    400.
0.   -2.    4.    400.
```

Computed solutions

EQUATION NO.	SOLUTION X
1	.450000E+03
2	.350000E+03
3	.275000E+03

Fortran

```
      PROGRAM NGELIM
C.....PROGRAM FOR SOLVING A SET OF SIMULTANEOUS
C.....LINEAR EQUATIONS  [A]{X} = {B}  USING
C.....NAIVE GAUSS ELIMINATION
      DIMENSION A(50,50), B(50), X(50)
      OPEN(UNIT=7, FILE='INPUT.DAT', STATUS='OLD')
      OPEN(UNIT=8, FILE= 'SOL.OUT', STATUS='NEW')
C.....READ TOTAL NUMBER OF EQUATIONS TO BE SOLVED:
      READ(7,*)  N
C.....READ MATRIX [A] AND VECTOR {B}:
      DO 10  IROW=1,N
      READ(7,*)  (A(IROW,ICOL), ICOL=1,N), B(IROW)
   10 CONTINUE
      CALL GAUSS(N, A, B, X)
      WRITE(8,100)
  100 FORMAT(/, 7X, 'EQUATION NO.',
     *          7X, 'SOLUTION X', /)
      DO 20  I=1,N
      WRITE(8,200)   I, X(I)
  200 FORMAT(I12, 8X, E16.6)
   20 CONTINUE
      STOP
      END
C-----------------------------------------------
      SUBROUTINE  GAUSS(N, A, B, X)
      DIMENSION  A(50,50), B(50), X(50)
C.....FORWARD ELIMINATION: PERFORM ACCORDING TO
C.....THE ORDER OF 'PRIME' FROM 1 TO N-1:
      DO 100  IP=1,N-1
C.....LOOP OVER EACH EQUATION STARTING FROM THE
C.....ONE THAT CORRESPONDS WITH THE ORDER OF
C.....'PRIME' PLUS ONE:
      DO 200  IE=IP+1,N
      RATIO = A(IE,IP)/A(IP,IP)
C.....COMPUTE NEW COEFF. OF THE EQ. CONSIDERED:
      DO 300  IC=IP+1,N
      A(IE,IC) = A(IE,IC) - RATIO*A(IP,IC)
  300 CONTINUE
      B(IE) = B(IE) - RATIO*B(IP)
  200 CONTINUE
C.....SET COEFF. ON LOWER LEFT PORTION TO ZERO:
      DO 400  IE=IP+1,N
      A(IE,IP) = 0.
  400 CONTINUE
  100 CONTINUE
C.....BACK SUBSTITUTION:
C.....COMPUTE SOLUTION OF THE LAST EQUATION:
      X(N) = B(N)/A(N,N)
C.....COMPUTE SOLUTIONS FROM EQUATION N-1 TO 1:
      DO 500  IE=N-1,1,-1
      SUM = 0.
      DO 600  IC=IE+1,N
      SUM = SUM + A(IE,IC)*X(IC)
  600 CONTINUE
      X(IE) = (B(IE) - SUM)/A(IE,IE)
  500 CONTINUE
      RETURN
      END
```

MATLAB

```
% PROGRAM NGELIM
% PROGRAM FOR COMPUTING A SET OF SIMULTANEOUS
% LINEAR EQUATIONS  [A]{X} = {B}  USING
% NAIVE GAUSS ELIMINATION
%
% READ TOTAL NUMBER OF EQUATIONS TO BE SOLVED:
% READ MATRIX [A] AND VECTOR {B} :
fid = fopen('input.dat', 'r');
n   = fscanf(fid,'%f',1);
a   = fscanf(fid,'%f',[n+1 n]);
a = a.';
b = squeeze(a(:,n+1));
a = squeeze(a(:,1:n));
fclose(fid);
x = gauss(n, a, b);
% WRITE THE SOLUTIONS:
fid = fopen('output.dat', 'w');
fprintf(fid,'\n   EQUATION NO.          SOLUTION X \n');
fprintf(    '\n   EQUATION NO.          SOLUTION X \n');
for i = 1:n
    fprintf(fid,' %8d          %14.6e\n', i, x(i));
    fprintf(    ' %8d          %14.6e\n', i, x(i));
end
fclose(fid);
```

```
function x = gauss(n, a, b)
% FORWARD ELIMINATION: PERFORM ACCORDING TO
%   THE ORDER OF 'PRIME' FROM 1 TO N-1:
for ip = 1:n-1
% LOOP OVER EACH EQUATION STARTING FROM THE
%   ONE THAT COORESPONDS WITH THE ORDER OF
%   'PRIME' PLUS ONE:
    for ie = ip+1:n
        ratio = a(ie,ip)/a(ip,ip);
% COMPUTE NEW COEFF. OF THE EQ. CONSIDERED:
        for ic = ip+1:n
            a(ie,ic) = a(ie,ic) - ratio*a(ip,ic);
        end
        b(ie) = b(ie) - ratio*b(ip);
    end
% SET COEFF. ON LOWER LEFT PORTION TO ZERO:
    for ie = ip+1:n
        a(ie,ip) = 0.;
    end
end
% BACK SUBSTITUTION:
% COMPUTE SOLUTION OF THE LAST EQUATION:
x(n) = b(n)/a(n,n);
% COMPUTE SOLUTIONS FROM EQUATION N-1 TO 1.
for ie = n-1:-1:1
    sum = 0.;
    for ic = ie+1:n
        sum = sum + a(ie,ic)*x(ic);
    end
    x(ie) = (b(ie) - sum)/a(ie,ie);
end
```

Figure 3.2 Computer program for solving a system of equations
by the Gauss elimination method.

3.4 Problems of Gauss Elimination Method

Example 3.4 shows that the process in the Gauss elimination method is simple and easy to understand. The corresponding computer program can be developed easily as shown in Fig. 3.2. However, the Gauss elimination method explained in the preceding section may have some pitfalls as will be explained in this section.

3.4.1 Division by zero.

The system of equations used in example 3.4 is

$$4x_1 - 4x_2 \quad\quad = \quad 400 \tag{3.11a}$$
$$-x_1 + 4x_2 - 2x_3 = \quad 400 \tag{3.11b}$$
$$\quad\quad -2x_2 + 4x_3 = \quad 400 \tag{3.11c}$$

In the process of forward elimination, Eq. (3.11a) is first divided by the coefficient x_1 that has the value of 4. If Eq. (3.11a) is interchanged with Eq. (3.11c) so that the system of equations becomes

$$\quad\quad -2x_2 + 4x_3 = \quad 400 \tag{3.11c}$$
$$-x_1 + 4x_2 - 2x_3 = \quad 400 \tag{3.11b}$$
$$4x_1 - 4x_2 \quad\quad = \quad 400 \tag{3.11a}$$

The coefficient of x_1 in the first equation is now zero. Division by zero thus creates problem in the forward elimination process. For general problems, it is possible that the coefficient $a_{ii}^{(i-1)}$ in Eq. (3.14) may be zero or close to zero. Small values of $a_{ii}^{(i-1)}$ can also lead to inaccurate solutions. The techniques to avoid these problems will be explained in section 3.5.

3.4.2 Round-off error.

Another problem that can occur in the process of the Gauss elimination method is the round-off error. Values are stored in computer with a limit amount of significant figures during computation. Numerical operations in the computational process of the Gauss elimination method can produce solutions with round-off error. The round-off error propagates as the number of operations increases. Such error can be understood clearly by studying the following example.

Example 3.5 Given the system of 2 equations as,

$$\begin{bmatrix} 2.000112 & 1.414214 \\ 1.414214 & 1.000102 \end{bmatrix} \begin{Bmatrix} x_1 \\ x_2 \end{Bmatrix} = \begin{Bmatrix} 0.521471 \\ 0.232279 \end{Bmatrix} \tag{3.17}$$

If the computer program as shown in Fig. 3.2 is used and executed on a typical computer, the computed solutions are $x_1 = 613.0448$ and $x_2 = -866.6558$. But if a calculator with 6 significant figures is used, the system of equations, Eq. (3.17), becomes

$$\begin{bmatrix} 2.00011 & 1.41421 \\ 1.41421 & 1.00010 \end{bmatrix} \begin{Bmatrix} x_1 \\ x_2 \end{Bmatrix} = \begin{Bmatrix} 0.521471 \\ 0.232279 \end{Bmatrix} \tag{3.18}$$

With the forward elimination process, Eq. (3.18) is changed to

$$\begin{bmatrix} 2.00011 & 1.41421 \\ 0 & 0.167263 \times 10^{-3} \end{bmatrix} \begin{Bmatrix} x_1 \\ x_2 \end{Bmatrix} = \begin{Bmatrix} 0.521471 \\ -0.136435 \end{Bmatrix}$$

After performing the back substitution, the computed solutions are $x_1 = 577.008$ and $x_2 = -815.691$. These solutions have an error of 6% as compared to the true solutions of Eq. (3.17). It is noted, in addition, that such a large error occurs from the fact that the square matrix on the left-hand side of the equation is an ill-conditioned matrix.

3.4.3 Ill-conditioned system.

A square matrix on the left-hand side of Eq. (3.17) is an ill-conditioned matrix because a small change of its coefficients in the matrix

$$[A] = \begin{bmatrix} 2.000112 & 1.414214 \\ 1.414214 & 1.000102 \end{bmatrix} \tag{3.19}$$

can cause a large change in the computed solutions. A matrix can be classified as an ill-conditioned matrix if its determinant is close to zero or very small as compared to its coefficients. In this example, the determinant of matrix $[A]$ is

$$\det[A] = (2.000112)(1.000102) - (1.414214)(1.414214)$$
$$= 0.000315$$

which is very small as compared to the matrix coefficients. For a practical engineering problem, a system of equations may consist of hundred thousand equations so that the size of matrix $[A]$ is very large. Finding its determinant is not practical because a large computational time is required. It is noted that, however, most engineering problems lead to the well-conditioned systems of equations. For example, the determinant of matrix $[A]$ is 32 in the example of heat transfer on the metal plate as shown in example 3.1.

In summary, the matrix $[A]$ is an ill-conditioned matrix if one or more of the following conditions exist:

(a) A small change in matrix $[A]$ coefficients causes a large change in the computed solutions.

(b) Values of the diagonal coefficients in matrix $[A]$ are small as compared to the off-diagonal ones.

(c) The determinant of matrix $[A]$ is close to zero or the product of $(\det[A]) \cdot (\det[A]^{-1})$ is far from unity.

(d) The product of $[A][A]^{-1}$ is somewhat different from the identity matrix.

3.5 Improved Gauss Elimination Method

As explained in the preceding section, some difficulties may be encountered while using the Gauss elimination method. One of the major difficulties is the division of coefficients by zero or values that are close to zero. Such difficulty can be avoid or alleviated by using the techniques of pivoting and scaling as described below.

3.5.1 Pivoting

If any coefficient a_{ii} along the diagonal line of the $[A]$ matrix is zero, the method of Gauss elimination method can not proceed. Or if such coefficients are close to zero, the method will produce solutions with error. Inaccurate solutions that occur in the later case can be seen clearly by considering the following example.

Example 3.6 Solve the following system of equations by using the standard Gauss elimination method,

$$\begin{bmatrix} 0.0003 & 3.0000 \\ 1.0000 & 1.0000 \end{bmatrix} \begin{Bmatrix} x_1 \\ x_2 \end{Bmatrix} = \begin{Bmatrix} 2.0001 \\ 1.0000 \end{Bmatrix} \qquad (3.20)$$

The exact solutions for the set of the two equations above are $x_1 = 1/3$ and $x_2 = 2/3$. It is noted that the value of the diagonal coefficient in the first equation is very small (0.0003) as compared to the other coefficients (3.0000 and 1.0000). After performing the forward elimination, the system of equations becomes

$$\begin{bmatrix} 0.0003 & 3.0000 \\ 0 & -9,999 \end{bmatrix} \begin{Bmatrix} x_1 \\ x_2 \end{Bmatrix} = \begin{Bmatrix} 2.0001 \\ -6,666 \end{Bmatrix} \qquad (3.21)$$

Then, by using the back substitution, the two solutions are obtained as

$$x_2 = 2/3$$

and $x_1 = (2.0001 - 3x_2)/0.0003$

The accuracy of the computed solution x_1 depends on the amount of the significant figures used during computation. For example, if a calculator with only 5 significant figures is used, the computed solutions are,

$$x_2 = 0.66667 \qquad \text{and} \qquad x_1 = 0.30000 \qquad (3.22)$$

The error of the computed solution x_1 is about 10%.

If the two equations in Eq. (3.20) are switched, i.e., their orders are interchanged such that,

$$\begin{bmatrix} 1.0000 & 1.0000 \\ 0.0003 & 3.0000 \end{bmatrix} \begin{Bmatrix} x_1 \\ x_2 \end{Bmatrix} = \begin{Bmatrix} 1.0000 \\ 2.0001 \end{Bmatrix} \qquad (3.23)$$

Then, by performing the forward elimination,

$$\begin{bmatrix} 1.0000 & 1.0000 \\ 0 & 2.9997 \end{bmatrix} \begin{Bmatrix} x_1 \\ x_2 \end{Bmatrix} = \begin{Bmatrix} 1.0000 \\ 1.9998 \end{Bmatrix}$$

and back substitution, the computed solutions are,

$$x_2 = 0.66667 \qquad \text{and} \qquad x_1 = 0.33333 \qquad (3.24)$$

In this case, the error of the computed solution x_1 reduces to 0.001%.

The example shows that accuracy of the solutions obtained from the Gauss elimination method depends on the values of the coefficients along the diagonal line of the matrix $[A]$. To obtain solutions with a higher accuracy, the orders of the equations must be changed such that the magnitudes of the coefficients along the matrix diagonal line are large as compared to its off-diagonal terms. To solve a set of n equations as shown in Eq. (3.12), the orders of equations must be rearranged. This is done by selecting the equation that has the largest coefficient and moving it up while performing the forward elimination process. As an example, before performing the third loop in the forward elimination as shown in Eq. (3.14), the equation within Eqs. (3.14c) – (3.14n) that has the largest coefficient of x_3 is selected and interchanged with the former Eq. (3.14c). Such process is called partial pivoting and presented in the subroutine (or function in MATLAB) PIVOT in the computer program in Fig. 3.3.

3.5.2 Scaling

The matrix $[A]$ in a system of equations generated from an engineering problem may consist of coefficients with a large difference in magnitudes. For example, a system of equations that occurs in the high-speed compressible flow analysis around the space shuttle in Fig. 1.3 may consist of unknowns of the density, velocity components and temperature of the flow field. The matrix $[A]$ of such problem consists of coefficients that are quite different in magnitudes. Large differences in the magnitudes of these coefficients create solution error as demonstrated in the following example.

Example 3.7   Solve the following system of equations by the Gauss elimination method,

$$\begin{bmatrix} 2 & 100,000 \\ 1 & 1 \end{bmatrix} \begin{Bmatrix} x_1 \\ x_2 \end{Bmatrix} = \begin{Bmatrix} 100,000 \\ 2 \end{Bmatrix} \tag{3.25}$$

The exact solutions of the above set of equations are $x_1 = 1.00002$ and $x_2 = 0.99998$.

If the standard Gauss elimination method is used, the forward elimination process yields,

$$\begin{bmatrix} 2 & 100,000 \\ 0 & -49,999 \end{bmatrix} \begin{Bmatrix} x_1 \\ x_2 \end{Bmatrix} = \begin{Bmatrix} 100,000 \\ -49,999 \end{Bmatrix} \tag{3.26}$$

After performing the back substitution, the computed solutions with three significant figures are,

$$x_2 = 1.00 \quad \text{and} \quad x_1 = 0.00 \tag{3.27}$$

for which the error of x_1 is 100%.

But, if the scaling process is first applied by dividing each equation by the largest value of all coefficients in that equation, Eq. (3.25) becomes,

$$\begin{bmatrix} 0.00002 & 1 \\ 1 & 1 \end{bmatrix} \begin{Bmatrix} x_1 \\ x_2 \end{Bmatrix} = \begin{Bmatrix} 1 \\ 2 \end{Bmatrix} \tag{3.28}$$

Then, by applying the pivoting technique as explained in the preceding section,

$$\begin{bmatrix} 1 & 1 \\ 0.00002 & 1 \end{bmatrix} \begin{Bmatrix} x_1 \\ x_2 \end{Bmatrix} = \begin{Bmatrix} 2 \\ 1 \end{Bmatrix}$$

With the forward elimination, the set of equations becomes,

$$\begin{bmatrix} 1 & 1 \\ 0 & 0.99998 \end{bmatrix} \begin{Bmatrix} x_1 \\ x_2 \end{Bmatrix} = \begin{Bmatrix} 2 \\ 0.99996 \end{Bmatrix}$$

After performing the back substitution, the computed solutions are

$$x_2 = 1.00 \quad \text{and} \quad x_1 = 1.00 \tag{3.29}$$

This example demonstrates that the scaling technique can increase accuracy of the computed solutions. A subroutine for performing the scaling technique can also be developed easily. Figure 3.3 shows a typical subroutine SCALE that can be included into the standard Gauss elimination program.

Example 3.8 Modify the Gauss elimination computer program as shown in Fig. 3.2 by including the capability of pivoting and scaling. Check accuracy of the computed solutions obtained from the modified computer program by using Eq. (3.10) in example 3.4.

The subroutines SCALE and PIVOT as shown in Fig. 3.3 are called by the subroutine GAUSS before and after the forward elimination loop. Details of each subroutine are presented in Fig. 3.3. Table 3.2 shows the computed solutions of Eq. (3.10) generated from the modified program.

Fortran

```
                    .
                    .
                    .
                    .
       SUBROUTINE  GAUSS(N, A, B, X)
       DIMENSION  A(50,50), B(50), X(50)
C.....PERFORM SCALING:
       CALL SCALE(N, A, B)
C.....FORWARD ELIMINATION: PERFORM ACCORDING TO
C.....THE ORDER OF 'PRIME' FROM 1 TO N-1:
       DO 100  IP=1,N-1
C.....PERFORM PARTIAL PIVOTING:
       CALL PIVOT(N, A, B, IP)
C.....LOOP OVER EACH EQUATION STARTING FROM THE
C.....ONE THAT CORRESPONDS WITH THE ORDER OF
C.....'PRIME' PLUS ONE:
       DO 200  IE=IP+1,N
                    .
                    .
                    .
                    .

C-----------------------------------------------
       SUBROUTINE SCALE(N, A, B)
       DIMENSION  A(50,50), B(50)
C.....PERFORM SCALING:
       DO 10  IE=1,N
       BIG = ABS(A(IE,1))
       DO 20  IC=2,N
       AMAX = ABS(A(IE,IC))
       IF(AMAX.GT.BIG)  BIG = AMAX
   20 CONTINUE
       DO 30  IC=1,N
       A(IE,IC) = A(IE,IC)/BIG
   30 CONTINUE
       B(IE) = B(IE)/BIG
   10 CONTINUE
       RETURN
       END
C-----------------------------------------------
       SUBROUTINE PIVOT(N, A, B, IP)
       DIMENSION  A(50,50), B(50)
C.....PERFORM PARTIAL PIVOTING:
       JP = IP
       BIG = ABS(A(IP,IP))
       DO 10  I=IP+1,N
       AMAX = ABS(A(I,IP))
       IF(AMAX.GT.BIG)  THEN
         BIG = AMAX
         JP  = I
       ENDIF
   10 CONTINUE
         IF(JP.NE.IP)  THEN
       DO 20  J=IP,N
       DUMY    = A(JP,J)
       A(JP,J) = A(IP,J)
       A(IP,J) = DUMY
   20 CONTINUE
       DUMY   = B(JP)
       B(JP)  = B(IP)
       B(IP)  = DUMY
         ENDIF
       RETURN
       END
```

MATLAB

```
function x = gauss(n, a, b)
%   PERFORM SCALING
[a, b] = scale(n, a, b);
%   FORWARD ELIMINATION: PERFORM ACCORDING TO
%   THE ORDER OF 'PRIME' FROM 1 TO N-1:
for ip = 1:n-1
%   PERFORM PARTIAL PIVOTING
[a, b] = pivot(n, a, b, ip);
%   LOOP OVER EACH EQUATION STARTING FROM THE
%   ONE THAT COORESPONDS WITH THE ORDER OF
%   'PRIME' PLUS ONE:
    for ie = ip+1:n
        ratio = a(ie,ip)/a(ip,ip);
%   COMPUTE NEW COEFF. OF THE EQ. CONSIDERED:
        for ic = ip+1:n
            a(ie,ic) = a(ie,ic) - ratio*a(ip,ic);
        end
        b(ie) = b(ie) - ratio*b(ip);
    end
%   SET COEFF. ON LOWER LEFT PORTION TO ZERO:
    for ie = ip+1:n
        a(ie,ip) = 0.;
    end
end
%   BACK SUBSTITUTION:
%   COMPUTE SOLUTION OF THE LAST EQUATION:
x(n) = b(n)/a(n,n);
%   COMPUTE SOLUTIONS FROM EQUATION N-1 TO 1:
for ie = n-1:-1:1
    sum = 0.;
    for ic = ie+1:n
        sum = sum + a(ie,ic)*x(ic);
    end
    x(ie) = (b(ie) - sum)/a(ie,ie);
end
% -----------------------------------------------
%   FUNCTION SCALING
function [a, b] = scale(n, a, b)
for ie = 1:n
    big = abs(a(ie,1));
    for ic = 2:n
        amax = abs(a(ie,ic));
        if amax > big
            big = amax;
        end
    end
    for ic = 1:n
        a(ie,ic) = a(ie,ic)/big;
    end
    b(ie) = b(ie)/big;
end
% -----------------------------------------------
%   FUNCTION PIVOTING
function [a, b] = pivot(n, a, b, ip)
jp = ip;
big = abs(a(ip,ip));
for i = ip+1:n
    amax = abs(a(i,ip));
    if amax > big
        big = amax;
        jp  = i;
    end
end
if jp ~= ip
    for j = ip:n
        dumy    = a(jp,j);
        a(jp,j) = a(ip,j);
        a(ip,j) = dumy;
    end
    dumy   = b(jp);
    b(jp)  = b(ip);
    b(ip)  = dumy;
end
```

Figure 3.3 Gauss elimination computer program after including the capability of scaling and pivoting.

Table 3.2 Example of input data file for Eq. (3.10) for the modified program and its computed solutions.

Input data for Eq. (3.10)				Computed solutions	
3				EQUATION NO.	SOLUTION X
4.	-4.	0.	400.	1	.450000E+03
-1.	4.	-2.	400.	2	.350000E+03
0.	-2.	4.	400.	3	.275000E+03

3.5.3 Tridiagonal system

The method of Gauss elimination can be used effectively to solve a special form of the system of equations,

$$\underset{(n\times n)\,(n\times 1)}{[A]\{X\}} = \underset{(n\times 1)}{\{B\}} \tag{3.4}$$

There are many engineering problems that yield the matrix $[A]$ with the non-zero terms only along the three main diagonal lines,

$$[A] = \begin{bmatrix} a_{11} & a_{12} & 0 & 0 & \cdots & 0 \\ a_{21} & a_{22} & a_{23} & 0 & \cdots & 0 \\ 0 & a_{32} & a_{33} & a_{34} & \cdots & 0 \\ \vdots & \vdots & \vdots & \vdots & \ddots & \vdots \\ 0 & 0 & 0 & 0 & a_{n,n-1} & a_{n,n} \end{bmatrix} \tag{3.30}$$

In this case, the system Eq. (3.4) is written in full as,

$$\begin{bmatrix} b_1 & c_1 & & & & & \\ a_2 & b_2 & c_2 & & & & \\ & a_3 & b_3 & c_3 & & & \\ & & \cdot & \cdot & \cdot & & \\ & & & \cdot & \cdot & \cdot & \\ & & & & \cdot & \cdot & \cdot \\ & & & & a_{n-1} & b_{n-1} & c_{n-1} \\ & & & & & a_n & b_n \end{bmatrix} \begin{Bmatrix} x_1 \\ x_2 \\ x_3 \\ \cdot \\ \cdot \\ \cdot \\ x_{n-1} \\ x_n \end{Bmatrix} = \begin{Bmatrix} d_1 \\ d_2 \\ d_3 \\ \cdot \\ \cdot \\ \cdot \\ d_{n-1} \\ d_n \end{Bmatrix} \tag{3.31}$$

The Gauss elimination method can be applied to solve such system of equations by eliminating the coefficients $a_2, a_3, \ldots, a_n$ in the forward elimination process and then performing the back substitution in order to obtain the solutions.

Example 3.9 Use the Gauss elimination method to solve the set of tridiagonal system in Eq. (3.10),

$$\begin{bmatrix} 4 & -4 & 0 \\ -1 & 4 & -2 \\ 0 & -2 & 4 \end{bmatrix} \begin{Bmatrix} x_1 \\ x_2 \\ x_3 \end{Bmatrix} = \begin{Bmatrix} 400 \\ 400 \\ 400 \end{Bmatrix} \tag{3.10}$$

The process starts from forward elimination by dividing the first equation by b_1, multiplying with a_2, and subtracting it from the second equation to obtain the new value of b_2 and d_2 as,

$$b_2' = b_2 - \frac{a_2}{b_1}c_1 = 4 - \frac{(-1)}{4}(-4) = 3$$

$$d_2' = d_2 - \frac{a_2}{b_1}d_1 = 400 - \frac{(-1)}{4}(400) = 500$$

It is noted that the value of c_2 is not altered because the coefficient above it is zero.

The forward elimination is then applied to the third equation to obtain,

$$\begin{bmatrix} 4 & -4 & 0 \\ 0 & 3 & -2 \\ 0 & 0 & 8 \end{bmatrix} \begin{Bmatrix} x_1 \\ x_2 \\ x_3 \end{Bmatrix} = \begin{Bmatrix} 400 \\ 500 \\ 2,200 \end{Bmatrix} \tag{3.32}$$

With back substitution starting from the last equation, the solutions of x_3, x_2 and x_1, are determined,

$$x_3 = 275 ; \qquad x_2 = 350 ; \qquad x_1 = 450$$

A computer program can be developed so that only the non-zero coefficients of a, b and c are stored as the three vectors for minimizing computer memory. The Gauss elimination process is performed in the program by using these three vectors as shown in Fig. 3.4.

Fortran

```
      SUBROUTINE  TRIDG(A, B, C, D, X, N)
C.....SOLVE TRIDIAGONAL SYSTEM OF N EQUATIONS
      DIMENSION  A(50), B(50), C(50), D(50), X(50)
C.....PERFORM FORWARD ELIMINATION:
      DO 10  I=2,N
      A(I) = A(I)/B(I-1)
      B(I) = B(I) - A(I)*C(I-1)
      D(I) = D(I) - A(I)*D(I-1)
   10 CONTINUE
C.....PERFORM BACKWARD SUBSTITUTION:
      X(N) = D(N)/B(N)
      DO 20  I=N-1,1,-1
      X(I) = (D(I) - C(I)*X(I+1))/B(I)
   20 CONTINUE
      RETURN
      END
```

MATLAB

```
function x = tridg(a, b, c, d, n)
%   SOLVE TRIDIAGONAL SYSTEM OF N EQUATIONS
%   PERFORM FORWARD ELIMINATION:
for i = 2:n
   a(i) = a(i)/b(i-1);
   b(i) = b(i) - a(i)*c(i-1);
   d(i) = d(i) - a(i)*d(i-1);
end
%   PERFORM BACKWARD SUBSTITUTION:
x(n) = d(n)/b(n);
for i = n-1:-1:1
   x(i) = (d(i) - c(i)*x(i+1))/b(i);
end
```

Figure 3.4 Subroutine for solving tridiagonal system of equations

3.6 Gauss-Jordan Method

The Gauss-Jordan method is an extension of the Gauss elimination method for solving a set of equations,

$$\begin{bmatrix} a_{11} & a_{12} & a_{13} \\ a_{21} & a_{22} & a_{23} \\ a_{31} & a_{32} & a_{33} \end{bmatrix} \begin{Bmatrix} x_1 \\ x_2 \\ x_3 \end{Bmatrix} = \begin{Bmatrix} b_1 \\ b_2 \\ b_3 \end{Bmatrix} \tag{3.33}$$

so that, after performing the elimination, it reduces into the form of,

$$\begin{bmatrix} 1 & 0 & 0 \\ 0 & 1 & 0 \\ 0 & 0 & 1 \end{bmatrix} \begin{Bmatrix} x_1 \\ x_2 \\ x_3 \end{Bmatrix} = \begin{Bmatrix} b_1^* \\ b_2^* \\ b_3^* \end{Bmatrix} \tag{3.34}$$

Because the square matrix on the left-hand side of Eq. (3.34) is an identity matrix, thus the equation gives the solutions directly as,

$$\begin{Bmatrix} x_1 \\ x_2 \\ x_3 \end{Bmatrix} = \begin{Bmatrix} b_1^* \\ b_2^* \\ b_3^* \end{Bmatrix}$$

Example 3.10 Solve Eq. (3.10) again by using the Gauss-Jordan method.

$$\begin{bmatrix} 4 & -4 & 0 \\ -1 & 4 & -2 \\ 0 & -2 & 4 \end{bmatrix} \begin{Bmatrix} x_1 \\ x_2 \\ x_3 \end{Bmatrix} = \begin{Bmatrix} 400 \\ 400 \\ 400 \end{Bmatrix} \tag{3.10}$$

By using Gauss elimination method as explained in example 3.4, the resulting matrix after applying forward elimination is,

$$\begin{bmatrix} 4 & -4 & 0 \\ 0 & 3 & -2 \\ 0 & 0 & 8/3 \end{bmatrix} \begin{Bmatrix} x_1 \\ x_2 \\ x_3 \end{Bmatrix} = \begin{Bmatrix} 400 \\ 500 \\ 2,200/3 \end{Bmatrix}$$

All equations are modified so that the coefficients along the main diagonal line of matrix $[A]$ are equal to one,

$$\begin{bmatrix} 1 & -1 & 0 \\ 0 & 1 & -2/3 \\ 0 & 0 & 1 \end{bmatrix} \begin{Bmatrix} x_1 \\ x_2 \\ x_3 \end{Bmatrix} = \begin{Bmatrix} 100 \\ 500/3 \\ 275 \end{Bmatrix}$$

The forward elimination is applied to eliminate the coefficient of x_2 in the first equation,

$$\begin{bmatrix} 1 & 0 & -2/3 \\ 0 & 1 & -2/3 \\ 0 & 0 & 1 \end{bmatrix} \begin{Bmatrix} x_1 \\ x_2 \\ x_3 \end{Bmatrix} = \begin{Bmatrix} 800/3 \\ 500/3 \\ 275 \end{Bmatrix}$$

Similarly, the coefficient of x_3 in the first equation is eliminated,

$$\begin{bmatrix} 1 & 0 & 0 \\ 0 & 1 & -2/3 \\ 0 & 0 & 1 \end{bmatrix} \begin{Bmatrix} x_1 \\ x_2 \\ x_3 \end{Bmatrix} = \begin{Bmatrix} 450 \\ 500/3 \\ 275 \end{Bmatrix}$$

After the coefficient of x_3 in the second equation is eliminated, the system of equations becomes,

$$\begin{bmatrix} 1 & 0 & 0 \\ 0 & 1 & 0 \\ 0 & 0 & 1 \end{bmatrix} \begin{Bmatrix} x_1 \\ x_2 \\ x_3 \end{Bmatrix} = \begin{Bmatrix} 450 \\ 350 \\ 275 \end{Bmatrix}$$

Thus, the solutions of the system of equations are,

$$\begin{Bmatrix} x_1 \\ x_2 \\ x_3 \end{Bmatrix} = \begin{Bmatrix} 450 \\ 350 \\ 275 \end{Bmatrix}$$

The above example illustrates that the Gauss-Jordan method leads to the solutions of the system of equations directly. It is noted that the difficulties that arise by using the Gauss elimination method also occur in the Gauss-Jordan method. The techniques of pivoting and scaling are thus needed to apply. In practice, the method is not used for solving a set of equations. This is because the method requires a large number of operations in the order of $(n^3 + n^2 - n)$, where n is the number of equations. Such the number of operations is greater than that required by the Gauss elimination method. However, the Gauss-Jordan method offers an advantage for finding the inverse of a matrix conveniently as will be explained in the next section.

3.7 Matrix Inversion Method

The Gauss-Jordan method can be used to find the inverse of a matrix. If a matrix with the size of (3×3), for example, is considered,

$$[A] = \begin{bmatrix} a_{11} & a_{12} & a_{13} \\ a_{21} & a_{22} & a_{23} \\ a_{31} & a_{32} & a_{33} \end{bmatrix} \tag{3.35}$$

The multiplication of the matrix $[A]$ and its matrix inverse $[A]^{-1}$ leads to,

$$\underset{(3\times3)}{[A]} \underset{(3\times3)}{[A]^{-1}} = \underset{(3\times3)}{[I]} \tag{3.36}$$

where $[I]$ is the identity matrix. If the matrix inverse consists of three column vectors as,

$$\underset{(3\times3)}{[A]^{-1}} = [\underset{(3\times1)}{\{X\}_1} \underset{(3\times1)}{\{X\}_2} \underset{(3\times1)}{\{X\}_3}] \tag{3.37}$$

then, by substituting it into Eq. (3.36),

$$\begin{bmatrix} a_{11} & a_{12} & a_{13} \\ a_{21} & a_{22} & a_{23} \\ a_{31} & a_{32} & a_{33} \end{bmatrix} \begin{bmatrix} \begin{Bmatrix} x_1 \\ x_2 \\ x_3 \end{Bmatrix}_1 & \begin{Bmatrix} x_1 \\ x_2 \\ x_3 \end{Bmatrix}_2 & \begin{Bmatrix} x_1 \\ x_2 \\ x_3 \end{Bmatrix}_3 \end{bmatrix} = \begin{bmatrix} 1 & 0 & 0 \\ 0 & 1 & 0 \\ 0 & 0 & 1 \end{bmatrix} \tag{3.38}$$

Equation (3.38) can be arranged into the three systems of equations as,

$$\begin{bmatrix} a_{11} & a_{12} & a_{13} \\ a_{21} & a_{22} & a_{23} \\ a_{31} & a_{32} & a_{33} \end{bmatrix} \begin{Bmatrix} x_1 \\ x_2 \\ x_3 \end{Bmatrix} = \begin{Bmatrix} 1 \\ 0 \\ 0 \end{Bmatrix}, \begin{Bmatrix} 0 \\ 1 \\ 0 \end{Bmatrix}, \begin{Bmatrix} 0 \\ 0 \\ 1 \end{Bmatrix} \tag{3.39}$$

These three systems of equations have the same square matrix $[A]$ on left-hand side but with three different vectors on the right-hand side. The Gauss-Jordan method can be applied to these three systems of equations simultaneously as shown in the example below.

Example 3.11 Solve Eq. (3.10) again by using the matrix inversion method.

$$\begin{bmatrix} 4 & -4 & 0 \\ -1 & 4 & -2 \\ 0 & -2 & 4 \end{bmatrix} \begin{Bmatrix} x_1 \\ x_2 \\ x_3 \end{Bmatrix} = \begin{Bmatrix} 400 \\ 400 \\ 400 \end{Bmatrix} \tag{3.10}$$

In this example, the matrix $[A]$ is

$$[A] = \begin{bmatrix} 4 & -4 & 0 \\ -1 & 4 & -2 \\ 0 & -2 & 4 \end{bmatrix}$$

To determine the matrix inverse $[A]^{-1}$, the following systems of equations are solved,

$$\begin{bmatrix} 4 & -4 & 0 \\ -1 & 4 & -2 \\ 0 & -2 & 4 \end{bmatrix} \begin{Bmatrix} x_1 \\ x_2 \\ x_3 \end{Bmatrix} = \begin{Bmatrix} 1 \\ 0 \\ 0 \end{Bmatrix}, \begin{Bmatrix} 0 \\ 1 \\ 0 \end{Bmatrix}, \begin{Bmatrix} 0 \\ 0 \\ 1 \end{Bmatrix} \tag{3.40}$$

By applying the Gauss-Jordan method as explained in section 3.6, the systems of equations below are obtained,

$$\begin{bmatrix} 1 & 0 & 0 \\ 0 & 1 & 0 \\ 0 & 0 & 1 \end{bmatrix} \begin{Bmatrix} x_1 \\ x_2 \\ x_3 \end{Bmatrix} = \begin{Bmatrix} 3/8 \\ 1/8 \\ 1/16 \end{Bmatrix}, \begin{Bmatrix} 1/2 \\ 1/2 \\ 1/4 \end{Bmatrix}, \begin{Bmatrix} 1/4 \\ 1/4 \\ 3/8 \end{Bmatrix} \tag{3.41}$$

Thus, the matrix inverse is,

$$[A]^{-1} = \begin{bmatrix} 3/8 & 1/2 & 1/4 \\ 1/8 & 1/2 & 1/4 \\ 1/16 & 1/4 & 3/8 \end{bmatrix} \tag{3.42}$$

Then, the solutions of Eq. (2.10) can be determined directly as,

$$\begin{Bmatrix} x_1 \\ x_2 \\ x_3 \end{Bmatrix} = \begin{bmatrix} 3/8 & 1/2 & 1/4 \\ 1/8 & 1/2 & 1/4 \\ 1/16 & 1/4 & 3/8 \end{bmatrix} \begin{Bmatrix} 400 \\ 400 \\ 400 \end{Bmatrix} = \begin{Bmatrix} 450 \\ 350 \\ 275 \end{Bmatrix} \tag{3.43}$$

From this example, the matrix inversion method can be used for finding the matrix inverse and solving for solutions of the system of equations. However, the number of operations required by this method is quite large (about $2n^3$) as compared to those needed by the other methods. Thus, the matrix inversion method is not employed in solving systems of equations for practical problems. However, the method can be used for testing the ill-conditioning of the matrix $[A]$ when its size is not too large as follows.

(a) Apply the scaling technique as explained in section 3.5.2 on matrix $[A]$ and then find the matrix inverse $[A]^{-1}$. If coefficients in the matrix inverse $[A]^{-1}$ are large as compared to the coefficients in the original matrix $[A]$, the matrix is considered as an ill-conditioned matrix.

(b) Multiply matrix $[A]$ by $[A]^{-1}$ to obtain a matrix. If that matrix is not close to the identity matrix $[I]$, the matrix $[A]$ is considered as an ill-conditioned matrix.

(c) Determine the matrix inverse $[A]^{-1}$ and find the matrix inverse of the matrix $[A]^{-1}$ again. If the resulting matrix is not close to the original matrix $[A]$, the matrix is considered as an ill-conditioned matrix.

3.8 Solving System of Linear Equations by MATLAB

In MATLAB, the slash symbol (/) is used for division. For example, 2/3 has the meaning of dividing 2 by 3. If this symbol is used for the matrix operation such as $[U]/[V]$, it means that the matrix $[U]$ is multiplied by the inverse of matrix $[V]$, i.e., $[U] \times [V]^{-1}$. The backslash symbol (\ ) in MATLAB is called the left division and is used for solving a system of equations. For example, $[U] \backslash [V]$ means that the inverse of matrix $[U]$ is multiplied by matrix $[V]$, i.e., $[U]^{-1} \times [V]$.

For a system of equations in the form,

$$[A]\{X\} = \{B\} \tag{3.4}$$

The vector $\{X\}$, which contains solutions of the system of equations, is obtained by multiplying the inverse of matrix $[A]$ on both sides of the equations,

$$[A]^{-1}[A]\{X\} = [A]^{-1}\{B\}$$

$$[I]\{X\} = [A]^{-1}\{B\}$$

$$\{X\} = [A]^{-1}\{B\}$$

The vector $\{X\}$ can be obtained easily by using the backslash symbol or left division between the matrices $[A]$ and $\{B\}$ as shown in the following example.

Example 3.11 Solve the following system of equations by using MATLAB,

$$\begin{bmatrix} 4 & -4 & 0 \\ -1 & 4 & -2 \\ 0 & -2 & 4 \end{bmatrix} \begin{Bmatrix} x_1 \\ x_2 \\ x_3 \end{Bmatrix} = \begin{Bmatrix} 400 \\ 400 \\ 400 \end{Bmatrix} \tag{3.10}$$

The procedures for obtaining the solutions of the vector $\{X\}$ are as follows,

```
>> A = [4 -4 0; -1 4 -2; 0 -2 4];
>> B = [400; 400; 400];
>> x = A\B

x =
     450.0000
     350.0000
     275.0000
```

In general, MATLAB uses the Gauss elimination method with partial pivoting for solving a system of equations when the left division is employed. Other solution methods, such as the LU decomposition and Cholesky decomposition, can also be selected. These methods are explained in details in the following sections.

3.9   LU Decomposition Method

One of the popular methods for solving a large system of equations is the LU decomposition method. The basic idea of this method is to decompose the matrix $[A]$ as the product of the two matrices $[L]$ and $[U]$. The matrix $[L]$ contains non-zero coefficients on the lower left portion of the matrix, while all other coefficients on the upper right portion are zero. The matrix $[U]$ contains the zero and non-zero coefficients similar to the matrix $[L]$ but in the opposite pattern. Procedures of the LU decomposition method can be explained by considering the system of equations,

$$[A]\{X\} = \{B\} \tag{3.4}$$

The first step is to decompose the matrix $[A]$ into the product of matrices $[L]$ and $[U]$ as,

$$[A] = [L][U] \tag{3.44}$$

For example, if the size of the matrix $[A]$ is (3x3), then

$$\underbrace{\begin{bmatrix} a_{11} & a_{12} & a_{13} \\ a_{21} & a_{22} & a_{23} \\ a_{31} & a_{32} & a_{33} \end{bmatrix}}_{[A]} = \underbrace{\begin{bmatrix} \ell_{11} & 0 & 0 \\ \ell_{21} & \ell_{22} & 0 \\ \ell_{31} & \ell_{32} & \ell_{33} \end{bmatrix}}_{[L]} \underbrace{\begin{bmatrix} 1 & u_{12} & u_{13} \\ 0 & 1 & u_{23} \\ 0 & 0 & 1 \end{bmatrix}}_{[U]} \tag{3.45}$$

where the matrix $[L]$ contains the non-zero coefficients only in the lower left portion. The matrix $[U]$ contains the non-zero coefficients only in the upper right portion with the values of one along the main diagonal line. By substituting Eq. (3.45) into Eq. (3.43),

$$[L]\underbrace{[U]\{X\}}_{\{Y\}} = \{B\} \tag{3.46}$$

Equation (3.46) can be written as,

$$[L]\{Y\} = \{B\} \tag{3.47}$$

which can be solved by the forward substitution to obtain the values in the vector $\{Y\}$. Then, the values in the vector $\{X\}$ can be determined by the back substitution from,

$$[U]\{X\} = \{Y\} \tag{3.48}$$

From the above explanation, the LU decomposition method for solving a system of equations $[A]\{X\} = \{B\}$ consists of the three main steps:

(a) Decompose the matrix $[A]$ into the two matrices $[L]$ and $[U]$,

(b) Solve the equations $[L]\{Y\} = \{B\}$ by forward substitution to obtain the vector $\{Y\}$,

(c) Then, solve the equations $[U]\{X\} = \{Y\}$ by back substitution to obtain the vector $\{X\}$ which contains the solutions of the system of equations.

The most time consuming process is the decomposition of the matrix $[A]$ into the two matrices $[L]$ and $[U]$. In many engineering problems, the matrix $[A]$ and the vector $\{B\}$ have their physical meanings. As an example of the stresses that occur in a bridge structure from the load of moving vehicles, the matrix $[A]$ represents the stiffness of the bridge structure while the vector $\{B\}$ contains the load of the moving vehicles. The vector $\{B\}$ changes with different loadings from the number of vehicles, while the matrix $[A]$ representing the bridge structure stiffness remains the same. This means, to solve for the solutions under different loadings, the matrix $[A]$ can be decomposed only once into the product of $[L][U]$. The vector $\{B\}$ is changed according to the loading conditions. Thus, the LU decomposition method offers an advantage in reducing the computational time where many sets of the solutions $\{X\}$ are needed from the different vectors $\{B\}$.

When the matrix $[A]$ is decomposed into the matrices $[L]$ and $[U]$ as shown in Eq. (3.45), the coefficients along the main diagonal line of the matrix $[L]$ can be any value while those along the main diagonal line of the matrix $[U]$ are set as one. With such setting, the method is called the Crout decomposition. On the other hand, if the coefficients along the main diagonal line of the matrix $[L]$ are set to be one, the method is called the Doolittle decomposition. Either the decomposition method can be used for solving the system of equations. Details of LU decomposition based on the Crout method are presented in the following example.

Example 3.13 Use the LU decomposition method to solve of the solutions of the system of equations,

$$\begin{bmatrix} 4 & -4 & 0 \\ -1 & 4 & -2 \\ 0 & -2 & 4 \end{bmatrix} \begin{Bmatrix} x_1 \\ x_2 \\ x_3 \end{Bmatrix} = \begin{Bmatrix} 400 \\ 400 \\ 400 \end{Bmatrix} \tag{3.10}$$

Here, the matrix $[A]$ is

$$[A] = \begin{bmatrix} a_{11} & a_{12} & a_{13} \\ a_{21} & a_{22} & a_{23} \\ a_{31} & a_{32} & a_{33} \end{bmatrix} = \begin{bmatrix} 4 & -4 & 0 \\ -1 & 4 & -2 \\ 0 & -2 & 4 \end{bmatrix}$$

The matrix $[A]$ can be decomposed into the matrices $[L]$ and $[U]$ by first writing,

$$\begin{bmatrix} \ell_{11} & 0 & 0 \\ \ell_{21} & \ell_{22} & 0 \\ \ell_{31} & \ell_{32} & \ell_{33} \end{bmatrix} \begin{bmatrix} 1 & u_{12} & u_{13} \\ 0 & 1 & u_{23} \\ 0 & 0 & 1 \end{bmatrix} = \begin{bmatrix} 4 & -4 & 0 \\ -1 & 4 & -2 \\ 0 & -2 & 4 \end{bmatrix} \tag{3.49}$$

From Eq. (3.49), the coefficients of the matrices $[L]$ and $[U]$ can be determined as follows:

$$\ell_{11} = a_{11} = 4 \tag{3.50a}$$

$$\ell_{21} = a_{21} = -1 \tag{3.50b}$$

$$\ell_{31} = a_{31} = 0 \tag{3.50c}$$

$$\ell_{11}u_{12} = a_{12} \qquad \to u_{12} = a_{12}/\ell_{11} = -1 \tag{3.51a}$$

$$\ell_{11}u_{13} = a_{13} \qquad \to u_{13} = a_{13}/\ell_{11} = 0 \tag{3.51b}$$

$$\ell_{21}u_{12} + \ell_{22} = a_{22} \qquad \to \ell_{22} = a_{22} - \ell_{21}u_{12} = 3 \tag{3.52a}$$

$$\ell_{31}u_{12} + \ell_{32} = a_{32} \qquad \to \ell_{32} = a_{32} - \ell_{31}u_{12} = -2 \tag{3.52b}$$

$$\ell_{21}u_{13} + \ell_{22}u_{23} = a_{23} \qquad \to u_{23} = \frac{a_{23} - \ell_{21}u_{13}}{\ell_{22}} = -\frac{2}{3} \tag{3.53a}$$

$$\ell_{31}u_{13} + \ell_{32}u_{23} + \ell_{33} = a_{33} \qquad \to \ell_{33} = a_{33} - \ell_{31}u_{13} - \ell_{32}u_{23} = \frac{8}{3} \tag{3.53b}$$

Therefore,

$$[A] = [L][U] = \begin{bmatrix} 4 & 0 & 0 \\ -1 & 3 & 0 \\ 0 & -2 & 8/3 \end{bmatrix} \begin{bmatrix} 1 & -1 & 0 \\ 0 & 1 & -2/3 \\ 0 & 0 & 1 \end{bmatrix}$$

Then, Eq. (3.10) can be written in the form,

$$\underbrace{\begin{bmatrix} 4 & 0 & 0 \\ -1 & 3 & 0 \\ 0 & -2 & 8/3 \end{bmatrix}}_{[L]} \underbrace{\underbrace{\begin{bmatrix} 1 & -1 & 0 \\ 0 & 1 & -2/3 \\ 0 & 0 & 1 \end{bmatrix}}_{[U]} \underbrace{\begin{Bmatrix} x_1 \\ x_2 \\ x_3 \end{Bmatrix}}_{\{X\}}}_{\{Y\}} = \underbrace{\begin{Bmatrix} 400 \\ 400 \\ 400 \end{Bmatrix}}_{\{B\}} \tag{3.54}$$

The system of equations $[L]\{Y\} = \{B\}$ is first solved by forward substitution to give the unknowns in vector $\{Y\}$,

$$\begin{bmatrix} 4 & 0 & 0 \\ -1 & 3 & 0 \\ 0 & -2 & 8/3 \end{bmatrix} \begin{Bmatrix} y_1 \\ y_2 \\ y_3 \end{Bmatrix} = \begin{Bmatrix} 400 \\ 400 \\ 400 \end{Bmatrix}$$

$$4y_1 = 400 \quad \to \quad y_1 = 100$$

$$-y_1 + 3y_2 = 400 \quad \to \quad y_2 = 500/3$$

$$-2y_2 + \frac{8}{3}y_3 = 400 \quad \to \quad y_3 = 275$$

Then, the system of equations $[U]\{X\} = \{Y\}$ is used to determine the unknowns in vector $\{X\}$ by back substitution as follows,

$$\begin{bmatrix} 1 & -1 & 0 \\ 0 & 1 & -2/3 \\ 0 & 0 & 1 \end{bmatrix} \begin{Bmatrix} x_1 \\ x_2 \\ x_3 \end{Bmatrix} = \begin{Bmatrix} 100 \\ 500/3 \\ 275 \end{Bmatrix}$$

$$x_3 = 275 \quad \rightarrow \quad x_3 = 275$$

$$x_2 - \frac{2}{3}x_3 = 500/3 \quad \rightarrow \quad x_2 = 350$$

$$x_1 - x_2 = 100 \quad \rightarrow \quad x_1 = 450$$

so that all solutions of Eq. (3.10) are obtained.

The procedures for determining coefficients in the matrices $[L]$ and $[U]$ from the matrix $[A]$ as presented in example 3.13 can be generalized for the system of n equations. Equations (3.50a) – (3.53b) can be written in the general forms as follows,

$$\ell_{i1} = a_{i1} \qquad\qquad i = 1, 2, \ldots, n \qquad (3.55a)$$

$$u_{1j} = \frac{a_{1j}}{\ell_{11}} \qquad\qquad j = 2, 3, \ldots, n \qquad (3.55b)$$

and for $j = 2, 3, \ldots, n-1$;

$$\ell_{ij} = a_{ij} - \sum_{k=1}^{j-1} \ell_{ik} u_{kj} \qquad\qquad i = j, j+1, \ldots, n \qquad (3.55c)$$

$$u_{jk} = \frac{a_{jk} - \sum_{i=1}^{j-1} \ell_{ji} u_{ik}}{\ell_{jj}} \qquad\qquad k = j+1, j+2, \ldots, n \qquad (3.55d)$$

with,

$$\ell_{nn} = a_{nn} - \sum_{k=1}^{n-1} \ell_{nk} u_{kn} \qquad\qquad (3.55e)$$

After the matrices $[L]$ and $[U]$ are obtained, the vectors $\{Y\}$ and $\{X\}$ can be determined from the formulas,

$$y_1 = \frac{b_1}{\ell_{11}}$$

$$y_i = \frac{b_i - \sum_{j=1}^{i-1} \ell_{ij} y_j}{\ell_{ii}} \qquad\qquad i = 2, 3, \ldots, n \qquad (3.56a)$$

and

$$x_n = y_n$$

$$x_i = y_i - \sum_{j=i+1}^{n} u_{ij} x_j \qquad\qquad i = n-1, n-2, \ldots, 1 \qquad (3.56b)$$

The procedures above are used to develop a corresponding computer program as shown in Fig. 3.5. Table 3.3 shows example of the input data file of Eq. (3.10) and the computed solutions generated from the program.

Fortran

```
PROGRAM LUDCOM
C.....PROGRAM FOR SOLVING A SET OF SIMULTANEOUS
C.....LINEAR EQUATIONS   [A]{X} = {B}
C.....USING NAIVE LU DECOMPOSITION
          DIMENSION  A(50,50), B(50), X(50)
          DIMENSION  AL(50,50), AU(50,50), Y(50)
          OPEN(UNIT=7, FILE='INPUT.DAT', STATUS='OLD')
          OPEN(UNIT=8, FILE= 'SOL.OUT', STATUS='NEW')
C.....READ TOTAL NUMBER OF EQUATIONS TO BE SOLVED:
          READ(7,*)  N
C.....READ MATRIX [A] AND VECTOR {B}:
          DO 10  IROW=1,N
          READ(7,*)  (A(IROW,ICOL), ICOL=1,N), B(IROW)
   10 CONTINUE
          CALL LU(N, A, B, X, AL, AU, Y)
          WRITE(8,100)
  100 FORMAT(/, 7X, 'EQUATION NO.',
       *              7X, 'SOLUTION X', /)
          DO 20  I=1,N
          WRITE(8,200)   I, X(I)
  200 FORMAT(I12, 8X, E16.6)
   20 CONTINUE
          STOP
          END
C-------------------------------------------------
          SUBROUTINE  LU(N, A, B, X, AL, AU, Y)
          DIMENSION  A(50,50), B(50), X(50)
          DIMENSION  AL(50,50), AU(50,50), Y(50)
C.....PERFORM DECOMPOSITION  [A] = [L][U]:
          DO 10  I=1,50
          DO 10  J=1,50
          AL(I,J) = 0.
          AU(I,J) = 0.
   10 CONTINUE
          DO 100  I=1,N
          AL(I,1) = A(I,1)
  100 CONTINUE
          DO 150  J=2,N
          AU(1,J) = A(1,J)/AL(1,1)
  150 CONTINUE
          DO 200  J=2,N-1
          DO 300  I=J,N
          SUM = 0.
          DO 350  K=1,J-1
          SUM = SUM + AL(I,K)*AU(K,J)
  350 CONTINUE
          AL(I,J) = A(I,J) - SUM
  300 CONTINUE
          DO 400  K=J+1,N
          SUM = 0.
          DO 450  I=1,J-1
          SUM = SUM + AL(J,I)*AU(I,K)
  450 CONTINUE
          AU(J,K) = (A(J,K) - SUM)/AL(J,J)
  400 CONTINUE
  200 CONTINUE
          SUM = 0.
          DO 500  K=1,N-1
          SUM = SUM + AL(N,K)*AU(K,N)
  500 CONTINUE
          AL(N,N) = A(N,N) - SUM
C.....PERFORM FORWARD PASS TO SOLVE [L]{Y} = {B}:
          Y(1) = B(1)/AL(1,1)
          DO 600  I=2,N
          SUM = 0.
          DO 650  J=1,I-1
          SUM = SUM + AL(I,J)*Y(J)
  650 CONTINUE
          Y(I) = (B(I) - SUM)/AL(I,I)
  600 CONTINUE
C.....PERFORM BACKWARD PASS TO SOLVE [U]{X} = {Y}:
          X(N) = Y(N)
          DO 700  I=N-1,1,-1
          SUM = 0.
          DO 750  J=I+1,N
          SUM = SUM + AU(I,J)*X(J)
  750 CONTINUE
          X(I) = Y(I) - SUM
  700 CONTINUE
          RETURN
          END
```

MATLAB

```
% PROGRAM LUDCOM
% PROGRAM FOR SOLVING A SET OF SIMULTANEOUS
% LINEAR EQUATIONS   [A]{X} = {B}
% USING NAIVE LU DECOMPOSITION
%
% READ TOTAL NUMBER OF EQUATIONS TO BE SOLVED:
% READ MATRIX [A] AND VECTOR {B} :
fid = fopen('input.dat', 'r');
n = fscanf(fid,'%f',1);
a = fscanf(fid,'%f',[n+1 n]);
a = a.';
b = squeeze(a(:,n+1));
a = squeeze(a(:,1:n));
fclose(fid);
x = LU(n, a, b);
% WRITE THE SOLUTIONS:
fid = fopen('output.dat', 'w');
fprintf(fid,'\n   EQUATION NO.          SOLUTION X \n');
fprintf(    '\n   EQUATION NO.          SOLUTION X \n');
for i = 1:n
      fprintf(fid,' %8d          %14.6e\n', i, x(i));
      fprintf(    ' %8d          %14.6e\n', i, x(i));
end
fclose(fid);


function x = LU(n, a, b)
% PERFORM DECOMPOSITION   [A] = [L][U]:
AL = zeros(n,n);
AU = zeros(n,n);
y  = zeros(n,1);
for i = 1:n
      AL(i,1) = a(i,1);
end
for j = 2:n
      AU(1,j) = a(1,j)/AL(1,1);
end
for j = 2:n-1
      for i = j:n
          sum = 0.;
          for k = 1:j-1
              sum = sum + AL(i,k)*AU(k,j);
          end
          AL(i,j) = a(i,j) - sum;
      end
      for k = j+1:n
          sum = 0.;
          for i = 1:j-1
              sum = sum + AL(j,i)*AU(i,k);
          end
          AU(j,k) = (a(j,k) - sum)/AL(j,j);
      end
end
for k = 1:n-1
      sum = sum + AL(n,k)*AU(k,n);
end
AL(n,n) = a(n,n) - sum;
% PERFORM FORWARD PASS TO SOLVE [L]{Y} = {B}
y(1) = b(1)/AL(1,1);
for i = 2:n
      sum = 0.;
      for j = 1:i-1
          sum = sum + AL(i,j)*y(j);
      end
      y(i) = (b(i) - sum)/AL(i,i);
end
% PERFORM BACKWARD PASS TO SOLVE [U]{X} = {Y}
x(n) = y(n);
for i = n-1:-1:1
      sum = 0.;
      for j = i+1:n
          sum = sum + AU(i,j)*x(j);
      end
      x(i) = y(i) - sum;
end
```

Figure 3.5 Computer program for solving system of equations
by the LU decomposition method.

Table 3.3 Example of input data file for Eq. (3.10) for the LU decomposition computer program as shown in Fig. 3.5 and its computed solutions.

Input data for Eq. (3.10)

Computational results

```
3
4.  -4.   0.   400.
-1.   4.  -2.   400.
0.  -2.   4.   400.
```

EQUATION NO.	SOLUTION X
1	.450000E+03
2	.350000E+03
3	.275000E+03

The number of numerical operations required by the LU decomposition method to solve a system of equations is close to that needed by the Gauss elimination method. However, the LU decomposition method requires more computer memory to store the two matrices $[L]$ and $[U]$ as compared to the Gauss elimination method. The required computer memory may be reduced by developing a more complex computer program.

3.10 MATLAB Function for LU Decomposition

MATLAB uses the built-in function lu to decompose the matrix $[A]$ into the two matrices $[L]$ and $[U]$. The syntax for this function is,

$$[L,U] = lu(A)$$

where L, U, and A represent the matrices $[L]$, $[U]$ and $[A]$, respectively.

Example 3.14 Solve the following system of equations by using MATLAB with the LU decomposition method,

$$\begin{bmatrix} 4 & -4 & 0 \\ -1 & 4 & -2 \\ 0 & -2 & 4 \end{bmatrix} \begin{Bmatrix} x_1 \\ x_2 \\ x_3 \end{Bmatrix} = \begin{Bmatrix} 400 \\ 400 \\ 400 \end{Bmatrix} \tag{3.10}$$

The two matrices $[A]$ and $\{B\}$ are first defined by using the commands,

```
>> A = [4 -4 0; -1 4 -2; 0 -2 4];
>> B = [400; 400; 400];
```

Then, the function lu is employed to decompose the matrix $[A]$ into the two matrices $[L]$ and $[U]$,

```
>> [L,U] = lu(A)

L =

       1.0000        0        0
      -0.2500   1.0000        0
            0  -0.6667   1.0000

U =

       4.0000   -4.0000        0
            0    3.0000  -2.0000
            0         0    2.6667
```

The computed matrices $[L]$ and $[U]$ can be verified by multiplying them together. The result must be equal to the original matrix $[A]$, i.e.,

```
>> L*U
ans =
        4      -4       0
       -1       4      -2
        0      -2       4
```

To solve the system of equations by the LU decomposition method, the left division is used in the two following steps,

```
>> Y = L\B
Y =
      400.0000
      500.0000
      733.3333
>> x = U\Y
x =
      450.0000
      350.0000
      275.0000
```

The above example shows that the roots of a system of equations can be obtained easily by using the LU decomposition method in MATLAB. Understanding the method procedure, however, is required prior to using the function `lu` built in MATLAB.

3.11 Cholesky Decomposition Method

All methods explained earlier in this chapter can be used to solve the system of equations $[A]\{X\} = \{B\}$ where the square matrix $[A]$ may be a symmetric or an asymmetric matrix. For many engineering problems, the matrix $[A]$ generated from the finite difference or finite element technique (will be explained in details in chapters 8 and 9, respectively) is normally a symmetric matrix. If the matrix $[A]$ is a symmetric matrix, the LU decomposition method can be modified so that its solving procedure is reduced. Such modified method is known as the Cholesky decomposition method. The method is often used for solving linear structural problems for which their matrices $[A]$ are symmetric. The Cholesky decomposition method starts from the system of equations,

$$[A]\{X\} = \{B\} \tag{3.4}$$

by decomposing the symmetric matrix $[A]$ into the form,

$$[A] = [L][L]^T \tag{3.57}$$

where $[L]$ is the matrix that contains all zero coefficients above its diagonal line. For example, if the system of Eq. (3.4) consists of 3 equations, the matrix $[A]$ is decomposed into the form,

$$\underbrace{\begin{bmatrix} a_{11} & a_{12} & a_{13} \\ a_{12} & a_{22} & a_{23} \\ a_{13} & a_{23} & a_{33} \end{bmatrix}}_{[A]} = \underbrace{\begin{bmatrix} \ell_{11} & 0 & 0 \\ \ell_{21} & \ell_{22} & 0 \\ \ell_{31} & \ell_{32} & \ell_{33} \end{bmatrix}}_{[L]} \underbrace{\begin{bmatrix} \ell_{11} & \ell_{21} & \ell_{31} \\ 0 & \ell_{22} & \ell_{32} \\ 0 & 0 & \ell_{33} \end{bmatrix}}_{[L]^T} \tag{3.58}$$

The coefficients in the matrix $[L]$ can be determined as follows,

$$\ell_{11}^2 = a_{11} \rightarrow \ell_{11} = \sqrt{a_{11}}$$

$$\ell_{11}\ell_{21} = a_{12} \rightarrow \ell_{21} = a_{12}/\ell_{11}$$

$$\ell_{11}\ell_{31} = a_{13} \rightarrow \ell_{31} = a_{13}/\ell_{11}$$

$$\ell_{21}^2 + \ell_{22}^2 = a_{22} \rightarrow \ell_{22} = \sqrt{a_{22} - \ell_{21}^2}$$

$$\ell_{21}\ell_{31} + \ell_{22}\ell_{32} = a_{23} \rightarrow \ell_{32} = (a_{23} - \ell_{21}\ell_{31})/\ell_{22}$$

$$\ell_{31}^2 + \ell_{32}^2 + \ell_{33}^2 = a_{33} \rightarrow \ell_{33} = \sqrt{a_{33} - \ell_{31}^2 - \ell_{32}^2}$$

By substituting Eq. (3.58) into Eq. (3.4),

$$\underbrace{\begin{bmatrix} \ell_{11} & 0 & 0 \\ \ell_{21} & \ell_{22} & 0 \\ \ell_{31} & \ell_{32} & \ell_{33} \end{bmatrix}}_{[L]} \underbrace{\underbrace{\begin{bmatrix} \ell_{11} & \ell_{21} & \ell_{31} \\ 0 & \ell_{22} & \ell_{32} \\ 0 & 0 & \ell_{33} \end{bmatrix}}_{[L]^T} \begin{Bmatrix} x_1 \\ x_2 \\ x_3 \end{Bmatrix}}_{\{Y\}} = \underbrace{\begin{Bmatrix} b_1 \\ b_2 \\ b_3 \end{Bmatrix}}_{\{B\}} \tag{3.59}$$

The vector $\{Y\}$ is determined by using forward substitution,

$$\begin{bmatrix} \ell_{11} & 0 & 0 \\ \ell_{21} & \ell_{22} & 0 \\ \ell_{31} & \ell_{32} & \ell_{33} \end{bmatrix} \begin{Bmatrix} y_1 \\ y_2 \\ y_3 \end{Bmatrix} = \begin{Bmatrix} b_1 \\ b_2 \\ b_3 \end{Bmatrix} \tag{3.60}$$

Then, the vector $\{X\}$ is determined by using back substitution,

$$\begin{bmatrix} \ell_{11} & \ell_{21} & \ell_{31} \\ 0 & \ell_{22} & \ell_{32} \\ 0 & 0 & \ell_{33} \end{bmatrix} \begin{Bmatrix} x_1 \\ x_2 \\ x_3 \end{Bmatrix} = \begin{Bmatrix} y_1 \\ y_2 \\ y_3 \end{Bmatrix} \tag{3.61}$$

The procedures for determining the coefficients in the matrix $[L]$ as shown in the above example can be generalized for the system of n equations as follows.

For the first row of matrix $[L]$,

$$\ell_{11} = \sqrt{a_{11}} \tag{3.62a}$$

For the next k^{th} row, where $k = 2, 3, \ldots, n$,

$$\ell_{ki} = \frac{a_{ik} - \sum_{j=1}^{i-1} \ell_{ij}\ell_{kj}}{\ell_{ii}} \qquad i = 1, 2, \ldots, k-1 \tag{3.62b}$$

with

$$\ell_{kk} = \sqrt{a_{kk} - \sum_{j=1}^{k-1} \ell_{kj}^2} \qquad (3.62c)$$

After the matrix $[L]$ is obtained, the vectors $\{Y\}$ and $\{X\}$ can be determined, respectively, as

$$y_1 = \frac{b_1}{\ell_{11}}$$

$$y_i = \frac{b_i - \sum_{j=1}^{i-1} \ell_{ij} y_j}{\ell_{ii}} \qquad i = 2, 3, ..., n \qquad (3.63)$$

and

$$x_n = \frac{y_n}{\ell_{nn}}$$

$$x_i = \frac{y_i - \sum_{j=i+1}^{n} \ell_{ji} x_j}{\ell_{ii}} \qquad i = n-1, n-2, ..., 1 \qquad (3.64)$$

Example 3.15 Develop a computer program to solve the system of n equations in the form of $[A]\{X\} = \{B\}$ where the matrix $[A]$ is symmetric. Verify the computer program by solving the following system of equations,

$$\begin{bmatrix} 4 & 3 & 1 \\ 3 & 5 & 2 \\ 1 & 2 & 6 \end{bmatrix} \begin{Bmatrix} x_1 \\ x_2 \\ x_3 \end{Bmatrix} = \begin{Bmatrix} 3,125 \\ 3,650 \\ 2,800 \end{Bmatrix} \qquad (3.65)$$

where the exact solutions are $x_1 = 450$, $x_2 = 350$ and $x_3 = 275$.

A computer program for solving the system of n equations in the form of $[A]\{X\} = \{B\}$ by using the Cholesky decomposition method can be developed as shown in Fig. 3.6. The program is verified by solving Eq. (3.65) with the computed solutions as shown in Table 3.4.

Table 3.4 Input data file of Eq. (3.65) for the computer program as shown in Fig. 3.6 and the computed solutions from the Cholesky decomposition method.

Input data for Eq. (3.65)				Computed solutions	
3				EQUATION NO.	SOLUTION X
4.	3.	1.	3125.	1	.450000E+03
3.	5.	2.	3650.	2	.350000E+03
1.	2.	6.	2800.	3	.275000E+03

Fortran

```
PROGRAM  CHOLESKY
C.....PROGRAM FOR SOLVING A SET OF SIMULTANEOUS

C.....LINEAR EQUATIONS  [A]{X} = {B}  USING
C.....CHOLESKY DECOMPOSITION IF [A] IS SYMMETRIC
      DIMENSION  A(50,50), B(50), X(50)
      DIMENSION  AL(50,50), Y(50)
      OPEN(UNIT=7, FILE='INPUT.DAT', STATUS='OLD')
      OPEN(UNIT=8, FILE= 'SOL.OUT', STATUS='NEW')
C.....READ TOTAL NUMBER OF EQUATIONS TO BE SOLVED:
      READ(7,*)  N
C.....READ MATRIX [A] AND VECTOR {B}:
      DO 10  IROW=1,N
      READ(7,*)  (A(IROW,ICOL), ICOL=1,N), B(IROW)
   10 CONTINUE
      CALL CHOLES(N, A, B, X, AL, Y)
      WRITE(8,100)
  100 FORMAT(/, 7X, 'EQUATION NO.',
     *          7X, 'SOLUTION X', /)
      DO 20  I=1,N
      WRITE(8,200)  I, X(I)
  200 FORMAT(I12, 8X, E16.6)
   20 CONTINUE
      STOP
      END
C------------------------------------------------
      SUBROUTINE  CHOLES(N, A, B, X, AL, Y)
      DIMENSION  A(50,50), B(50), X(50)
      DIMENSION  AL(50,50), Y(50)
C.....PERFORM DECOMPOSITION  [A] = [L][LT]:
      DO 10  I=1,50
      DO 10  J=1,50
      AL(I,J) = 0.
   10 CONTINUE
      AL(1,1) = SQRT(A(1,1))
      DO 100  K=2,N
      DO 200  I=1,K-1
      SUM = 0.
      IF(I.EQ.1)  GO TO 350
      DO 300  J=1,I-1
      SUM = SUM + AL(I,J)*AL(K,J)
  300 CONTINUE
  350 CONTINUE
      AL(K,I) = (A(K,I) - SUM)/AL(I,I)
  200 CONTINUE
      SUM = 0.
      DO 400  J=1,K-1
      SUM = SUM + AL(K,J)*AL(K,J)
  400 CONTINUE
      AL(K,K) = SQRT(A(K,K) - SUM)
  100 CONTINUE
C.....PERFORM FORWARD PASS:
      Y(1) = B(1)/AL(1,1)
      DO 500  I=2,N
      SUM = 0.
      DO 550  J=1,I-1
      SUM = SUM + AL(I,J)*Y(J)
  550 CONTINUE
      Y(I) = (B(I) - SUM)/AL(I,I)
  500 CONTINUE
C.....PERFORM BACKWARD PASS:
      X(N) = Y(N)/AL(N,N)
      DO 600  I=N-1,1,-1
      SUM = 0.
      DO 650  J=I+1,N
      SUM = SUM + AL(J,I)*X(J)
  650 CONTINUE
      X(I) = (Y(I) - SUM)/AL(I,I)
  600 CONTINUE
      RETURN
      END
```

MATLAB

```
% PROGRAM CHOLESKY
% PROGRAM FOR SOLVING A SET OF SIMULTANEOUS
% LINEAR EQUATIONS  [A]{X} = {B}
% USING CHOLESKY DECOMPOSITION IF [A] IS SYMMETRIC
%
% READ TOTAL NUMBER OF EQUATIONS TO BE SOLVED:
% READ MATRIX [A] AND VECTOR {B} :
fid = fopen('input.dat', 'r');
n = fscanf(fid,'%f',1);
a = fscanf(fid,'%f',[n+1 n]);
a = a.';
b = squeeze(a(:,n+1));
a = squeeze(a(:,1:n));
fclose(fid);
x = choles(n, a, b);
% WRITE THE SOLUTIONS:
fid = fopen('output.dat', 'w');
fprintf(fid,'\n   EQUATION NO.        SOLUTION X \n');
fprintf(    '\n   EQUATION NO.        SOLUTION X \n');
for i = 1:n
    fprintf(fid,'  %8d        %14.6e\n', i, x(i));
    fprintf(    '  %8d        %14.6e\n', i, x(i));
end
fclose(fid);
```

```
function x = choles(n, a, b)
%  PERFORM DECOMPOSITION  [A] = [L][U]:
AL = zeros(n,n);
y = zeros(n,1);
AL(1,1) = sqrt(a(1,1));
disp(a)
disp(b)
for k = 2:n
    for i = 1:k-1
        sum = 0.;
        if(i ~= 1)
            for j = 1:i-1
                sum = sum + AL(i,j)*AL(k,j);
            end
        end
        AL(k,i) = (a(k,i) - sum)/AL(i,i);
    end
    sum = 0.;
    for j = 1:k-1
        sum = sum + AL(k,j)*AL(k,j);
    end
    AL(k,k) = sqrt(a(k,k) - sum);
end
%  PERFORM FORWARD PASS
y(1) = b(1)/AL(1,1);
for i = 2:n
    sum = 0.;
    for j = 1:i-1
        sum = sum + AL(i,j)*y(j);
    end
    y(i) = (b(i) - sum)/AL(i,i);
end
%  PERFORM BACKWARD PASS
x(n) = y(n)/AL(n,n);
for i = n-1:-1:1
    sum = 0.;
    for j = i+1:n
        sum = sum + AL(j,i)*x(j);
    end
    x(i) = (y(i) - sum)/AL(i,i);
end
```

Figure 3.6 Computer program for solving a system of n equations by using the Cholesky decomposition method.

3.12 MATLAB **Function for Cholesky Decomposition**

MATLAB uses the built-in function `chol` to decompose the matrix $[A]$. The syntax for this function is

$$U = chol(A)$$

where `U` and `A` represents the matrices $[L]^T$ and $[A]$ as shown in Eq. (3.57), respectively.

Example 3.16 Solve the following system of equations by using the MATLAB function `chol` for the Cholesky decomposition method.

$$\begin{bmatrix} 4 & 3 & 1 \\ 3 & 5 & 2 \\ 1 & 2 & 6 \end{bmatrix} \begin{Bmatrix} x_1 \\ x_2 \\ x_3 \end{Bmatrix} = \begin{Bmatrix} 3,125 \\ 3,650 \\ 2,800 \end{Bmatrix}$$

The matrices $[A]$ and $\{B\}$ are first defined by the commands,

```
>> A = [4 3 1; 3 5 2; 1 2 6];
>> B = [3125; 3650; 2800];
```

The function `chol` is then used to decompose matrix $[A]$,

```
>> U = chol(A)

U =

        2.0000      1.5000      0.5000
             0      1.6583      0.7538
             0           0      2.2764
```

The result from decomposing can be verified by,

```
>> U'*U

ans =

        4      3      1
        3      5      2
        1      2      6
```

Finally, the left division command is applied to obtain the solutions as follows,

```
>> Y = U'\B

Y =

    1.0e+003 *

        1.5625
        0.7877
        0.6260

>> x = U\Y

x =

        450.0000
        350.0000
        275.0000
```

The computed solutions above are the exact solutions of the problem.

3.13 Jacobi Iteration Method

All methods for solving a system of equations explained so far are called the direct technique because the solutions are obtained directly from certain computational procedures. The technique, however, requires substantial computational time and memory when solving a large set of equations. Another group of methods, called the iterative technique, are sometimes used to solve a set of equations that arises from some types of engineering problems.

The simplest iterative technique is the Jacobi iteration method. The method can be understood easily by considering the following system of equations,

$$\begin{bmatrix} a_{11} & a_{12} & a_{13} \\ a_{21} & a_{22} & a_{23} \\ a_{31} & a_{32} & a_{33} \end{bmatrix} \begin{Bmatrix} x_1 \\ x_2 \\ x_3 \end{Bmatrix} = \begin{Bmatrix} b_1 \\ b_2 \\ b_3 \end{Bmatrix} \tag{3.66}$$

The basic idea behind this method is to rewrite each equation so that its unknown appears on the left-hand side of that equation,

$$x_1 = \frac{b_1 - a_{12} x_2 - a_{13} x_3}{a_{11}} \tag{3.67a}$$

$$x_2 = \frac{b_2 - a_{21} x_1 - a_{23} x_3}{a_{22}} \tag{3.67b}$$

$$x_3 = \frac{b_3 - a_{31} x_1 - a_{32} x_2}{a_{33}} \tag{3.67c}$$

The computational procedure starts from using a set of guess values x_1, x_2, x_3 on the right-hand side of Eq. (3.67a-c) in order to compute the new values x_1, x_2, x_3. The process is repeated until the updated values of x_1, x_2, x_3 converge to the solutions within the specified tolerance ε,

$$\left| \frac{x_i^{k+1} - x_i^k}{x_i^{k+1}} \right| \times 100\% < \varepsilon \tag{3.68}$$

where the subscript i is the equation number and the superscript k is the kth iteration.

The system of equation (3.10) is solved again but by using the Jacobi iteration method as shown in the example below.

Example 3.17 Use the Jacobi iteration method to solve Eq. (3.10),

$$\begin{bmatrix} 4 & -4 & 0 \\ -1 & 4 & -2 \\ 0 & -2 & 4 \end{bmatrix} \begin{Bmatrix} x_1 \\ x_2 \\ x_3 \end{Bmatrix} = \begin{Bmatrix} 400 \\ 400 \\ 400 \end{Bmatrix} \tag{3.10}$$

The three equations in Eq. (3.10) are,

$$4x_1 - 4x_2 = 400 \tag{3.69a}$$

$$-x_1 + 4x_2 - 2x_3 = 400 \tag{3.69b}$$

$$- 2x_2 + 4x_3 = 400 \tag{3.69c}$$

These three equations are written such that their unknowns appear on the left-hand-side of the equations,

$$x_1 = 100 + x_2 \tag{3.70a}$$

$$x_2 = 100 + \frac{1}{4}x_1 + \frac{1}{2}x_3 \tag{3.70b}$$

$$x_3 = 100 + \frac{1}{2}x_2 \tag{3.70c}$$

Then, the iteration numbers are included to indicate the old and new values of the solutions,

$$x_1^{k+1} = 100 + x_2^k \tag{3.71a}$$

$$x_2^{k+1} = 100 + \frac{1}{4}x_1^k + \frac{1}{2}x_3^k \tag{3.71b}$$

$$x_3^{k+1} = 100 + \frac{1}{2}x_2^k \tag{3.71c}$$

where the superscript k represents the kth iteration.

For example, if the initial guess values are $x_1 = x_2 = x_3 = 100$, then the new values of x_1, x_2, x_3 in Eq. (3.71a-c) are

$$x_1 = 100 + 100 \qquad\qquad = 200 \tag{3.72a}$$

$$x_2 = 100 + \frac{1}{4}(100) + \frac{1}{2}(100) = 175 \tag{3.72b}$$

$$x_3 = 100 + \frac{1}{2}(100) \qquad\qquad = 150 \tag{3.72c}$$

The differences between the old and new values are determined according to Eq. (3.68) as,

$$\text{Difference of } x_1 = \left|\frac{100 - 200}{200}\right| \times 100\% = 50.00\% \tag{3.73a}$$

$$\text{Difference of } x_2 = \left|\frac{100 - 175}{175}\right| \times 100\% = 42.86\% \tag{3.73b}$$

$$\text{Difference of } x_3 = \left|\frac{100 - 150}{150}\right| \times 100\% = 33.33\% \tag{3.73c}$$

The process is repeated until the differences of the old and new values of x_1, x_2, x_3 are less than a specified tolerance ε, such as $\varepsilon = 0.05\%$.

Figure 3.7 shows a computer program for solving the system of equations in this example. The initial values of x_1, x_2, x_3 used in the program are all 100 and the specified tolerance is $\varepsilon = 0.05\%$. The computed solutions at different iterations are shown in Table 3.5. The table shows that the solutions converge to the exact solution within 19 iterations. Figure 3.7 and Table 3.5 demonstrate that a computer program using the Jacobi iteration method can be developed so that the solutions from a set of equations are obtained easily.

Fortran

```
      PROGRAM  JACOBI
C.....JACOBI ITERATION METHOD FOR EXAMPLE 3.13
      DIMENSION  XOLD(3), XNEW(3)
      TOL = 0.05
      DO 10  I=1,3
      XOLD(I) = 100.
   10 CONTINUE
      WRITE(6,50)
   50 FORMAT(2X, 'ITERATION NO.',
     *      6X, 'X1', 8X, 'X2', 8X, 'X3', /)
      DO 100  ITER=1,500
      XNEW(1) = 100. + XOLD(2)
      XNEW(2) = 100. + 0.25*XOLD(1) + 0.50*XOLD(3)
      XNEW(3) = 100. + 0.50*XOLD(2)
      WRITE(6,500)  ITER, (XOLD(I), I=1,3)
  500 FORMAT(I8, 6X, 3F10.0)
      IFLAG = 0
      DO 200  I=1,3
      EPS = ABS((XNEW(I) - XOLD(I))*100./XNEW(I))
      IF(EPS.GE.TOL)  IFLAG = 1
  200 CONTINUE
      IF(IFLAG.EQ.0)  GO TO 400
      DO 300  I=1,3
      XOLD(I) = XNEW(I)
  300 CONTINUE
  100 CONTINUE
  400 CONTINUE
      STOP
      END
```

MATLAB

```
% PROGRAM JACOBI
% JACOBI ITERATION METHOD FOR EXAMPLE 3.13
tol = 0.05;
for i = 1:3
    xold(i) = 100.;
end
fprintf('\n  ITERATION NO.      X1       X2      ...
X3\n');
for iter = 1:500
    xnew(1) = 100. + xold(2);
    xnew(2) = 100. + 0.25*xold(1) + 0.5*xold(3);
    xnew(3) = 100. + 0.50*xold(2);
    fprintf('%8d%10.0f%10.0f%10.0f\n',iter,xold(1), ...
xold(2),xold(3));
    iflag = 0;
    for i = 1:3
        eps = abs((xnew(i) - xold(i))*100./xnew(i));
        if(eps >= tol)
            iflag = 1;
        end
    end
    if iflag == 0
        break
    end
    for i = 1:3
        xold(i) = xnew(i);
    end
end
```

Figure 3.7 Computer program for solving the system of Eq. (3.10) by the Jacobi iteration method.

Table 3.5 The computed solutions of Eq. (3.10) at each iteration by using the Jacobi iteration method.

ITERATION NO.	X1	X2	X3
1	100.	100.	100.
2	200.	175.	150.
3	275.	225.	188.
4	325.	263.	213.
5	363.	288.	231.
6	388.	306.	244.
7	406.	319.	253.
8	419.	328.	259.
9	428.	334.	264.
10	434.	339.	267.
11	439.	342.	270.
12	442.	345.	271.
13	445.	346.	272.
14	446.	347.	273.
15	447.	348.	274.
16	448.	349.	274.
17	449.	349.	274.
18	449.	349.	275.
19	449.	350.	275.

3.14 Gauss-Seidel Iteration Method

The Gauss-Seidel method is similar to the Jacobi method explained in the preceding section but can increase the solution convergence rate. The Gauss-Seidel method reduces the number of iterations by using the updated values of x_1, x_2, x_3 right after they were computed. The example below shows the Gauss-Seidel method for solving the same set of Eq. (3.10).

Example 3.18   Use the Gauss-Seidel iteration method to solve the set of equations,

$$\begin{bmatrix} 4 & -4 & 0 \\ -1 & 4 & -2 \\ 0 & -2 & 4 \end{bmatrix} \begin{Bmatrix} x_1 \\ x_2 \\ x_3 \end{Bmatrix} = \begin{Bmatrix} 400 \\ 400 \\ 400 \end{Bmatrix} \tag{3.10}$$

The same equations as shown in Example 3.17 can be used for the Gauss-Seidel iteration method except Eqs. (3.71b-c). The newly computed values of x_1 and x_2 are used immediately in the calculation for the new x_2 and x_3 in Eqs. (3.74b-c), respectively.

$$x_1^{k+1} = 100 + x_2^k \tag{3.74a}$$

$$x_2^{k+1} = 100 + \frac{1}{4}x_1^{k+1} + \frac{1}{2}x_3^k \tag{3.74b}$$

$$x_3^{k+1} = 100 + \frac{1}{2}x_2^{k+1} \tag{3.74c}$$

For example, if the initial guess values of $x_1 = x_2 = x_3 = 100$, then the new values of x_1, x_2, and x_3 can be determined from Eq. (3.74a-c) as,

$$x_1 = 100 + 100 \qquad\qquad = 200 \tag{3.75a}$$

$$x_2 = 100 + \frac{1}{4}(200) + \frac{1}{2}(100) = 200 \tag{3.75b}$$

$$x_3 = 100 + \frac{1}{2}(200) \qquad\qquad = 200 \tag{3.75c}$$

The process is repeated until the differences between the old and new values of x_1, x_2 and x_3 are less than the specified tolerance ε.

Figure 3.8 shows a computer program developed for this example. The initial values of x_1, x_2, x_3 are set to 100 with the specified tolerance of $\varepsilon = 0.05\%$. The computed solutions obtained from this program are shown in Table 3.6. The table shows that the variables x_1, x_2, x_3 converge to the final solutions within 12 iterations as compared to 19 iterations by the Jacobi method.

Fortran

```
      PROGRAM  GSEIDEL
C.....GAUSS-SEIDEL ITERATION METHOD FOR EX 3.14
      DIMENSION  XOLD(3), XNEW(3)
      TOL = 0.05
      DO 10  I=1,3
      XOLD(I) = 100.
   10 CONTINUE
      WRITE(6,50)
   50 FORMAT(2X, 'ITERATION NO.',
     *       6X, 'X1', 8X, 'X2', 8X, 'X3', /)
      DO 100  ITER=1,500
      XNEW(1) = 100. + XOLD(2)
      XNEW(2) = 100. + 0.25*XNEW(1) + 0.50*XOLD(3)
      XNEW(3) = 100. + 0.50*XNEW(2)
      WRITE(6,500)  ITER, (XOLD(I), I=1,3)
  500 FORMAT(I8, 6X, 3F10.0)
      IFLAG = 0
      DO 200  I=1,3
      EPS = ABS((XNEW(I) - XOLD(I))*100./XNEW(I))
      IF(EPS.GE.TOL)  IFLAG = 1
  200 CONTINUE
      IF(IFLAG.EQ.0)  GO TO 400
      DO 300  I=1,3
      XOLD(I) = XNEW(I)
  300 CONTINUE
  100 CONTINUE
  400 CONTINUE
      STOP
      END
```

MATLAB

```
%  PROGRAM GSEIDEL
%  GAUSS-SEIDEL ITERATION METHOD FOR EXAMPLE 3.14
tol = 0.05;
for i = 1:3
    xold(i) = 100.;
end
fprintf('\n   ITERATION NO.       X1          X2       ...
X3\n');
for iter = 1:500
    xnew(1) = 100. + xold(2);
    xnew(2) = 100. + 0.25*xnew(1) + 0.5*xold(3);
    xnew(3) = 100. + 0.50*xnew(2);
    fprintf('%8d%10.0f%10.0f%10.0f\n',iter,xold(1), ...
xold(2),xold(3));
    iflag = 0;
    for i = 1:3
        eps = abs((xnew(i) - xold(i))*100./xnew(i));
        if(eps >= tol)
            iflag = 1;
        end
    end
    if iflag == 0
        break
    end
    for i = 1:3
        xold(i) = xnew(i);
    end
end
```

Figure 3.8 Computer program for solving the system of Eq. (3.10) by the Gauss-Seidel iteration method.

Table 3.6 The computed solutions of Eq. (3.10) at each iteration by using the Gauss-Seidel iteration method.

ITERATION NO.	X1	X2	X3
1	100.	100.	100.
2	200.	200.	200.
3	300.	275.	238.
4	375.	313.	256.
5	413.	331.	266.
6	431.	341.	270.
7	441.	345.	273.
8	445.	348.	274.
9	448.	349.	274.
10	449.	349.	275.
11	449.	350.	275.
12	450.	350.	275.

3.15 Successive Over-relaxation Method

The successive over-relaxation method is modified from Gauss-Seidel method in order to further increase the solution convergence rate. The new solutions are obtained by weighting between the newly computed and the old values as,

$$x_i^{k+1} = \omega x_i^{k+1*} + (1-\omega)x_i^k \tag{3.76}$$

where the superscript k represent the kth iteration, x_i^{k+1*} is the newly computed value as explained in the Gauss-Seidel method and ω is the weighting factor which has a value between 0 and 2. If $\omega = 1$, the method becomes the Gauss-Seidel method. If the value of ω is between 1 and 2, the computed solutions are weighted by the new values more than the old values, which is called over-relaxation. If ω is between 0 and 1, the method is called under-relaxation that may help a nearly diverged solution to converge. In general, the value of ω should be selected between 1 and 2 to increase the convergence rate. It is noted that the appropriate value of ω varies depending on the problems.

To understand the successive over-relaxation method clearly, Eq. (3.10) is solved again as shown in the following example.

Example 3.19 Use the successive over-relaxation method to solve Eq. (3.10),

$$\begin{bmatrix} 4 & -4 & 0 \\ -1 & 4 & -2 \\ 0 & -2 & 4 \end{bmatrix} \begin{Bmatrix} x_1 \\ x_2 \\ x_3 \end{Bmatrix} = \begin{Bmatrix} 400 \\ 400 \\ 400 \end{Bmatrix} \tag{3.10}$$

From the Gauss-Seidel method, the new values of x_1, x_2, x_3 at the $(k+1)^{th}$ iteration are

$$x_1^{k+1} = 100 + x_2^k \tag{3.74a}$$

$$x_2^{k+1} = 100 + \frac{1}{4}x_1^{k+1} + \frac{1}{2}x_3^k \tag{3.74b}$$

$$x_3^{k+1} = 100 + \frac{1}{2}x_2^{k+1} \tag{3.74c}$$

With the successive over-relaxation, the new values of x_1, x_2, x_3 are

$$x_1^{k+1} = \omega\left(100 + x_2^k\right) + (1-\omega)x_1^k \tag{3.77a}$$

$$x_2^{k+1} = \omega\left(100 + \frac{1}{4}x_1^{k+1} + \frac{1}{2}x_3^k\right) + (1-\omega)x_2^k \tag{3.77b}$$

$$x_3^{k+1} = \omega\left(100 + \frac{1}{2}x_2^{k+1}\right) + (1-\omega)x_3^k \tag{3.77c}$$

Figure 3.9 shows a computer program developed with the initial value x_1, x_2, x_3 of 100, the weighting factor $\omega = 1.2$ and the specified tolerance $\varepsilon = 0.05\%$. The computed solutions obtained from the computer program are shown in Table 3.7. The results show that the converged solutions are obtained within 6 iterations as compared to 12 and 19 iterations by the Gauss-Seidel and Jacobi methods, respectively.

Fortran

```
          PROGRAM  SOR
C.....SUCCESSIVE OVER RELAXATION FOR EXAMPLE 3.15
          DIMENSION  XOLD(3), XNEW(3)
          W = 1.2
          TOL = 0.05
          DO 10  I=1,3
          XOLD(I) = 100.
   10 CONTINUE
          WRITE(6,50)
   50 FORMAT(2X, 'ITERATION NO.',
     *          6X, 'X1', 8X, 'X2', 8X, 'X3', /)
          DO 100  ITER=1,500
          XNEW(1) = 100. + XOLD(2)
          XNEW(1) = W*XNEW(1) + (1.-W)*XOLD(1)
          XNEW(2) = 100. + 0.25*XNEW(1) + 0.50*XOLD(3)
          XNEW(2) = W*XNEW(2) + (1.-W)*XOLD(2)
          XNEW(3) = 100. + 0.50*XNEW(2)
          XNEW(3) = W*XNEW(3) + (1.-W)*XOLD(3)
          WRITE(6,500)  ITER, (XOLD(I), I=1,3)
  500 FORMAT(I8, 6X, 3F10.0)
          IFLAG = 0
          DO 200  I=1,3
          EPS = ABS((XNEW(I) - XOLD(I))*100./XNEW(I))
          IF(EPS.GE.TOL)  IFLAG = 1
  200 CONTINUE
          IF(IFLAG.EQ.0)  GO TO 400
          DO 300  I=1,3
          XOLD(I) = XNEW(I)
  300 CONTINUE
  100 CONTINUE
  400 CONTINUE
          STOP
          END
```

MATLAB

```
% PROGRAM SOR
% SUCCESSIVE OVER RELEXATION FOR EXAMPLE 3.15
w = 1.2
tol = 0.05;
for i = 1:3
    xold(i) = 100.;
end
fprintf('\n   ITERATION NO.      X1        X2     ...
X3\n');
for iter = 1:500
    xnew(1) = 100. + xold(2);
    xnew(1) = w*xnew(1) + (1-w)*xold(1);
    xnew(2) = 100. + 0.25*xnew(1) + 0.5*xold(3);
    xnew(2) = w*xnew(2) + (1-w)*xold(2);
    xnew(3) = 100. + 0.50*xnew(2);
    xnew(3) = w*xnew(3) + (1-w)*xold(3);
    fprintf('%8d%10.0f%10.0f%10.0f\n',iter,xold(1), ...
xold(2),xold(3));
    iflag = 0;
    for i = 1:3
        eps = abs((xnew(i) - xold(i))*100./xnew(i));
        if(eps >= tol)
            iflag = 1;
        end
    end
    if iflag == 0
        break
    end
    for i = 1:3
        xold(i) = xnew(i);
    end
end
```

Figure 3.9 Computer program for solving the system of Eq. (3.10) by the successive over-relaxation method.

Table 3.7 The computed solutions of Eq. (3.10) at each iteration by using the successive over-relaxation method.

ITERATION NO.	X1	X2	X3
1	100.	100.	100.
2	220.	226.	236.
3	347.	320.	265.
4	435.	345.	274.
5	448.	350.	275.
6	450.	350.	275.

3.16 Conjugate Gradient Method

One of the most popular methods for solving a system of equations based on the iterative technique is the conjugate gradient method. The method is used for solving the system of equations in the form

$$\underset{(n \times n)}{[A]} \underset{(n \times 1)}{\{X\}} = \underset{(n \times 1)}{\{B\}} \tag{3.4}$$

where the matrix $[A]$ is a symmetric and positive definite matrix. Definitions of a positive definite matrix will be explained later. The method provides a converged solution within n iterations if the system of equations consists of n equations. In addition, the method performs effectively when the matrix $[A]$ is a sparse matrix.

In order to understand the concept of the method, an example for the system of two equations ($n = 2$) is considered,

$$\begin{bmatrix} 2 & 1 \\ 1 & 3 \end{bmatrix} \begin{Bmatrix} x_1 \\ x_2 \end{Bmatrix} = \begin{Bmatrix} 4 \\ -3 \end{Bmatrix} \tag{3.78}$$

where $[A]$ is a symmetric matrix with its dimensions of (2×2). Equation (3.78) are written in the form of two algebraic equations as,

$$2x_1 + x_2 = 4 \tag{3.79}$$

$$x_1 + 3x_2 = -3 \tag{3.80}$$

These two equations can be plotted and represented by the two straight lines as shown in Fig. 3.10. The two lines intersect at the point where $x_1 = 3$ and $x_2 = -2$ which are the solutions of the system of equations.

$$\{X\} = \begin{Bmatrix} x_1 \\ x_2 \end{Bmatrix} = \begin{Bmatrix} 3 \\ -2 \end{Bmatrix} \tag{3.81}$$

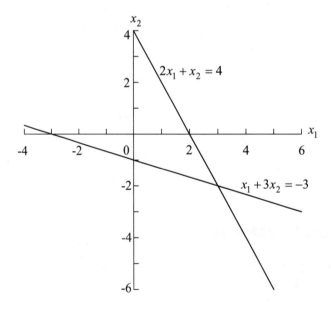

Figure 3.10 Two straight lines represented by Eqs. (3.79 - 3.80) and
the intersection point at $x_1 = 3$ and $x_2 = -2$.

The procedures to find the solutions of Eq. (3.78) by using the conjugate gradient method start from the quadratic function in the form,

$$f(x_1, x_2) = \frac{1}{2} \underset{(1\times2)}{\lfloor X \rfloor} \underset{(2\times2)}{[A]} \underset{(2\times1)}{\{X\}} - \underset{(1\times2)}{\lfloor B \rfloor} \underset{(2\times1)}{\{X\}}$$ (3.82)

or

$$f(x_1, x_2) = \frac{1}{2} \lfloor x_1 \; x_2 \rfloor \begin{bmatrix} 2 & 1 \\ 1 & 3 \end{bmatrix} \begin{Bmatrix} x_1 \\ x_2 \end{Bmatrix} - \lfloor 4 \; -3 \rfloor \begin{Bmatrix} x_1 \\ x_2 \end{Bmatrix}$$ (3.83)

The function $f(x_1, x_2)$ in Eq. (3.83) is plotted by using contour lines as shown in Fig. 3.11. The figure shows that the quadratic function $f(x_1, x_2)$ is minimum at the location where x_1 and x_2 are the solutions of the system of equations.

If the function $f(x_1, x_2)$ is plotted in three dimensions, the shape of the function is similar to a blunt cone as shown in Fig. 3.12. The location where $x_1 = 3$ and $x_2 = -2$ is at the bottom of the blunt cone.

At the bottom location of the blunt cone with the solutions of x_1 and x_2, the gradient of the quadratic function is zero. Since Eq. (3.83) is,

$$f(x_1, x_2) = \frac{1}{2}\left(2x_1^2 + 2x_1x_2 + 3x_2^2\right) - 4x_1 + 3x_2$$ (3.84)

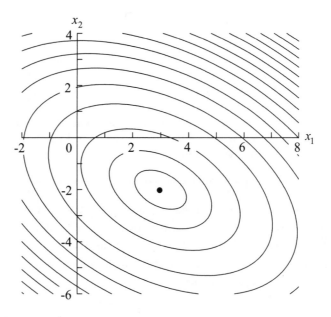

Figure 3.11 Contour plot of the quadratic function, $f(x_1, x_2)$.

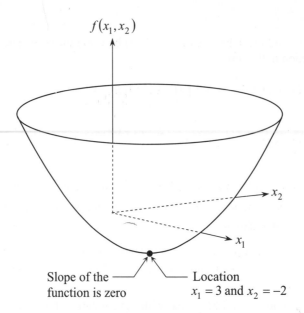

$$f(x_1, x_2)$$

$$x_2$$

$$x_1$$

Slope of the ——————— Location
function is zero $x_1 = 3$ and $x_2 = -2$

Figure 3.12 Shape of the quadratic function leading to
the concept of the conjugate gradient method.

the first derivatives with respect to x_1 and x_2 representing zero gradient at this location are,

$$\frac{\partial f}{\partial x_1} = 2x_1 + x_2 - 4 = 0 \tag{3.85a}$$

$$\frac{\partial f}{\partial x_2} = x_1 + 3x_2 + 3 = 0 \tag{3.85b}$$

The results are the original system of Eq. (3.78). The process shows that the solutions of system of equations can be determined from the condition of zero derivatives of the quadratic function.

To find the solutions of the system of equations by the conjugate gradient method, the quadratic function must vary as a blunt cup shape similar to that shown in Fig. (3.12). Such blunt cup shape occurs when the matrix $[A]$ is positive definite, i.e.,

$$\lfloor X \rfloor [A]\{X\} > 0 \tag{3.86}$$

where $\{X\}$ is a non-zero vector. Verifying that a matrix $[A]$ is a positive definite by using Eq. (3.86) is cumbersome if its dimensions are large. An alternative way for verifying a positive definite matrix $[A]$ is to observe its diagonal coefficients. If they are all positive, the matrix $[A]$ is likely a positive definite. Another way is to determine its determinant as follows. For the matrix $[A]$ with its dimensions of $(n \times n)$, the following determinants,

$$D_1 = |a_{11}| \tag{3.87a}$$

$$D_2 = \begin{vmatrix} a_{11} & a_{12} \\ a_{12} & a_{22} \end{vmatrix} \tag{3.87b}$$

$$D_3 = \begin{vmatrix} a_{11} & a_{12} & a_{13} \\ a_{12} & a_{22} & a_{23} \\ a_{13} & a_{23} & a_{33} \end{vmatrix} \tag{3.87c}$$

up to

$$D_n = \begin{vmatrix} a_{11} & a_{12} & \cdots & a_{1n} \\ a_{12} & a_{22} & \cdots & a_{2n} \\ \vdots & \vdots & \ddots & \vdots \\ a_{1n} & a_{2n} & \cdots & a_{nn} \end{vmatrix} \tag{3.87n}$$

must be greater than zero. For example, the matrix $[A]$ in Eq. (3.78) is positive definite because its $D_1 = 2$ and $D_2 = 5$.

To illustrate the procedure of the conjugate gradient method for solving the system of n equations in Eq. (3.4), the quadratic function

$$f(x_1, x_2, \ldots, x_n) = \frac{1}{2} \underset{(1 \times n)}{\lfloor X \rfloor} \underset{(n \times n)}{[A]} \underset{(n \times 1)}{\{X\}} - \underset{(1 \times n)}{\lfloor B \rfloor} \underset{(n \times 1)}{\{X\}} \tag{3.88}$$

is used. Their first derivatives with respect to $x_1, x_2, \ldots, x_n$ are determined and set to zero as,

$$\frac{\partial f}{\partial \{X\}} = [A]\{X\} - \{B\} = 0 \tag{3.89}$$

The result is identical Eq. (3.4) if the matrix $[A]$ in Eq. (3.89) is positive definite.

Because the conjugate gradient method is an iterative technique, a set of initial guess values in the vector $\{X\}$ is needed. Such set of values may represent a point on the surface of the blunt cone. The idea of the conjugate gradient method is to move from that point to the bottom location of the blunt cone representing the solutions of the system of equations. For a system of n equations, the shape of function f is a complex shape in hyperspace $(x_1, x_2, \ldots, x_n)$ which can not be displayed in three dimensions. The iterative process updates the vector $\{X\}$ by using its old solutions to determine the new solutions at the $(k+1)^{th}$ iteration as follows,

$$\{X\}^{k+1} = \{X\}^k + \lambda_k \{D\}^k \tag{3.90}$$

where $\{D\}^k$ is called the *search direction vector*. It is the vector that searches for the direction to the bottom location of the blunt cone. The scalar quantity λ_k represents the step size of the vector $\{D\}^k$ which changes at each iteration.

The value of λ_k for the kth iteration can be determined by substituting Eq. (3.90) into Eq. (3.88), taking the first derivative with respect to λ_k and setting them to zero as,

$$\frac{\partial f}{\partial \lambda_k} = \lfloor D \rfloor^k [A]\{D\}^k \lambda_k + \lfloor D \rfloor^k \{R\}^k = 0 \tag{3.91}$$

where
$$\{R\}^k = [A]\{X\}^k - \{B\} \tag{3.92}$$

is the residual vector. The vector approaches zero as the computed solutions converge to the exact solutions. Equation (3.91) can be written for determining the value λ_k as,

$$\lambda_k = -\frac{\lfloor D \rfloor^k \{R\}^k}{\lfloor D \rfloor^k [A]\{D\}^k} \tag{3.93}$$

It should be again noted that Eq. (3.93) is valid if and only if the matrix $[A]$ is positive definite.

During the iteration process, if the vector $\{X\}^k$ at the kth iteration is not the final solution, then the residual vector $\{R\}^k$ from Eqs. (3.89) and (3.92) is not zero,

$$\frac{\partial f}{\partial \{X\}^k} = [A]\{X\}^k - \{B\} = \{R\}^k \tag{3.94}$$

The residual vector $\{R\}^k$ represents the gradient of function f for the solutions in the vector $\{X\}^k$. Thus, if the vector $\{X\}^k$ contains the solutions of the system of equations at the bottom of the blunt cone, the residual vector $\{R\}^k$ representing the gradient of function must be zero.

At the first iteration ($k = 0$), the computational process starts from the initial guess vector $\{X\}^0$ and the search direction vector $\{D\}^0$. The search direction vector $\{D\}^0$ is first assigned in the opposite direction of the residual vector $\{R\}^0$ as,

$$\{D\}^0 = -\{R\}^0 \tag{3.95}$$

A new search direction vector $\{D\}$ is then determined from

$$\{D\}^{k+1} = -\{R\}^{k+1} + \alpha_k \{D\}^k \tag{3.96}$$

where α_k is the value that makes the vector $\{D\}^{k+1}$ to be $[A]$-conjugated with the vector $\{D\}^k$, i.e.,

$$\{D\}^{k+1} [A]\{D\}^k = 0 \tag{3.97}$$

By substituting Eq. (3.96) into Eq. (3.97) and arranging the equation for determining the value of α_k,

$$\alpha_k = \frac{\lfloor R \rfloor^{k+1} [A]\{D\}^k}{\lfloor D \rfloor^k [A]\{D\}^k} \tag{3.98}$$

It is noted that the word "conjugate" in the conjugate gradient method came from the requirement that if a vector $\{u\}$ is $[A]$-conjugated with another vector $\{v\}$, then

$$\lfloor u \rfloor [A]\{v\} = 0 \tag{3.99}$$

where the matrix $[A]$ must be symmetric and positive definite.

In conclusion, the procedures of the conjugate gradient method for solving the system of n equations,

$$[A]\{X\} = \{B\}$$
$$\underset{(n \times n)\ (n \times 1)}{} \qquad \underset{(n \times 1)}{}$$

(3.4)

where the matrix $[A]$ is symmetric and positive definite are as follows,

1. Assume initial vector $\{X\}^0$ and calculate the residual vector,

$$\{R\}^0 = [A]\{X\}^0 - \{B\}$$

2. Assign the search direction vector,

$$\{D\}^0 = -\{R\}^0$$

3. Start the iteration process for $k = 0, 1, 2, \ldots, n$ to determine,

$$\lambda_k = -\frac{\lfloor D \rfloor^k \{R\}^k}{\lfloor D \rfloor^k [A]\{D\}^k}$$

$$\{X\}^{k+1} = \{X\}^k + \lambda_k \{D\}^k$$

$$\{R\}^{k+1} = [A]\{X\}^{k+1} - \{B\}$$

In each iteration, determine and examine the error using $Error = \sqrt{\lfloor R \rfloor^{k+1}\{R\}^{k+1}} < \varepsilon$, where ε represents the acceptable error. If the condition is true, stop the iteration process. If not, the iteration process continues by determining,

$$\alpha_k = \frac{\lfloor R \rfloor^{k+1}[A]\{D\}^k}{\lfloor D \rfloor^k [A]\{D\}^k}$$

$$\{D\}^{k+1} = -\{R\}^{k+1} + \alpha_k \{D\}^k$$

and go back to step 3 for the new k^{th} iteration.

The above procedures are developed as a computer program as shown in Fig. 3.13 with detailed computations as shown in the example below.

Example 3.20   Use the conjugate gradient method to solve the system of equations,

$$\begin{bmatrix} 4 & -4 & 0 \\ -4 & 4 & -2 \\ 0 & -2 & 4 \end{bmatrix} \begin{Bmatrix} x_1 \\ x_2 \\ x_3 \end{Bmatrix} = \begin{Bmatrix} 400 \\ -950 \\ 400 \end{Bmatrix}$$

(3.100)

Compare the computed solutions with the exact solutions of $x_1 = 450$, $x_2 = 350$ and $x_3 = 275$.

Fortran

```
      SUBROUTINE CG(N, A, B, X)
      DIMENSION  A(50,50), B(50), X(50)
      DIMENSION  R(50), D(50)
C.....ASSIGN INITIAL VALUES IN VECTOR X:
      XZERO = 0.
      DO 10  I=1,N
      X(I) = XZERO
   10 CONTINUE
C.....ASSIGN TOLERANCE FOR STOPPING CRITERION:
      TOL = 0.0001
C.....COMPUTE INITIAL RESIDUAL & SEARCH DIRECTION:
      DO 20  I=1,N
      SUM = 0.
      DO 30  J=1,N
      SUM = SUM + A(I,J)*X(J)
   30 CONTINUE
      R(I) = SUM - B(I)
      D(I) = -R(I)
   20 CONTINUE
C.....ENTER THE ITERATION LOOP:
      DO 500  K=1,N+1
      UP = 0.
      DO 40  I=1,N
      UP = UP + D(I)*R(I)
   40 CONTINUE
      DOWN = 0.
      DO 50  I=1,N
      SUM = 0.
      DO 60  J=1,N
      SUM = SUM + A(I,J)*D(J)
   60 CONTINUE
      DOWN = DOWN + D(I)*SUM
   50 CONTINUE
      ALAM = -UP/DOWN
      DO 70  I=1,N
      X(I) = X(I) + ALAM*D(I)
   70 CONTINUE
      DO 80  I=1,N
      SUM = 0.
      DO 90  J=1,N
      SUM = SUM + A(I,J)*X(J)
   90 CONTINUE
      R(I) = SUM - B(I)
   80 CONTINUE
      RES = 0.
      DO 100  I=1,N
      RES = RES + R(I)*R(I)
  100 CONTINUE
      RES = SQRT(RES)
      IF(RES.LT.TOL)  RETURN
      UP = 0.
      DOWN = 0.
      DO 110  I=1,N
      SUM = 0.
      DO 120  J=1,N
      SUM = SUM + A(I,J)*D(J)
  120 CONTINUE
      UP = UP + R(I)*SUM
      DOWN = DOWN + D(I)*SUM
  110 CONTINUE
      ALPHA = UP/DOWN
      DO 130  I=1,N
      D(I) = -R(I) + ALPHA*D(I)
  130 CONTINUE
  500 CONTINUE
      RETURN
      END
```

MATLAB

```
function x = cg(n, a, b, x)
% ASSIGN INITIAL VALUES IN VECTOR X:
xzero = 0.;
x(1:n) = xzero;
% ASSIGN TOLERANCE FOR STOPPING CRITERION:
tol = 0.0001;
% COMPUTE INITIAL RESIDUAL & SEARCH DIRECTION:
for i = 1:n
    sum = 0.;
    for j = 1:n
        sum = sum + a(i,j)*x(j);
    end
    r(i) = sum - b(i);
    d(i) = -r(i);
end
% ENTER THE ITERATION LOOP:
for k = 1:n+1
    up = 0.;
    for i = 1:n
        up = up + d(i)*r(i);
    end
    down = 0.;
    for i = 1:n
        sum = 0.;
        for j = 1:n
            sum = sum + a(i,j)*d(j);
        end
        down = down + d(i)*sum;
    end
    alam = -up/down;
    for i = 1:n
        x(i) = x(i) + alam*d(i);
    end
    for i = 1:n
        sum = 0.;
        for j = 1:n
            sum = sum + a(i,j)*x(j);
        end
        r(i) = sum - b(i);
    end
    res = 0.;
    for i = 1:n
        res = res + r(i)*r(i);
    end
    res = sqrt(res);
    if res < tol
        break
    end
    up = 0.;
    down = 0.;
    for i = 1:n
        sum = 0.;
        for j = 1:n
            sum = sum + a(i,j)*d(j);
        end
        up = up + r(i)*sum;
        down = down + d(i)*sum;
    end
    alpha = up/down;
    for i = 1:n
        d(i) = -r(i) + alpha*d(i);
    end
end
```

Figure 3.13 Computer program for solving the system of equations
by using the conjugate gradient method.

Solution   Assume the initial guess values in the vector $\{x\}^0 = \begin{Bmatrix} 100 \\ 100 \\ 100 \end{Bmatrix}$, then

$$\{R\}^0 = [A]\{X\}^0 - \{B\} = \begin{bmatrix} 4 & -4 & 0 \\ -4 & 4 & -2 \\ 0 & -2 & 4 \end{bmatrix}\begin{Bmatrix} 100 \\ 100 \\ 100 \end{Bmatrix} - \begin{Bmatrix} 400 \\ -950 \\ 400 \end{Bmatrix} = \begin{Bmatrix} -400 \\ 750 \\ -200 \end{Bmatrix}$$

and assign $\{D\}^0 = -\{R\}^0 = \begin{Bmatrix} 400 \\ -750 \\ 200 \end{Bmatrix}$

Start the iteration process with $k = 0$,

$$\lambda_0 = -\frac{\lfloor D \rfloor^0 \{R\}^0}{\lfloor D \rfloor^0 [A]\{D\}^0} = 0.1260$$

$$\{X\}^1 = \{X\}^0 + \lambda_0\{D\}^0 = \begin{Bmatrix} 150.4132 \\ 5.4752 \\ 125.2066 \end{Bmatrix}$$

$$\{R\}^1 = [A]\{X\}^1 - \{B\} = \begin{Bmatrix} 179.7521 \\ 119.8347 \\ 89.8760 \end{Bmatrix}$$

$$Error = \sqrt{\lfloor R \rfloor^1 \{R\}^1} = 233.9848$$

$$\alpha_0 = \frac{\lfloor R \rfloor^1 [A]\{D\}^0}{\lfloor D \rfloor^0 [A]\{D\}^0} = 0.0718$$

$$\{D\}^1 = -\{R\}^1 + \alpha_0\{D\}^0 = \begin{Bmatrix} -151.0314 \\ -173.6861 \\ -75.5157 \end{Bmatrix}$$

Start the iteration process with $k = 1$,

$$\lambda_1 = -\frac{\lfloor D \rfloor^1 \{R\}^1}{\lfloor D \rfloor^1 [A]\{D\}^1} = -1.9836$$

$$\{X\}^2 = \{X\}^1 + \lambda_1\{D\}^1 = \begin{Bmatrix} 450.0000 \\ 350.0000 \\ 275.0000 \end{Bmatrix}$$

The obtained solutions are equal to the exact solutions. The residual vector is,

$$\{R\}^2 = [A]\{X\}^2 - \{B\} = \begin{Bmatrix} 0 \\ 0 \\ 0 \end{Bmatrix}$$

So that $Error = \sqrt{\lfloor R \rfloor^2 \{R\}^2} = 0$ and the computation stops.

If the computation continues, then

$$\alpha_1 = \frac{\lfloor R \rfloor^2 [A]\{D\}^1}{\lfloor D \rfloor^1 [A]\{D\}^1} = 0$$

and

$$\{D\}^2 = -\{R\}^2 + \alpha_1 \{D\}^1 = \begin{Bmatrix} 0 \\ 0 \\ 0 \end{Bmatrix}.$$

In addition, if the initial guess values in the vector $\lfloor X \rfloor^0 = \lfloor 0 \ \ 0 \ \ 0 \rfloor$ are used, the computational results from conjugate gradient method will converge to the solutions within 3 iterations $(k = 2)$.

As the number of equations increases, the computational time from more numerical operations also increases. The number of operations required by the conjugate gradient method, however, can be reduced. For example, the residual vector $\{R\}$ at the $(k+1)^{th}$ iteration in Eq. (3.94) can be determined from the residual vector $\{R\}$ and the product of $[A]\{D\}$ from the kth iteration. In addition, the residual vectors $\{R\}$ at different iterations have the orthogonal property, i.e.,

$$\lfloor R \rfloor^m \{R\}^n = 0 \qquad\qquad \text{when } m \neq n \qquad\qquad (3.101)$$

The property helps reducing the number of operations between matrices.

In conclusion, the modified conjugate gradient method that can reduce the computational time consists of determining the parameters as follows,

$$\lambda_k = \frac{\lfloor R \rfloor^k \{R\}^k}{\lfloor D \rfloor^k [A]\{D\}^k} \qquad\qquad (3.102)$$

and

$$\alpha_k = \frac{\lfloor R \rfloor^{k+1} \{R\}^{k+1}}{\lfloor R \rfloor^k \{R\}^k} \qquad\qquad (3.103)$$

Then, the following procedures are employed,

1.   Assume the initial vector $\{X\}^0$ and calculate the residual vector from,
$$\{R\}^0 = [A]\{X\}^0 - \{B\}$$

2.   Determine the search direction vector from,
$$\{D\}^0 = -\{R\}^0$$

3.   Calculate the residual from,
$$\delta^0 = \lfloor R \rfloor^0 \{R\}^0$$

4.   Start the iteration process from $k = 0, 1, 2, \ldots, n$ by calculating,
$$\{U\}^k = [A]\{D\}^k$$
$$\lambda_k = \frac{\delta^k}{\lfloor D \rfloor^k \{U\}^k}$$
$$\{X\}^{k+1} = \{X\}^k + \lambda_k \{D\}^k$$

$$\{R\}^{k+1} = \{R\}^k + \lambda_k \{U\}^k$$

$$\delta^{k+1} = \lfloor R \rfloor^{k+1} \{R\}^{k+1}$$

Examine the error using $Error = \sqrt{\lfloor R \rfloor^{k+1} \{R\}^{k+1}} < \varepsilon$, where ε is the acceptable error. If it is true, the computation stops. If not, the iteration process continues by calculating,

$$\alpha_k = \frac{\delta^{k+1}}{\delta^k}$$

$$\{D\}^{k+1} = -\{R\}^{k+1} + \alpha_k \{D\}^k$$

and then go back to the new k^{th} iteration in step 4.

The process explained above is used to develop a computer program as shown in Fig. 3.14.

Fortran

```
      SUBROUTINE CGNEW(N, A, B, X)
      DIMENSION  A(50,50), B(50), X(50)
      DIMENSION  R(50), D(50), U(50)
C.....ASSIGN INITIAL VALUES IN VECTOR X:
      XZERO = 0.
      DO 10  I=1,N
      X(I) = XZERO
   10 CONTINUE
C.....ASSIGN TOLERANCE FOR STOPPING CRITERION:
      TOL = 0.0001
C.....COMPUTE INITIAL RESIDUAL & SEARCH DIRECTION:
      DO 20  I=1,N
      SUM = 0.
      DO 30  J=1,N
      SUM = SUM + A(I,J)*X(J)
   30 CONTINUE
      R(I) = SUM - B(I)
      D(I) = -R(I)
   20 CONTINUE
      DEL = 0.
      DO 40  I=1,N
      DEL = DEL + R(I)*R(I)
   40 CONTINUE
C.....ENTER THE ITERATION LOOP:
      DO 500  K=1,N+1
      DO 50  I=1,N
      U(I) = 0.
      DO 60  J=1,N
      U(I) = U(I) + A(I,J)*D(J)
   60 CONTINUE
   50 CONTINUE
      DOWN = 0.
      DO 70  I=1,N
      DOWN = DOWN + D(I)*U(I)
   70 CONTINUE
      ALAM = DEL/DOWN
      DO 80  I=1,N
      X(I) = X(I) + ALAM*D(I)
      R(I) = R(I) + ALAM*U(I)
   80 CONTINUE
      DEL1 = 0.
      DO 90  I=1,N
      DEL1 = DEL1 + R(I)*R(I)
   90 CONTINUE
      IF(DEL1.LT.TOL)  RETURN
      ALPHA = DEL1/DEL
      DO 100  I=1,N
      D(I) = -R(I) + ALPHA*D(I)
  100 CONTINUE
      DEL = DEL1
  500 CONTINUE
      RETURN
      END
```

MATLAB

```
function x = cgnew(n, a, b, x)
%  ASSIGN INITIAL VALUES IN VECTOR X:
xzero = 0.;
x(1:n) = xzero;
%  ASSIGN TOLERANCE FOR STOPPING CRITERION:
tol = 0.0001;
%  COMPUTE INITIAL RESIDUAL & SEARCH DIRECTION:
for i = 1:n
    sum = 0.;
    for j = 1:n
        sum = sum + a(i,j)*x(j);
    end
    r(i) = sum - b(i);
    d(i) = -r(i);
end
del = 0.;
for i = 1:n
    del = del + r(i)*r(i);
end
%  ENTER THE ITERATION LOOP:
for k = 1:n+1
    for i = 1:n
        u(i) = 0.
        for j = 1:n
            u(i) = u(i) + a(i,j)*d(j);
        end
    end
    down = 0.;
    for i = 1:n
        down = down + d(i)*u(i);
    end
    alam = del/down;
    for i = 1:n
        x(i) = x(i) + alam*d(i);
        r(i) = r(i) + alam*u(i);
    end
    del2 = 0.;
    for i = 1:n
        del2 = del2 + r(i)*r(i);
    end
    if del2 < tol
        break
    end
    alpha = del2/del;
    for i = 1:n
        d(i) = -r(i) + alpha*d(i);
    end
    del = del2;
end
```

Figure 3.14 Computer program for solving the system of equations by the modified conjugate gradient method.

Example 3.21 Use the modified conjugate gradient method in Fig. 3.14 to solve the system of equations,

$$\begin{bmatrix} 4 & -4 & 0 \\ -1 & 4 & -2 \\ 0 & -2 & 4 \end{bmatrix} \begin{Bmatrix} x_1 \\ x_2 \\ x_3 \end{Bmatrix} = \begin{Bmatrix} 400 \\ 400 \\ 400 \end{Bmatrix} \tag{3.10}$$

Note that the matrix $[A]$ on left-hand side of Eq. (3.10) is an asymmetric matrix.
The exact solutions for the above system of equations are $x_1 = 450$, $x_2 = 350$ and $x_3 = 275$.

<u>Solution</u>   The system of equations in Eq. (3.10) is in the form,

$$[P]\{X\} = \{Q\} \tag{3.104}$$

where $[P]$ is an asymmetric matrix. Equation (3.104) can be modified by pre-multiplying matrix $[P]^T$ on both sides so that the resulting matrix on the left-hand side of the equation becomes a symmetric matrix,

$$[P]^T[P]\{X\} = [P]^T\{Q\} \tag{3.105}$$

After pre-multiplying by matrix $[P]^T$, the new system of equations is

$$\begin{bmatrix} 17 & -20 & 2 \\ -20 & 36 & -16 \\ 2 & -16 & 20 \end{bmatrix} \begin{Bmatrix} x_1 \\ x_2 \\ x_3 \end{Bmatrix} = \begin{Bmatrix} 1{,}200 \\ -800 \\ 800 \end{Bmatrix} \tag{3.106}$$

which is in the form of $[A]\{X\} = \{B\}$

The modified conjugate gradient method can then be applied by first assuming the initial guess vector, for example,

$$\{X\}^0 = \begin{Bmatrix} 100 \\ 100 \\ 100 \end{Bmatrix}$$

So that

$$\{R\}^0 = [A]\{X\}^0 - \{B\} = \begin{bmatrix} 17 & -20 & 2 \\ -20 & 36 & -16 \\ 2 & -16 & 20 \end{bmatrix} \begin{Bmatrix} 100 \\ 100 \\ 100 \end{Bmatrix} - \begin{Bmatrix} 1{,}200 \\ -800 \\ 800 \end{Bmatrix} = \begin{Bmatrix} -1{,}300 \\ 800 \\ -200 \end{Bmatrix}$$

Then

$$\{D\}^0 = -\{R\}^0 = \begin{Bmatrix} 1{,}300 \\ -800 \\ 200 \end{Bmatrix}$$

and

$$\delta^0 = \lfloor R \rfloor^0 \{R\}^0 = 2{,}370{,}000$$

By starting the iteration process at $k = 0$,

$$\{U\}^0 = [A]\{D\}^0 = \begin{Bmatrix} 38,500 \\ -58,000 \\ 19,400 \end{Bmatrix}$$

$$\lambda_0 = \frac{\delta^0}{\lfloor D \rfloor^0 \{U\}^0} = 0.0236$$

$$\{X\}^1 = \{X\}^0 + \lambda_0\{D\}^0 = \begin{Bmatrix} 130.7087 \\ 81.1024 \\ 104.7244 \end{Bmatrix}$$

$$\{R\}^1 = \{R\}^0 + \lambda_0\{U\}^0 = \begin{Bmatrix} -390.5512 \\ -570.0787 \\ 258.2677 \end{Bmatrix}$$

$$\delta^1 = \lfloor R \rfloor^1 \{R\}^1 = 544,222.1689$$

$$Error = \sqrt{\lfloor R \rfloor^1 \{R\}^1} = 737.7142$$

$$\alpha_0 = \frac{\delta^1}{\delta^0} = 0.2296$$

$$\{D\}^1 = -\{R\}^1 + \alpha_0\{D\}^0 = \begin{Bmatrix} 689.0697 \\ 386.3750 \\ -212.3418 \end{Bmatrix}$$

For the iteration process at $k = 1$,

$$\{U\}^1 = [A]\{D\}^1 = \begin{Bmatrix} 3,562.0 \\ 3,525.6 \\ -9,050.7 \end{Bmatrix}$$

$$\lambda_1 = \frac{\delta^1}{\lfloor D \rfloor^1 \{U\}^1} = 0.0948$$

$$\{X\}^2 = \{X\}^1 + \lambda_1\{D\}^1 = \begin{Bmatrix} 196.0579 \\ 117.7450 \\ 84.5866 \end{Bmatrix}$$

$$\{R\}^2 = \{R\}^1 + \lambda_1\{U\}^1 = \begin{Bmatrix} -52.7418 \\ -235.7237 \\ -600.0730 \end{Bmatrix}$$

$$\delta^2 = \lfloor R \rfloor^2 \{R\}^2 = 418,434.9655$$

$$Error = \sqrt{\lfloor R \rfloor^2 \{R\}^2} = 646.8655$$

$$\alpha_1 = \frac{\delta^2}{\delta^1} = 0.7689$$

$$\{D\}^2 \; = \; -\{R\}^2 + \alpha_1\{D\}^1 \; = \; \begin{Bmatrix} 582.5454 \\ 532.7951 \\ 436.8102 \end{Bmatrix}$$

For the iteration process at $k = 2$,

$$\{U\}^2 \; = \; [A]\{D\}^2 \; = \; \begin{Bmatrix} 120.9903 \\ 540.7525 \\ 1{,}376.5731 \end{Bmatrix}$$

$$\lambda_2 \; = \; \frac{\delta^2}{\lfloor D \rfloor^2 \{U\}^2} \; = \; 0.435918$$

$$\{X\}^3 \; = \; \{X\}^2 + \lambda_2\{D\}^2 \; = \; \begin{Bmatrix} 450.0000 \\ 350.0000 \\ 275.0000 \end{Bmatrix}$$

$$\{R\}^3 \; = \; \{R\}^2 + \lambda_2\{U\}^2 \; = \; \begin{Bmatrix} 0 \\ 0 \\ 0 \end{Bmatrix}$$

$$\delta^3 \; = \; \lfloor R \rfloor^3\{R\}^3 \; = \; 0$$

$$Error \; = \; \sqrt{\lfloor R \rfloor^3\{R\}^3} \; = \; 0$$

This example demonstrates that the computed solutions from the modified conjugate gradient method converge to the exact solutions of $x_1 = 450$, $x_2 = 350$ and $x_3 = 275$ with the error of $\delta = 0$.

For practical problems that are governed by nonlinear differential equations, the derived matrix $[A]$ is usually an asymmetric and non-positive definite matrix. In this case, the conjugate gradient method presented herein must be modified. The modification leads to the new methods such as the generalized minimal residual method or GMRES, the bi-conjugate gradient method, the squared conjugate gradient method and the quasi-minimal residual method, etc. Details for these methods are beyond the scope of this book but can be found in research literatures.

3.17 Closure

In this chapter, various methods for solving a set of system of equations, $[A]\{X\} = \{B\}$, are presented. These methods are classified into two groups of the direct and iterative techniques. The direct technique consists of the Cramer's rule, the Gauss-elimination method, the Gauss-Jordan method, the matrix inversion method, the LU decomposition method and the Cholesky method. The Cramer's rule should not be used in general because it requires a large number of operations as compared to the others. The Guass-elimination and the LU decomposition methods are popular and widely used for solving practical problems. The Cholesky decomposition method is often used for solving many engineering problems where the matrix $[A]$ is symmetric.

The iterative technique consists of the Jacobi method, the Gauss-Seidel method, the successive over-relaxation method and the conjugate gradient method. The conjugate gradient method is considered as a popular method today for solving large size problems because of its high computational efficiency.

The computational procedures and examples presented in this chapter can help readers to understand the concepts of various methods easily. The accompanied computer programs of each method can also be used for solving a large set of system of equations. Some of these computer programs can be implemented directly into the finite difference or finite element software for analyzing practical engineering problems.

Exercises

1. Solve the following system of equations by using the Cramer's rule,

$$-2x_1 + 3x_2 + x_3 = 9$$
$$3x_1 + 4x_2 - 5x_3 = 0$$
$$x_1 - 2x_2 + x_3 = -4$$

2. Solve the following system of equations by using the Cramer's rule,

$$\begin{bmatrix} 1 & 1 & 3 \\ 5 & 3 & 1 \\ 2 & 3 & 1 \end{bmatrix} \begin{Bmatrix} x_1 \\ x_2 \\ x_3 \end{Bmatrix} = \begin{Bmatrix} 2 \\ 3 \\ -1 \end{Bmatrix}$$

3. Solve the following system of equations by using the Cramer's rule,

$$\begin{bmatrix} 2 & 3 & 5 \\ 3 & 1 & -2 \\ 1 & 3 & 4 \end{bmatrix} \begin{Bmatrix} x_1 \\ x_2 \\ x_3 \end{Bmatrix} = \begin{Bmatrix} 0 \\ -2 \\ -3 \end{Bmatrix}$$

4. Solve the system of equations in Problems 1-3 by using the left division technique in MATLAB. Verify the computed solutions by substituting them back into the system of equations.

5. Show detailed computational procedure for solving the following system of equations,

$$\begin{bmatrix} 1 & 2 & 4 \\ 4 & -1 & 1 \\ 2 & 5 & 2 \end{bmatrix} \begin{Bmatrix} x_1 \\ x_2 \\ x_3 \end{Bmatrix} = \begin{Bmatrix} 11 \\ 8 \\ 3 \end{Bmatrix}$$

by using
   (a) the Gauss elimination method with scaling and pivoting.
   (b) the LU decomposition method.
Verify the computed solutions by substituting them back into the system of equations.

6.   Solve the following system of equations by using the Gauss elimination method,

$$3x_1 - 2x_2 + 8x_3 = 21$$
$$-5x_1 + 6x_2 + 4x_3 = 19$$
$$8x_1 + 3x_2 - 5x_3 = -1$$

        (a)   without pivoting and scaling.

        (b)   with pivoting and scaling.

        (c)   verify the computed solutions by using the computer program in Fig. 3.3.

7.   Solve the following system of equations by using the Gauss elimination method,

$$\begin{bmatrix} 1 & 2 & 1 & 4 \\ 4 & 2 & 0 & 3 \\ 3 & 0 & 2 & 1 \\ 0 & 3 & 1 & 1 \end{bmatrix} \begin{bmatrix} x_1 \\ x_2 \\ x_3 \\ x_4 \end{bmatrix} = \begin{Bmatrix} 12.700 \\ 11.880 \\ 18.218 \\ 17.100 \end{Bmatrix}$$

Verify the computed solutions with Eq. (2.59).

8.   Show detailed computational procedure for solving the following system of equations,

$$\begin{bmatrix} 4 & 3 & 2 & 1 \\ 3 & 4 & 3 & 2 \\ 2 & 3 & 4 & 3 \\ 1 & 2 & 3 & 4 \end{bmatrix} \begin{bmatrix} x_1 \\ x_2 \\ x_3 \\ x_4 \end{bmatrix} = \begin{Bmatrix} 1 \\ 1 \\ -1 \\ -1 \end{Bmatrix}$$

by using: (a) the Gauss elimination method, (b) the LU decomposition method, and (c) the Cholesky decomposition method. Verify the solutions by using one of the computer programs presented in this chapter.

9.   Solve the system of equations shown in Problem 5 by using the matrix inversion that is based on the Gauss Jordan method. Show detailed computational procedure. Explain advantages and disadvantages of the method.

10.   Solve the following system of equations,

$$\begin{bmatrix} 0 & -1 & 3 \\ 4 & 0 & -1 \\ -2 & 5 & 0 \end{bmatrix} \begin{Bmatrix} x_1 \\ x_2 \\ x_3 \end{Bmatrix} = \begin{Bmatrix} 3 \\ 14 \\ 7 \end{Bmatrix}$$

by using: (a) the Gauss elimination method, (b) the LU decomposition method, (c) the Gauss-Seidel iteration method, and (d) the conjugate gradient method. Explain difficulties that arise in each method and suggest how to overcome them.

11. Use the Gauss elimination method to solve the following system of equations,

$$\begin{bmatrix} 1 & 2 & 3 & 4 \\ 2 & 2 & 3 & 4 \\ 3 & 3 & 3 & 4 \\ 4 & 4 & 4 & 4 \end{bmatrix} \begin{Bmatrix} x_1 \\ x_2 \\ x_3 \\ x_4 \end{Bmatrix} = \begin{Bmatrix} 1.234 \\ 2.234 \\ 3.334 \\ 4.444 \end{Bmatrix}$$

Compare the computed solutions with those obtained from the Gauss-Seidel method that uses the stopping tolerance of $\varepsilon = 0.000001\%$. Then explain advantages and disadvantages of both methods.

12. The exact solutions of the following system of equations,

$$\begin{bmatrix} 1 & 1/2 & 1/3 \\ 1/2 & 1/3 & 1/4 \\ 1/3 & 1/4 & 1/5 \end{bmatrix} \begin{Bmatrix} x_1 \\ x_2 \\ x_3 \end{Bmatrix} = \begin{Bmatrix} 11/6 \\ 13/12 \\ 47/60 \end{Bmatrix}$$

are $x_1 = x_2 = x_3 = 1$. Use the Gauss elimination method to solve for the solutions. Explain why the computed solutions contain errors. Then suggest methods to reduce such errors and use them to improve the solutions.

13. Modify the Gauss elimination method to solve the tri-diagonal system of equations in the form,

$$\begin{bmatrix} b_1 & c_1 & 0 & 0 & 0 & 0 \\ a_2 & b_2 & c_2 & 0 & 0 & 0 \\ 0 & \ddots & \ddots & \ddots & 0 & 0 \\ 0 & 0 & \ddots & \ddots & \ddots & 0 \\ 0 & 0 & 0 & a_{n-1} & b_{n-1} & c_{n-1} \\ 0 & 0 & 0 & 0 & a_n & b_n \end{bmatrix} \begin{Bmatrix} x_1 \\ x_2 \\ x_3 \\ \vdots \\ x_{n-1} \\ x_n \end{Bmatrix} = \begin{Bmatrix} d_1 \\ d_2 \\ d_3 \\ \vdots \\ d_{n-1} \\ d_n \end{Bmatrix}$$

Show detailed computational procedure with a corresponding computer program.

14. Solve the following tri-diagonal system of equations

$$\begin{bmatrix} 4 & -3 & & & \\ -3 & 2 & 1 & & \\ & 1 & 3 & -1 & \\ & & -1 & 5 & -4 \\ & & & -4 & 2 \end{bmatrix} \begin{Bmatrix} x_1 \\ x_2 \\ x_3 \\ x_4 \\ x_5 \end{Bmatrix} = \begin{Bmatrix} -2 \\ 4 \\ 7 \\ -3 \\ -6 \end{Bmatrix}$$

Verify the computed solutions by using the computer program in Fig. 3.4.

15. Use the Gauss Jordan method to find the inverses of the following matrices,

(a) $\begin{bmatrix} 1 & 1 & 1 \\ 1 & -1 & -1 \\ 2 & 1 & 7 \end{bmatrix}$;   (b) $\begin{bmatrix} 1 & 1 & 1 \\ 1 & 2 & 1 \\ 3 & 3 & 7 \end{bmatrix}$

Then verify the results with those obtained by using the Cramer's rule.

16. Find the inverse of the matrix,

$$[A] = \begin{bmatrix} 1 & 1/2 & 1/3 \\ 1/2 & 1/3 & 1/4 \\ 1/3 & 1/4 & 1/5 \end{bmatrix}$$

by using the Gauss Jordan method as explained in section 3.7. Examine whether the matrix is ill-conditioned by using the three criteria as explained at the end of section 3.7.

17. Use the Cholesky decomposition computer program as shown in Fig. 3.6 to solve the following system of equations,

$$\begin{bmatrix} 4 & -4 & 0 \\ -4 & 4 & -2 \\ 0 & -2 & 4 \end{bmatrix} \begin{Bmatrix} x_1 \\ x_2 \\ x_3 \end{Bmatrix} = \begin{Bmatrix} 400 \\ -950 \\ 400 \end{Bmatrix}$$

which has exact solutions of $x_1 = 450$, $x_2 = 350$ and $x_3 = 275$. Explain difficulties that may arise by using the method and how to overcome them.

18. Study the expressions for obtaining the coefficients in the matrices $[L]$ and $[U]$ as shown in Eqs. (3.55a-e) of the LU decomposition method. Prepare these expressions for the system of 4 equations. Then, follow Eqs. (3.56a-b) to write expressions for the forward and back substitution procedures. Use the system of 4 equations in Problem 7 to verify the expressions derived.

19. Add the pivoting and scaling capability as explained in sections 3.5.1-2 to the computer program that uses the LU decomposition method for solving the system of equations shown in Fig. 3.5. Then use the modified program to solve the system of equations in Problem 17.

20. Develop a compute program for solving the system of nonlinear equations by the Newton-Raphson method. During the iteration procedure, use the Gauss elimination method as shown by the computer program in Fig. 3.3 to solve the system of equations. Verify the developed computer program by solving Example 2.12 and compare the solutions with those shown in Table 2.8.

21. Develop a computer program to solve the following system of equations,

$$4x + y^2 + z = 11$$
$$x + 4y + z^2 = 18$$
$$x^2 + y + 4z = 15$$

by using the Gauss-Seidel iteration and conjugate gradient methods. Use the initial guess values of $x = y = z = 1$ with the acceptable error of $\varepsilon = 0.000001\%$.

22. Develop a computer program to solve the following system of equations,

$$\begin{bmatrix} 3 & 2 & 0 & 0 \\ 2 & 3 & 2 & 0 \\ 0 & 2 & 3 & 2 \\ 0 & 0 & 2 & 3 \end{bmatrix} \begin{Bmatrix} x_1 \\ x_2 \\ x_3 \\ x_4 \end{Bmatrix} = \begin{Bmatrix} 12 \\ 17 \\ 14 \\ 7 \end{Bmatrix}$$

by using: (a) the Jacobi iteration method, (b) the Gauss-Seidel iteration method, (c) the successive over-relaxation method with $1.20 \le \omega \le 1.40$ for every $\Delta\omega = 0.01$, and (d) the conjugate gradient method with the acceptable error of $\varepsilon = 0.001\%$.

23. Study the convergence rates for the following system of equations,

$$\begin{bmatrix} 10 & 7 & 8 & 7 \\ 7 & 5 & 6 & 5 \\ 8 & 6 & 10 & 9 \\ 7 & 5 & 9 & 10 \end{bmatrix} \begin{Bmatrix} x_1 \\ x_2 \\ x_3 \\ x_4 \end{Bmatrix} = \begin{Bmatrix} 32 \\ 23 \\ 33 \\ 31 \end{Bmatrix}$$

by using: (a) the Gauss-Seidel iteration method, (b) the successive over-relaxation method with $\omega = 1.25, 1.50, 1.75$, and (c) the conjugate gradient method with the initial guess values of $x_1 = x_2 = x_3 = x_4 = 0$. Compare the computed solutions with those obtained from the Gauss elimination method.

24. Study the convergence rate for solving the system of Eq. (3.10) by employing the successive over-relaxation method. Use the weighting factor ω that varies from 0 to 2 with the increment of 0.05. Then, suggest the optimal value of ω for solving this system of equations.

25. Solve the system of equations in Problem 12 again but by using the conjugate gradient method. Use appropriate initial guess values to obtain the converged solutions.

26. Solve the following system of equations by using the conjugate gradient method,

$$\begin{bmatrix} 2 & 4 & 1 \\ 4 & 3 & 2 \\ 1 & 2 & 4 \end{bmatrix} \begin{Bmatrix} x_1 \\ x_2 \\ x_3 \end{Bmatrix} = \begin{Bmatrix} 13 \\ 16 \\ 17 \end{Bmatrix}$$

with the initial guess values of $x_1 = x_2 = x_3 = 1$. Show the computational procedures in details.

27.   Matrix $[A]$ in the following system of equations is an asymmetric matrix,

$$\begin{bmatrix} 1 & 2 & 3 \\ -2 & 1 & 4 \\ 1 & -3 & 2 \end{bmatrix} \begin{Bmatrix} x_1 \\ x_2 \\ x_3 \end{Bmatrix} = \begin{Bmatrix} 8 \\ 4 \\ -1 \end{Bmatrix}$$

Use the conjugate gradient method to solve for the solutions with the initial guess values of $x_1 = x_2 = x_3 = 1$. Show the computational procedures in details.

28.   Solve the following system of equations,

$$\begin{bmatrix} 17 & 1 & 4 & 3 & -1 & 2 & 3 & -7 \\ 2 & 10 & -1 & 1 & -2 & 1 & 1 & -4 \\ -1 & 1 & -8 & 2 & -5 & 2 & -1 & 1 \\ 2 & 4 & 1 & -11 & 1 & 3 & 4 & -1 \\ 1 & 3 & 1 & 7 & -15 & 1 & -2 & 4 \\ -2 & 1 & 7 & -1 & 2 & 12 & -1 & 8 \\ 3 & 4 & 5 & 1 & 2 & 8 & -19 & 2 \\ 5 & 1 & 1 & 1 & 1 & 1 & -7 & 10 \end{bmatrix} \begin{Bmatrix} x_1 \\ x_2 \\ x_3 \\ x_4 \\ x_5 \\ x_6 \\ x_7 \\ x_8 \end{Bmatrix} = \begin{Bmatrix} 71 \\ 43 \\ -11 \\ -37 \\ -61 \\ 52 \\ -73 \\ 21 \end{Bmatrix}$$

by using the computer programs of:
   (a)   the Gauss elimination method
   (b)   the LU decomposition method
   (c)   the conjugate gradient method
presented in this chapter. Provide comments on the accuracy of the solutions obtained from the three methods.

29.   Solve the following system of equations,

$$\begin{bmatrix} 8 & -2 & 1 & 0 & \cdots & \cdots & \cdots & \cdots & \cdots & 0 \\ -2 & 8 & -2 & 1 & 0 & \cdots & \cdots & \cdots & \cdots & 0 \\ 1 & -2 & 8 & -2 & 1 & 0 & \cdots & \cdots & \cdots & 0 \\ 0 & 1 & -2 & 8 & -2 & 1 & 0 & \cdots & \cdots & 0 \\ \cdots & \cdots & \cdots & \cdots & \cdots & \cdots & \cdots & \cdots & \cdots & \cdots \\ \cdots & \cdots & \cdots & \cdots & \cdots & \cdots & \cdots & \cdots & \cdots & \cdots \\ 0 & \cdots & \cdots & 0 & 1 & -2 & 8 & -2 & 1 & 0 \\ 0 & \cdots & \cdots & \cdots & 0 & 1 & -2 & 8 & -2 & 1 \\ 0 & \cdots & \cdots & \cdots & \cdots & 0 & 1 & -2 & 8 & -2 \\ 0 & \cdots & \cdots & \cdots & \cdots & \cdots & 0 & 1 & -2 & 8 \end{bmatrix} \begin{Bmatrix} x_1 \\ x_2 \\ x_3 \\ x_4 \\ \cdots \\ \cdots \\ x_{47} \\ x_{48} \\ x_{49} \\ x_{50} \end{Bmatrix} = \begin{Bmatrix} 100 \\ 100 \\ 100 \\ 100 \\ \cdots \\ \cdots \\ 100 \\ 100 \\ 100 \\ 100 \end{Bmatrix}$$

by using the computer programs of:

(a) the Gauss elimination method

(b) the LU decomposition method

(c) the Cholesky decomposition method

(d) the conjugate gradient method

presented in this chapter. Explain advantages and disadvantages of each method for solving such system of equations.

30. Solve the system of equations in Problem 28 again but by using the function `lu` with the left division technique in MATLAB. Compare the computed solutions with those obtained in Problem 28.

31. Solve the system of equations in Problem 29 again but using the function `lu` or `chol` with the left division technique in MATLAB. Compare the computed solutions with those obtained in Problem 29.

32. In the determination of the currents and voltages in an electrical circuit in Fig. P3.32, the Kirchhoff's current rule (summation of currents at any point is zero) and the Kirchhoff's voltage rule (summation of voltages in any closed-circuit loop is zero) are needed.

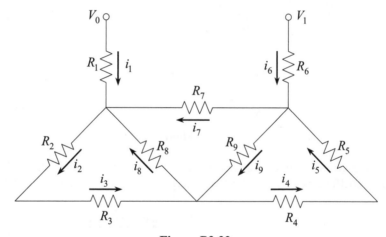

Figure P3.32.

By applying the Kirchhoff's current and voltage rules on the circuit, the following system of equations is obtained,

$$i_2 - i_3 = 0 \qquad\qquad i_2 R_2 + i_3 R_3 + i_8 R_8 = 0$$

$$i_4 - i_5 = 0 \qquad\qquad i_7 R_7 - i_8 R_8 - i_9 R_9 = 0$$

$$i_1 + i_7 + i_8 - i_2 = 0 \qquad\qquad i_9 R_9 + i_4 R_4 + i_5 R_5 = 0$$

$$i_3 + i_9 - i_8 - i_4 = 0 \qquad\qquad i_1 R_1 - i_7 R_7 - i_6 R_6 = V_0 - V_1$$

$$i_6 + i_5 - i_7 - i_9 = 0$$

If V_0 = 150 Volt, V_1 = 0 Volt, R_1 = R_8 = R_9 = 10 Ω, R_2 = R_4 = 5 Ω, R_3 = R_6 = 15 Ω and R_5 = R_7 = 20 Ω, use MATLAB to solve the system of equations. Compare the solutions with those obtained by using the Gauss elimination method.

33. Develop a system of equations by apply the Kirchhoff's current and voltage rules on the electrical circuit in Fig. P3.33. Then, use MATLAB to solve such system of equations to obtain the currents passing through each resistor. Compare the computed solutions with those obtained from using the Gauss-Seidel iteration method.

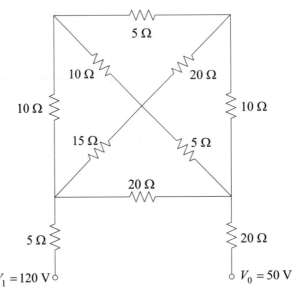

Figure P3.33.

34. A moving system of three weights connected by the two cords is shown in Fig. P3.34. The friction coefficients between the two weights on the floor are 0.3 and 0.5 as shown in the figure.

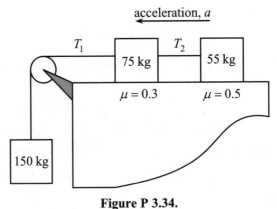

Figure P 3.34.

By applying the Newton's second law $\left(\Sigma \vec{F} = m\vec{a}\right)$ on the three weights, the following system of equations is obtained,

$$
\begin{aligned}
150a + T_1 &= 1{,}471.500 \\
75a - T_1 + T_2 &= -220.725 \\
55a - T_2 &= -269.775
\end{aligned}
$$

Use MATLAB to solve the system of equations for the system acceleration a and the tensions T_1 and T_2. Compare the computed solutions with those obtained by using the Gauss elimination method.

35.   A moving system consisting of three weights connected by cords is shown in Fig. P3.35. Apply the Newton's second law $\left(\Sigma \vec{F} = m\vec{a}\right)$ on the three weights to derive a system of three equations. Then, use MATLAB to solve for the system acceleration a and the cord tensions T_1 and T_2.

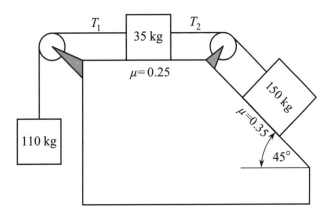

Figure P3.35.

Chapter

4

Interpolation and Extrapolation

4.1 Introduction

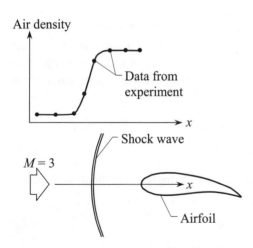

Figure 4.1 Shock wave and air density distribution in front of an airfoil.

Most of data obtained from experiment are always at discrete points. For examples, the temperatures on the solar heating panel are measured at specific positions or the pressures over an airfoil are collected at discrete locations. Often, the data at the other locations that have not been measured from the experiment are needed. Interpolation and extrapolation techniques are thus required to provide information at these locations.

Figure 4.1 shows experimental data of the air density from a supersonic Mach 3 flow over an airfoil. The data are collected at some discrete locations. It would be better if these data are fitted to yield a continuous function. The function can be used to provide information of the density at the other locations that have not been measured.

Another example that requires fitting a set of data is the prediction of the temperature through the thickness of a hot water pipe. The analytical solution of the temperature distribution through the pipe thickness is in the form of the Bessel function. Values of the Bessel function is not easy to calculate so that they appear in form of a table in most of the heat transfer textbooks. Table 4.1 shows typical values of the Bessel function that varies with x. These values are at discrete points from $x = 2.0$ to 4.0 with an increment of 0.2. If the Bessel function at $x = 2.5784$ is needed, the data in the table must be interpolated. Many interpolation methods are explained in this chapter. These interpolation methods are basis for studying higher-level numerical methods, such as the finite difference and finite element methods, that will be explained in the later chapters.

Table 4.1   Values of the Bessel function at discrete points x.

x	2.0	2.2	2.4	2.6	2.8	3.0
$J_0(x)$	0.2239	0.1104	0.0025	-0.0968	-0.1850	-0.2601
x	3.0	3.2	3.4	3.6	3.8	4.0
$J_0(x)$	-0.2601	-0.3202	-0.3643	-0.3918	-0.3992	-0.3971

The widely used interpolation methods are: (1) the Newton's divided differences, (2) the Lagrange polynomials, and (3) the spline interpolation. These methods are presented and explained in details herein.

4.2 Newton's Divided Differences

One of the most popular methods for interpolating a set of data points is the Newton's divided differences. The main idea of the method is to create a polynomial function $f(x)$ from the given data points and use that function to interpolate the values at any other locations. The simplest form of such function is linear as explained in the following section.

4.2.1 Linear interpolation

From Table 4.1, if the value of the Bessel function at any point x between the two points $x_0 = 2.0$ and $x_1 = 4.0$ is needed, the easiest way to estimate such value is to create a linear function or the first-order polynomial between these two data points as shown by the dashed line in Fig. 4.2. The dashed line is then used to estimate the value of Bessel function at the point x from the values of the function at x_0 and x_1.

The first-order polynomial or linear function can be derived by writing it in the form

$$f(x) = C_0 + C_1(x - x_0) \tag{4.1}$$

where C_0 and C_1 are unknown constants that can be determined by using the conditions at the points x_0 and x_1 as follows,

at $\quad x = x_0$: $\qquad f(x_0) = C_0 + 0 = C_0 \tag{4.2a}$

at $\quad x = x_1$: $\qquad f(x_1) = C_0 + C_1(x_1 - x_0)$

$$= f(x_0) + C_1(x_1 - x_0)$$

$$C_1 = \frac{f(x_1) - f(x_0)}{x_1 - x_0} \tag{4.2b}$$

Thus,
$$f(x) = f(x_0) + (x - x_0)\frac{f(x_1) - f(x_0)}{x_1 - x_0} \tag{4.3}$$

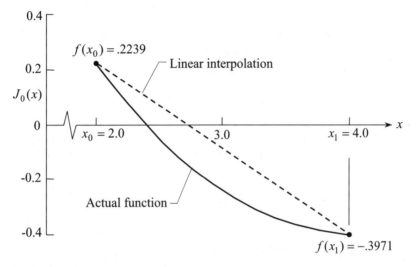

Figure 4.2 Use of a linear function to interpolate the Bessel function values between $x_0 = 2.0$ and $x_1 = 4.0$.

Example 4.1 The values of the Bessel function $f(x_0 = 2.0) = 0.2239$ and $f(x_1 = 4.0) = -0.3971$ are given at two points in Table 4.1 as shown in Fig. 4.2. Estimate the value of the Bessel function at the point $x = 3.2$ by using the linear interpolation. Compare the estimated value with the exact solution shown in Table 4.1.

From Eq. (4.3), the estimated value of the Bessel function at the point $x = 3.2$ by using the linear interpolation is

$$f(x = 3.2) = 0.2239 + (3.2 - 2.0)\frac{-0.3971 - 0.2239}{4.0 - 2.0}$$

$$= 0.2239 - 0.3726$$

$$f(3.2) = -0.1487 \tag{4.4}$$

The estimated value has the true error of

$$\varepsilon = \frac{-0.3202 + 0.1487}{-0.3202} \times 100\% = 53.56\% \tag{4.5}$$

4.2.2 Quadratic interpolation

A quadratic interpolation can be derived by using the same procedure as for the linear interpolation explained in the preceding section. The quadratic interpolation is based on the second-order polynomial. Their coefficients are determined by using the three data values at x_0, x_1 and x_2. The quadratic function that passes through the three data values is written in the form

$$f(x) = C_0 + C_1(x - x_0) + C_2(x - x_0)(x - x_1) \qquad (4.6)$$

where C_0, C_1 and C_2 are unknown constants which can be determined by using the three data values at x_0, x_1 and x_2. If the three data values are $f(x_0 = 2.0) = 0.2239$, $f(x_1 = 3.0) = -0.2601$ and $f(x_2 = 4.0) = -0.3971$, the unknown constants can be determined as follows

$$\text{at} \quad x = x_0 \; : \qquad f(x_0) = C_0 + 0 + 0 = C_0 \qquad (4.7a)$$

$$\text{at} \quad x = x_1 \; : \qquad f(x_1) = C_0 + C_1(x_1 - x_0)$$

$$= f(x_0) + C_1(x_1 - x_0)$$

$$C_1 = \frac{f(x_1) - f(x_0)}{x_1 - x_0} \qquad (4.7b)$$

$$\text{at} \quad x = x_2 \; : \qquad f(x_2) = C_0 + C_1(x_2 - x_0) + C_2(x_2 - x_0)(x_2 - x_1)$$

$$C_2 = \frac{\dfrac{f(x_2) - f(x_1)}{x_2 - x_1} - \dfrac{f(x_1) - f(x_0)}{x_1 - x_0}}{x_2 - x_0} \qquad (4.7c)$$

These values of C_0, C_1 and C_2 are then substituted into Eq. (4.6) to obtain the quadratic interpolation.

Example 4.2 Three data values of the Bessel function $f(x_0 = 2.0) = 0.2239$, $f(x_1 = 3.0) = -0.2601$ and $f(x_2 = 4.0) = -0.3971$ are given in Table 4.1 as shown in Fig. 4.3. Estimate the value of the Bessel function at $x = 3.2$ by using the quadratic interpolation. Compare the estimated value with exact solution in Table 4.1.

From the given three data points, the constants C_1, C_2 and C_3 are determined from Eq. (4.7a-c) as

$$C_0 = 0.2239$$

$$C_1 = \frac{-0.2601 - 0.2239}{3.0 - 2.0} = -0.4840$$

$$C_2 = \frac{\dfrac{-0.3971 + 0.2601}{4.0 - 3.0} - \dfrac{-0.2601 - 0.2239}{3.0 - 2.0}}{4.0 - 2.0} = 0.1735$$

By substituting these constants into Eq. (4.6), the quadratic interpolation is

$$f(x) = 0.2239 - 0.4840(x-2.0) + 0.1735(x-2.0)(x-3.0)$$

Then, the Bessel function at $x = 3.2$ can be determined as

$$f(3.2) = 0.2239 - 0.4840(1.2) + 0.1735(1.2)(0.2)$$
$$= 0.2239 - 0.5808 + 0.0416$$

$$f(3.2) = -0.3153 \qquad (4.8)$$

The true error as compared to the exact solution is

$$\varepsilon = \frac{-0.3202 + 0.3153}{-0.3202} \times 100\% = 1.53\% \qquad (4.9)$$

which is less than the error obtained from the linear interpolation in Example 4.1.

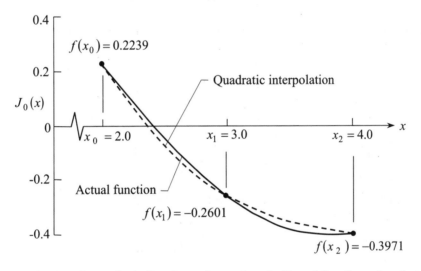

Figure 4.3 Use of a quadratic function to interpolate the Bessel function values between $x_0 = 2.0$ and $x_2 = 4.0$.

4.2.3 n^{th}-order Polynomial interpolation

The procedure explained in the preceding sections can be used to create the nth-order polynomial interpolation passing through the $n+1$ data values at x_0, x_1, x_2,..., x_n as shown in Fig. 4.4.

The nth-order polynomial interpolation as shown in Fig. 4.4 can be written in a general form as

$$f(x) = C_0 + C_1(x-x_0) + C_2(x-x_0)(x-x_1) + C_3(x-x_0)(x-x_1)(x-x_2)$$
$$+ ... + C_n(x-x_0)(x-x_1)...(x-x_{n-1}) \qquad (4.10)$$

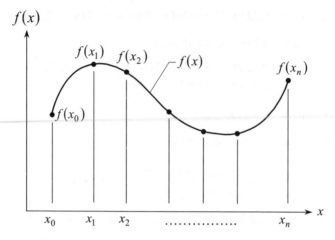

Figure 4.4 The nth-order polynomial interpolation passing through $n+1$ data values.

where the coefficients C_i, $i = 0, 1, 2,..., n$ can be determined and are obtained in the forms

$$C_0 = f(x_0) \tag{4.11a}$$

$$C_1 = f[x_1, x_0] \tag{4.11b}$$

$$C_2 = f[x_2, x_1, x_0] \tag{4.11c}$$

$$\vdots \qquad \vdots \qquad\qquad\qquad \vdots$$

$$C_n = f[x_n, x_{n-1}, ..., x_1, x_0] \tag{4.11n}$$

The square parentheses shown in Eq. (4.11) represent the divided differences. For example, the first divided difference is determined from

$$f[x_i, x_j] = \frac{f(x_i) - f(x_j)}{x_i - x_j} \tag{4.12}$$

The result is the same as the value of C_1 in Eq. (4.7b). The second divided difference is

$$f[x_i, x_j, x_k] = \frac{f[x_i, x_j] - f[x_j, x_k]}{x_i - x_k} \tag{4.13}$$

which is in the same form of Eq. (4.7c) for determining the value of C_2. Thus, the nth divided difference can be written in the form

$$f[x_n, x_{n-1}, ..., x_1, x_0] = \frac{f[x_n, x_{n-1}, ..., x_1] - f[x_{n-1}, ..., x_1, x_0]}{x_n - x_0} \tag{4.14}$$

The values of divided differences as shown in Eqs. (4.12) - (4.14) can be determined sequentially as shown in Table 4.2.

Table 4.2 Sequential steps to determine the values of the divided differences.

i	x	$f(x)$	Divided differences First	Second	Third
0	x_0	$f(x_0)$	$f[x_1, x_0]$	$f[x_2, x_1, x_0]$	$f[x_3, x_2, x_1, x_0]$
1	x_1	$f(x_1)$	$f[x_2, x_1]$	$f[x_3, x_2, x_1]$	
2	x_2	$f(x_2)$	$f[x_3, x_2]$		
3	x_3	$f(x_3)$			

The divided differences in the first row of Table 4.2 represent the values of the constants C_i, $i = 0, 1, 2,..., n$ as shown in Eq. (4.11). By substituting these divided differences into Eq. (4.10), the general form of the nth-order polynomial interpolation is

$$f(x) = f(x_0) + (x - x_0)f[x_1, x_0] + (x - x_0)(x - x_1)f[x_2, x_1, x_0]$$
$$+ ... + (x - x_0)(x - x_1)...(x - x_{n-1})f[x_n, x_{n-1}, ..., x_0] \tag{4.15}$$

Equation (4.15) is called the Newton's divided-difference interpolating polynomials. The equation can be used to develop a corresponding computer program directly.

Figure 4.5 shows a computer program to determine a value from a set of data points by using the Newton's divided-difference interpolating polynomials in Eq. (4.15).

Fortran

```
      PROGRAM  NEWDIV
C.....PROGRAM FOR COMPUTING  F(X)  AT A GIVEN X
C.....USING NEWTON'S DIVIDED-DIFFERENCE
C.....INTERPOLATING POLYNOMIALS
      DIMENSION  X(10), FX(10,10)
C.....READ NUMBER OF DATA SETS, DATA OF X AND FX:
      READ(5,*)  N
      DO 10  IROW=1,N
      READ(5,*)  X(IROW), FX(IROW,1)
   10 CONTINUE
C.....COMPUTE DIVIDED-DIFFERENCE COEFFICIENTS:
      M = N
      DO 20  ICOL=2,N
      M = M - 1
      DO 30  IROW=1,M
      FX(IROW,ICOL) =
     1    FX(IROW+1,ICOL-1) - FX(IROW,ICOL-1)
      FX(IROW,ICOL) =
     1    FX(IROW,ICOL)/(X(IROW+ICOL-1) - X(IROW))
   30 CONTINUE
   20 CONTINUE
C.....COMPUTE DESIRED F(X) AT THE GIVEN X VALUE:
      READ(5,*)  XX
      FF  = FX(1,1)
      FAC = 1.
      DO 40  I=2,N
      FAC = FAC*(XX - X(I-1))
      FF  = FF + FX(1,I)*FAC
   40 CONTINUE
      WRITE(6,100)  XX, FF
  100 FORMAT(' VALUE OF F(X) AT X =', E10.4,
     *       ' IS', E16.7              )
      STOP
      END
```

MATLAB

```
%   PROGRAM NEWDIV
%   PROGRAM FOR COMPUTING F(X) AT A GIVEN X
%   USING NEWTON'S DIVIDED-DIFFERENCE
%   INTERPOLATING POLYNOMIALS
%   READ NUMBER OF DATA SETS
n = input('\nEnter number of data: ');
for irow = 1:n
    x(irow)      = input('\nEnter point x:      ');
    fx(irow,1) = input('Enter value f(x):  ');
end
%   COMPUTE DIVIDED-DIFFERENCE COEFFICIENTS:
m = n;
for icol = 2:n
    m = m-1;
    for irow = 1:m
        fx(irow,icol) = fx(irow+1,icol-1) - fx(irow,icol-1);
        fx(irow,icol) = fx(irow,icol)/(x(irow+icol-1)-x(irow));
    end
end
%   COMPUTE DESIRED F(X) AT THE GIVEN X VALUE:*)
xx = input('\nEnter point x to find f(x):  ');
ff = fx(1,1);
fac = 1.;
for i = 2:n
    fac = fac*(xx - x(i-1));
    ff = ff + fx(1,i)*fac;
end
fprintf('\nVALUE OF F(X) AT X = %14.6f IS %14.6e', xx, ff)
```

Figure 4.5 A computer program for determining an interpolated value from a given set of data points by using the Newton's divided-difference method.

Example 4.3 Employ the computer program in Fig. 4.5 of the Newton's divided-difference method to estimate the value of the Bessel function at $x = 3.2$. Use all data points in Table 4.1 except the data at $x = 3.2$. Then, determine the true error by comparing the estimated value with the exact solution.

Figure 4.6 shows the input data needed by the computer program. The input data consists of all data values that appear in Table 4.1 except the value at $x = 3.2$. The estimated value obtained from the computer program is -0.3201. Thus, the true error is

$$\varepsilon = \frac{-0.3202 + 0.3201}{-0.3202} \times 100\% = 0.03\% \tag{4.16}$$

The estimated value at $x = 3.2$ obtained from using the values at the ten data points is quite accurate with the error of only 0.03%. It is noted that the errors of the estimated values by using the linear and quadratic interpolations in Examples 4.1 and 4.2 are 53.56% and 1.53%, respectively.

```
10
2.0    .2239
2.2    .1104
2.4    .0025
2.6   -.0968
2.8   -.1850
3.0   -.2601
3.4   -.3643
3.6   -.3918
3.8   -.3992
4.0   -.3971
   3.2

VALUE OF F(X) AT X = .3200E+01 IS    -.3201119E+00
```

Figure 4.6 Input data from Table 4.1 for the computer program in Fig. 4.5 that uses the Newton's divided-difference method and the computed value.

The above examples show that Newton's divided-difference method is simple and easy to understand. The computational procedure of the method can be used to develop a computer program directly. The main idea of the method is to derive a function that passes through all data points. The function is then used to determine values at locations within these data points. The general form of the Newton's divided-difference interpolating polynomials consists of the set of coefficient C_i as shown in Eq. (4.10). These coefficients can be determined by using the sequential steps as explained in Table 4.2. In the following section, another method that can be used to interpolate values from the given data points is presented. The method is also simple and represents a basis to study high-level numerical methods for solving practical problems.

4.3 Lagrange Interpolating Polynomials

The Lagrange polynomials are widely used to interpolate values from a set of data points. The first-order Lagrange polynomial is the simplest one which uses a linear function as explained below.

4.3.1 Linear interpolation

A straight line is used to connect the two data values of $f(x_0)$ and $f(x_1)$ at x_0 and x_1, respectively, as shown in Fig. 4.7.

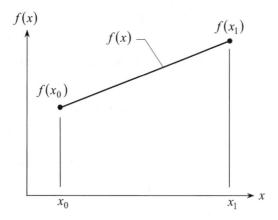

Figure 4.7 Linear interpolation between points x_0 and x_1.

The straight line is represented by a linear function in the form

$$f(x) = ax + b \tag{4.17}$$

where a and b are constants and can be determined by using the data values at point x_0 and x_1 as follows

at $\quad x = x_0 \quad :$ $\qquad\qquad f(x_0) \;=\; ax_0 + b \tag{4.18a}$

at $\quad x = x_1 \quad :$ $\qquad\qquad f(x_1) \;=\; ax_1 + b \tag{4.18b}$

Equation (4.18b) is first subtracted from Eq. (4.18a) to yield

$$f(x_1) - f(x_0) \;=\; a(x_1 - x_0)$$

$$a \;=\; \frac{f(x_1) - f(x_0)}{x_1 - x_0}$$

Then, by substituting a into Eq. (4.18a) to give

$$b \;=\; f(x_0) - \frac{f(x_1) - f(x_0)}{x_1 - x_0} x_0$$

Thus, Eq. (4.17) becomes

$$f(x) \;=\; \frac{f(x_1) - f(x_0)}{x_1 - x_0} x \;+\; f(x_0) - \frac{f(x_1) - f(x_0)}{x_1 - x_0} x_0$$

The above equation can be rewritten as

$$f(x) \;=\; \left(\frac{x_1 - x}{x_1 - x_0}\right) f(x_0) + \left(\frac{x_0 - x}{x_0 - x_1}\right) f(x_1) \tag{4.19}$$

or,

$$f(x) \;=\; L_0(x) f(x_0) \;+\; L_1(x) f(x_1) \tag{4.20}$$

where $\qquad L_0(x) = \dfrac{x_1 - x}{x_1 - x_0}$ and $L_1(x) = \dfrac{x_0 - x}{x_0 - x_1}$ $\qquad$ (4.21)

The functions $L_0(x)$ and $L_1(x)$ are called the Lagrange interpolation functions. Their distributions are shown in Fig. 4.8.

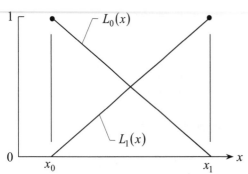

Figure 4.8 Linear Lagrange interpolation functions.

From Fig. 4.8,

$$L_0(x) = \begin{cases} 1 & ; \ x = x_0 \\ 0 & ; \ x = x_1 \end{cases} \qquad (4.22a)$$

and

$$L_1(x) = \begin{cases} 0 & ; \ x = x_0 \\ 1 & ; \ x = x_1 \end{cases} \qquad (4.22b)$$

Or, it can be concluded that

$$L_i(x) = \begin{cases} 1 & ; \ x = x_i \\ 0 & ; \ \text{at other points} \end{cases} \qquad (4.23)$$

Example 4.4 Use linear Lagrange interpolation function to estimate the value of the Bessel function at $x = 3.2$. The values of the Bessel function at $x_0 = 2.0$ and $x_1 = 4.0$ are $f(x_0) = 0.2239$ and $f(x_1) = -0.3971$, respectively. Compare the estimated value with the exact solution in Table 4.1.

From Eq. (4.21), the linear Lagrange interpolation functions are

$$L_0(x) = \frac{x_1 - x}{x_1 - x_0} = \frac{4.0 - x}{4.0 - 2.0} \qquad (4.24a)$$

$$L_1(x) = \frac{x_0 - x}{x_0 - x_1} = \frac{2.0 - x}{2.0 - 4.0} \qquad (4.24b)$$

Then, the linear Lagrange interpolation is

$$f(x) = \left(\frac{4.0 - x}{4.0 - 2.0} \right) f(x_0) + \left(\frac{2.0 - x}{2.0 - 4.0} \right) f(x_1) \qquad (4.25)$$

By substituting the values of $f(x_0)$ and $f(x_1)$ into Eq. (4.25), the estimated value at $x = 3.2$ is

$$f(x = 3.2) = \left(\frac{4.0 - 3.2}{4.0 - 2.0} \right)(0.2239) + \left(\frac{2.0 - 3.2}{2.0 - 4.0} \right)(-0.3971)$$

$$= 0.08956 - 0.23826$$

$$f(3.2) = -0.1487 \qquad (4.26)$$

The result is identical to that obtained from the Newton's divided-difference method in Example 4.1 with the true error of 53.56%.

4.3.2 Quadratic interpolation

A quadratic function is used to fit a set of three data $f(x_0)$, $f(x_1)$, $f(x_2)$ at x_0, x_1, x_2, respectively. The quadratic function is expressed in the form as shown in Fig. 4.9.

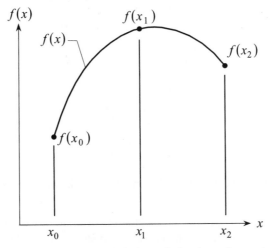

Figure 4.9 Quadratic interpolation passing through the three data points at x_0, x_1 and x_2.

$$f(x) = ax^2 + bx + c \tag{4.27}$$

where a, b and c are constants and can be determined from the three data at x_0, x_1 and x_2 as

at $\quad x = x_0$: $\qquad f(x_0) = ax_0^2 + bx_0 + c \tag{4.28a}$

at $\quad x = x_1$: $\qquad f(x_1) = ax_1^2 + bx_1 + c \tag{4.28b}$

at $\quad x = x_2$: $\qquad f(x_2) = ax_2^2 + bx_2 + c \tag{4.28c}$

The procedure to determine the constants a, b and c is left as an exercise at the end of the chapter. After these constants are determined and substituted into Eq. (4.27), the quadratic Lagrange interpolation is obtained in the form

$$f(x) = L_0(x)f(x_0) + L_1(x)f(x_1) + L_2(x)f(x_2) \tag{4.29}$$

where

$$L_0(x) = \frac{(x_2 - x)(x_1 - x)}{(x_2 - x_0)(x_1 - x_0)} \tag{4.30a}$$

$$L_1(x) = \frac{(x_2 - x)(x_0 - x)}{(x_2 - x_1)(x_0 - x_1)} \tag{4.30b}$$

$$L_2(x) = \frac{(x_1 - x)(x_0 - x)}{(x_1 - x_2)(x_0 - x_2)} \tag{4.30c}$$

A typical Lagrange interpolation function (L_i) as shown in Eq. (4.30a-c) is equal to unity at x_i and zero at other points as shown in Fig. 4.10.

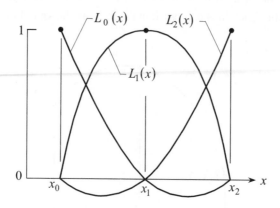

Figure 4.10 Quadratic Lagrange interpolation functions.

Example 4.5 Use the quadratic Lagrange interpolation to estimate the value of the Bessel function at $x = 3.2$. The three values of the Bessel function are given as $f(x_0 = 2.0) = 0.2239$, $f(x_1 = 3.0) = -0.2601$ and $f(x_2 = 4.0) = -0.3971$. Compare the estimated value with exact solution in Table 4.1.

From Eq. (4.30a-c), the quadratic Lagrange interpolation functions are

$$L_0(x) = \frac{(x_2 - x)(x_1 - x)}{(x_2 - x_0)(x_1 - x_0)} = \frac{(4-x)(3-x)}{(4-2)(3-2)} = \frac{(4-x)(3-x)}{2}$$

$$L_1(x) = \frac{(x_2 - x)(x_0 - x)}{(x_2 - x_1)(x_0 - x_1)} = \frac{(4-x)(2-x)}{(4-3)(2-3)} = -(4-x)(2-x)$$

$$L_2(x) = \frac{(x_1 - x)(x_0 - x)}{(x_1 - x_2)(x_0 - x_2)} = \frac{(3-x)(2-x)}{(3-4)(2-4)} = \frac{(3-x)(2-x)}{2}$$

Then, the quadratic Lagrange interpolation in Eq. (4.29) is

$$f(x) = \frac{(4-x)(3-x)}{2}f(x_0) - (4-x)(2-x)f(x_1) + \frac{(3-x)(2-x)}{2}f(x_2)$$

By substituting the data values of $f(x_0)$, $f(x_1)$ and $f(x_2)$ into the above equation, the estimated value at $x = 3.2$ is

$$f(x = 3.2) = \frac{(4-3.2)(3-3.2)}{2}(0.2239) - (4-3.2)(2-3.2)(-0.2601)$$

$$+ \frac{(3-3.2)(2-3.2)}{2}(-0.3971)$$

$$= -0.017912 - 0.249696 - 0.047652$$

$$f(3.2) = -0.3153 \qquad\qquad (4.31)$$

The estimated value is identical to that obtained from the Newton's divided-difference method in Example 4.2 with the true error of 1.53%.

4.3.3 Polynomial interpolation

By understanding the linear and quadratic Lagrange interpolations explained in the preceding sections, the nth-order Lagrange interpolation as shown in Fig. 4.4 can be derived in the same manner. The general form of the nth-order Lagrange interpolation can be written in the form

$$f(x) = \sum_{i=0}^{n} L_i(x) f(x_i) \tag{4.32}$$

where

$$L_i(x) = \prod_{\substack{j=0 \\ j=i}}^{n} \frac{x - x_j}{x_i - x_j} \tag{4.33}$$

The symbol $\prod$ represents the multiplication operator. For example,

$$\prod_{k=1}^{3} (k+1) = (1+1)(2+1)(3+1) = (2)(3)(4) = 24 \tag{4.34}$$

The use of the Lagrange interpolation function $L_i(x)$ in the form of Eq. (4.33) can be explained by using an example with three data points (i.e., $n = 2$). In this case, the function $L_1(x)$ is

$$L_1(x) = \prod_{\substack{j=0 \\ j \neq 1}}^{2} \frac{x - x_j}{x_1 - x_j} = \frac{(x - x_0)(x - x_2)}{(x_1 - x_0)(x_1 - x_2)}$$

which is identical to that shown in Eq. (4.30b).

The Lagrange interpolation function in Eq. (4.33) is equal to unity at x_i and zero at the other points. Figure 4.11 shows the profile of a typical Lagrange interpolation function that has such properties.

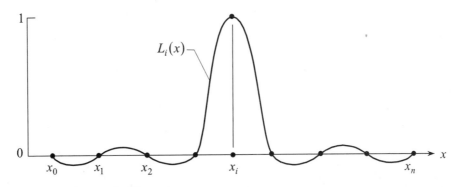

Figure 4.11 Profile of a typical Lagrange interpolation function.

From Examples 4.4 and 4.5, the linear and quadratic Lagrange interpolation functions can be derived easily from a few data points. For a general problem with many data points, the derivation of the Lagrange interpolation function is tedious and a computer program for determining them is needed. Figure 4.12 shows a computer program for determining the Lagrange interpolation functions for $n+1$ data points. The program, in addition, estimates the interpolated value from a given set of data points.

Fortran

```
      PROGRAM  LAGPOL
C.....PROGRAM FOR COMPUTING  F(X)  AT A GIVEN  X
C.....USING LAGRANGE INTERPOLATION
      DIMENSION  X(10), FX(10)
C.....READ NUMBER OF DATA SETS, DATA OF X AND FX:
      READ(5,*)  N
      DO 10  I=1,N
      READ(5,*)  X(I), FX(I)
 10 CONTINUE
C.....COMPUTE DESIRED F(X) AT THE GIVEN X VALUE:
      READ(5,*)  XX
      YY = 0.
      DO 20  I=1,N
      AL = 1.
      DO 30  J=1,N
        IF(J.NE.I)   THEN
          AL = AL*(XX-X(J))/(X(I)-X(J))
        ENDIF
 30 CONTINUE
      YY = YY + AL*FX(I)
 20 CONTINUE
      WRITE(6,100)  XX, YY
100 FORMAT(' VALUE OF F(X) AT X =', E10.4,
     *       ' IS', E16.7              )
      STOP
      END
```

MATLAB

```
%   PROGRAM LAGPOL
%   PROGRAM FOR COMPUTING F(X) AT A GIVEN X
%   USING LAGRANGE INTERPOLATION
%   READ NUMBER OF DATA SETS
n = input('\nEnter number of data: ');
for irow = 1:n
    x(irow)  = input('\nEnter point x:      ');
    fx(irow) = input('Enter value f(x):  ');
end
%   COMPUTE DESIRED F(X) AT THE GIVEN X VALUE:*)
xx  = input('\nEnter point x to find f(x):  ');
yy = 0.;
for i = 1:n
    AL = 1.;
    for j = 1:n
        if j ~= i
            AL = AL*(xx-x(j))/(x(i)-x(j));
        end
    end
    yy = yy + AL*fx(i);
end
fprintf('\nVALUE OF F(X) AT X = %14.6f IS %14.6e', xx, yy)
```

Figure 4.12 Computer program for determining the Lagrange interpolation functions and the interpolated value from a given set of data.

Example 4.6 Use the computer program for determining the Lagrange interpolation functions as shown in Fig. 4.12 to estimate the value of the Bessel function at $x = 3.2$ from all data points in Table 4.1 except at $x = 3.2$. Compare the estimated value with that obtained from the Newton's divided-difference method in Example 4.3.

Figure 4.13 shows the input data needed for the computer program in Fig. 4.12. The data are from Table 4.1 except the one at $x = 3.2$. The computed value from the computer program is -0.3201 which is identical to that from the Newton's divided-difference method in Example 4.3.

```
    10
   2.0    .2239
   2.2    .1104
   2.4    .0025
   2.6   -.0968
   2.8   -.1850
   3.0   -.2601
   3.4   -.3643
   3.6   -.3918
   3.8   -.3992
   4.0   -.3971
      3.2

VALUE OF F(X) AT X = .3200E+01 IS  -.3201119E+00
```

Figure 4.13 Example of input data from Table 4.1 for using with the computer program in Fig. 4.12 and the computed value from the Lagrange interpolation method.

It should be noted that some patterns of data points cannot be fitted well by using the higher-order polynomials. For example, the data points that are taken from the Runge's function which is represented by

$$f(x) \quad = \quad \frac{1}{1+25x^2} \tag{4.35}$$

If the fourth-order Lagrange interpolation is used to fit the five data (at $x = -1, -0.5, 0, 0.5, 1$) taken from the Runge's function, its distribution is shown in Fig. 4.14. The figure shows that the distribution of the fourth-order Lagrange interpolation is quite different from the actual profile of the Runge's function. In this case, it is obvious that the Lagrange interpolation should not be used to represent the distribution of the Runge's function. In the next section, another interpolation method that can satisfactory fit some special set of data or functions will be explained.

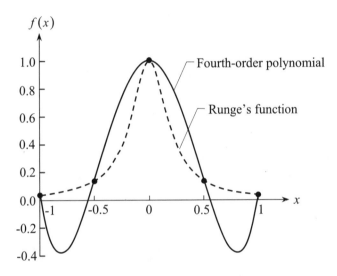

Figure 4.14 Comparison between the distribution fitted by the fourth-order polynomial and the actual profile of the Runge's function.

4.4 Spline Interpolations

The basic idea of the Newton's divided-difference and Lagrange interpolation methods is to create a polynomial function that passes through all data points. However, the distribution of such a single polynomial function may not realistically represent the true behavior of the problem. For example, a set of eight data points for the air density measured from an experiment of a supersonic flow over an airfoil is shown in Fig. 4.15. Figure 4.15(a) shows a single polynomial function that passes through all eight data points. The distribution of the fitted polynomial is not realistic with up and down deviation between these data points. A more realistic distribution is obtained by using many functions to fit these data points as shown in Fig. 4.15(b). In this section, the spline interpolation method that can produce more realistic distributions of some particular data behavior is explained.

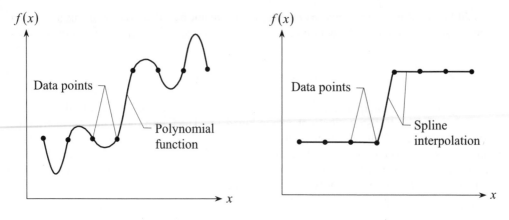

(a) Distribution of polynomial function.   (b) Distribution of spline interpolation.

Figure 4.15 Distributions of a polynomial function and spline interpolation to fit a set of data distributed as a step function pattern.

4.4.1 Linear spline

The linear spline is the simplest spline function that uses a straight line to connect two data points. Figure 4.16 shows the three linear splines that are used to connect the four data points.

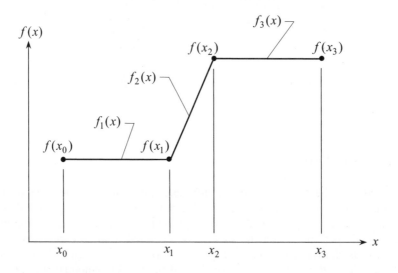

Figure 4.16 Three linear splines that connect four data points.

These three linear splines are

$$f_1(x) = f(x_0) + m_1(x - x_0) \qquad ; \qquad x_0 \le x \le x_1 \qquad (4.36a)$$

$$f_2(x) = f(x_1) + m_2(x - x_1) \qquad ; \qquad x_1 \le x \le x_2 \qquad (4.36b)$$

$$f_3(x) = f(x_2) + m_3(x - x_2) \qquad ; \qquad x_2 \le x \le x_3 \qquad (4.36c)$$

where
$$m_i = \frac{f(x_i) - f(x_{i-1})}{x_i - x_{i-1}} \qquad i = 1, 2, 3 \qquad (4.36d)$$

are their slopes. A computer program for creating the n linear splines from the given $n+1$ data points is easy to develop and is left as an exercise.

4.4.2 Quadratic Spline

The quadratic spline uses a quadratic function to represent distribution between the two data points. Figure 4.17 shows the three quadratic splines that pass through the four data points. These three quadratic splines are

$$f_1(x) = a_1 x^2 + b_1 x + c_1 \qquad ; \qquad x_0 \le x \le x_1 \qquad (4.37a)$$

$$f_2(x) = a_2 x^2 + b_2 x + c_2 \qquad ; \qquad x_1 \le x \le x_2 \qquad (4.37b)$$

$$f_3(x) = a_3 x^2 + b_3 x + c_3 \qquad ; \qquad x_2 \le x \le x_3 \qquad (4.37c)$$

where a_i, b_i, c_i, $i = 1, 2, 3$ are unknown constants. These unknowns are determined by using the conditions as follows.

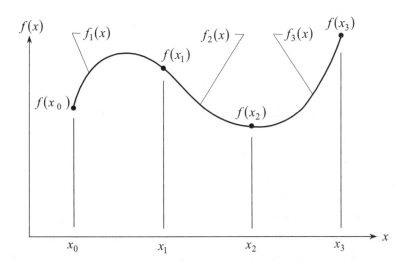

Figure 4.17 Three quadratic splines that connect four data points.

(a) At any internal point, the values of the two quadratic splines connected at that point must be equal. For example, the conditions at point x_1 in Fig. 4.17 are

$$f_1(x_1) \quad = \quad a_1 x_1^2 + b_1 x_1 + c_1 \quad = \quad f(x_1) \tag{4.38a}$$

and

$$f_2(x_1) \quad = \quad a_2 x_1^2 + b_2 x_1 + c_2 \quad = \quad f(x_1) \tag{4.38b}$$

Thus, the two internal points x_1 and x_2 in Fig. 4.17 yield 4 conditions.

(b) The first quadratic spline must pass through the first data point x_0,

$$f_1(x_0) \quad = \quad a_1 x_0^2 + b_1 x_0 + c_1 \quad = \quad f(x_0) \tag{4.39a}$$

and the third quadratic spline must pass through the last data point x_3,

$$f_3(x_3) \quad = \quad a_3 x_3^2 + b_3 x_3 + c_3 \quad = \quad f(x_3) \tag{4.39b}$$

so that two additional conditions are produced.

(c) At any internal point, the slopes of the two quadratic splines connected at that point must be equal. For example, the condition at point x_1 is

$$f_1'(x_1) \quad = \quad f_2'(x_1) \tag{4.40a}$$

$$2a_1 x_1 + b_1 \quad = \quad 2a_2 x_1 + b_2 \tag{4.40b}$$

The requirement for equal slopes yields two conditions for the two internal points in Fig. 4.17.

The total number of conditions for the four data points using the three quadratic splines is $4+2+2 = 8$. Since there are 9 unknowns, the coefficient a_1 may be assigned to be zero which means $f_1(x)$ is a linear spline. The eight unknowns can then be solved from the system of eight equations

$$\begin{bmatrix} (8\times 8) \end{bmatrix} \begin{Bmatrix} b_1 \\ c_1 \\ a_2 \\ b_2 \\ c_2 \\ a_3 \\ b_3 \\ c_3 \end{Bmatrix} = \begin{Bmatrix} (8\times 1) \end{Bmatrix} \tag{4.41}$$

For a general problem with $n + 1$ data points, the total of n quadratic splines of $f_1(x)$, $f_2(x)$, ..., $f_n(x)$ is needed. Since each quadratic spline has 3 unknowns, so the total number of unknowns is $3n$. However, the number of equations that can be created are $2(n-1)+2+(n-1) = 3n-1$ which is one condition fewer than the number of unknowns. If the first quadratic spline is assumed to be linear by setting its coefficient a_1 as zero, then the number of unknowns reduces to $3n-1$ so that they can be solved from $3n-1$ equations.

A computer program for determining the quadratic splines can be developed to create the system of Eq. (4.41) from the conditions as shown in Eqs. (4.38) - (4.40). This system of equations can be solved for the unknown coefficients by a procedure for solving a system of equations explained in Chapter 3. The quadratic splines are then obtained after substituting the computed coefficients into Eqs. (4.37a-c).

4.4.3 Cubic spline

The procedure for creating a cubic spline is similar to the linear and quadratic splines as explained in the preceding sections. Distribution of the quadratic spline is either in a convex or concave shape. The cubic spline is widely used because its distribution can be in both the convex and concave shapes between two data points.

If three cubic splines are to be created from the four data points in Fig. 4.17, the three cubic functions connecting the three pairs of data points are

$$f_1(x) = a_1 x^3 + b_1 x^2 + c_1 x + d_1 \quad ; \quad x_0 \le x \le x_1 \quad\quad (4.42a)$$

$$f_2(x) = a_2 x^3 + b_2 x^2 + c_2 x + d_2 \quad ; \quad x_1 \le x \le x_2 \quad\quad (4.42b)$$

$$f_3(x) = a_3 x^3 + b_3 x^2 + c_3 x + d_3 \quad ; \quad x_2 \le x \le x_3 \quad\quad (4.42c)$$

where a_i, b_i, c_i, d_i, $i = 1, 2, 3$ are unknown coefficients. The twelve unknown coefficients of the three cubic functions can be determined from the following conditions.

(a) At any internal point, the values of the two cubic splines connected at that point must be equal. For example, the conditions at point x_1 in Fig. 4.17 are

$$f_1(x_1) = a_1 x_1^3 + b_1 x_1^2 + c_1 x_1 + d_1 = f(x_1) \quad\quad (4.43a)$$

and
$$f_2(x_1) = a_2 x_1^3 + b_2 x_1^2 + c_2 x_1 + d_2 = f(x_1) \quad\quad (4.43b)$$

Thus, the two internal points x_1 and x_2 in Fig. 4.17 yield four conditions.

(b) The first cubic spline must pass through the first data point x_0,

$$f_1(x_0) = a_1 x_0^3 + b_1 x_0^2 + c_1 x_0 + d_1 = f(x_0) \quad\quad (4.44a)$$

and the last cubic spline must pass through the last data point, x_3,

$$f_3(x_3) = a_3 x_3^3 + b_3 x_3^2 + c_3 x_3 + d_3 = f(x_3) \quad\quad (4.44b)$$

so that two additional conditions are produced.

(c) At any internal point, the slopes of the two cubic splines connected at that point must be equal. For example, the condition at point x_1 is

$$f_1'(x_1) = f_2'(x_1) \quad\quad (4.45a)$$

$$3a_1x_1^2 + 2b_1x_1 + c_1 \;=\; 3a_2x_1^2 + 2b_2x_1 + c_2 \qquad\qquad (4.45b)$$

The requirement for equal slopes yields two conditions for the two internal points in Fig. 4.17.

(d)   At any internal point, the second derivatives of the two cubic splines connected at that point must be equal. For example, the condition at point x_1 is

$$f_1''(x_1) \;=\; f_2''(x_1) \qquad\qquad (4.46a)$$

$$6a_1x_1 + 2b_1 \;=\; 6a_2x_1 + 2b_2 \qquad\qquad (4.46b)$$

The requirement for equal second-derivatives yields two conditions for the two internal points in Fig. 4.17.

(e)   The conditions of zero second-derivative for the cubic splines at the first and last data points may be assigned to produce two more conditions of

$$f_1''(x_0) \;=\; f_3''(x_3) \;=\; 0 \qquad\qquad (4.46c)$$

From the set of four data points connected by the three cubic splines in Fig. 4.17, there is the total of 12 unknown coefficients. These unknown coefficients are determined from the $4 + 2 + 2 + 2 = 12$ conditions from Eqs. (4.43)-(4.46). The twelve equations can be written in matrix form as

$$\begin{bmatrix} & & \\ & (12\times12) & \\ & & \end{bmatrix} \begin{Bmatrix} a_1 \\ b_1 \\ c_1 \\ d_1 \\ \vdots \\ a_3 \\ b_3 \\ c_3 \\ d_3 \end{Bmatrix} = \begin{Bmatrix} \\ (12\times1) \\ \end{Bmatrix} \qquad\qquad (4.47)$$

For a set of $n+1$ data points to be fitted by n cubic splines $f_1(x)$, $f_2(x)$,..., $f_n(x)$, there are $4n$ unknown coefficients to be determined. These unknown coefficients are solved from a set of $2(n-1) + 2 + (n-1) + (n-1) + 2 = 4n$ equations according to the conditions explained in Eqs. (4.43) - (4.46).

Solving a set of $4n$ equations to produce n cubic splines may not be simple, especially for the case of a large value of n. An alternative approach explained below solves only $n-1$ equations in order to produce n cubic splines. The idea behind such approach is to assume a linear distribution for the second derivative of the spline function $f_i(x)$. Such linear distribution can be written in form of the Lagrange interpolation as

$$f_i''(x) = f''(x_{i-1})\frac{x-x_i}{x_{i-1}-x_i} + f''(x_i)\frac{x-x_{i-1}}{x_i-x_{i-1}} \tag{4.48}$$

where $i = 1, 2, \ldots, n$. The spline function $f_i(x)$ can then be obtained by performing integration twice. The integrations lead to two integrating constants which can be determined from the two conditions of

at point $\qquad x = x_{i-1}$: $\qquad f_i(x_{i-1}) = f(x_{i-1})$ $\qquad$ (4.49a)

and at point $\qquad x = x_i$: $\qquad f_i(x_i) = f(x_i)$ $\qquad$ (4.49b)

Once the two integrating constants are determined and substituted back, the final spline function is obtained in the form,

$$f_i(x) = \frac{f''(x_{i-1})}{6(x_i-x_{i-1})}(x_i-x)^3 + \frac{f''(x_i)}{6(x_i-x_{i-1})}(x-x_{i-1})^3$$

$$+ \left[\frac{f(x_{i-1})}{(x_i-x_{i-1})} - \frac{(x_i-x_{i-1})f''(x_{i-1})}{6}\right](x_i-x)$$

$$+ \left[\frac{f(x_i)}{(x_i-x_{i-1})} - \frac{(x_i-x_{i-1})f''(x_i)}{6}\right](x-x_{i-1}) \tag{4.50}$$

The right-hand side of the derived spline function in Eq. (4.50) consists of the two unknown second-derivatives at x_{i-1} and x_i. It is noted that the first derivative of the two spline functions connected at an interior point must be equal,

$$f_{i-1}'(x_i) = f_i'(x_i) \tag{4.51}$$

The first derivatives of the spline functions $f_{i-1}(x)$ and $f_i(x)$ can be derived from Eq. (4.50). These derivatives are then required to be equal according to Eq. (4.51) which leads to

$$(x_i-x_{i-1})f''(x_{i-1}) + 2(x_{i+1}-x_{i-1})f''(x_i) + (x_{i+1}-x_i)f''(x_{i+1})$$

$$= \frac{6}{x_{i+1}-x_i}(f(x_{i+1})-f(x_i)) + \frac{6}{x_i-x_{i-1}}(f(x_{i-1})-f(x_i)) \tag{4.52}$$

Equation (4.52) consists of the three unknowns which are the second derivatives of the spline functions at x_{i-1}, x_i and x_{i+1}, where $i = 1, 2, \ldots, n$. If the second derivatives of the spline functions for the first and last data points are assumed to be zero,

$$f''(x_0) = f''(x_n) = 0 \tag{4.53}$$

then, Eq. (4.52) yields a set of simultaneous equations with $n-1$ unknowns of $f''(x_1)$, $f''(x_2)$, $\ldots\ldots$, $f''(x_{n-1})$ in the form

$$\begin{bmatrix} x & x & & & & & \\ x & x & x & & & & \\ & x & x & x & & & \\ & & \ddots & \ddots & \ddots & & \\ & & & \ddots & \ddots & \ddots & \\ & & & & x & x & x \\ & & & & & x & x \end{bmatrix} \begin{Bmatrix} f''(x_1) \\ f''(x_2) \\ f''(x_3) \\ \vdots \\ \vdots \\ f''(x_{n-2}) \\ f''(x_{n-1}) \end{Bmatrix} = \begin{Bmatrix} x \\ x \\ x \\ \vdots \\ \vdots \\ x \\ x \end{Bmatrix} \tag{4.54}$$

where the symbol x represents the non-zero coefficient in the matrix. The square matrix on left-hand side of Eq. (4.54) is a tridiagonal matrix because all elements in the matrix are zero except those along the three main diagonals. Such system of equations can be solved conveniently by using the methods explained in Chapter 3.

After solving the system of equations in Eq. (4.54), the second derivatives are substituted into Eq. (4.50) to yield the desired cubic spline functions between x_{i-1} and x_i. The following example demonstrates the performance of the cubic spline functions for fitting a set of special data as compared to that from the high-order polynomial interpolation.

Example 4.7   Develop a computer program that employs the cubic spline interpolation as explained in Eqs. (4.48) - (4.54) to fit the data given in the table below. Compare the result with that obtained from the use of high-order Lagrange interpolation. It is noted that the result from the use of high-order Lagrange interpolation can be obtained from the computer program in Fig. 4.12.

x	0.00	0.10	0.30	0.44	0.56	0.60	0.70	0.85	1.00
$f(x)$	0.1	0.1	0.1	0.1	0.5	0.5	0.5	0.5	0.5

Figure 4.18 shows the computer program for cubic spline interpolation. The procedure in the program starts from reading the data and the location needed for the interpolated result. Next, the program forms the system of equations as shown in Eq. (4.54) and solves such system of equations to obtain the second derivatives for all data points. Finally, the values of the computed second derivatives are substituted into Eq. (4.50) to yield the cubic spline interpolation.

Figure 4.19 shows the comparison between the distributions obtained from the cubic spline interpolation and high-order Lagrange polynomial. The figure shows that the cubic spline interpolation provides realistic distribution while the high-order Lagrange polynomial yields oscillated distribution. The example also suggests that users should understand nature of the data before applying an appropriate interpolating method.

Fortran

```
      PROGRAM  CUBSPLN
C.....A CUBIC SPLINE INTERPOLATING PROGRAM
      DIMENSION  X(10), FX(10)
      DIMENSION  A(10), B(10), C(10), D(10), E(10)
C.....READ NUMBER OF DATA SETS, DATA OF X AND FX,
C.....THEN THE VALUE OF X THAT F(X) IS NEEDED:
      READ(5,*)  N
      DO 10  I=1,N
      READ(5,*)  X(I), FX(I)
   10 CONTINUE
      READ(5,*)  XX
C.....FORM UP TRIDIAGONAL SYSTEM OF N EQUATIONS:
      DO 20  I=2,N-1
      A(I) = X(I) - X(I-1)
      B(I) = 2.*(X(I+1) - X(I-1))
      C(I) = X(I+1) - X(I)
      D(I) = 6.*(FX(I+1)-FX(I))/(X(I+1)-X(I))
     1     + 6.*(FX(I-1)-FX(I))/(X(I)-X(I-1))
   20 CONTINUE
      B(1) = 1.
      C(1) = 0.
      D(1) = 0.
      A(N) = 0.
      B(N) = 1.
      D(N) = 0.
C.....SOLVE TRIDIAGONAL SYSTEM OF N EQUATIONS FOR
C.....2ND DERIVATIVES, RETURN SOLUTION IN E( ):
C.....[STANDARD TRIDIAGONAL SYSTEM SOLVER - N EQS]
C.....COMPUTE  F(X)  AT THE GIVEN  X:
      DO 30  I=2,N
      A(I) = A(I)/B(I-1)
      B(I) = B(I) - A(I)*C(I-1)
   30 CONTINUE
      DO 35  I=2,N
      D(I) = D(I) - A(I)*D(I-1)
   35 CONTINUE
      E(N) = D(N)/B(N)
      DO 40  I=N-1,1,-1
      E(I) = (D(I) - C(I)*E(I+1))/B(I)
   40 CONTINUE
      DO 50  I=2,N
      IF((XX.GE.X(I-1)).AND.(XX.LE.X(I)))   THEN
      D1 = X(I) - XX
      D2 = XX - X(I-1)
      DD = X(I) - X(I-1)
      T1 = E(I-1)*D1*D1*D1/(6.*DD)
      T2 = E(I)*D2*D2*D2/(6.*DD)
      T3 = (FX(I-1)/DD - E(I-1)*DD/6.)*D1
      T4 = (FX(I)/DD - E(I)*DD/6.)*D2
      FF = T1 + T2 + T3 + T4
      ENDIF
   50 CONTINUE
      WRITE(6,100)  XX, FF
  100 FORMAT(' VALUE OF F(X) AT X =', E10.4,
     *        ' IS', E16.7                    )
      STOP
      END
```

MATLAB

```
%   PROGRAM CUBSPLN
%   A CUBIC SPLINE INTERPOLATING PROGRAM
%   READ NUMBER OF DATA SETS, DATA OF X AND FX
%   THEN THE VALUE OF X THAT F(X) IS NEEDED:
n = input('\nEnter number of data: ');
for irow = 1:n
    x(irow)  = input('\nEnter point x:     ');
    fx(irow) = input('Enter value f(x): ');
end
xx  = input('\nEnter point x to find f(x): ');
%   FORM UP TRIDIAGONAL SYSTEM OF N EQUATIONS: *)
for i = 2:n-1
    a(i) = x(i) - x(i-1);
    b(i) = 2.0*(x(i+1) - x(i-1));
    c(i) = x(i+1) - x(i);
    d(i) = 6.0*(fx(i+1)-fx(i))/(x(i+1)-x(i))+ 6.0* ...
(fx(i-1)-fx(i))/(x(i)-x(i-1));
end
b(1) = 1.0;
c(1) = 0.0;
d(1) = 0.0;
a(n) = 0.0;
b(n) = 1.0;
d(n) = 0.0;
%   SOLVE TRIDIAGONAL SYSTEM OF N EQUATIONS FOR
%   2ND DERIVATIVES, RETURN SOLUTION IN E( ):
%   [STANDARD TRIDIAGONAL SYSTEM SOLVER - N EQS]
%   COMPUTE  F(X)  AT THE GIVEN  X:
for i = 2:n
    a(i) = a(i)/b(i-1);
    b(i) = b(i) - a(i)*c(i-1);
end
for i = 2:n
    d(i) = d(i) - a(i)*d(i-1);
end
e(n) = d(n)/b(n);
for i = n-1:-1:1
    e(n) = (d(i) - c(i)*e(i+1))/b(i);
end
for i = 2:n
    if ((xx >= x(i-1)) & (xx <= x(i)))
        d1 = x(i) - xx;
        d2 = xx - x(i-1);
        dd = x(i) - x(i-1);
        t1 = e(i-1)*d1*d1*d1/(6.0*dd);
        t2 = e(i)*d2*d2*d2/(6.0*dd);
        t3 = (fx(i-1)/dd-e(i-1)*dd/6.0)*d1;
        t4 = (fx(i)/dd-e(i)*dd/6.0)*d2;
        ff = t1 + t2 + t3 + t4;
    end
end
fprintf('\nVALUE OF F(X) AT X = %10.4f IS %14.6e', xx, ff);
```

Figure 4.18 Computer program for cubic spline interpolation method.

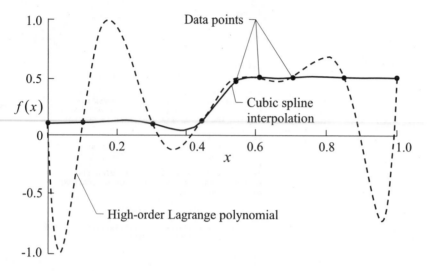

Figure 4.19 Comparison between the distributions obtained from using the cubic spline interpolation and higher-order Lagrange polynomial.

4.5 MATLAB Functions for Interpolations

Two built-in MATLAB functions that can be used for interpolation are explained herein. The two functions are `spline` and `interp1`. The `spline` function is based on the cubic spline interpolation method explained in the preceding section. The `interp1` function allows the user to select a method for the interpolation.

The `spline` function has the syntax as follow,

$$yy = spline(x, y, xx)$$

where   x and y   are the locations and values of the data
       xx   is the location to perform interpolation
       yy   is the interpolated value at the location xx

Example 4.8   Employ the `spline` function to fit the Runge's function,

$$f(x) = \frac{1}{1 + 25x^2} \qquad (4.35)$$

by using nine data at $x = -1.0, -0.75, -0.5, -0.25, 0.0, 0.25, 0.5, 0.75$ and 1.0. Plot to compare the computed distribution with the actual distribution of the Runge's function.

The values at the specified nine locations can be generated by using the

```
>> x = [-1:0.25:1];
>> y = 1./(1+25*x.^2);
```

Then, the `spline` function is employed to determine the values of `yy` at `xx` starting from -1 to 1 with the interval of 0.125 as follows

```
>> xx = [-1:0.125:1];
>> yy = spline(x,y,xx);
```

Figure 4.20 shows the comparison of the distribution from the values by the `spline` function (dashed line) and the actual distribution of the Runge's function (solid line). The comparison shows that the distribution from the values by the `spline` function agrees very well with the Runge's function.

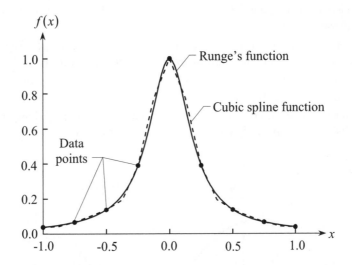

Figure 4.20 Comparison between the distribution from the values by the `spline` function and the actual distribution of the Rung's function.

The second useful function for interpolation in MATLAB is `interp1`. The syntax of this function is

$$yy = interp1(x, y, xx, 'method')$$

where   `x` and `y`   are the locations and values of the data

       `xx`   is the location to perform interpolation

       `yy`   is the interpolated value at the location `xx`

       `'method'`   is the option for interpolation:

             `'nearest'`   The option interpolates the value at the location `xx` by using the values from the nearest locations

             `'linear'`   The option uses linear interpolation to determine the value at the location `xx`

             `'spline'`   The option uses cubic spline interpolation to determine the value at the location `xx`

The default `'method'` is `'linear'` if one of the three options above is not selected.

4.6 Extrapolation

All methods in the preceding sections explain the interpolation procedures. These methods create a function $f(x)$ passing through all of the data at x_0, x_1, x_2,..., x_n. The function is then used to interpolate a value at the desired location between x_0 and x_n. Accuracy of the interpolated value depends on the number of the given data points and its location. The accuracy of the interpolated value also increases if the desired location is closed to a given data point.

The basic idea of the extrapolation is to use the function created from an interpolation method to estimate the value outside the range of the given data points. Figure 4.21 shows an example of the interpolation and extrapolation from the three data points at x_0, x_1 and x_2. For interpolation, a quadratic function can be created to estimate the value within the given range. The figure shows that the quadratic function can provide accurate interpolated value between x_0 and x_2. The same quadratic function, however, may yield an extrapolated value with a large error at a location outside the range of x_0 and x_2. The figure suggests that extrapolation must be performed with care, especially if the number of data points is a few. The following two examples show that the accuracy of the extrapolated value can be improved by increasing the number of data points.

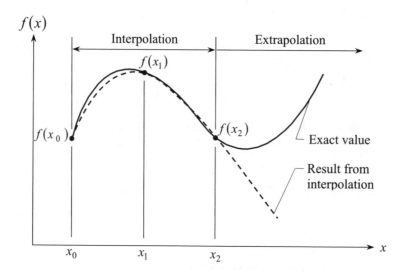

Figure 4.21 Comparison between interpolation and extrapolation.

Example 4.9 The values of the Bessel's function at three locations are given by $f(x_0 = 2.0) = 0.2239$, $f(x_1 = 2.6) = -0.0968$ and $f(x_2 = 3.2) = -0.3202$. Determine the value of the Bessel's function at point $x = 4.0$ by performing extrapolation.

From the given three data points, the computer program for the Newton's divided-difference method as shown in Fig. 4.5 can be used to extrapolate the value of the Bessel's function at $x = 4.0$. The extrapolated value is $f(x = 4.0) = -0.4667$. The extrapolated value has the true error as compared to the exact solution in Table 4.1 as

$$\varepsilon = \frac{-0.3971+0.4667}{-0.3971}\times100\% = -17.53\% \tag{4.55}$$

The error from the extrapolated value in Eq. (4.55) is relatively large. The same extrapolation is performed again but by using more data points as shown in the following example.

Example 4.10   The values of the Bessel's function at five locations are given by $f(x_0 = 2.0) =$ 0.2239, $f(x_1 = 2.4) = 0.0025$, $f(x_2 = 2.8) = -0.1850$, $f(x_3 = 3.2) = -0.3202$ and $f(x_4 = 3.6) =$ −0.3918. Determine the value of the Bessel's function at point $x = 4.0$ by performing extrapolation.

From the values of the Bessel's function at five locations, the computer program in Fig. 4.5 can be used again to determine the value of the Bessel's function at point $x = 4$. The extrapolated value is $f(x = 4.0) = -0.3956$. The extrapolated value has the true error as compared to the exact solution in Table 4.1 as

$$\varepsilon = \frac{-0.3971+0.3956}{-0.3971}\times100\% = -0.38\% \tag{4.56}$$

The error from this later Example 4.10 is quite small as compared to that from Example 4.9. These two examples demonstrate that the error of the extrapolated value can be decreased by increasing the number of the data points.

4.7 Closure

In this chapter, the interpolation and extrapolation methods are presented. Both methods estimate values at locations from a set of data points. The interpolation methods presented herein are the Newton's divided differences, Lagrange polynomial and spline interpolations. The Newton's divided differences and Lagrange interpolations create the nth-order polynomial that pass through $n+1$ data points. Linear, quadratic and general n^{th}-order polynomials are derived and explained in details with examples. If the given set of data points behaves properly, accurate interpolated values can be obtained by using high-order polynomials.

For some special sets of data points, high-order polynomials may yield oscillated distributions leading to inaccurate interpolated values. In this case, the spline interpolation methods may be used. The most popular spline interpolation method is to employ a cubic function to represent the data distribution between two data points. Derivation of the cubic spline interpolation is explained and presented in details with examples. For these interpolation methods, their corresponding computer programs are also developed so that the interpolated values from a large set of data points can be obtained conveniently.

Extrapolation to estimate values outside the given range of data by using the Newton's divided differences and Lagrange polynomials is presented at the end of the chapter. The methods and associate examples suggest that the extrapolation must be performed with care. The extrapolated values may be inaccurate from a set of few data points or their locations are far away from the given data locations. It is noted that some of the interpolation methods, such as the use of the Lagrange polynomials, are the basis for studying the finite difference and finite element methods in the later chapters.

Exercises

1.   Relation between the applied force P and displacement u of the three-bar truss structure in Fig. P4.1 is nonlinear according to the data below

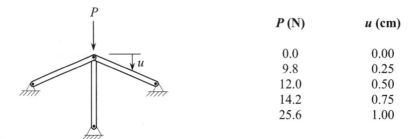

P (N)	u (cm)
0.0	0.00
9.8	0.25
12.0	0.50
14.2	0.75
25.6	1.00

Figure P4.1

   (a)   Plot the data to show the nonlinear relation between the force and displacement.
   (b)   Derive the Lagrange polynomial for the given data points.
   (c)   Plot the Lagrange polynomial obtained in (b) and compare with the data in (a).
   (d)   Use the derived Lagrange polynomial to estimate the forces at the displacements of 0.17, 0.62 and 1.25 cm.

2.   From an experiment, the heat conduction coefficient k of an aluminum material varies with the temperature T as shown in the table. Derive the Lagrange interpolation polynomial to estimate values of the heat conduction coefficient at 50°C, 250 °C and 500 °C.

T (°C)	-100	0	100	200	300	400
k (W/m-°C)	215	202	206	215	228	249

3.   Values of the gravitational acceleration g depend on the altitude y as shown in table. Use the Newton's divided difference method to estimate the value of gravitational accelerations at the altitudes of 5,000 m, 42,000 m and 90,000 m. Plot the distribution of the derived interpolation function together with the given data.

y (m)	0	20,000	40,000	60,000	80,000
g (m/sec^2)	9.8100	9.7487	9.6879	9.6278	9.5682

4.   The air density ρ depends on the temperature T as shown in the table. Employ the computer program of the Newton's divided-difference method in Fig. 4.5 to estimate values of the air density at 250 K, 800 K and 3,000 K. Then, modify the computer program to determine the air density at the temperature at every 10 K from 100 K to 2,500 K. Plot the computed air density distribution that varies with the temperature.

T (K)	100	200	300	500	700	1,000	1,500	2,000	2,500
ρ (kg/m^3)	3.6010	1.7684	1.1774	0.7048	0.5030	0.3524	0.2355	0.1762	0.1394

5. Show that the Lagrange interpolation functions in Eqs. (4.20) and (4.29) are equivalent to the interpolation functions obtained from Newton's divided-difference method in Eqs. (4.1) and (4.6).

6. The set of data in the table below represents the displacement u of a spring from the applied force F. Employ the computer program of the Newton's divided differences or Lagrange interpolation method as shown in Figs. 4.5 and 4.12 to estimate the force needed to displace the spring to 0.3 m and 0.4 m. Then, plot the function obtained from the computer program along with the given data points.

u (m)	0.107	0.172	0.238	0.351	0.388	0.417	0.432	0.441
F (kN)	100	200	300	400	500	600	700	800

7. Temperature drop from wind chill depends on the wind speed. A set of the temperature data that varies with the wind speed is shown in the table. Use the Newton's divided-difference method to estimate the temperature at the wind speed of 35 km/hr. From the given set of data, is it possible for the temperature to drop lower than -50 °C. If it is possible, find the wind speed that causes such temperature.

Wind speed (km/hr)	0	10	20	30	40	50
Temperature (°C)	-12	-23	-31	-36	-38	-39

8. Show detailed procedure for deriving the Lagrange quadratic interpolation functions in Eqs. (4.29) - (4.30).

9. Develop a computer program for the linear interpolation as explained in section 4.4.1. The program should be able to solve any problem with at least 100 data points. Verify the computer program with the set of data in Example 4.7.

10. Use the cubic spline interpolation method explained in section 4.4.3 to derive a set of functions that fit the data as shown in the table. Explain the computational steps in details and determine the value of $f(x = 5)$. Then, compare the computed value with that obtained from the computer program in Fig. 4.18.

x	1	4	6	9	10
$f(x)$	4	9	15	7	3

11. Develop a computer program for the quadratic spline interpolation as explained in section 4.4.2. The program should be able to solve any problem with at least 100 data points. Verify the computer program with the set of data in Example 4.7.

12. From the data in the table as shown below, derive the interpolation functions by using

x	f(x)
2	9.5
4	8.0
6	10.5
8	39.5
10	72.0

   (a) the Newton's differences method

   (b) the Lagrange polynomial method

   (c) the cubic spline method

   Plot the functions obtained from (a)-(c) together with the data points. Then, determine values of these functions at $x = 7$.

13. Fit the data in Problem 12 again but by using the quadratic spline interpolation as explained in section 4.4.2. Show the computational steps in details.

14. Solve Problem 1 again but by using the cubic spline interpolation method as explained in section 4.4.3. Verify the solution by using the computer program in Fig. 4.18. Plot the distributions of the interpolations obtained from these two problems. Then, provide comments on the advantages and disadvantages of each method.

15. From the data in the table as shown below, derive the interpolation functions by using

x	f(x)
0	-7
1	-3
2	6
3	25
4	62
5	129

   (a) the Newton's differences method

   (b) the Lagrange polynomial method

   (c) the cubic spline method

   Determine the values from each function at $x = 2.5$ and 4.5. Then, compare the computed values with those obtained from the computer programs in Figs. 4.5, 4.12 and 4.18.

16. Use the computer programs for the Lagrange polynomial and cubic spline methods as shown in Figs. 4.12 and 4.18 to determine the interpolation functions from the set of data points in Problem 15. Plot the functions obtained and discuss the advantages and disadvantages of each method.

17. Solve Problem 6 again but by using the cubic spline method as explained in section 4.4.3. Plot the distribution of the derived functions and compare them with that obtained from Problem 6.

18. The Runge's function is given by $f(x) = 1/(1+25x^2)$ for $-1 \le x \le 1$ as shown in Fig. 4.14. Create a set of data points from this function in the given range with an increment of $\Delta x = 0.2$. Then, fit the set of data with the tenth-order Lagrange polynomial function. Plot and compare the derived polynomial function with the Runge's function. Discuss the accuracy of the derived Lagrange polynomial function.

19. Solve Problem 18 again but by using the computer program for the cubic spline interpolation in Fig. 4.18. Plot to compare the result with the function obtained from Problem 18.

20. The Runge's function is given by $f(x) = 1/(1+25x^2)$ for $-5 \le x \le 5$. Create a set of data points from this function in the given range with $\Delta x = 1$. Then, derive the interpolation function by using the Newton's divided-difference method. Plot the distribution of the derived function and compare it with the Runge's function.

21. Solve Problem 20 again but by using the computer program for the cubic spline interpolation in Fig. 4.18. Plot to compare the result with the function obtained from Problem 20.

22. The Runge's function is given by $f(x) = 1/(1 + 25x^2)$ for $-2 \le x \le 2$. Create a set of data points from this function in the given range with $\Delta x = 1$ and then derive the interpolation function by using the Lagrange polynomial and cubic spline interpolation methods. Plot the derived functions to compare with the Runge's function. Discuss the accuracy of these derived functions for representing the Runge's function.

23. The data in the table below are generated from the function $f(x) = 100/x^2$. Use an extrapolation method to estimate the value of the function at $x = 5.7$. Compare the estimated value with the exact solution. Suggest ways to improve the solution accuracy from the extrapolation.

x	1	2	3	4	5
$f(x)$	100.000	25.000	11.111	6.250	4.000

24. Data in the table below represent the distribution of the air density ρ in front of an airfoil which moves at a supersonic speed as shown in Fig. 4.1. Use the computer program in Fig. 4.18 to derive the cubic spline interpolation function. Plot to compare such function with the given data and discuss the result. Provide comments if the Lagrange polynomial interpolation method is used to fit the same set of data.

x	ρ	x	ρ
1.000	4.2	1.200	15.2
1.020	4.8	1.204	18.7
1.040	5.1	1.208	23.5
1.060	5.2	1.212	28.9
1.080	5.3	1.216	34.0
1.100	5.5	1.220	38.3
1.120	5.8	1.228	42.7
1.140	6.1	1.236	45.3
1.160	6.5	1.244	46.2
1.180	7.4	1.250	46.4
1.189	9.1	1.300	46.4
1.196	12.9	1.400	46.4

25. From the table shown in Problem 4, use the MATLAB function interp1 with spline option to interpolate the air density values from 100 K to 2,500 K at every 10 K. Compare the result obtained from MATLAB with those from Problem 4.

26. Solve Problem 6 again but by using the function spline in MATLAB. Compare the result with that obtained from the Lagrange polynomial interpolation.

27. From the table in Problem 10, use functions spline and interp1 in MATLAB with linear option to interpolate the value of function at $x = 5$. Compare the result with that obtained from Problem 10.

28. From an experiment of air flow in a pipe, the average air velocity $\bar{u}$ depends the distance y from the pipe surface as shown in the table below.

$\bar{u}$	0.318	0.300	0.264	0.228	0.221	0.179	0.152	0.140
y	0.0075	0.0071	0.0061	0.0055	0.0051	0.0041	0.0034	0.0030

Use the MATLAB function `spline` to determine the average velocities $\bar{u}$ from $y = 0.0030$ to 0.0075 with the increment at every 0.0001. Plot and compare the computed results with the data in the table.

29. Performance curve of a pump is obtained from a set of data between the head and flow rate. The table below shows a set of data for the head (H) and flow rate (Q).

H (m)	45.1	42.7	39.6	35.1	30.5	22.9	15.2
Q (m³/hr)	0	180	270	360	450	540	630

Use a function in MATLAB to determine value of the head at the flow rate of 400 m³/hr.

Chapter
5

Least-Squares Regression

5.1 Introduction

Most of the material properties used in engineering design and analysis were obtained from experiments. For example, the material elasticity curve is needed in the stress analysis for designing a new product. Such curve is obtained from a function that best fits the experimental data. In chapter 4, procedures for deriving an interpolating function that matches the given data points exactly are presented. Use of the interpolating function is suitable for few data points that vary smoothly. For a large set of data points, especially when the data deviate considerably, an interpolating function may not provide accurate representation of the overall data behavior.

Derivation of an approximate function that best fits a given set of data points by using the least-squares regression is explained in this chapter. The method minimizes the sum of the squares of the differences between the data values and the values of the approximate function. To ease understanding of the method, a set of data for the wind velocities measured at different elevations of a building is shown in Table 5.1. The data indicate that the measured wind velocity increases with the building elevation.

Table 5.1 The measured wind velocities at different elevations of the building.

Building elevation, x (m)	Wind velocity, y (m/sec)
10	2.2
15	4.6
20	4.2
25	7.0
30	6.6
35	9.2

The data of the wind velocities at different elevations of the building as shown in Table 5.1 are plotted in Fig. 5.1. Due to the deviation of the measured wind velocities at different elevations, the function obtained from the interpolation method can not sufficiently represent the realistic behavior of the phenomenon as shown in the figure. Figure 5.1 also shows two approximate functions fitted by using eye-ball. These approximate functions do not represent the best fitted function for such set of data. The least-squares regression methods that are explained in this chapter provide the best fitted function for any set of data.

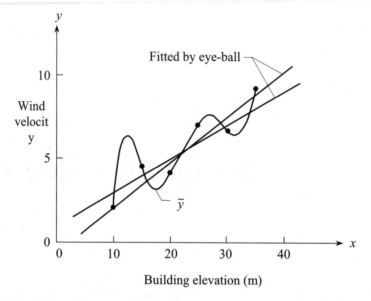

Figure 5.1 Data of the wind velocity that varies with the building elevation fitted by using eye-ball and interpolating function.

Several least-squares regression methods are presented herein. These methods are: (1) the linear regression, (2) the linear regression for nonlinear data, (3) the polynomial regression, and (4) the multiple regression. These methods are explained in details with illustrated examples and computer programs.

5.2 Linear Regression

Linear regression is a simple method for fitting a set of data that tends to vary linearly. Figure 5.2 shows a set of data with n data points, $x_i, y_i, i = 1, 2, ..., n$. The fitted function is assumed in the form,

$$g(x) = a_0 + a_1 x \qquad (5.1)$$

where a_0 and a_1 are the unknown coefficients to be determined later.

As shown in Fig. 5.2, the data y_i at a typical location x_i differs from the value of the fitted function $g(x)$ as $d(x_i)$. The idea behind the least-squares method is to minimize the squares of the differences between the data values and the function values. The total error the occurs from all n data points is

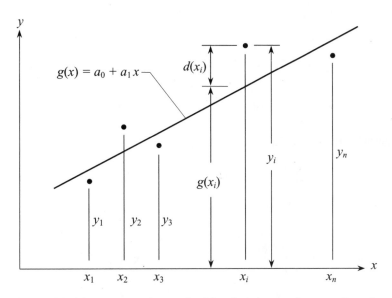

Figure 5.2 Linear regression method for data that tend to vary linearly.

$$E = \sum_{i=1}^{n} [d(x_i)]^2 \tag{5.2}$$

It is noted that by squaring the differences, the positive differences are not cancelled the negative differences. Equation (5.2) can also be written as

$$E = \sum_{i=1}^{n} [y_i - g(x_i)]^2 \tag{5.3}$$

By substituting Eq. (5.1) with $x = x_i$, Eq. (5.3) becomes

$$E = \sum_{i=1}^{n} [y_i - (a_0 + a_1 x_i)]^2 \tag{5.4}$$

The least-squares method is based on results from calculus demonstrating that a function has a minimum value when its partial derivatives are zero. Thus, by performing minimization of the function E in Eq. (5.4) with respect to the unknown coefficients a_0 and a_1, two equations are obtained as

$$\frac{\partial E}{\partial a_0} = 0 \tag{5.5a}$$

and

$$\frac{\partial E}{\partial a_1} = 0 \tag{5.5b}$$

The partial derivative in Eq. (5.5a) yields

$$2\sum_{i=1}^{n} [y_i - (a_0 + a_1 x_i)](-1) = 0$$

$$\sum_{i=1}^{n} y_i - \sum_{i=1}^{n} a_0 - \sum_{i=1}^{n} a_1 x_i = 0$$

$$n a_0 + \left(\sum_{i=1}^{n} x_i\right) a_1 = \sum_{i=1}^{n} y_i \tag{5.6a}$$

Similarly, the partial derivative in Eq. (5.5b) gives

$$2\sum_{i=1}^{n} [y_i - (a_0 + a_1 x_i)](-x_i) = 0$$

$$\sum_{i=1}^{n} x_i y_i - \sum_{i=1}^{n} a_0 x_i - \sum_{i=1}^{n} a_1 x_i^2 = 0$$

$$\left(\sum_{i=1}^{n} x_i\right) a_0 + \left(\sum_{i=1}^{n} x_i^2\right) a_1 = \sum_{i=1}^{n} x_i y_i \tag{5.6b}$$

Equations (5.6a) and (5.6b) can be written together in matrix form as

$$\begin{bmatrix} n & \sum_{i=1}^{n} x_i \\ \sum_{i=1}^{n} x_i & \sum_{i=1}^{n} x_i^2 \end{bmatrix} \begin{Bmatrix} a_0 \\ a_1 \end{Bmatrix} = \begin{Bmatrix} \sum_{i=1}^{n} y_i \\ \sum_{i=1}^{n} x_i y_i \end{Bmatrix} \tag{5.7}$$

By using the Cramer's rule explained in section 3.2, the two unknown coefficients a_0 and a_1 can be determined as

$$a_0 = \frac{\left(\sum_{i=1}^{n} y_i\right)\left(\sum_{i=1}^{n} x_i^2\right) - \left(\sum_{i=1}^{n} x_i y_i\right)\left(\sum_{i=1}^{n} x_i\right)}{n\left(\sum_{i=1}^{n} x_i^2\right) - \left(\sum_{i=1}^{n} x_i\right)^2} \tag{5.8a}$$

$$a_1 = \frac{n\left(\sum_{i=1}^{n} x_i y_i\right) - \left(\sum_{i=1}^{n} x_i\right)\left(\sum_{i=1}^{n} y_i\right)}{n\left(\sum_{i=1}^{n} x_i^2\right) - \left(\sum_{i=1}^{n} x_i\right)^2} \tag{5.8b}$$

The fitted function $g(x)$ is then obtained by substituting the computed coefficients a_0 and a_1 in Eqs. (5.8a-b) back into Eq. (5.1).

Example 5.1 Employ the linear regression method to establish the best fitted function for the set of the wind velocity and building elevation data in Table 5.1.

     The data in Table 5.1 are used to calculate values in Table 5.2 for determining the coefficients a_0 and a_1 of the fitted function according to Eqs. (5.8a-b) as

$$a_0 = \frac{(33.8)(3,475)-(870)(135)}{6(3,475)-(135)^2} = \frac{5}{2,625} = 0.001904$$

$$a_1 = \frac{6(870)-(135)(33.8)}{6(3,475)-(135)^2} = \frac{657}{2,625} = 0.250286$$

Table 5.2 Values required for calculating the two coefficients a_0 and a_1 of the fitted function $g(x)$ in Example 5.1.

x_i	y_i	x_i^2	$x_i y_i$
10	2.2	100	22
15	4.6	225	69
20	4.2	400	84
25	7.0	625	175
30	6.6	900	198
35	9.2	1,225	322
$\sum$ 135	33.8	3,475	870

Then, the fitted function $g(x)$ according to Eq. (5.1) is

$$g(x) = 0.001904 + 0.250286\,x \qquad (5.9)$$

Distribution of the fitted function $g(x)$ is plotted to compare with the given data as shown in Fig. 5.3.

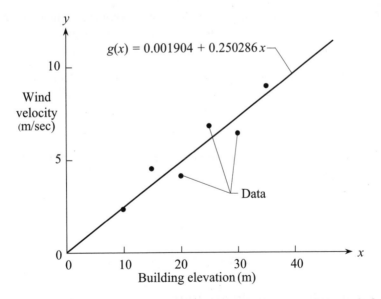

Figure 5.3 Comparison between the fitted function and data of the wind velocity that varies with building elevation.

Example 5.2 Develop a linear regression computer program to determine the two coefficients of the fitted function for n data points. Validate the program by using the data in Example 5.1.

Figure 5.4 shows a linear regression computer program for determining the two coefficients of the fitted function for n data points, x_i, y_i, $i = 1, 2, \ldots, n$. Table 5.3 shows the input data required by the program and output of the coefficients a_0 and a_1 for the set of data in Example 5.1.

Fortran

```
       PROGRAM  LREGRES
C.....A LINEAR REGRESSION PROGRAM
       DIMENSION  X(100), Y(100)
C.....READ NUMBER OF DATA SETS, DATA OF X AND Y:
       READ(5,*)  N
       DO 10  I=1,N
       READ(5,*)  X(I), Y(I)
    10 CONTINUE
C.....COMPUTE SUMMATION TERMS:
       SUMX  = 0.
       SUMY  = 0.
       SUMX2 = 0.
       SUMXY = 0.
       DO 20  I=1,N
       SUMX  = SUMX  + X(I)
       SUMY  = SUMY  + Y(I)
       SUMX2 = SUMX2 + X(I)*X(I)
       SUMXY = SUMXY + X(I)*Y(I)
    20 CONTINUE
C.....SOLVE FOR COEFFICIENTS:
       DET = N*SUMX2 - SUMX*SUMX
       A0  = (SUMY*SUMX2 - SUMXY*SUMX)/DET
       A1  = (N*SUMXY - SUMX*SUMY)/DET
       WRITE(6,100)  A0, A1
   100 FORMAT(' COEFFICIENT A0  =', E14.6,
      *     /, ' COEFFICIENT A1  =', E14.6)
       STOP
       END
```

MATLAB

```
%   PROGRAM LREGRES
%   A LINEAR REGRESSION PROGRAM
%   READ NUMBER OF DATA SETS, DATA OF X AND Y:
n = input('\nEnter number of data: ');
for irow = 1:n
    x(irow) = input('\nEnter value of x:  ');
    y(irow) = input('Enter value of y:  ');
end
%   COMPUTE SUMMATION TERMS:
sumx  = 0.0;
sumy  = 0.0;
sumx2 = 0.0;
sumxy = 0.0;
for i = 1:n
    sumx  = sumx  + x(i);
    sumy  = sumy  + y(i);
    sumx2 = sumx2 + x(i)*x(i);
    sumxy = sumxy + x(i)*y(i);
end
%   SOLVE FOR COEFFICIENTS:
det = n*sumx2 - sumx*sumx;
A0 = (sumy*sumx2 - sumxy*sumx)/det;
A1 = (n*sumxy - sumx*sumy)/det;
fprintf('\nCOEFFICIENT A0 = %14.6e', A0)
fprintf('\nCOEFFICIENT A1 = %14.6e', A1)
```

Figure 5.4 Linear regression computer program for determining the two coefficients a_0 and a_1 of the fitted linear function from n data points.

Table 5.3 Input and output data of the linear regression computer program in Fig. 5.4 for the set of data in Example 5.1.

Input data		Output data		
6		COEFFICIENT A0	=	.190375E-02
10	2.2	COEFFICIENT A1	=	.250286E+00
15	4.6			
20	4.2			
25	7.0			
30	6.6			
35	9.2			

5.3 Linear Regression for Nonlinear Data

Most of data obtained from experiment distribute nonlinearly. These data may be fitted by using the polynomial regression method that will be explained in the next section. The polynomial regression procedure is similar to that of the linear regression except more unknown coefficients are needed to determine. There are sets of data, however, that distribute in some certain patterns. One of the patterns is in the form of the power equation as shown in Fig. 5.5(a). The general form of the power equation is

$$\bar{y} = a\bar{x}^b \tag{5.10}$$

The linear regression method presented in the preceding section can be employed to fit such set of data. The advantage for using the linear regression method to fit the data distribute in the pattern of the power equation is that only two unknown coefficients are needed to determine. The procedure starts from linearlization of the power Eq. (5.10) by taking its logarithm to yield

$$\log \bar{y} = \log a + b \log \bar{x} \tag{5.11}$$

The result is in the form

$$y = a_0 + a_1 x \tag{5.12}$$

which is in the same form as Eq. (5.1) as shown in Fig. 5.5(b) where

$$x = \log \bar{x} \tag{5.13a}$$

$$y = \log \bar{y} \tag{5.13b}$$

$$a_0 = \log a \tag{5.13c}$$

$$a_1 = b \tag{5.13d}$$

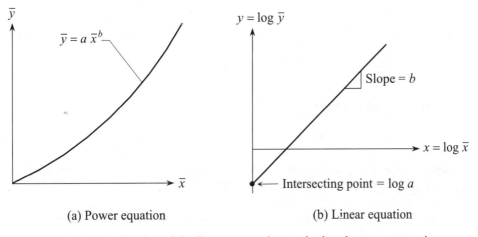

(a) Power equation   (b) Linear equation

Figure 5.5 Application of the linear regression method to the power equation.

With the fitted function in the linearized form of Eq. (5.12), the linear regression method as explained in section 5.2 can be applied. After the two unknowns coefficients a_0 and a_1 are solved, Eqs. (5.13c-d) are used to obtain the coefficients a and b of the power Eq. (5.10).

$\bar{x}$	1	2	3	4	5
$\bar{y}$	0.1	0.7	0.9	1.7	2.1

Example 5.3 Apply the linear regression method to derive the coefficients a and b of the power equation,

$$\bar{y} = a\bar{x}^b$$

by using the set of data given in the table.

Table 5.4 shows the data and their logarithmic values. These values are used to determine the two coefficients a_0 and a_1 of Eq. (5.12) according to Eq. (5.8a-b) as follows,

$$a_0 = \frac{(-0.649)(1.170)-(0.294)(2.079)}{5(1.170)-(2.079)^2} = \frac{-1.371}{1.528} = -0.897$$

$$a_1 = \frac{5(0.294)-(2.079)(-0.649)}{5(1.170)-(2.079)^2} = \frac{2.819}{1.528} = 1.845$$

Table 5.4 The given data and their logarithmic values for applying linear regression method to derive the power equation in Example 5.3.

$\bar{x}_i$	$\bar{y}_i$	$x_i = \log \bar{x}_i$	$y_i = \log \bar{y}_i$	x_i^2	$x_i y_i$
1	0.1	0.000	-1.000	0.000	0.000
2	0.7	0.301	-0.155	0.091	-0.047
3	0.9	0.477	-0.046	0.228	-0.022
4	1.7	0.602	0.230	0.362	0.138
5	2.1	0.699	0.322	0.489	0.225
Σ		2.079	-0.649	1.170	0.294

Then, the linearized equation in form of Eq. (5.12) is

$$y = -0.897 + 1.845x \tag{5.14}$$

The computed coefficients a_0 and a_1 are used to determine the coefficients a and b of the power equation according to Eqs. (5.13c-d) as

$$a_0 = \log a \quad \rightarrow \quad a = 0.127 \tag{5.15a}$$

$$a_1 = b \quad \rightarrow \quad b = 1.845 \tag{5.15b}$$

Thus, the fitted power equation is

$$\bar{y} = 0.127\bar{x}^{1.845} \tag{5.16}$$

Distribution of the fitted power Eq. (5.16) is plotted and compared with the given data points as shown in Fig. 5.6.

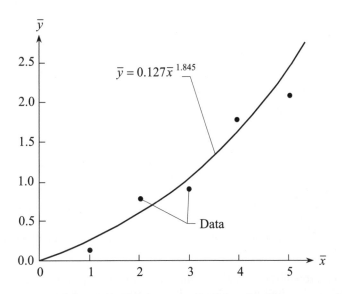

Figure 5.6 Comparison of the fitted power equation using linear regression with the given data in Example 5.3.

There are other types of equations that the linear regression method can be applied to fit a given set of data that distribute in certain patterns. One of the patterns is in form of the exponential model as

$$\bar{y} = a e^{b\bar{x}} \tag{5.17}$$

The exponential model in Eq. (5.17) can also be linearized by taking its natural logarithm to yield

$$\ln \bar{y} = \ln a + b\bar{x} \ln e$$

Since $\ln e = 1$, then

$$\ln \bar{y} = \ln a + b\bar{x} \tag{5.18}$$

Equation (5.18) can be written in the linear form as

$$y = a_0 + a_1 x \tag{5.12}$$

where

$$y = \ln \bar{y} \quad ; \quad x = \bar{x}$$
$$a_0 = \ln a \quad ; \quad a_1 = b \tag{5.19}$$

Figure 5.7(a) shows the distribution of the exponential model according to Eq. (5.17) while Fig. 5.17(b) shows the linear distribution of Eq. (5.12). The unknown coefficients a_0 and a_1 of the linearized Eq. (5.12) and the coefficients a and b can be determined in the same way as those of the power equation.

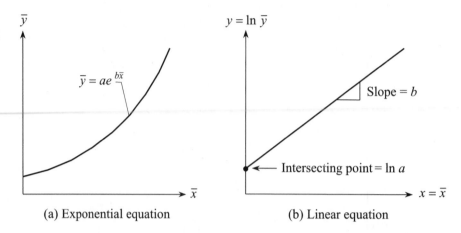

(a) Exponential equation   (b) Linear equation

Figure 5.7 Application of the linear regression to the exponential equation.

The saturation-growth-rate equation is another nonlinear equation which is often used to fit a growth data behavior with a limiting condition. The equation is in the form

$$\bar{y} = a\,\frac{\bar{x}}{b+\bar{x}} \tag{5.20}$$

or

$$\frac{b+\bar{x}}{a\bar{x}} = \frac{1}{\bar{y}}$$

$$\frac{1}{\bar{y}} = \frac{1}{a} + \frac{b}{a}\frac{1}{\bar{x}} \tag{5.21}$$

Equation (5.21) can be written in the linear form as

$$y = a_0 + a_1 x \tag{5.12}$$

where

$$y = 1/\bar{y} \quad ; \quad x = 1/\bar{x}$$

$$a_0 = 1/a \quad ; \quad a_1 = b/a \tag{5.22}$$

Figure 5.8(a-b) shows the distributions the saturation-growth-rate equation and the linear equation with its slope and intersection point. Again, the unknown coefficients a_0 and a_1 of the linearized Eq. (5.12) and the coefficients a and b can be determined in the same way as those of the power and exponential equations.

The application of the linear regression method to establish functions for fitting sets of nonlinear data by using the power, exponential and saturation-growth-rate equations is simple. The fitted functions are accurate when the data distributions are in the patterns that could be represented by such equations. If the data distributions are in other patterns, different forms of the fitted functions should be used. The following section explains the polynomial regression method to derived fitted functions for the more general data patterns.

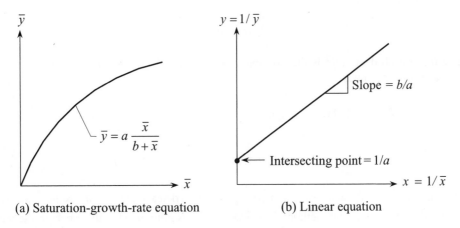

(a) Saturation-growth-rate equation   (b) Linear equation

Figure 5.8 Application of the linear regression to the saturation-growth-rate equation.

5.4 Polynomial Regression

The polynomial regression method is suitable to derive a function for fitting a set of data scattered in a polynomial pattern. Figure 5.9 shows a typical fitted polynomial for a set of n data points of x_i, y_i, $i = 1, 2, \ldots, n$. A polynomial of order m can be written in the form

$$g(x) = a_0 + a_1 x + a_2 x^2 + \ldots + a_m x^m \tag{5.23}$$

where $a_0, a_1, a_2, \ldots, a_m$ are the unknown coefficients. These coefficients are determined by first writing the total error E as the sum of the error squares for all data points as

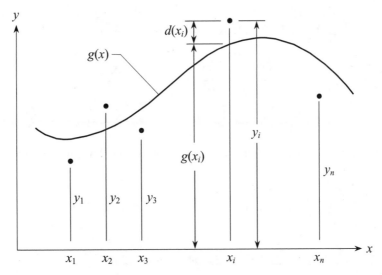

Figure 5.9 Fitting n data points by the polynomial regression method.

$$E = \sum_{i=1}^{n} [d(x_i)]^2 \tag{5.24}$$

Since the error at each data point is the difference between the data value and the polynomial value, then, the total error E becomes

$$E = \sum_{i=1}^{n} [y_i - g(x_i)]^2$$

$$E = \sum_{i=1}^{n} [y_i - (a_0 + a_1 x + a_2 x^2 + ... + a_m x^m)]^2 \tag{5.25}$$

The $m+1$ unknown coefficients of $a_0, a_1, a_2, ..., a_m$ are determined in the same way as those explained in the linear regression method. The total error E is minimized with respect to the unknown coefficients leading to a set of $m+1$ simultaneous equations as

$$\left.\begin{array}{c} \dfrac{\partial E}{\partial a_0} = 0 \\[2mm] \dfrac{\partial E}{\partial a_1} = 0 \\[2mm] \dfrac{\partial E}{\partial a_2} = 0 \\[1mm] \vdots \qquad \vdots \\[2mm] \dfrac{\partial E}{\partial a_m} = 0 \end{array}\right\} \quad m+1 \text{ equations} \tag{5.26}$$

For example, the minimization of the total error E with respect to the unknown a_0 in the first equation of Eq. (5.26) yields

$$2\sum_{i=1}^{n} [y_i - (a_0 + a_1 x_i + a_2 x_i^2 + ... + a_m x_i^m)](-1) = 0$$

$$\sum_{i=1}^{n} y_i - \sum_{i=1}^{n} a_0 - \sum_{i=1}^{n} a_1 x_i - \sum_{i=1}^{n} a_2 x_i^2 - ... - \sum_{i=1}^{n} a_m x_i^m = 0$$

$$n a_0 + \left(\sum_{i=1}^{n} x_i\right) a_1 + \left(\sum_{i=1}^{n} x_i^2\right) a_2 + ... + \left(\sum_{i=1}^{n} x_i^m\right) a_m = \sum_{i=1}^{n} y_i$$

Similarly, the minimization of the total error E with respect to the unknown a_1 in the second equation of Eq. (5.26) yields

$$2\sum_{i=1}^{n} [y_i - (a_0 + a_1 x_i + a_2 x_i^2 + ... + a_m x_i^m)](-x_i) = 0$$

$$\sum_{i=1}^{n} x_i y_i - \sum_{i=1}^{n} a_0 x_i - \sum_{i=1}^{n} a_1 x_i^2 - \sum_{i=1}^{n} a_2 x_i^3 - ... - \sum_{i=1}^{n} a_m x_i^{m+1} = 0$$

$$\left(\sum_{i=1}^{n} x_i\right) a_0 + \left(\sum_{i=1}^{n} x_i^2\right) a_1 + \left(\sum_{i=1}^{n} x_i^3\right) a_2 + ... + \left(\sum_{i=1}^{n} x_i^{m+1}\right) a_m = \sum_{i=1}^{n} x_i y_i$$

The minimization of the other equations in Eq. (5.26) can be performed in the same fashion leading to a set of $m+1$ simultaneous equations in the matrix form as

$$\begin{bmatrix} n & \sum_{i=1}^{n} x_i & \sum_{i=1}^{n} x_i^2 & \cdots & \sum_{i=1}^{n} x_i^m \\ \sum_{i=1}^{n} x_i & \sum_{i=1}^{n} x_i^2 & \sum_{i=1}^{n} x_i^3 & \cdots & \sum_{i=1}^{n} x_i^{m+1} \\ \sum_{i=1}^{n} x_i^2 & \sum_{i=1}^{n} x_i^3 & \sum_{i=1}^{n} x_i^4 & \cdots & \sum_{i=1}^{n} x_i^{m+2} \\ \vdots & \vdots & \vdots & \ddots & \vdots \\ \sum_{i=1}^{n} x_i^m & \sum_{i=1}^{n} x_i^{m+1} & \sum_{i=1}^{n} x_i^{m+2} & \cdots & \sum_{i=1}^{n} x_i^{2m} \end{bmatrix} \begin{Bmatrix} a_0 \\ a_1 \\ a_2 \\ \vdots \\ a_m \end{Bmatrix} = \begin{Bmatrix} \sum_{i=1}^{n} y_i \\ \sum_{i=1}^{n} x_i y_i \\ \sum_{i=1}^{n} x_i^2 y_i \\ \vdots \\ \sum_{i=1}^{n} x_i^m y_i \end{Bmatrix} \quad (5.27)$$

The square $(m+1)\times(m+1)$ matrix on the left-hand side and the $m+1$ vector on the right-hand side of Eq. (5.27) are known. Thus, the $m+1$ unknown coefficients of $a_0, a_1, a_2, ..., a_m$ can be solved from the set of simultaneous Eq. (5.27) by using any direct method explained in chapter 3.

Example 5.4 Develop a computer program to derive a polynomial of order m for fitting a set of n data. Then, apply the program to establish a third-order polynomial for fitting the data of the water specific heat that varies with the temperature as shown in Table 5.5. Plot to compare distribution of the fitted polynomial with the given data.

Table 5.5 Data of water specific heat c_p (kJ/kg·°C) that varies with the temperature T(°C).

T	c_p	T	c_p
0	1.00762	55	0.99919
5	1.00392	60	0.99967
10	1.00153	65	1.00024
15	1.00000	70	1.00091
20	0.99907	75	1.00167
25	0.99852	80	1.00253
30	0.99826	85	1.00351
35	0.99818	90	1.00461
40	0.99828	95	1.00586
45	0.99849	100	1.00721
50	0.99878		

Figure 5.10 shows a polynomial regression computer program for determining the unknown coefficients of the fitted m-order polynomial for n data points. The program generates a set of $m+1$ simultaneous equations that is solved by calling the subroutine GAUSS explained in chapter 3. Details of the subroutine GAUSS (not included herein) is shown in Fig. 3.2.

Fortran

```
      PROGRAM PREGRES
C.....A POLYNOMIAL REGRESSION PROGRAM
      DIMENSION X(100), Y(100)
      DIMENSION A(10,10), B(10), XX(10)
C.....READ NUMBER OF DATA SETS, DATA OF X AND Y:
      OPEN(UNIT=7, FILE='INPUT.DAT', STATUS='OLD')
      READ(7,*) N
      DO 10 I=1,N
      READ(7,*) X(I), Y(I)
   10 CONTINUE
C.....READ ORDER OF POLYNOMIAL NEEDED:
      READ(7,*) M
      DO 50 IR=1,10
      B(IR) = 0.
      DO 50 IC=1,10
      A(IR,IC) = 0.
   50 CONTINUE
C.....COMPUTE SQUARE MATRIX ON LHS AND
C.....VECTOR ON RHS OF SYSTEM EQUATIONS:
      DO 100 IR=1,M+1
      DO 200 IC=1,M+1
      K = IR + IC - 2
      DO 300 I=1,N
      A(IR,IC) = A(IR,IC) + X(I)**K
  300 CONTINUE
  200 CONTINUE
      DO 400 I=1,N
      B(IR) = B(IR) + Y(I)*(X(I)**(IR-1))
  400 CONTINUE
  100 CONTINUE
C.....CALL SUBROUTINE FOR SOLVING SYSTEM EQS:
      MP1 = M + 1
      CALL GAUSS(MP1, A, B, XX)
C.....PRINT OUT POLYNOMIAL COEFFICIENTS:
      WRITE(6,500)

  500 FORMAT(/,
     *  ' COEFFICIENTS OF FITTED POLYNOMIAL ARE:')
      DO 600 I=1,M+1
      IM1 = I-1
      WRITE(6,700) IM1, XX(I)
  700 FORMAT(' A(', I1, ') =', E13.7)
  600 CONTINUE
      STOP
      END
```

MATLAB

```
%   PROGRAM PREGRES
%   A POLYNOMIAL REGRESSION PROGRAM
%   READ NUMBER OF DATA SETS, DATA OF X AND Y:
fid = fopen('input.dat', 'r');
n   = fscanf(fid,'%f',1);
[x, y] = textread('input.dat','%f %7.5f', ...
'headerlines',1);
fclose(fid);
%   READ ORDER OF POLINOMIAL NEEDED:
m = input('Enter order of polinomial:    ');
for ir = 1:m+1
    b(ir) = 0.0;
    for ic = 1:m+1
        a(ir,ic) = 0.0;
    end
end
%   COMPUTE SQUARE MATRIX ON LHS AND
%   VECTOR ON RHS OF SYSTEM EQUATIONS:
for ir = 1:m+1
    for ic = 1:m+1
        k = ir + ic -2;
        for i = 1:n
            a(ir,ic) = a(ir,ic) + x(i)^k;
        end
    end
    for i = 1:n
        b(ir) = b(ir) + y(i)*(x(i)^(ir-1));
    end
end
%   CALL SUBROUTINE FOR SOLVING SYSTEM EQS:
mp1 = m + 1;
xx = gauss(mp1, a, b);
%   PRINT OUT POLINOMIAL COEFFICIENTS:
fprintf('\nCOEFFICIENTS OF FITTED POLINOMIAL ARE:')
for i = 1:m+1
    im1 = i-1;
    fprintf('\n A(%1d) = %13.7e', im1, xx(i));
end
```

Figure 5.10 Polynomial regression computer program for determining the $m+1$ coefficients of the fitted m-order polynomial from n data points. The program calls the subroutine GAUSS shown in Fig. 3.2.

The computer program starts from reading an input file that contains the total of $n = 21$ data for this example. The program then establishes a third-order polynomial ($m = 3$) that has $m+1 = 4$ unknown coefficients. These coefficients are solved from a set of $m+1 = 4$ simultaneous equations as shown below

$$\begin{bmatrix} 0.2100000E+02 & 0.1050000E+04 & 0.7175000E+05 & 0.5512500E+07 \\ 0.1050000E+04 & 0.7175000E+05 & 0.5512500E+07 & 0.4516662E+09 \\ 0.7175000E+05 & 0.5512500E+07 & 0.4516662E+09 & 0.3854156E+11 \\ 0.5512500E+07 & 0.4516662E+09 & 0.3854156E+11 & 0.3382122E+13 \end{bmatrix} \begin{Bmatrix} a_0 \\ a_1 \\ a_2 \\ a_3 \end{Bmatrix}$$

$$= \begin{Bmatrix} 0.2102805E+02 \\ 0.1051999E+04 \\ 0.7195141E+05 \\ 0.5531869E+07 \end{Bmatrix}$$

The computed 4 coefficients are

$$a_0 = 0.1006448E+01 \qquad a_1 = -0.4988565E-03$$
$$a_2 = 0.8460584E-05 \qquad a_3 = -0.3457691E-07 \qquad (5.28)$$

Thus, the fitted third-order polynomial is

$$c_p = (0.1006448E+01) + (-0.4988565E-03)\,T$$
$$+ (0.8460584E-05)\,T^2 + (-0.3457691E-07)\,T^3 \qquad (5.29)$$

Distribution of the fitted polynomial in Eq. (5.29) is compared with the given data as shown in Fig. 5.11.

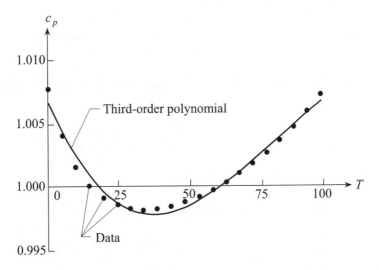

Figure 5.11 Comparison of the fitted third-order polynomial and the given data of the water specific heat that varies with temperature.

5.5 MATLAB Functions for Least-Squares Regression

One of the MATLAB functions used in the least-squares regression method is the `polyfit` function. The command for using the function is

$$p = polyfit(x,y,n)$$

where `x` and `y`   are the data
      `p`   is the vector containing the computed coefficients
      `n`   is the order of the fitted polynomial

Example 5.5 Employ the `polyfit` function to determine the coefficients of the linear function for fitting the data of the wind velocity and building height as given in Table 5.1.

The data in Table 5.1 are first assign to store in the variables x and y by using the commands

```
>> x = [10 15 20 25 30 35];
>> y = [2.2 4.6 4.2 7.0 6.6 9.2];
```

The `polyfit` function is then applied by using the first-order polynomial for linear regression as

```
>> a = polyfit(x,y,1)

a =

     0.2503    0.0019
```

The computed coefficients are 0.2503 and 0.0019. The first coefficient of 0.2503 is the slope of the fitted function while the second coefficient of 0.0019 is the intersection point on the y-axis. The computed coefficients are identical to those obtained in Example 5.1.

Example 5.6   Employ the `polyfit` function to determine the coefficients of the third-order polynomial for fitting the water specific heat that varies with the temperature by using the set of data in Table 5.5.

Similar to Example 5.5, the data in Table 5.5 are first assigned to store in the variables x and y by using the commands

```
>> x = [0 5 10 15 20 25 30 35 40 45 50 55 60 65 70 75 80 85 90 95 100];
>> y = [1.00762 1.00392 1.00153 1 0.99907 0.99852 0.99826 0.99818 ...
0.99828 0.99849 0.99878 0.99919 0.99967 1.00024 1.00091 1.00167 ...
1.00253 1.00351 1.00461 1.00586 1.00721];
```

Then, the `polyfit` function with the polynomial order of 3 is used to fit the data,

```
>> format long
>> a = polyfit(x,y,3)

a =

  -0.00000003453390    0.00000845381042   -0.00049857114790
1.00644556747600
```

The computed coefficients are identical to those obtained in Example 5.4.

Another valuable function that can be used with the `polyfit` function is `polyval`. The function determines a value between data points from the fitted function. The command for using the `polyval` function is

$$z = \text{polyval}(p,x)$$

where z is the computed value of the function at x location
        p is the vector containing the coefficients of the fitted polynomial
        x is the location for determining the function value

For example, the coefficients of the fitted polynomial in Example 5.6 can be used to determine the polynomial value at $x = 47$ by the commands,

```
>> format short
>> z = polyval(a,47)

z =

    0.9981
```

5.6 Multiple Regression

The preceding sections explain the least-squares regression methods to establish the fitted functions for a set of data. In these methods, the fitted function y depends on a single variable x. For practical problems, the fitted function y may depend on many independent variable x. As an example of a drag force measurement on a car surface in a wind tunnel, the wind pressure y on the car surface varies with the car length x_1, i.e.,

$$y \;=\; y(x_1) \tag{5.30}$$

In addition, the wind pressure varies with the car width x_2, so that

$$y \;=\; y(x_1, x_2) \tag{5.31}$$

Furthermore, the wind pressure on the car surface also depends on the wind speed x_3,

$$y \;=\; y(x_1, x_2, x_3) \tag{5.32}$$

Thus, in general, the fitted function y are dependent of many variables $(x_1, x_2, x_3, ..., x_k)$ such that it can be written as

$$y \;=\; y(x_1, x_2, x_3, ..., x_k) \tag{5.33}$$

where k is the number of the independent variables.

5.6.1 Linear

If the data pattern of each independent variable x_j, $j = 1, 2, ..., k$ (for k independent variables) distributes linearly, the fitted g function can be assumed in the form,

$$g \;=\; a_0 + a_1 x_1 + a_2 x_2 + ... + a_k x_k \tag{5.34}$$

where a_j, $j = 0, 1, 2, ..., k$ are the unknown coefficients to be determined. The least-squares method is applied by first writing the total error E, which is the summation of the squares of the differences between the fitted function values and the data values, as

$$E \;=\; \sum_{i=1}^{n} [y_i - (a_0 + a_1 x_{1i} + a_2 x_{2i} + ... + a_k x_{ki})]^2 \tag{5.35}$$

The total error E in Eq. (5.35) is then minimized with respect to the unknown coefficients leading to a set of $k+1$ simultaneous equations as

$$\left.\begin{array}{c} \dfrac{\partial E}{\partial a_0} = 0 \\[2mm] \dfrac{\partial E}{\partial a_1} = 0 \\[2mm] \dfrac{\partial E}{\partial a_2} = 0 \\[1mm] \vdots \qquad \vdots \\[1mm] \dfrac{\partial E}{\partial a_k} = 0 \end{array}\right\} \; k+1 \text{ equations} \tag{5.36}$$

Details of the minimization process is omitted herein and left as an exercise. The minimization process in Eq. (5.36) leads to a set of $k+1$ simultaneous equations written in matrix form as

$$\begin{bmatrix} n & \sum\limits_{i=1}^{n} x_{1i} & \sum\limits_{i=1}^{n} x_{2i} & \cdots & \sum\limits_{i=1}^{n} x_{ki} \\[2mm] \sum\limits_{i=1}^{n} x_{1i} & \sum\limits_{i=1}^{n} x_{1i} x_{1i} & \sum\limits_{i=1}^{n} x_{1i} x_{2i} & \cdots & \sum\limits_{i=1}^{n} x_{1i} x_{ki} \\[2mm] \sum\limits_{i=1}^{n} x_{2i} & \sum\limits_{i=1}^{n} x_{1i} x_{2i} & \sum\limits_{i=1}^{n} x_{2i} x_{2i} & \cdots & \sum\limits_{i=1}^{n} x_{2i} x_{ki} \\[2mm] \vdots & \vdots & \vdots & \ddots & \vdots \\[2mm] \sum\limits_{i=1}^{n} x_{ki} & \sum\limits_{i=1}^{n} x_{1i} x_{ki} & \sum\limits_{i=1}^{n} x_{2i} x_{ki} & \cdots & \sum\limits_{i=1}^{n} x_{ki} x_{ki} \end{bmatrix} \begin{Bmatrix} a_0 \\ a_1 \\ a_2 \\ \vdots \\ a_k \end{Bmatrix} = \begin{Bmatrix} \sum\limits_{i=1}^{n} y_i \\[2mm] \sum\limits_{i=1}^{n} x_{1i} y_i \\[2mm] \sum\limits_{i=1}^{n} x_{2i} y_i \\[2mm] \vdots \\[2mm] \sum\limits_{i=1}^{n} x_{ki} y_i \end{Bmatrix} \tag{5.37}$$

where the square $(k+1)\times(k+1)$ matrix on the left-hand side and the $(k+1)\times 1$ vector on the right-hand side of the Eq. (5.37) are known. Equation (5.37) is then solved to obtain the unknown coefficients $a_0, a_1, a_2, ..., a_k$ for $k+1$ values by using any direct method described in chapter 3.

Example 5.7 Develop a computer program using the multiple regression method to obtain a fitted function g for a set of data that varies with k independent variables. Test the program on a set of 6 data ($n = 6$) that varies with 2 independent variables ($k = 2$) generated from the equation

$$y = 1 + 2x_1 + 3x_2 \tag{5.38}$$

The 6 data generated from Eq. (5.38) are shown in the table below

i	x_{1i}	x_{2i}	y_i
1	0	0	1
2	0	1	4
3	1	0	3
4	1	2	9
5	2	1	8
6	2	2	11

The generated data from the tested function in the table above lead to the values in the matrices of the simultaneous Eq. (5.37) as

i	x_{1i}	x_{2i}	y_i	$x_{1i}x_{1i}$	$x_{1i}x_{2i}$	$x_{2i}x_{2i}$	$x_{1i}y_i$	$x_{2i}y_i$
1	0	0	1	0	0	0	0	0
2	0	1	4	0	0	1	0	4
3	1	0	3	1	0	0	3	0
4	1	2	9	1	2	4	9	18
5	2	1	8	4	2	1	16	8
6	2	2	11	4	4	4	22	22
Σ	6	6	36	10	8	10	50	52

So that the set of simultaneous equations is

$$\begin{bmatrix} 6 & 6 & 6 \\ 6 & 10 & 8 \\ 6 & 8 & 10 \end{bmatrix} \begin{Bmatrix} a_0 \\ a_1 \\ a_2 \end{Bmatrix} = \begin{Bmatrix} 36 \\ 50 \\ 52 \end{Bmatrix} \tag{5.39}$$

The set of simultaneous equations is solved to give the unknown coefficients of

$$a_0 = 1, \quad a_1 = 2, \quad a_2 = 3 \tag{5.40}$$

Thus, the fitted function according to Eq. (5.34) is

$$\begin{aligned} g &= a_0 + a_1 x_1 + a_2 x_2 \\ &= 1 + 2x_1 + 3x_2 \end{aligned} \tag{5.41}$$

which is identical to the tested function that was used to generate the data. The distribution of the fitted function is in form of a flat plane as shown in Fig. 5.12. With this fitted function, other values that are not on the given data points can be determined. For example, the value at $x_1 = x_2 = 1$ is

$$g(x_1 = 1, x_2 = 1) = 1 + 2(1) + 3(1) = 6$$

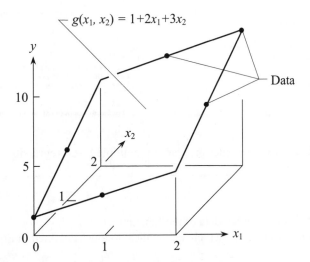

Figure 5.12 Distribution of the fitted function g obtained from the multiple linear regression of the data y that varies with the two independent variables x_1 and x_2.

Figure 5.13 shows a computer program for the multiple linear regression method. The program starts from reading the set of n input data points. Each data point contains the values $x_{1i}, x_{2i}, ..., x_{ki}$ of the k independent variables together with the value y_i. With these data, the program establishes a set of simultaneous equations in the form of Eq. (5.37). The set of simultaneous equations is then solved for the unknown coefficients $a_j, j = 0, 1, 2, ..., k$ by calling the subroutine GAUSS shown in Fig. 3.2.

Fortran

```
       PROGRAM  MREGRES
C.....A MULTIPLE LINEAR REGRESSION PROGRAM
       DIMENSION  X(100,10), Y(100)
       DIMENSION  A(10,10), B(10), XX(10)
C.....READ NUMBER OF DATA SETS N,
C.....NUMBER OF INDEPENDENT VARIABLES K,
C.....AND DATA OF X(I,K) AND Y(I):
       OPEN(UNIT=7, FILE='INPUT.DAT', STATUS='OLD')
       READ(7,*)  N, K
       DO 10  I=1,N
       READ(7,*)   (X(I,J), J=1,K), Y(I)
    10 CONTINUE
       DO 50  IR=1,10
       B(IR) = 0.
       DO 50  IC=1,10
       A(IR,IC) = 0.
    50 CONTINUE
C.....COMPUTE SQUARE MATRIX ON LHS AND
C.....VECTOR ON RHS OF SYSTEM EQUATIONS:
C.....CALL SUBROUTINE FOR SOLVING SYSTEM EQS:
       DO 100  I=1,N
       DO 200  IR=1,K+1
       IF(IR.EQ.1)  FR = 1.
       IF(IR.GT.1)  FR = X(I,IR-1)
       DO 300  IC=1,K+1
       IF(IC.EQ.1)  FC = 1.
       IF(IC.GT.1)  FC = X(I,IC-1)
       A(IR,IC) = A(IR,IC) + FR*FC
   300 CONTINUE
       B(IR) = B(IR) + FR*Y(I)
   200 CONTINUE
   100 CONTINUE
       KP1 = K + 1
       CALL GAUSS(KP1, A, B, XX)
C.....PRINT OUT COEFFICIENTS:
       WRITE(6,500)
   500 FORMAT(/,
      *  ' COEFFICIENTS OF FITTED FUNCTION ARE:')
       DO 600  I=1,K+1
       IM1 = I-1
       WRITE(6,700)  IM1, XX(I)
   700 FORMAT(' A(', I1, ') =', E13.7)
   600 CONTINUE
       STOP
       END
```

MATLAB

```
%  PROGRAM MREGRES
%  A MULTIPLE LINEAR REGRESSION PROGRAM
%  READ NUMBER OF DATA SETS N,
%  NUMBER OF INDEPENDENT VARIABLES K,
%  AND DATA OF X(I,K) AND Y(I):
fid = fopen('input.dat', 'r');
n = fscanf(fid,'%f',1);
k = fscanf(fid,'%f',1);
x = fscanf(fid,'%f',[k n]);
x = x';
y = fscanf(fid,'%f',[n]);
b = zeros(k+1,1);
a = zeros(k+1,k+1);
%  COMPUTE SQUARE MATRIX ON LHS AND
%  VECTOR ON RHS OF SYSTEM EQUATIONS:
%  CALL SUBROUTINE FOR SOLVING SYSTEM EQS:
for i = 1:n
    for ir = 1:k+1
        if ir == 1
            fr = 1.;
        end
        if ir > 1
            fr = x(i,ir-1);
        end
        for ic = 1:k+1
            if ic == 1
                fc = 1.;
            end
            if ic > 1
                fc = x(i,ic-1);
            end
            a(ir,ic) = a(ir,ic) + fr*fc;
        end
        b(ir) = b(ir) + fr*y(i);
    end
end
kp1 = k + 1;
xx = gauss(kp1, a, b);
%  PRINT OUT COEFFICIENTS:
fprintf('\nCOEFFICIENTS OF FITTED FUNCTION ARE:')
for i = 1:k+1
    im1 = i-1;
    fprintf('\n A(%1d) = %13.7e', im1, xx(i));
end
```

Figure 5.13 Computer program for multiple linear regression method.

Table 5.6 shows the 6 input data of the function y in Eq. (5.38) for Example 5.7. The table also shows the 3 computed coefficients which are identical to those obtained in Eq. (5.40).

Table 5.6 Input and output data of the multiple linear regression computer program
in Fig. 5.13 for the set of data in Example 5.7.

Input data			Output data
6	2		COEFFICIENTS OF FITTED FUNCTION ARE:
0	0	1	A(0) = .1000000E+01
0	1	4	A(1) = .2000000E+01
1	0	3	A(2) = .3000000E+01
1	2	9	
2	1	8	
2	2	11	

Understanding the behavior of the data pattern can improve the accuracy of the fitted function. For example, the power functions should be used if the data pattern distributes in the power form. In this case, the fitted function y may be assumed to depend on the 3 independent variables x_1, x_2 and x_3 as

$$\bar{y} = a \bar{x}_1^b \bar{x}_2^c \bar{x}_3^d \tag{5.42}$$

The coefficients a, b, c, d in Eq. (5.42) are unknowns to be determined. The fitted function in Eq. (5.42) is then linearized by taking its logarithm to yield

$$\log \bar{y} = \log a + b \log \bar{x}_1 + c \log \bar{x}_2 + d \log \bar{x}_3 \tag{5.43}$$

which can be written in the form

$$y = a_0 + a_1 x_1 + a_2 x_2 + a_3 x_3 \tag{5.44}$$

Herein, by matching Eqs. (5.44) and (5.43),

$$y = \log \bar{y} \quad ; \quad a_1 = b \quad ; \quad a_2 = c \quad ; \quad a_3 = d$$

$$a_0 = \log a \quad ; \quad x_1 = \log \bar{x}_1 \quad ; \quad x_2 = \log \bar{x}_2 \quad ; \quad x_3 = \log \bar{x}_3 \tag{5.45}$$

Minimization of total error E of the fitted function y in Eq. (5.44) leads to a set of 4 simultaneous equations in the form of Eq. (5.37) with the 4 unknowns of a_0, a_1, a_2 and a_3. Once these unknowns are determined, the coefficients a, b, c, d can then be obtained from Eq. (5.45). The example below shows the application of the multiple regression method with a fitted power equation for a practical problem.

Example 5.8 Insulating tiles are used as the thermal protection system for the space shuttle. These tiles, which have the size of approximately 20×20 cm and thickness of 6 to 10 cm, are placed beneath the shuttle body and wings. The tiles are subjected to high aerodynamic heating rate during descending at hypersonic speed through the earth atmosphere. Proper gaps must be provided between these tiles to allow their expansions from high temperature. Typical tile layout is shown in Fig. 5.14 with the gap width of w. With the layout of the tiles, there was a concern that a hot gas from the high-speed flow may pass through the gap and creates an excessive heating rate at location A on the adjacent tile as shown in the figure. An experiment was established in a high-speed wind tunnel to measure the heating rate, especially at that location. Table 5.7 shows the measured heating rate data at location A which varies with several flow parameters.

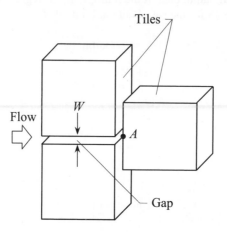

Figure 5.14 Typical layout of space shuttle tiles with gap.

Table 5.7 Measured heating rates at location A in Fig. 5.14 from hypersonic flow past shuttle tiles.

q	δ	w	Re	P
2.53	1.13	0.18	14.2×10^5	0.02
2.49	1.01	0.18	4.0×10^5	0.03
2.15	1.01	0.10	8.3×10^5	0.03
1.95	1.01	0.10	4.2×10^5	0.03
3.80	1.01	0.30	8.3×10^5	0.03
2.00	1.01	0.30	4.2×10^5	0.03
3.45	1.01	0.41	8.8×10^5	0.03
2.99	1.01	0.41	4.4×10^5	0.03

Study of the data distribution has shown that the heating rate at location A of the tile in Fig. 5.14 varies with the four parameters in the form

$$q \;=\; a\left(\frac{\delta}{w}\right)^{b} Re^{c}\, P^{d} \tag{5.46}$$

where q is the heating rate, δ is the boundary layer thickness, Re is the Reynolds number and P is the pressure.

To determine the coefficients a, b, c and d, Eq. (5.46) is first linearized by taking its logarithm to yield

$$\log q \;=\; \log a \,+\, b \log\left(\frac{\delta}{w}\right) \,+\, c \log Re \,+\, d \log P \tag{5.47}$$

which can be further written in the form

$$y \;=\; a_0 \,+\, a_1 x_1 \,+\, a_2 x_2 \,+\, a_3 x_3 \tag{5.48}$$

Then, the given data in Table 5.7 are transformed so that they can be used with the multiple linear regression program in Fig. 5.13 as follows

$y = \log q$	$x_1 = \log(\delta/w)$	$x_2 = \log(Re)$	$x_3 = \log(P)$
0.40312	0.79781	6.15229	-1.69897
0.39620	0.74905	5.60206	-1.52288
0.33244	1.00432	5.91908	-1.52288
0.29003	1.00432	5.62325	-1.52288
0.57978	0.52720	5.91908	-1.52288
0.30103	0.52720	5.62325	-1.52288
0.53782	0.39154	5.99445	-1.52288
0.47567	0.39154	5.64345	-1.52288

With the transformed data above, the computer program in Fig. 5.13 yields the values for the coefficients a_0, a_1, a_2 and a_3 as

$$a_0 = -0.400726 \quad ; \quad a_2 = 0.326507$$
$$a_1 = -0.272717 \quad ; \quad a_3 = 0.581139 \tag{5.49}$$

Then, by using the relations in Eq. (5.45), the unknown coefficients of the fitted function in Eq. (5.42) are

$$a = 0.397442 \quad ; \quad c = 0.326507$$
$$b = -0.272717 \quad ; \quad d = 0.581139 \tag{5.50}$$

Thus, the fitted function for the heating rate is

$$q = (0.397442)\left(\frac{\delta}{w}\right)^{-0.272717} Re^{0.326507} \, P^{0.581139} \tag{5.51}$$

The fitted function can be use to estimate the heating rate for other conditions different from the given data. For example, the heating rate at $\delta = 1.01$ cm, $w = 0.2$ cm, $Re = 4.0 \times 10^5$ and $P = 0.03$ is

$$q = (0.397442)\left(\frac{1.01}{0.2}\right)^{-0.272717} \left(4.0 \times 10^5\right)^{0.326507} (0.03)^{0.581139}$$

$$= 2.25$$

5.6.2 Polynomial

The multiple linear regression method for which the fitted function y varies linearly with the k independent variables of x_j, $j = 1, 2, ..., k$ was presented in the preceding section. The method can be extended to the cases when the data y varies nonlinearly in the form of polynomials. For example, if the data tend to vary with the cubic and quadratic distributions along x_1 and x_2, respectively, as shown in Fig. 5.15, the fitted function may be assumed in the form

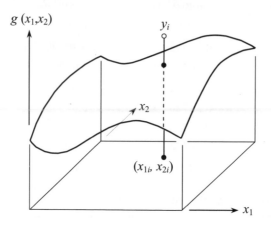

Figure 5.15 A fitted function g that varies with cubic and quadratic distributions along x_1 and x_2, and a typical data y_i at (x_{1i}, x_{2i}).

$$g(x_1, x_2) = \left(b_0 + b_1 x_1 + b_2 x_1^2 + b_3 x_1^3\right)\left(c_0 + c_1 x_2 + c_2 x_2^2\right) \tag{5.52}$$

where b_0, b_1, b_2, b_3 and c_0, c_1, c_2 are the unknown coefficients. The fitted function in Eq. (5.52) can be expanded to yield

$$g(x_1, x_2) = a_0 + a_1 x_1 + a_2 x_1^2 + a_3 x_1^3 + a_4 x_2 + a_5 x_1 x_2 + a_6 x_1^2 x_2$$
$$+ a_7 x_1^3 x_2 + a_8 x_2^2 + a_9 x_1 x_2^2 + a_{10} x_1^2 x_2^2 + a_{11} x_1^3 x_2^2 \tag{5.53}$$

where a_j, $j = 0, 1, \ldots, 11$ are the unknown coefficients from the products of b_j and c_j in Eq. (5.52). These unknown coefficients can be determined by the least-squares method as explained in the preceding sections. The total error E is first written as summation of the squares of the difference between the fitted function values and the data values as

$$E = \sum_{i=1}^{n}\left[y_i - \left(a_0 + a_1 x_1 + a_2 x_1^2 + \ldots + a_{11} x_1^3 x_2^2\right)\right]^2 \tag{5.54}$$

Minimization of the total error E with respect to the 12 unknown constants is then performed leading to a set of 12 simultaneous equations

$$\left.\begin{array}{l} \dfrac{\partial E}{\partial a_0} = 0 \\[2mm] \dfrac{\partial E}{\partial a_1} = 0 \\[2mm] \dfrac{\partial E}{\partial a_2} = 0 \\[1mm] \quad\vdots \quad\quad \vdots \\[2mm] \dfrac{\partial E}{\partial a_{11}} = 0 \end{array}\right\} \;\; 12 \text{ equations} \tag{5.55}$$

Detailed derivation of the simultaneous equations is omitted herein so that it is left for an exercise. The set of 12 simultaneous equations can be written in matrix form as

$$
\begin{bmatrix}
n & \sum_{i=1}^{n} x_{1i} & \sum_{i=1}^{n} x_{1i}^2 & \cdots & \sum_{i=1}^{n} x_{1i}^3 x_{2i}^2 \\[2mm]
\sum_{i=1}^{n} x_{1i} & \sum_{i=1}^{n} x_{1i}^2 & \sum_{i=1}^{n} x_{1i}^3 & \cdots & \sum_{i=1}^{n} x_{1i}^4 x_{2i}^2 \\[2mm]
\sum_{i=1}^{n} x_{1i}^2 & \sum_{i=1}^{n} x_{1i}^3 & \sum_{i=1}^{n} x_{1i}^4 & \cdots & \sum_{i=1}^{n} x_{1i}^5 x_{2i}^2 \\[2mm]
\vdots & \vdots & \vdots & \ddots & \vdots \\[2mm]
\sum_{i=1}^{n} x_{1i}^3 x_{2i}^2 & \sum_{i=1}^{n} x_{1i}^4 x_{2i}^2 & \sum_{i=1}^{n} x_{1i}^5 x_{2i}^2 & \cdots & \sum_{i=1}^{n} x_{1i}^6 x_{2i}^4
\end{bmatrix}
\begin{Bmatrix}
a_0 \\[2mm] a_1 \\[2mm] a_2 \\[2mm] \vdots \\[2mm] a_{11}
\end{Bmatrix}
=
\begin{Bmatrix}
\sum_{i=1}^{n} y_i \\[2mm]
\sum_{i=1}^{n} x_{1i} y_i \\[2mm]
\sum_{i=1}^{n} x_{1i}^2 y_i \\[2mm]
\vdots \\[2mm]
\sum_{i=1}^{n} x_{1i}^3 x_{2i}^3 y_i
\end{Bmatrix}
\tag{5.56}
$$

where the (12×12) matrix on the left-hand side and the (12×1) vector on the right-hand side of the simultaneous equations above are known. The simultaneous equations are then solved for the 12 unknown coefficients $a_0, a_1, a_2, \ldots, a_{11}$. The fitted function $g(x_1, x_2)$ is then obtained by substituting these computed coefficients back into Eq. (5.53).

The multiple regression methods explained in this section suggest that the form for the fitted function should be selected properly. Distribution behavior of the data must be studied prior to selecting an appropriate function for representing them. With the appropriate function, the least-squares method can be applied and the coefficients of the fitted function are then obtained straight forwardly. It should also be noted that the minimization process presented in this chapter is the fundamental of many advanced numerical methods for analyzing engineering problems.

5.7 Closure

The least-squares method to derive the best fitted function for a given set of data is presented. The method starts from an assumed function that may be in the form of polynomials, power, exponential or some other types. These functions contain the unknown coefficients that are to be determined. The total error between the fitted function and the actual data is then constructed. Such total error is defined as the summation of the squares of the differences between the function values and the data values. The total error is minimized with respect to the unknown coefficients leading to a set of simultaneous equations. The set of simultaneous equation is solved for the unknown coefficients. The computed coefficients are substituted back into the assumed function resulting in the best fitted function.

Several regression methods are presented to fit sets of data that distribute linearly and in the form of higher-order polynomials. The methods are called the linear and polynomial regression methods, respectively. The linear regression method for fitting sets of nonlinear data that distribute in the forms of the power, exponential and saturation-growth-rate equations is also presented. Application of the linear regression method to fit the nonlinear data in the mentioned forms helps reducing complexity of the formulation and the computational effort.

The regression methods above are used to derive the best fitted functions for problems that have only single independent variable. For problems with several independent variables, the multiple regression method is used. The multiple regression method follows the same procedure but leads to more unknown coefficients.

The essential step in the least-squares method is the minimization process. The process seeks the best fitted function by performing partial derivatives of the total error with respect to the problem unknowns and set them to zero. Understanding such process is essential to study other advanced numerical methods for analyzing practical engineering problems.

Exercises

1.   Use the least-square regression method to establish a linear function that best fits the data in the table below.

x	0	1	2	4	6	7	9	10	12
y	1	6	10	23	32	39	49	56	65

Compare the computed coefficients of fitted function with those obtained from the computer program in Fig. 5.4. Plot to compare distribution of the fitted function with the data.

2.   Use the least-square regression method to establish a linear function that best fits the data in the table below.

x	1,000	1,500	2,000	2,500	3,000	3,500
y	1,457	1,282	932	788	465	264

Compare the computed coefficients of fitted function with those obtained from the computer program in Fig. 5.4. Plot to compare distribution of the fitted function with the data.

3.   The data below represent the thermal expansion of a metal that depends on the temperature.

Temperature, °C	40	50	60	70	80	90	100	110
Expansion, %	1.1	1.3	1.3	1.5	1.7	1.9	2.0	2.3

Use the linear regression method to derive the fitted function and estimate values of the thermal expansion at 63°C and 95°C.

4. In the test of a car braking system, the stopping distances are found to depend on the car speeds as shown in the table below.

Velocity, km/hr	10	15	20	30	40	50	60	70	80
Stopping distance, m	5	9	15	18	22	30	35	38	43

Use the linear regression method to derive the best fitted function and estimate the stopping distance when the car speed is 65 km/hr.

5. The expressway toll-fee is determined according to the driving distance. The table below shows the exit numbers, distances and toll-fees.

Exit Number	Distance (Miles)	Toll-fee (Dollars)	Exit Number	Distance (Miles)	Toll-fee (Dollars)
2	10	0.65	8	68	2.45
3	24	1.00	8A	74	2.65
4	35	1.35	9	83	3.05
5	43	1.70	10	87	3.15
6	50	2.05	11	90	3.45
7	53	2.10	12	95	3.80
7A	59	2.25	13	115	4.60

If the expressway authority needs to construct two new exit ways which are 5A and 12A at the distances of 46 and 107 miles, respectively, determine the toll-fees at these new exit ways by using the linear regression method.

6. Distribution of the data given in the table is in the exponential form as

$$\bar{y} = a e^{b\bar{x}}$$

Apply the linear regression method for the nonlinear data to determine the constants a and b of the fitted exponential function above. Then, compare the distribution of fitted function with the data in the table.

$\bar{x}$	0	1	2	3	4	5	6
$\bar{y}$	1.2	1.9	2.1	2.8	3.2	4.1	4.9

7. Distribution of the data as shown in the table is in the saturation-growth-rate form as

$$\bar{y} = a \frac{\bar{x}}{b+\bar{x}}$$

Apply the least-squares method to determine the coefficients a and b of the saturation-growth-rate equation above. Repeat the problem by using the second-order polynomial regression. Plot to compare distributions of the two fitted functions with the given data.

$\bar{x}$	0	1	2	3	4	5	6
$\bar{y}$	0.02	0.17	0.29	0.34	0.41	0.43	0.47

8. The friction coefficient f of a laminar flow in a tube is found to vary with the Reynolds number Re in the form

$$f = a\, Re^b$$

Apply the linear regression method to the nonlinear data as shown in the table to determine the unknown constants a and b. Then, use the fitted function to estimate the fiction coefficients at the Reynolds numbers of 752 and 1,427

Re	500	1,000	1,500	2,000
f	0.0320	0.0160	0.0107	0.0080

9. The atmospheric pressure P (mm. Hg.) is found to vary with the altitude h (feet) above the sea level in the form

$$P = \alpha e^{-\beta h}$$

Determine the coefficients α and β of the fitted function above from the data in the table below by using the linear regression method. Then, use the fitted function to estimate the pressure at the altitude of 1,250 feet.

P (mm. Hg.)	29.9	29.4	29.0	28.4	27.7
h (feet)	0	500	1,000	1,500	2,000

10. The stress-strain (σ - ε) data obtained from testing a concrete column are shown in the table below. The data is best fitted by from the relation

$$\sigma = a\varepsilon e^{-b\varepsilon}$$

Apply the linear regression method for the nonlinear data to determine the coefficients a and b of the fitted equation above. Then, plot to compare distribution of the fitted function with the data.

σ (MPa)	7.1	9.7	11.8	14.4	16.7	19.0	20.7	19.7	18.5
$\varepsilon \times 10^3$	0.265	0.400	0.500	0.700	0.950	1.360	2.080	2.450	2.940

11. Use the polynomial regression method to fit the data in the table below with the second-order polynomial.

x	0	1	3	4	6	8	9	10	11	12
y	1	-7	-17	-19	-17	-7	1	11	23	37

Show the derivation in details. Compare the computed polynomial coefficients with those obtained from the computer program in Fig. 5.10. Then, plot to compare distribution of the fitted polynomial with the data.

12. Fit the data in problem 11 again by using the third-order polynomial. Plot to compare the distributions of the second- and third-order polynomials. Give reasons and comments if the two distributions are the same.

13. From an experiment, the thermal conduction coefficient k of an aluminum material is found to vary with the temperature T as shown in the table. Apply the least-squares method to establish the three fitted functions of the first-, second- and third-order polynomials. Plot to compare distributions of the three fitted polynomials with the experimental data. Then, determine the total error that occurs from each fitted polynomial.

T (°C)	-100	0	100	200	300	400
k (W/m-°C)	215	202	206	215	228	249

14. In Example 5.4, the third-order polynomial is used to fit the water specific heat that varies with the temperature. Employ the computer program in Fig. 5.10 to fit the data in this example again by using the second- and fourth-order polynomials. Plot to compare distributions of the three fitted polynomials. Also, determine the total error that occurs from each fitted polynomial.

15. The air specific heat is found to vary with the temperature in the form of the second-order polynomial according to the data in the table. Use the polynomial regression computer program in Fig. 5.10 to determine the coefficients of the fitted polynomial. Plot to compare distribution of the fitted polynomial with the data. Then, use the fitted polynomial to estimate the values of the specific heat at 1,000 °C 1,500 °C and 2,000 °C.

T (°C)	700	1,200	1,700	2,200	2,700
C_p (kJ/kg-°C)	1.1427	1.2059	1.2522	1.2815	1.2938

16. The table below shows the data of the stress σ that varies with the strain ε. Plot distribution of the data and fit them by an appropriate polynomial. Then, plot to compare distribution of the fitted polynomial with the data. Also, determine the total error of the fitted polynomial from the given data.

σ(MPa)	$\varepsilon \times 10^3$	σ(MPa)	$\varepsilon \times 10^3$
57.7	0.15	383.0	1.66
123.5	0.52	423.0	1.86
191.8	0.76	465.8	2.08
236.0	1.01	497.5	2.27
267.7	1.12	530.6	2.56
309.1	1.42	576.2	2.86
354.0	1.52	613.4	3.19

17. Apply the least-squares regression method to fit data in the table below by using the fourth-order polynomial. Verify the computed polynomial coefficients by comparing with those obtained from the computer program in Fig. 5.10. Then, plot to compare distribution of the fitted polynomial with the given data.

x	1	2	3	4	5	6	7	8	9	10
y	0	2	18	45	100	190	310	505	761	1,127

18. Solve Problem 2 again but by using the MATLAB function `polyfit` with $n = 1$ to establish a linear function. Then, plot to compare the fitted function with the given data.

19. Solve Problem 4 again but by using the MATLAB function `polyfit` with $n = 1$ to establish a linear function. Then, employ the MATLAB function `polyval` to estimate the stopping distance at the car speed of 65 km/hr.

20. Employ the MATLAB `polyfit` function to fit the data in Problem 17 by using the fifth-order polynomial. Then, plot to compare distribution of the fitted polynomial with the given data. Provide comments on the total error of the fitted polynomial as compared to that obtained in Problem 17.

21. From the experimental data as shown in the table below, the pressure head H of a water pump is found to vary with the flow rate Q in the form

$$H = A - BQ^2$$

Apply the linear regression method to fit the nonlinear data and determine the coefficients A and B of the fitted equation above. Then, use the fitted equation to estimate the pressure head at the flow rate of $Q = 260$ m^3/h. Solve the problem again but by using the MATLAB functions `polyfit` and `polyval`.

H(m)	40.5	37.4	32.9	28.3	23.2	16.8	13.3
Q(m^3/h)	0	115	183	228	274	319	342

22. In the assessment of a water pump performance, the flow rate Q is found to vary with the input power P as shown in the table below.

P (kW)	81	78	72	67	64	56	51
Q (m³/h)	387	349	310	272	231	192	153

Use the MATLAB function `polyfit` to establish an appropriate polynomial for fitting the data above. Then, plot to compare distribution of the fitted polynomial with the given data.

23. The stress-strain (σ-ε) data from the uni-axial tensile testing of a material are shown in the table below.

σ (MPa)	0	229	318	343	360	373
ε (mm/mm)	0	0.0007	0.0012	0.0017	0.0022	0.0027

Plot the distribution of the tested data. Then, employ the MATLAB `polyfit` function with an appropriate polynomial order to fit the data. Plot to compare distribution of the fitted polynomial with the given data. Determine the total error of the fitted polynomial from the data and provide comments on how to reduce such error.

24. Show the derivation of the simultaneous equations in Eq. (5.37) for the multiple linear regression in details. Note that the fitted function y varies linearly with the independent variables $x_i, i = 1, 2, ..., k$.

25. From the equation of a flat plane, $z = ax + by + c$, determine its values a, b and c to best fit the following data

x	0.4	1.2	3.4	4.1	5.7	7.2	9.3
y	0.7	2.1	4.0	4.9	6.3	8.1	8.9
z	0.03	0.93	3.06	3.35	4.87	5.76	8.92

Then, used the fitted function to estimate the value of z at $x = 3.6$ and $y = 7.1$.

26. Use the multiple linear regression method as explained in section 5.6 to fit the data in the table below. Show detailed calculation and plot to compare distribution of the fitted function with the given data.

x_1	1	0	2	3	4	2	1
x_2	0	1	4	2	1	3	6
x_3	1	3	1	2	5	3	4
y	4	-5	-6	0	-1	-7	-20

27.   Use the multiple linear regression method explained in section 5.6 to fit the data with 4 independent variables as shown in the table below. Verify the computed coefficients of the fitted function with those obtained from the computer program in Fig. 5.13. Plot to compare distribution of the fitted function with the given data.

x_1	3	1	4	0	2	5	1	2
x_2	2	5	1	2	3	4	0	1
x_3	4	2	3	4	1	0	2	3
x_4	0	1	4	3	2	1	2	3
y	-7	7	11	2	11	15	1	5

28.   Develop a multiple regression computer program to establish the fitted function y that varies in the forms of the polynomials with order m_1 and m_2 along the independent variables x_1 and x_2 using, respectively. Test the program by using a set of 10 data points ($n = 10$) that are generated from the equation

$$g(x_1, x_2) = \left(1 + 2x_1 + 3x_1^2\right)\left(2 + 3x_2 + 4x_2^2\right)$$

Verify the computed coefficients obtained from the program with the coefficients in the equation above.

29.   From an experiment of heat transfer measurement in a heat exchanger, the Nusselt number Nu is found to vary with the Reynolds number Re and the Prandtl number Pr in the form

$$Nu = \alpha\, Re^{\beta}\, Pr^{\gamma}\, r^{\delta}$$

where r is the ratio of the fluid viscosities at the average and wall temperatures. Apply the multiple regression method to determine the values of α, β, γ and δ for the data in the table below.

Nu	*Re*	*Pr*	*r*
277.0	49,000.0	2.30	0.947
348.0	68,600.0	2.28	0.954
421.0	84,800.0	2.27	0.959
223.0	34,200.0	2.32	0.943
177.0	22,900.0	2.36	0.936
114.8	1,321.0	246.0	0.592
95.9	931.0	247.0	0.583
68.3	518.0	251.0	0.579
49.1	346.0	273.0	0.290
56.0	122.9	1,518.0	0.294
39.9	54.0	1,590.0	0.279
47.0	84.6	1,521.0	0.267
94.2	1,249.0	107.4	0.724
99.9	1,021.0	186.0	0.612
83.1	465.0	414.0	0.512
35.9	54.8	1,302.0	0.273

30.   Use the multiple regression method to derive the set of simultaneous equations in Eq. (5.56). The set of simultaneous equations is derived to determine the coefficients of a fitted function that varies in the form of the third- and second-order polynomials along the independent variables x_1 and x_2, respectively. Develop a corresponding computer program to determine the coefficients of the fitted function by using the data in the table below. The data represent values of the measured temperatures in degree Kelvin at different locations along the outer circumference of a cylinder. The cylinder is placed in a high-speed wind tunnel and is subjected to a hot air flow at the speed of Mach 8. Herein, x_1 represents the angular location in degrees along the cylinder outer circumference and x_2 denotes the time in seconds.

		x_2					
		0	20	40	60	80	100
	0	560	700	840	950	1,000	1,080
	5	560	680	790	900	950	1,050
	10	560	820	920	1,300	1,400	1,500
	15	560	710	960	1,180	1,320	1,440
	20	560	740	1,080	1,270	1,430	1,530
	25	560	760	1,160	1,270	1,390	1,480
x_1	30	560	720	1,140	1,220	1,340	1,410
	35	560	700	1,060	1,140	1,220	1,310
	40	560	700	980	1,080	1,150	1,220
	45	560	670	900	960	1,050	1,110
	50	560	660	860	920	990	1,060
	55	560	650	810	870	920	980
	60	560	630	760	820	860	910

Chapter

6

Numerical Integration and Differentiation

6.1 Introduction

Integration and differentiation always arise in the process of solving scientific and engineering problems. Most of the functions that occur during solving practical problems are complicated and can not be integrated analytically. Numerical integration is thus required to provide approximate solutions. Several numerical integration methods are firstly presented in this chapter. Numerical differentiation methods are explained later at the end of the chapter. Numerical integration and differentiation methods presented in this chapter are the basis to study higher-level numerical methods in the later chapters for solving practical problems.

Performing integration by using integrating formulas has been taught in high-school and early years of undergraduate level. Simple functions can be integrated easily by using standard integrating formulas. For example, the integral of the polynomial function below is,

$$I = \int_0^2 \left(2x^3 - 5x^2 + 3x + 1\right) dx = \left[\frac{2x^4}{4} - \frac{5x^3}{3} + \frac{3x^2}{2} + x\right]_0^2 = 2\frac{2}{3} \qquad (6.1)$$

Or, the integral of a logarithmic function is,

$$I = \int_1^2 \ln x \, dx = \left[x \ln x - x\right]_1^2 = 0.386294 \qquad (6.2)$$

Integration of some functions yields complicated result, such as

$$\int_0^b \frac{dx}{1+x^4} = \frac{1}{4\sqrt{2}} \ln\left(\frac{b^2 + b\sqrt{2} + 1}{b^2 - b\sqrt{2} + 1}\right) - \frac{1}{2\sqrt{2}} \tan^{-1}\left(\frac{b\sqrt{2}}{b^2 - 1}\right) \tag{6.3}$$

Results of the integrations above are exact and can be used directly.

There are numerous functions that are complicated and occur during solving practical problems. These functions can not be integrated analytically to obtain exact solutions. For example, integration of an error function below is needed in the process for solving the transient temperature response from conduction heat transfer in a bar,

$$I = \int_a^b e^{-x^2} dx \tag{6.4}$$

where a and b are constants. A numerical method is needed to provide solution for the integration of the error function above. In some other problems, such as the behavior of a swinging pendulum, integration of the function shown in Eq. (6.5) is required.

$$K(\theta) = \int_0^{\pi/2} \frac{dx}{\sqrt{1 - \sin^2 \frac{\theta}{2} \sin^2 x}} \tag{6.5}$$

Equation (6.5) is called the elliptic integral of the first kind where θ is the swinging angle. A numerical method is again needed to provide the integral solution. The solutions at different angles are normally tabulated and presented in textbooks so that they can be used conveniently.

Numerical methods presented in this chapter can be applied to integrate a given function easily and conveniently. The function or *integrand* may be in any form, simple or complex function, as shown in the examples above. The general form of the integration is

$$I = \int_a^b f(x)dx \tag{6.6}$$

The integral in Eq. (6.6) means that the function f at the x location is multiplied by dx and summed over the interval $x = a$ to $x = b$. It should be noted that the integral sign $\int$ has the style of the letter S representing the meaning of *Summation*.

From the explanation above, the integration is similar to the multiplication between the height of the function f at the x location and the length dx which leads to a long skinny area as shown in Fig. 6.1(a). These areas are then summed together to yield the total area under the curve of the given function. By developing a computer program and using a very small value of dx, the area under the curve can be determined accurately and effectively. Mathematically, the integral obtained from such process approaches the exact solution as $dx \to 0$ as shown in Fig. 6.1(b). Determination for the area under the curve, by using several numerical integration methods, is presented in this chapter. Some of these methods provide high solution accuracy while the others are quite simple to use.

Numerical integration methods presented herein are: (1) the trapezoidal rule with single and multiple segments, (2) the Simpson's rule with single and multiple segments, (3) the Romberg integration, and (4) the Gauss quadrature. Multiple integration and numerical differentiation are then explained. The last two topics are important for solving practical problems and are the basis for studying other higher-level numerical methods in the later chapters.

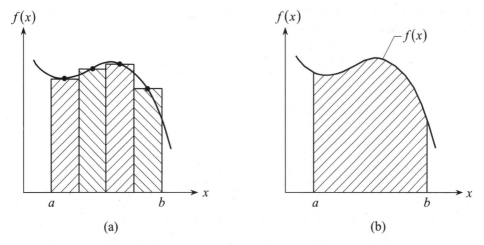

Figure 6.1 Integration of a function represented by area under curve.

6.2 Trapezoidal Rule

The procedure to determine an integral by using the trapezoidal rule is simple and easy to understand. The integral is approximated by the trapezoidal area as shown by Fig. 6.2.

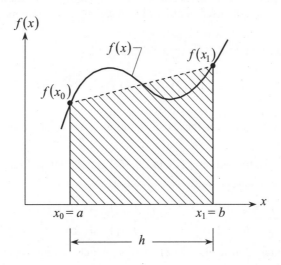

Figure 6.2 Approximate integral by the trapezoidal rule.

Figure 6.2 shows distribution of a function $f(x)$ in the interval $a \le x \le b$. The integral of the function within such interval is

$$I \;=\; \int_a^b f(x)\,dx \tag{6.7}$$

The integral solution is represented by the area under the function $f(x)$. The area may be approximates by the trapezoidal area under the dashed line as,

$$I \approx (x_1 - x_0) \frac{f(x_0) + f(x_1)}{2}$$

$$= \frac{h}{2} \left[f(x_0) + f(x_1) \right] \tag{6.8}$$

From the figure, $x_1 - x_0 = b - a = h$, then

$$I = \frac{b-a}{2} \left[f(x_0) + f(x_1) \right] \tag{6.9}$$

It is noted that the dashed line is the first-order Lagrange polynomial as shown in Eq. (4.20). If Eq. (4.20) is substituted into Eq. (6.7),

$$I \approx \int_a^b \left[\frac{x_1 - x}{x_1 - x_0} f(x_0) + \frac{x_0 - x}{x_0 - x_1} f(x_1) \right] dx \tag{6.10}$$

and the integration is performed, then

$$I = \frac{h}{2} \left[f(x_0) + f(x_1) \right] \tag{6.11}$$

The integral is identical to that obtained in Eq. (6.8). The above procedure also suggests that high order Lagrange polynomials can be used in order to provide more accurate integral solutions.

Example 6.1 Use the trapezoidal rule to estimate the integral

$$I = \int_0^2 f(x) \, dx = \int_0^2 \left(2x^3 - 5x^2 + 3x + 1 \right) dx \tag{6.12}$$

Compare the computed solution with the exact solution. Also determine the true error and true percentage error.

Distribution of the given function $f(x)$ in Eq. (6.12) is shown in Fig. 6.3.

Herein,

$$x_0 = a = 0 \quad ; \quad f(x_0) = 0 - 0 + 0 + 1 = 1$$
$$x_1 = b = 2 \quad ; \quad f(x_1) = 16 - 20 + 6 + 1 = 3$$

Then, the approximate integral from Eq. (6.8) or (6.9) which is represented by the area under the dashed line is

$$I = \frac{h}{2} [f(x_0) + f(x_1)]$$

$$= \frac{2-0}{2} (1+3)$$

$$I = 4 \tag{6.13}$$

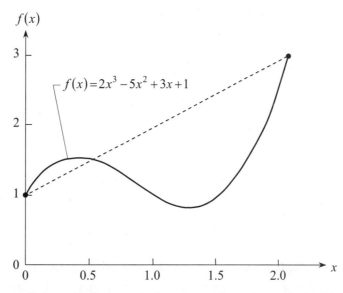

Figure 6.3 Use of trapezoidal rule to obtain an approximate integral which is the area under the dashed line.

It is noted that the exact integral is

$$I = \int_0^2 \left(2x^3 - 5x^2 + 3x + 1\right) dx = \left[\frac{2x^4}{4} - \frac{5x^3}{3} + \frac{3x^2}{2} + x\right]_0^2$$

$$= \frac{8}{3} = 2.666667 \tag{6.14}$$

Thus, the true error is

$$E_t = \frac{8}{3} - 4 = -1.333333 \tag{6.15}$$

and the true percentage error is

$$\varepsilon_t = \frac{\frac{8}{3} - 4}{\frac{8}{3}} \times 100\% = -50\% \tag{6.16}$$

Since the exact integral is not available in general, the accuracy of the computed integral obtained from the trapezoidal rule is not known. The error that arises by using the trapezoidal rule, which will be shown later, is

$$E_t = -\frac{1}{12} f''(\xi)(b - a)^3 \tag{6.17}$$

where ξ is a value between the integration limits a and b. Equation (6.17) implies that the trapezoidal rule provides exact integral if the given function $f(x)$ is linear (f' is constant and f'' vanishes). Understanding the error statement in Eq. (6.17) can lead to an improved integral solution. Such understanding is a basis of the Romberg integration method that will be explained in Section 6.7 for providing more accurate integrals.

The integration error that arises by using the trapezoidal rule as shown in Eq. (6.17) can be derived from the Taylor's series. The Taylor's series as shown in Eq. (2.29) is firstly written as,

$$f(x) = f(x_0) + f'(x_0)(x-x_0) + \frac{f''(x_0)}{2!}(x-x_0)^2 + \dots$$

$$+ \frac{f^{(n)}(x_0)}{n!}(x-x_0)^n + R_n \tag{6.18}$$

where R_n denotes the remainder that consists the remaining terms of the infinite series. The remainder can be written in the form

$$R_n = \frac{f^{(n+1)}(\xi)}{(n+1)!}(x-x_0)^{n+1} \tag{6.19}$$

where ξ is an unknown value between x_0 to x. For example, the function $f(x)$ can be determined if the function and its first derivative at point a are known. So that Eq. (6.18) becomes

$$f(x) = f(a) + f'(a)(x-a) + R_1 \tag{6.20}$$

where R_1 is the remainder that consists of the second to the infinite terms. In this case, the remainder from Eq. (6.19) is

$$R_1 = \frac{f''(\xi)}{2!}(x-x_0)^2 \tag{6.21}$$

Equations (6.20) and (6.21) imply that the exact value of $f(x)$ can be obtained if a value of ξ is selected properly.

The polynomial function of order n passing through $n+1$ data points was derived in Eq. (4.10) of Chapter 4. To fit a function $f(x)$ by a polynomial function, the remainder similar to Eq. (6.18) must be included as follow

$$f(x) = C_0 + C_1(x-x_0) + C_2(x-x_0)(x-x_1) + \dots$$
$$+ C_n(x-x_0)(x-x_1)\dots(x-x_{n-1}) + R_n \tag{6.22}$$

where the coefficients $C_i, i=0,1,2,\dots,n$ are determined by applying the conditions at the locations $x_0, x_1, x_2, \dots, x_n$. These coefficients are identical to those shown in Eq. (4.11) as follows,

$$C_0 = f(x_0) \tag{6.23a}$$

$$C_1 = \frac{f(x_1)-f(x_0)}{h} = \frac{\Delta f(x_0)}{h} \tag{6.23b}$$

$$C_2 = \frac{f(x_2)-2f(x_1)+f(x_0)}{2h^2} = \frac{\Delta^2 f(x_0)}{2!h^2} \tag{6.23c}$$

through C_n, where the symbol Δ refers to the forward differencing. In this case, the remainder in Eq. (6.22) is

$$R_n = \frac{f^{(n+1)}(\xi)}{(n+1)!}(x-x_0)(x-x_1)...(x-x_{n-1})(x-x_n) \tag{6.24}$$

which is in the same form as that of the Taylor's series in Eq. (6.19). Inclusion of the remainder to the expression of Eq. (6.22) leads to the exact value of function $f(x)$ if the value of ξ between x_0 and x_n is selected properly.

As an example of a simple case shown in Fig. 6.2, Eq. (6.22) with $x_0 = a$ and $x_1 = b$ is

$$f(x) = C_0 + C_1(x-x_0) + R_1$$

$$= f(a) + \frac{\Delta f(a)}{h}(x-a) + \frac{f''(\xi)}{2!}(x-a)(x-b) \tag{6.25}$$

By substituting Eq. (6.25) which is the exact expression for any function $f(x)$ into the integral Eq. (6.7) and performing integration,

$$I = \int_a^b \left[f(a) + \frac{\Delta f(a)}{h}(x-a) + \frac{f''(\xi)}{2!}(x-a)(x-b) \right] dx$$

$$= \left[f(a)x + \frac{\Delta f(a)}{h}\left(\frac{x^2}{2}-ax\right) + \frac{f''(\xi)}{2!}\left(\frac{x^3}{3}-\frac{ax^2}{2}-\frac{bx^2}{2}+abx\right) \right]_a^b$$

$$I = f(a)(b-a) + \frac{\Delta f(a)}{h}\frac{(b-a)^2}{2} + \frac{f''(\xi)}{2!}\left(-\frac{(b-a)^3}{6}\right)$$

Since $(b-a)=h$ and $\Delta f(a)=f(b)-f(a)$ from Eq. (6.23b), then

$$I = f(a)h + \frac{f(b)-f(a)}{h}\frac{h^2}{2} - \frac{1}{12}f''(\xi)h^3$$

$$I = \frac{h}{2}(f(b)+f(a)) - \frac{1}{12}f''(\xi)h^3 \tag{6.26}$$

The first term in Eq. (6.26) is the approximate integral by using the trapezoidal rule while the second term represents the error. Thus, the error that occurs from the trapezoidal rule is

$$E_t = -\frac{1}{12}f''(\xi)h^3 = -\frac{1}{12}f''(\xi)(b-a)^3 \tag{6.27}$$

Example 6.2 The exact integral of the function

$$f(x) = 2x^3 - 5x^2 + 3x + 1 \tag{6.28}$$

from the limit $a = 0$ to $b = 2$ as shown in Example 6.1 is 2.666667. The approximate solution by using the trapezoidal rule is 4 and the exact error is -1.333333. If the exact integral is not available, the error may be determined as follows.

From Eq. (6.27), the error from using the trapezoidal rule is

$$E_t = -\frac{1}{12}f''(\xi)h^3 \tag{6.27}$$

Herein, $h = b - a = 2$ and the derivatives of the function $f(x)$ are

$$f'(x) = 6x^2 - 10x + 3 \tag{6.29a}$$

$$f''(x) = 12x - 10 \tag{6.29b}$$

If the location ξ is chosen at $x = 1$ which is at the middle of the interval between $a = 0$ to $b = 2$, then

$$f''(\xi) = 12(1) - 10 \qquad = 2$$

Thus, the error from Eq. (6.27) is

$$E_a = -\frac{1}{12}(2)(2)^3 \qquad = -1.333333 \tag{6.30}$$

The approximate error as shown in Eq. (6.30) is exact. The example demonstrates that the exact error can be obtained if the location ξ is chosen properly. Because the location ξ for the exact error is not known, the average value for second derivative of the function is determined instead as

$$f''(\xi) = \int_0^2 (12x - 10)\,dx = \left[\frac{12x^2}{2} - 10x\right]_0^2$$

$$= 24 - 20 \qquad = 4$$

Thus, the approximate error is

$$E_a = -\frac{1}{12}(4)(2)^3 \qquad = -2.666667 \tag{6.31}$$

which has the same sign and order of magnitude as the exact error. It is noted that understanding of the approximate error above is useful to evaluate the accuracy of the computed solution by other methods in the later sections.

6.3 Composite Trapezoidal Rule

The trapezoidal method presented in the preceding section can provide an improved integral solution if the integration interval is divided into many segments so that the trapezoidal rule is applied to each segment. The computed areas from these segments are then combined to yield the integral solution for the entire interval. The procedure is called *composite* or *multiple-application trapezoidal rule* as shown by Fig. 6.4.

Figure 6.4 shows distribution of a typical function $f(x)$ between the interval $a \le x \le b$. The interval from a to b is divided into n segments with equal width of h,

$$h = \frac{b - a}{n} \tag{6.32}$$

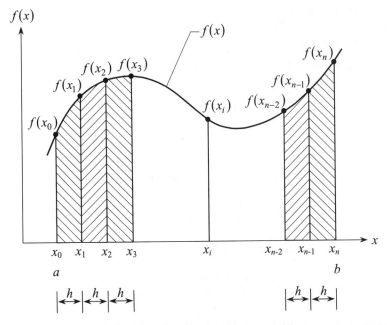

Figure 6.4 Approximate integral determination by composite trapezoidal rule.

The coordinates at both ends of the segments are

$$x_i \quad = \quad x_0 + ih \qquad\qquad i = 0, 1, 2, \ldots, n \qquad\qquad (6.33)$$

Integration of the function $f(x)$ between $a \le x \le b$ is performed by first dividing the entire interval into n segments starting from the segments with $x_0 \le x \le x_1$, $x_1 \le x \le x_2$ until $x_{n-1} \le x \le x_n$ as,

$$I \quad = \quad \int_a^b f(x)dx$$

$$I \quad = \quad \int_{x_0}^{x_1} f(x)dx \; + \; \int_{x_1}^{x_2} f(x)dx \; + \; \ldots \; + \; \int_{x_{n-1}}^{x_n} f(x)dx \qquad\qquad (6.34)$$

Then, the trapezoidal rule is applied to each segment that has the width of h as follows,

$$I \quad \approx \quad \frac{h}{2}(f(x_0) + f(x_1)) + \frac{h}{2}(f(x_1) + f(x_2)) + \ldots + \frac{h}{2}(f(x_{n-1}) + f(x_n))$$

$$= \quad \frac{h}{2}(f(x_0) + 2f(x_1) + 2f(x_2) + \ldots + 2f(x_{n-1}) + f(x_n))$$

$$I \quad = \quad \frac{h}{2}\left(f(x_0) + f(x_n) + 2\sum_{i=1}^{n-1} f(x_i)\right) \qquad\qquad (6.35)$$

Equation (6.35) can be used to develop a corresponding computer program directly. The program can be employed to find approximate integral of a given function conveniently. A more accurate solution is obtained by simply increasing the number of segments n in the program.

The error arisen by using the composite trapezoidal rule that divides the entire interval from the limit a to b into n segments is

$$E_t = -\frac{1}{12}\left(\frac{b-a}{n}\right)^3 \sum_{i=1}^{n} f''(\xi_i) \qquad (6.36)$$

where $f''(\xi_i)$ is the second derivative of the function at location ξ_i of the segment i. The second derivative value of the function is different from segment to segment. Their average value is

$$\overline{f}'' \cong \frac{1}{n}\left(\sum_{i=1}^{n} f''(\xi_i)\right) \qquad (6.37)$$

Thus, the total approximate error according to Eq. (6.36) becomes

$$E_a = -\frac{1}{12}\frac{(b-a)^3}{n^2}\overline{f}'' \qquad (6.38)$$

The term n^2 on the numerator of Eq. (6.38) implies that the total error will reduce four times if number of the segments is double. Understanding the behavior of the error reduction above can help verifying the accuracy of the computed integral solutions.

Example 6.3 Apply the composite trapezoidal rule to estimate the integral

$$I = \int_0^2 f(x)\,dx = \int_0^2 \left(2x^3 - 5x^2 + 3x + 1\right)dx \qquad (6.12)$$

which is in the same form as Example 6.1 but by developing a computer program. The program can vary the number of segments so that accuracy of the computed solutions can be studied by comparing them with the exact solution.

To clearly demonstrate the use of the composite trapezoidal rule, the entire interval from $a = 0$ to $b = 2$ is divided into 4 segments ($n = 4$) as shown in Fig. 6.5. The width h of each segment according to Eq. (6.32) is

$$h = \frac{b-a}{n} = \frac{2-0}{4} = 0.5 \qquad (6.39)$$

The values of the function $f(x)$ at the ends of the four segments are

$$f(x_0 = 0) \;=\; 0 - 0 + 0 + 1 \qquad\qquad = 1$$
$$f(x_1 = 0.5) \;=\; 0.25 - 1.25 + 1.5 + 1 \qquad = 1.5$$
$$f(x_2 = 1.0) \;=\; 2 - 5 + 3 + 1 \qquad\qquad = 1$$
$$f(x_3 = 1.5) \;=\; 6.75 - 11.25 + 4.5 + 1 \qquad = 1$$
$$f(x_4 = 2.0) \;=\; 16 - 20 + 6 + 1 \qquad\qquad = 3$$

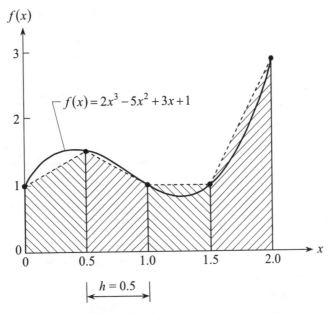

Figure 6.5 Use of the composite trapezoidal rule to approximate the integral in Example 6.3 by dividing the entire interval into 4 segments.

Thus, the approximate integral, according to Eq. (6.35), by using four segments ($n = 4$) is,

$$I = \frac{h}{2}\left(f(x_0) + f(x_4) + 2\sum_{i=1}^{4-1} f(x_i)\right) = \frac{0.5}{2}(1 + 3 + 2(1.5 + 1 + 1))$$

$$I = 2.75 \tag{6.40}$$

Table 6.1 shows the approximate integrals obtained from the computer program in Fig. 6.6 by using the composite trapezoidal rule. The table indicates that the error is reduced by four times as the number of segments is double. The behavior of the error reduction can be observed from the solutions when $n = 2$ and $n = 4$ or when $n = 5$ and $n = 10$.

Table 6.1 Approximate solutions of the integral in Example 6.3 by using the composite trapezoidal computer program in Fig. 6.6 as compared to the exact solution of $I = 2.666667$.

n	h	I	ε_t (%)
2	1.0000	3.000000	−12.5
3	0.6667	2.814815	−5.6
4	0.5000	2.750000	−3.1
5	0.4000	2.720000	−2.0
6	0.3333	2.703704	−1.4
7	0.2857	2.693878	−1.0
8	0.2500	2.687500	−0.8
9	0.2222	2.683128	−0.6
10	0.2000	2.680000	−0.5

Figure 6.6 shows a computer program to determine approximate integral of a given function $f(x)$ from a to b by using the composite trapezoidal rule. The number of segments is input by the

user. The program can be modified to integrate other functions by simply changing the function statement and the integration limits declared in the program.

Fortran

```
      PROGRAM  TRAPEZ
C.....A MULTIPLE-SEGMENT TRAPEZOIDAL PROGRAM
C.....FOR ESTIMATING INTEGRAL OF F(X)
      A = 0.
      B = 2.
C.....READ NUMBER OF SEGMENTS REQUIRED:
      READ(5,*)  N
      H = (B - A)/N
      SUM = 0.
      X = A + H
      DO 10  I=1,N-1
      FX = FUNC(X)
      SUM = SUM + FX
      X = X + H
   10 CONTINUE
      FX0 = FUNC(A)
      FXN = FUNC(B)
      SOL = (FX0 + FXN + 2.*SUM)*H/2.
      WRITE(6,100)  N, SOL
  100 FORMAT(' INTEGRAL OF F(X) USING', I3,
     *        ' SEGMENTS IS', F10.6)
      STOP
      END
C----------------------------------------------
      FUNCTION FUNC(X)
      FUNC = 2.*X*X*X - 5.*X*X + 3.*X + 1.
      RETURN
      END
```

MATLAB

```
%    PROGRAM TRAPEZ
%    A MULTIPLE-SEGMENT TRAPEZOIDAL PROGRAM
%    FOR ESTIMATING INTEGRAL OF F(X)
%-----------------------------------------------
func = inline('2.*x*x*x - 5.*x*x + 3.*x + 1.', 'x');
%-----------------------------------------------
a = 0.;
b = 2.;
%    READ NUMBER OF SEGMENTS REQUIRED:
n = input('\nEnter the number of segments: ');
h = (b - a)/n;
sum = 0.;
x = a + h;
for i = 1:n-1
    fx  = func(x);
    sum = sum + fx;
    x = x + h;
end
fx0 = func(a);
fxn = func(b);
sol = (fx0 + fxn + 2.*sum)*h/2.;
fprintf('INTEGRAL OF F(X) USING %3d SEGMENTS IS ...
%10.6f',n,sol)
```

Figure 6.6 A computer program to determine the integral in Example 6.3 by using the composite trapezoidal rule.

6.4 Simpson's Rule

From the trapezoidal rule explained in section 6.2, the integral value is estimated by using the area under a straight line (dashed line connecting points a and b in Fig. 6.2). Because the distribution of a function to be integrated is arbitrary, the area under the straight line is, in general, not accurate for representing the integral value. In this section, the Simpson's rule is presented for which the integral value is determined from the area under the second-order polynomial (dashed line in Fig. 6.7).

Figure 6.7 shows the distribution of a function $f(x)$ between $a \le x \le b$. The objective is to determine the integral

$$I = \int_a^b f(x)dx \tag{6.41}$$

The integral value is determined by approximating the function $f(x)$ in the form of a second-order polynomial. By substituting the second-order Lagrange polynomial as shown in Eq. (4.29) into Eq. (6.41)

$$I \approx \int_a^b \left[\frac{(x-x_1)(x-x_2)}{(x_0-x_1)(x_0-x_2)} f(x_0) + \frac{(x-x_0)(x-x_2)}{(x_1-x_0)(x_1-x_2)} f(x_1) \right.$$
$$\left. + \frac{(x-x_0)(x-x_1)}{(x_2-x_0)(x_2-x_1)} f(x_2) \right] dx \tag{6.42}$$

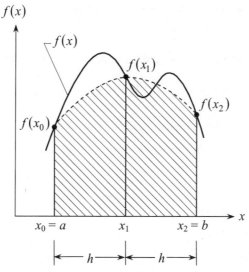

Figure 6.7 Use of the Simpson's 1/3 rule to obtain approximate integral which is the area under the dashed line.

The values of the function $f(x_0), f(x_1), f(x_2)$ can be determined at the locations x_0, x_1, x_2. If $x_2 - x_1 = x_1 - x_0 = h$, the approximate integral from Eq. (6.42) is

$$I \approx \frac{h}{3}[f(x_0) + 4f(x_1) + f(x_2)] \tag{6.43}$$

where

$$h = \frac{b-a}{2} \tag{6.44}$$

Equation (6.43) is called the Simpson's one-third rule. It is noted that the term "one-third" refers to the presence of factor 1/3 in front of the expression. Equation (6.43) can also be written in a more general form as

$$I \approx \frac{(b-a)}{6}[f(x_0) + 4f(x_1) + f(x_2)] \tag{6.45}$$

The error of the integral value obtained by using the Simpson's 1/3 rule can be derived in the same manner as that for the trapezoidal rule. Derivation of the error expression is omitted herein so that it will be used as an exercise. The error of the integral value from the Simpson's 1/3 rule is

$$E_t = -\frac{1}{90}h^5 f^{(4)}(\xi) \tag{6.46}$$

By substituting $h = (b-a)/2$ from Eq. (6.44), then

$$E_t = -\frac{(b-a)^5}{2,880} f^{(4)}(\xi) \tag{6.47}$$

where ξ is a value between the integration limit a to b. The error that occurs from the Simpson's 1/3 rule in Eq. (6.47) is less than that in Eq. (6.27) of the trapezoidal rule. Equation (6.47) also implies that the Simpson's 1/3 rule can provide exact integral value if the function to be integrated is a polynomial of third order or lower. Obtaining such exact integral value by using the Simpson's 1/3 rule is demonstrated in the following example.

Example 6.4 Determine the integral in Example 6.1 again but by using the Simpson's 1/3 rule.

$$I = \int_0^2 f(x)\,dx = \int_0^2 \left(2x^3 - 5x^2 + 3x + 1\right)dx \tag{6.12}$$

Compare the computed integral value with the exact solution of 2.666667.

Herein, the width $h = x_2 - x_1 = x_1 - x_0$ as shown in Eq. (6.44) is

$$h = \frac{b-a}{2} = \frac{2-0}{2} = 1$$

and the values of the function at the three locations are

$$f(x_0 = 0) = 0 - 0 + 0 + 1 = 1$$
$$f(x_1 = 1) = 2 - 5 + 3 + 1 = 1$$
$$f(x_2 = 2) = 16 - 20 + 6 + 1 = 3$$

By substituting these values into Eq. (6.43), the approximate integral is

$$I = \frac{1}{3}[1 + 4(1) + 3] = \frac{8}{3} = 2.666667$$

which is equal to the exact solution.

The Simpson's 1/3 rule won't provide exact integral value if the polynomial is of fourth order or higher. For example,

$$I = \int_0^2 \left(x^4 + 2x^3 - 5x^2 + 3x + 1\right)dx \tag{6.48}$$

for which the exact solution is 9.066667. In this case, the Simpson's 1/3 rule yields

$$I = \frac{1}{3}[1 + 4(2) + 19] = 9.333333 \tag{6.49}$$

The computed integral has the true error of -0.266667 or -2.9%.

6.5 Composite Simpson's Rule

The Simpson's 1/3 rule can provide more accurate integral value if the integration limits from a to b is divided into many segments. The idea is similar to the composite trapezoidal rule explained in section 6.3. The procedure can be understood clearly by considering Fig. 6.8.

Figure 6.8 shows distribution of a function $f(x)$ within the interval $a \le x \le b$. If the interval from a to b is divided into n segments, then, the width h of each segment is

$$h = \frac{b-a}{n} \tag{6.50}$$

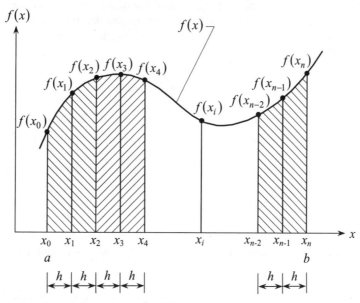

Figure 6.8 Approximate integral determination by composite Simpson's rule.

where the coordinates at both ends of each interval are

$$x_i = x_0 + ih \qquad\qquad i = 0, 1, 2, ..., n \qquad\qquad (6.51)$$

Since the general form for the integral of a function $f(x)$ between $a \le x \le b$ is

$$I = \int_a^b f(x)dx \qquad\qquad (6.52)$$

The integral in Eq. (6.52) is firstly divided into $n/2$ sub-integrals with their integration limits between $x_0 \le x \le x_2$, $x_2 \le x \le x_4$, to $x_{n-2} \le x \le x_n$ as

$$I = \int_{x_0}^{x_2} f(x)dx + \int_{x_2}^{x_4} f(x)dx + ... + \int_{x_{n-2}}^{x_n} f(x)dx \qquad\qquad (6.53)$$

The Simpson's rule in Eq. (6.43) is then applied to each sub-integral as follow

$$I \approx \frac{h}{3}[f(x_0)+4f(x_1)+f(x_2)] + \frac{h}{3}[f(x_2)+4f(x_3)+f(x_4)]$$

$$+ ... + \frac{h}{3}[f(x_{n-2})+4f(x_{n-1})+f(x_n)]$$

$$= \frac{h}{3}[f(x_0) + 4f(x_1) + 2f(x_2) + 4f(x_3) + 2f(x_4) + ...$$

$$+ 2f(x_{n-2}) + 4f(x_{n-1}) + f(x_n)]$$

$$I = \frac{h}{3}\left[f(x_0) + f(x_n) + 4\sum_{i=1,3,5}^{n-1} f(x_i) + 2\sum_{i=2,4,6}^{n-2} f(x_i)\right] \qquad\qquad (6.54)$$

It should be noted that the number of segments must be an even value because the Simpson's rule is applied to $n/2$ segments. Such constraint must be implemented into the computer program that employs the composite Simpson's rule. The integral error obtained by using the composite Simpson's rule (left as an exercise) is

$$E_a = -\frac{(b-a)^5}{180n^4} \overline{f}^{(4)} \tag{6.55}$$

where $\overline{f}^{(4)}$ is the average value of the fourth-order derivative of the function from all segments. The expression is in the similar form as that for the composite trapezoidal rule as shown by Eq. (6.37).

Example 6.5 Develop a computer program that uses the composite Simpson's rule to estimate the integral

$$I = \int_0^2 f(x)\,dx = \int_0^2 \left(2x^3 - 5x^2 + 3x + 1\right)dx \tag{6.12}$$

Note that the same integral was previously evaluated in Example 6.1 with the exact solution of 2.666667.

Fortran

```
      PROGRAM  SIMPSON
C.....A MULTIPLE-SEGMENT SIMPSON'S 1/3 PROGRAM
C.....FOR ESTIMATING INTEGRAL OF F(X)
      A = 0.
      B = 2.
C.....READ NUMBER OF SEGMENTS REQUIRED:
      READ(5,*)  N
      M = N - (N/2)*2
      IF(M.EQ.0)  GO TO 5
      WRITE(6,50)
   50 FORMAT(' NUMBER OF SEGMENTS MUST BE EVEN')
      STOP
    5 CONTINUE
      H = (B - A)/N
      SUM1 = 0.
      X = A + H
      DO 10  I=1,N-1,2
      FX = FUNC(X)
      SUM1 = SUM1 + FX
      X = X + 2.*H
   10 CONTINUE
      SUM2 = 0.
      X = A + 2.*H
      DO 20  I=2,N-2,2
      FX = FUNC(X)
      SUM2 = SUM2 + FX
      X = X + 2.*H
   20 CONTINUE
      FX0 = FUNC(A)
      FXN = FUNC(B)
      SOL = (FX0 + FXN + 4.*SUM1 + 2.*SUM2)*H/3.
      WRITE(6,100)  N, SOL
  100 FORMAT(' INTEGRAL OF F(X) USING', I3,
     *       ' SEGMENTS IS', F10.6)
      STOP
      END
C-----------------------------------------------------
      FUNCTION FUNC(X)
      FUNC = 2.*X*X*X - 5.*X*X + 3.*X + 1.
      RETURN
      END
```

MATLAB

```
%  PROGRAM  SIMPSON
%  A MULTIPLE-SEGMENT SIMPSON'S 1/3 PROGRAM
%  FOR ESTIMATING INTEGRAL OF F(X)
%----------------------------------------------------
func = inline('x^4 + 2.*x^3 - 5.*x^2 + 3.*x + 1.', ...
'x');
%----------------------------------------------------
a = 0.;
b = 2.;
%  READ NUMBER OF SEGMENTS REQUIRED:
n = input('\nEnter the number of segments: ');
m = n - fix(n/2)*2;
if m~= 0
    disp(' Number of segments must be even')
    break
end
h = (b - a)/n;
sum1 = 0.;
x = a + h;
for i = 1:2:n-1
    fx = func(x);
    sum1 = sum1 + fx;
    x = x + 2.*h;
end
sum2 = 0.;
x = a + 2.*h;
for i = 2:2:n-2
    fx = func(x);
    sum2 = sum2 + fx;
    x = x + 2.*h;
end
fx0 = func(a);
fxn = func(b);
sol = (fx0 + fxn + 4.*sum1 + 2.*sum2)*h/3.;
fprintf('INTEGRAL OF F(X) USING %3d SEGMENTS IS ...
%10.6f',n,sol)
```

Figure 6.9 A computer program to determine the integral in Example 6.5 by using the composite Simpson's rule.

The computer program for determining the integral in this example is presented in Fig. 6.9. Users need to input an even number of the segments n. A warning statement will appear if an odd number of the segments n is input while executing the program. For this example, the program always gives the exact solution of 2.666667 for any input even number n.

Example 6.6 Modify the computer program developed for Example 6.5 to determine the integral in Eq. (6.48) which is

$$I = \int_0^2 (x^4 + 2x^3 - 5x^2 + 3x + 1) dx \tag{6.48}$$

Investigate the solution improvement by increasing the number of segments n. Compare the computed integral values with the exact solution of 9.066667.

Table 6.2 presents the computed integral values obtained from using different numbers of the segments n. The table shows that the computed integral value approaches the exact solution as the number of the segments n increases.

Table 6.2 Computed integral values by using the composite Simpson's rule with different number of segments n for Example 6.6. The exact integral value is $I = 9.066667$.

n	h	I	ε_t (%)
2	1.0000	9.333333	-2.9412
4	0.5000	9.083333	-.1838
6	0.3333	9.069960	-.0363
8	0.2500	9.067708	-.0115
10	0.2000	9.067093	-.0047
20	0.1000	9.066696	-.0003

6.6 Newton-Cotes Formulas

The trapezoidal rule in section 6.2 estimates the integral value from the area under a straight line or first-order polynomial. Accuracy of the estimated integral value is increased by using the Simpson's 1/3 rule as explained in section 6.4. The Simpson's 1/3 rule estimates the integral value from the area under the second-order polynomial. Thus, a more accurate integral value can be obtained by determining the area under the higher order polynomial.

Figure 6.10 shows the estimation of the integral from the area under the third-order polynomial. Integration of a function $f(x)$ from the limits a to b is

$$I = \int_a^b f(x) dx \tag{6.41}$$

The integral value is approximated by the area under the third-order polynomial as shown by the dashed line in Fig. 6.10. The third-order polynomial can be derived from the general form of the Lagrange polynomial in Eq. (4.32) when $n = 3$. By substituting such third-order polynomial into Eq. (6.41) and performing integration, the approximate integral value is

$$I \approx \frac{3h}{8} [f(x_0) + 3f(x_1) + 3f(x_2) + f(x_3)] \tag{6.56}$$

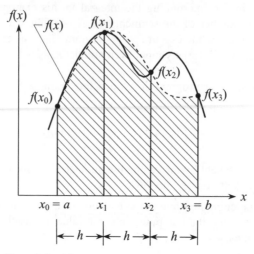

Figure 6.10 Use of the Simpson's 3/8 rule to obtain approximate integral
which is the area under the dashed line.

where, in this case

$$h = \frac{b-a}{3} \tag{6.57}$$

Equation (6.56) is called the Simpson's three-eighth rule because the factor of 3/8 appears in front of the expression. It is noted that Eq. (6.56) can be written in a more general form as

$$I \approx \frac{(b-a)}{8}[f(x_0) + 3f(x_1) + 3f(x_2) + f(x_3)] \tag{6.58}$$

The integral error from the Simpson's 3/8 rule can be derived in the same manner as that for the trapezoidal rule. The integral error from the Simpson's 3/8 rule is

$$E_t = -\frac{3}{80}h^5 f^{(4)}(\xi) \tag{6.59}$$

By substituting $h = (b-a)/3$ from Eq. (6.57) into Eq. (6.59)

$$E_t = -\frac{(b-a)^5}{6,480} f^{(4)}(\xi) \tag{6.60}$$

where ξ is a value between the integration limits a to b. The integral error as shown in Eq. (6.60) obtained from the Simpson's 3/8 rule is less than that from the Simpson's 1/3 rule in Eq. (6.47). The integral value from the Simpson's 3/8 rule is more accurate because the function is determined at the four locations of x_0, x_1, x_2 and x_3 as compared to the three locations used in the Simpson's 1/3 rule.

Example 6.7 Employ the Simpson's 3/8 rule to estimate the integral in Example 6.1

$$I = \int_0^2 f(x)\,dx = \int_0^2 \left(2x^3 - 5x^2 + 3x + 1\right)dx \tag{6.12}$$

Compare the computed integral value with the exact solution of 2.666667.

The width h to be used in the Simpson's 3/8 rule from Eq. (6.57) for this example is

$$h = \frac{b-a}{3} = \frac{2-0}{3} = \frac{2}{3} \tag{6.61}$$

and the values of the function $f(x)$ at the four locations are

$$
\begin{aligned}
f(x_0 = 0) &= & 0 - & & 0 + 0 + 1 &= 1 \\
f(x_1 = 2/3) &= & \frac{16}{27} - & \frac{20}{9} + 2 + 1 &= \frac{37}{27} \\
f(x_2 = 4/3) &= & \frac{128}{27} - & \frac{80}{9} + 4 + 1 &= \frac{23}{27} \\
f(x_3 = 2) &= & 16 - & 20 + 6 + 1 &= 3
\end{aligned}
$$

By substituting these values into Eq. (6.56), the approximate integral obtained from the Simpson's 3/8 rule is

$$I = \left(\frac{3}{8}\right)\left(\frac{2}{3}\right)\left[1 + 3\left(\frac{37}{27}\right) + 3\left(\frac{23}{27}\right) + 3\right] = \frac{8}{3} \tag{6.62}$$

which is equal to the exact solution of 2.666667. The result confirms that the Simpson's 3/8 rule can provide exact integral as implied by Eq. (6.60) if the integrand is a polynomial of third order or less. However, if the integrand is a polynomial of fourth order or higher, the Simpson's 3/8 rule can not provide exact integral as shown in the following example.

Example 6.8 Use the Simpson's 3/8 rule to estimate the integral in Eq. (6.48)

$$I = \int_0^2 \left(x^4 + 2x^3 - 5x^2 + 3x + 1\right) dx \tag{6.48}$$

Compare the computed integral value with the exact solution of 9.066667. It is noted that the integral value obtained from the Simpson's 1/3 rule in Eq. (6.49) is 9.333333 with the true error of -2.9%.

The width h according to Eq. (6.57) for the Simpson's 3/8 rule is

$$h = \frac{b-a}{3} = \frac{2-0}{3} = \frac{2}{3} \tag{6.63}$$

and the values of function $f(x)$ at the four locations are

$$
\begin{aligned}
f(x_0 = 0) &= & 0 + & 0 - & 0 + 0 + 1 &= 1 \\
f(x_1 = 2/3) &= & \frac{16}{81} + & \frac{16}{27} - & \frac{20}{9} + 2 + 1 &= \frac{127}{81} \\
f(x_2 = 4/3) &= & \frac{256}{81} + & \frac{128}{27} - & \frac{80}{9} + 4 + 1 &= \frac{325}{81} \\
f(x_3 = 2) &= & 16 + & 16 - & 20 + 6 + 1 &= 19
\end{aligned}
$$

By substituting these values into Eq. (6.56) of the Simpson's 3/8 rule, the computed integral solution is

$$I = \left(\frac{3}{8}\right)\left(\frac{2}{3}\right)\left[1 + 3\left(\frac{127}{81}\right) + 3\left(\frac{325}{81}\right) + 19\right] = 9.185185 \qquad (6.64)$$

The computed integral solution has the true error from the exact solution of 9.066667 - 9.185185 = -0.118518 or -1.3%. The error is less than -2.9% which is produced by the Simpson's 1/3 rule.

    The trapezoidal rule, Simpson's 1/3 rule and Simpson's 3/8 rule use the first-, second- and third-order Lagrange polynomial, respectively for estimating the integral. Higher order Lagrange polynomials can be used to derive more accurate integral solution. The different orders of the Lagrange polynomials lead to a family of the *Newton-Cotes integration formulas* as follows.

(a) For $n = 1$ with 2 points (trapezoidal rule),

$$I = \frac{b-a}{2}[f(x_0) + f(x_1)] \qquad (6.65)$$

The error is $-(1/12)h^3 f^{(2)}(\xi)$ where $h = b - a$.

(b) For $n = 2$ with 3 points (Simpson's 1/3 rule),

$$I = \frac{(b-a)}{6}\left[f(x_0) + 4f(x_1) + f(x_2)\right] \qquad (6.66)$$

The error is $-(1/90)h^5 f^{(4)}(\xi)$ where $h = (b-a)/2$.

(c) For $n = 3$ with 4 points (Simpson's 3/8 rule),

$$I = \frac{(b-a)}{8}\left[f(x_0) + 3f(x_1) + 3f(x_2) + f(x_3)\right] \qquad (6.67)$$

The error is $-(3/80)h^5 f^{(4)}(\xi)$ where $h = (b-a)/3$.

(d) For $n = 4$ with 5 points (Boole's rule),

$$I = \frac{(b-a)}{90}\left[7f(x_0) + 32f(x_1) + 12f(x_2) + 32f(x_3) + 7f(x_4)\right] \qquad (6.68)$$

The error is $-(8/945)h^7 f^{(6)}(\xi)$ where $h = (b-a)/4$.

(e) For $n = 5$ with 6 points,

$$I = \frac{(b-a)}{288}\left[19f(x_0) + 75f(x_1) + 50f(x_2) + 50f(x_3) + 75f(x_4) + 19f(x_5)\right] \qquad (6.69)$$

The error is $-(275/12{,}096)h^7 f^{(6)}(\xi)$ where $h = (b-a)/5$.

(f) For $n = 6$ with 7 points,

$$I = \frac{(b-a)}{840}\left[41f(x_0) + 216f(x_1) + 27f(x_2) + 272f(x_3) + 27f(x_4) \right.$$

$$\left. + 216f(x_5) + 41f(x_6)\right] \qquad (6.70)$$

The error is $-(9/1,400)\, h^9 f^{(8)}(\xi)$ where $h = (b-a)/6$.

(g) For $n = 7$ with 8 points,

$$I = \frac{(b-a)}{17,280}\Big[751 f(x_0) + 3,577 f(x_1) + 1,323 f(x_2) + 2,989 f(x_3)$$

$$+ 2,989 f(x_4) + 1,323 f(x_5) + 3,577 f(x_6) + 751 f(x_7)\Big] \tag{6.71}$$

The error is $-(8,183/518,400)\, h^9 f^{(8)}(\xi)$ where $h = (b-a)/7$.

Even though the Newton-Cotes formulas with high numbers of n or points can provide accurate integral solution, they are rarely used in practice. The composite trapezoidal and Simpson's rules are used instead because they are simple. Accurate integral solution can be obtained by employing the composite trapezoidal or Simpson's rule with a large number of segments. The computer programs for the composite trapezoidal and Simpson's rules as shown in Figs. 6.6 and 6.9 can be used to serve for this purpose conveniently.

6.7 Romberg Integration

From the explanation in section 6.3 of the composite trapezoidal rule, the exact integral value I consists of two parts,

$$I = I(h) + E(h) \tag{6.72}$$

The first part is the integral value $I(h)$ obtained from the composite trapezoidal rule such as that shown in Eq. (6.35). The accuracy of the computed integral value depends on the width h of the segments within the integration limits a to b. The second part is the error $E(h)$ as shown in Eq. (6.38)

$$E = -\frac{1}{12}\frac{(b-a)^3}{n^2}\,\overline{f}'' \tag{6.38}$$

where $\overline{f}''$ is the average second-derivative from all the segments n as expressed in Eq. (6.37). The error statement in Eq. (6.38) suggests that the integral error E will reduce four times if the number of the segments n is double. Such idea leads the development of the Romberg integration method that can improve the computed integral accuracy. The method divides the integration limits from a to b twice so that the errors obtained from each case are used to produce a more accurate integral solution. The method is based on the *Richardson's extrapolation technique* for which the two integral estimates can lead to a more accurate integral value.

If the composite trapezoidal rule is applied to integrate a given function twice by using the two different widths of h_1 and h_2, then the exact integral condition in Eq. (6.72) gives

$$I(h_1) + E(h_1) = I(h_2) + E(h_2) \tag{6.73}$$

Since $h = (b-a)/n$, Eq. (6.38) can be written in term of h as

$$E = -\frac{b-a}{12} h^2 \overline{f''} \tag{6.74}$$

If the average second-derivatives from using the two different widths are assumed to be equal, then the ratio of the errors from the two cases depends on the widths h_1 and h_2 as

$$\frac{E(h_1)}{E(h_2)} = \frac{h_1^2}{h_2^2} \tag{6.75}$$

or,

$$E(h_1) = E(h_2)\left(\frac{h_1}{h_2}\right)^2 \tag{6.76}$$

By substituting Eq. (6.76) into Eq. (6.73),

$$I(h_1) + E(h_2)\left(\frac{h_1}{h_2}\right)^2 = I(h_2) + E(h_2)$$

the error from using the width h_2 can be written in form of the two integral estimates and the two widths as

$$E(h_2) = \frac{I(h_2) - I(h_1)}{(h_1/h_2)^2 - 1} \tag{6.77}$$

From Eq. (6.72), the exact integral by using the width h_2 is

$$I = I(h_2) + E(h_2) \tag{6.78}$$

By substituting Eq. (6.77) into Eq. (6.78), a new integral value is obtained from the two integral values that were previously estimated by using the widths h_1 and h_2 as

$$I = I(h_2) + \frac{I(h_2) - I(h_1)}{(h_1/h_2)^2 - 1} = \left(1 + \frac{h_2^2}{h_1^2 - h_2^2}\right) I(h_2) - \left(\frac{h_2^2}{h_1^2 - h_2^2}\right) I(h_1) \tag{6.79}$$

Determination of the new integral solution in Eq. (6.79) can be understood clearly if the width h_2 is a half of the width h_1 as

$$h_2 = h_1/2 \tag{6.80}$$

Then, Eq. (6.79) becomes

$$I = \frac{4}{3} I(h_2) - \frac{1}{3} I(h_1) \tag{6.81}$$

Example 6.9 The single and composite trapezoidal rules were used to determine the integral

$$I = \int_0^2 f(x)\,dx = \int_0^2 \left(2x^3 - 5x^2 + 3x + 1\right) dx \tag{6.12}$$

as shown in Examples 6.1 and 6.3. Their solutions are tabulated as shown below

n	h	I	ε_t (%)
1	2.0	4.00	-50.0
2	1.0	3.00	-12.5
4	0.5	2.75	-3.0

From the table, a new integral value can be determined from two previous integral values when $n = 1$ and $n = 2$ ($h_1 = 2.00$ and $h_2 = 1.00$) according to Eq. (6.81) as

$$I = \frac{4}{3}(3.00) - \frac{1}{3}(4.00) = 2.666667 \tag{6.82}$$

The new integral value obtained in Eq. (6.82) is exact as compared to the exact solution in Eq. (6.14).

Similarly, another new integral value can be obtained from the two previous integral values when $n = 2$ and $n = 4$ ($h_1 = 1.00$ and $h_2 = 0.50$) according to Eq. (6.81) as

$$I = \frac{4}{3}(2.75) - \frac{1}{3}(3.00) = 2.666667 \tag{6.83}$$

In general, the integrand $f(x)$ is not in form of the polynomials, the Romberg integration must be repeated to produce an accurate integral solution. As shown in Eq. (6.74), the error from the trapezoidal rule varies with h^2 or $O(h^2)$. It can be shown that the error is of order $O(h^4)$ after the second application of Romberg integration. The same procedure as shown in Eqs. (6.75) - (6.80) can be repeated so that the new integral solution after the second application of Romberg integration is

$$I = \frac{16}{15}I_M - \frac{1}{15}I_L \tag{6.84}$$

where I_M is the more accurate integral solution and I_L is the less accurate integral solution. The newer integral solution in Eq. (6.84) has the error of order $O(h^6)$. If the Romberg integration technique is applied again, the newer integral solution is

$$I = \frac{64}{63}I_M - \frac{1}{63}I_L \tag{6.85}$$

The new integral solutions obtained from the applications of the Romberg integration as shown in Eqs. (6.81), (6.84), (6.85) can be written in a more general form as

$$I = \frac{2^{2k}I_M - I_L}{2^{2k} - 1} \tag{6.86}$$

where $k = 1, 2, 3, \ldots$ represents the kth application of the Romberg integration. Equation (6.86) can be written in another form for convenient computer programming as

$$I = \frac{4^k I_M - I_L}{4^k - 1} \tag{6.87}$$

The example below shows the use of Eq. (6.87) of the Romberg integration method in details.

Example 6.10 Use the Romberg integration method to determine the integral

$$I = \int_0^{\pi/2} \sin x \, dx \tag{6.88}$$

Apply the method until the relative error is less than 0.0001%. Note that the exact solution of the integral value is 1.

The trapezoidal rule is first applied to integrate the function $f(x) = \sin x$ from the limits $a = 0$ to $b = \pi/2$ by using the numbers of intervals $n = 1$ and 2 with the widths of $h_1 = \pi/2$ and $h_2 = \pi/4$, respectively. The applications lead to the two integral values of $I(h_1) = I_L = 0.7853981643$ and $I(h_2) = I_M = 0.9480594490$. The Romberg integration according to Eq. (6.87) is applied with $k = 1$ to give

$$I = \frac{4^1(0.9480594490) - (0.7853981643)}{4^1 - 1} = 1.0022798780 \tag{6.89}$$

The procedure can be shown in a tabulated form for ease of understanding as

n		$k = 1$
1	0.7853981643 →	1.0022798780
2	0.9480594490	

The relative error computed from the previous more accurate solution is

$$\varepsilon_a = \left| \frac{1.0022798780 - 0.9480594490}{1.0022798780} \right| \times 100\% = 5.4097\% \tag{6.90}$$

Since the computed error is higher than the specified allowable error, the composite trapezoidal rule is applied again using $n = 4$ with the width $h = \pi/8$. The application yields the corresponding integral value of 0.9871158010. Then, the table becomes

n		$k = 1$		$k = 2$
1	0.7853981643 →	1.0022798780 →		0.9999915655
2	0.9480594490 →	1.0001345850		
4	0.9871158010			

where the new integral solution after applying the second Romberg integration $k = 2$ is determined from Eq. (6.87) as

$$I = \frac{4^2(1.0001345850) - (1.0022798780)}{4^2 - 1} = 0.9999915655 \tag{6.91}$$

The new integral solution has the relative error of

$$\varepsilon_a = \left| \frac{0.9999915655 - 1.0001345850}{0.9999915655} \right| \times 100\% = 0.0143\% \tag{6.92}$$

which is still higher than the specified allowable error. The composite trapezoidal rule is applied again with $n = 8$ and the width $h = \pi/16$ leading to the integral value of 0.9967851719. The Romberg integration when $k = 3$ is used and the table is updated as

n		$k=1$	$k=2$	$k=3$
1	0.7853981643	1.0022798780	0.9999915655	1.0000000090
2	0.9480594490	1.0001345850	0.9999998771	
4	0.9871158010	1.0000082960		
8	0.9967851719			

where the newer integral solution of 1.0000000090 is determined from Eq. (6.87) when $k = 3$ as

$$I = \frac{4^3(0.9999998771) - (0.9999915655)}{4^3 - 1} = 1.0000000090 \tag{6.93}$$

With the newer integral solution, the relative error is

$$\varepsilon_a = \left| \frac{1.0000000090 - 0.9999998771}{1.0000000090} \right| \times 100\% = 0.00001\% \tag{6.94}$$

Since the relative error is now less than the specified allowable error of 0.0001%, the Romberg integration process is terminated with the final integral solution of 1.0000000090.

The Romberg integration process as explained in Example 6.10 can be used to develop a computer program such as that shown in Fig. 6.11. The program can be used to integrate an arbitrary function $f(x)$ from the given integration limits a to b and the specified relative error. It is noted that the program stores the integral values in the format as shown below.

$$R_{11} \rightarrow R_{12} \rightarrow R_{13} \rightarrow R_{14}$$
$$R_{21} \rightarrow R_{22} \rightarrow R_{23}$$
$$R_{31} \rightarrow R_{32}$$
$$R_{41}$$

In the above format, R_{14} is the final integral solution.

Fortran

```
      PROGRAM  ROMBERG
C.....A ROMBERG INTEGRATING PROGRAM FOR ESTIMATING
C.....INTEGRAL OF F(X) WITHIN SPECIFIED % ERROR
      DIMENSION R(10,10)
      PI = 4.*ATAN(1.)
      A = 0.
      B = PI/2.
      EPS = .0001
C.....COMPUTE R(1,1):
      FX0 = FUNC(A)
      FXN = FUNC(B)
      R(1,1) = (FX0+FXN)*(B-A)/2.

C.....LOOP OVER NUMBER OF ROMBERG APPLICATIONS:
      DO 100  I=1,9
      N = 2**I
      CALL TRAP(A, B, N, AREA)
      R(I+1,1) = AREA
      DO 200  IC=2,I+1
      K = IC - 1
      IR = 2 + I - IC
      R(IR,IC) = ((4**K)*R(IR+1,K) - R(IR,K))/
     1            (4**K - 1.)
  200 CONTINUE
      ERR = 100.*(R(1,K+1)-R(2,K))/R(1,K+1)
      ERR = ABS(ERR)
      IF(ERR.LE.EPS)  GO TO 300
  100 CONTINUE
  300 CONTINUE
      WRITE(6,400)  R(1,K+1), ERR
  400 FORMAT(' FINAL INTEGRAL VALUE =', E16.10, /,
     *  ' WITH RELATIVE ERROR  =', E12.6, ' %')
      STOP
      END
C-------------------------------------------------
      SUBROUTINE  TRAP(A, B, N, AREA)
C.....MULTIPLE-SEGMENT TRAPEZOIDAL RULE
      H = (B - A)/N
      SUM = 0.
      X = A + H
      DO 10  I=1,N-1
      FX = FUNC(X)
      SUM = SUM + FX
      X = X + H
   10 CONTINUE
      FX0 = FUNC(A)
      FXN = FUNC(B)
      AREA = (FX0 + FXN + 2.*SUM)*H/2.
      RETURN
      END
C-------------------------------------------------
      FUNCTION  FUNC(X)
      FUNC = SIN(X)
      RETURN
      END
```

MATLAB

```
%   PROGRAM  ROMBERG
%   A ROMBERG INTEGRATING PROGRAM FOR ESTIMATING
%   INTEGRAL OF F(X) WITHIN SPECIFIED % ERROR
%------------------------------------------------
func = inline('sin(x)','x');
%------------------------------------------------
a = 0.;
b = pi/2.;
es = 0.0001;
%   COMPUTE R(1,1):
fx0 = func(a);
fxn = func(b);
r(1,1) = (fx0 + fxn)*(b - a)/2.;
%   LOOP OVER NUMBER OF ROMBERG APPLICATIONS:
for i = 1:9
    n = 2^i;
    area = trap(a, b, n);
    r(i+1,1) = area;
    for ic = 2:i+1
        k  = ic - 1;
        ir = 2 + i - ic;
        r(ir,ic) = ((4^k)*r(ir+1,k) - r(ir,k))/ ...
(4^k - 1.);
    end
    err = 100.*(r(1,k+1) - r(2,k))/r(1,k+1);
    err = abs(err);
    if err <= es
        continue
    end
end
fprintf('\nFINAL INTEGRAL VALUE = %16.10e', r(1,k+1));
fprintf('\nWITH RELATIVE ERROR  = %12.6e' , err);
```

```
function area = trap(a, b, n)
%   MULTIPLE-SEGMENT TRAPEZOIDAL RULE
%----------------------------------------
func = inline('sin(x)','x');
%----------------------------------------
h = (b - a)/n;
sum = 0.;
x = a + h;
for i = 1:n-1
    fx = func(x);
    sum = sum + fx;
    x = x + h;
end
fx0 = func(a);
fxn = func(b);
area = (fx0 + fxn + 2.*sum)*h/2.;
```

Figure 6.11 Computer program for determining the integral solution in Example 6.10 by using the Romberg integration method.

6.8 Gauss Integration

Guass integration is one of the most popular integration methods widely used in science and engineering computation. The method is sometimes called Gauss quadrature because the integral is determined from the approximate quadrilateral area under the function to be integrated. Determining the quadrilateral area by using the Gauss integration method is similar to that for the trapezoidal rule as explained in section 6.2. However, the quadrilateral area obtained from the Gauss integration method is more accurate in representing the integral as depicted in Fig. 6.12.

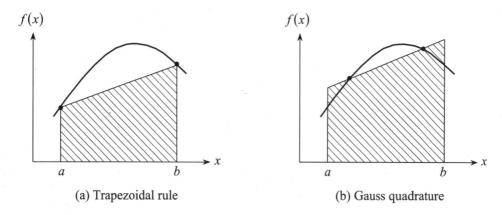

(a) Trapezoidal rule   (b) Gauss quadrature

Figure 6.12 Integral determination from the quadrilater area by the trapezoidal rule and Gauss quadrature.

Figure 6.12 shows the difference in the integral estimation from the area under the function $f(x)$ by using the trapezoidal rule and Guass quadrature. If the distribution of the function $f(x)$ to be integrated is in the concave form as shown in Fig. 6.12(a), the trapezoidal rule will produce a considerable error because the quadrilateral area is determined from the two values of the function $f(x)$ evaluated at the integration limits a and b. The error is substantial if the integrating interval from a to b is large. Figure 6.12(b) shows that a more accurate integral solution can be obtained if the quadrilateral area is determined from the two values of the function $f(x)$ at the two locations away from the two integration limits a and b. The question is, thus, what should be the two locations that can provide a more accurate integral solution.

It is noted that the Newton-Cotes formulas as shown in Eqs. (6.65) - (6.71) can be written in a general form as

$$I = \int_a^b f(x)\,dx \cong \sum_{i=1}^n W_i\, f(x_i) \tag{6.95}$$

For example, the integral formula for the trapezoidal rule in Eq. (6.65) can be written in the form

$$I = W_1\, f(x_1) + W_2\, f(x_2) \tag{6.96}$$

where W_1 and W_2 may be thought as the weights that correspond to the functions evaluated at two appropriate locations. In this case,

$$W_1 \;=\; W_2 \;=\; \frac{b-a}{2} \qquad\qquad (6.97a)$$

$$f(x_1) \;=\; f(a) \qquad ; \qquad f(x_2) \;=\; f(b) \qquad\qquad (6.97b)$$

The integral formula in Eq. (6.96) with the weights and locations in Eqs. (6.97a-b) produce the quadrilateral area as shown in Fig. 6.12(a).

To develop the Gauss integration formula, the integral form as shown in Eq. (6.96) is used. However, the integration limits are from -1 to +1 in the ξ-coordinate direction as shown in Fig. 6.13 so that the developed formula can be used in general. Thus, the integral statement is written in the form

$$I \;=\; \int_{-1}^{1} f(\xi)\,d\xi$$

$$=\; W_1\,f(\xi_1) + W_2\,f(\xi_2) \qquad\qquad (6.98)$$

where the weights W_1, W_2 and the locations ξ_1, ξ_2 are unknowns.

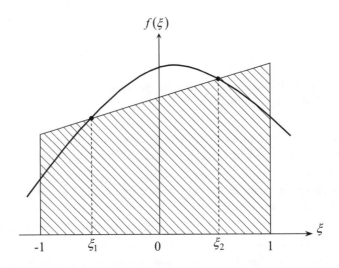

Figure 6.13 Gauss integration in the ξ-coordinate direction.

The four unknowns are to be determined from the four appropriate conditions. The four conditions are such that the exact integral must be obtained by using Eq. (6.98) if the function $f(\xi)$ is constant, linear, parabola or cubic, i.e.,

$$f(\xi) \;=\; 1, \xi, \xi^2, \xi^3 \qquad\qquad (6.99)$$

The function $f(\xi)$ in the four forms of Eq. (6.99) lead to the four conditions according to Eq. (6.98) as follows,

$$W_1 + W_2 \ = \ \int_{-1}^{1} 1 \, d\xi \ = \ 2 \qquad (6.100\text{a})$$

$$W_1 \xi_1 + W_2 \xi_2 \ = \ \int_{-1}^{1} \xi \, d\xi \ = \ 0 \qquad (6.100\text{b})$$

$$W_1 \xi_1^2 + W_2 \xi_2^2 \ = \ \int_{-1}^{1} \xi^2 \, d\xi \ = \ \frac{2}{3} \qquad (6.100\text{c})$$

$$W_1 \xi_1^3 + W_2 \xi_2^3 \ = \ \int_{-1}^{1} \xi^3 \, d\xi \ = \ 0 \qquad (6.100\text{d})$$

Eqs. (6.100a-d) are nonlinear with the four unknowns of W_1, W_2, ξ_1 and ξ_2. These unknowns can be determined by starting from Eq. (6.100b)

$$W_2 \ = \ -\frac{W_1 \xi_1}{\xi_2} \qquad (6.101\text{a})$$

By substituting W_2 from Eq. (6.101a) into Eq. (6.100d),

$$W_1 \xi_1^3 - W_1 \xi_1 \xi_2^2 \ = \ 0$$

or,

$$\xi_1^2 \ = \ \xi_2^2$$

But ξ_1 is not equal to ξ_2, then

$$\xi_1 \ = \ -\xi_2 \qquad (6.101\text{b})$$

By substituting ξ_1 into Eq. (6.101a), the weight $W_1 = W_2$. With the use of Eq. (6.100a), it is found that

$$W_1 \ = \ W_2 \ = \ 1 \qquad (6.101\text{c})$$

Then, by substituting Eqs. (6.101b-c) into Eq. (6.100c),

$$\xi_1^2 + \xi_2^2 \ = \ \frac{2}{3}$$

to give

$$\xi_1 \ = \ -\frac{1}{\sqrt{3}} \qquad (6.101\text{d})$$

In summary, the weights W_1, W_2 and the two locations ξ_1, ξ_2 are

$$W_1 \ = \ W_2 \ = \ 1 \qquad (6.102\text{a})$$

$$\xi_1 \ = \ -\frac{1}{\sqrt{3}} \ = \ -0.5773502692 \qquad (6.102\text{b})$$

$$\xi_2 \ = \ +\frac{1}{\sqrt{3}} \ = \ +0.5773502692 \qquad (6.102\text{c})$$

The locations ξ_1 and ξ_2 are called the Gauss point locations. With the weights and locations in Eqs. (6.102a-c), the integral formula for the two-point Gauss integration in Eq. (6.98) is

$$I = (1)f\left(-\frac{1}{\sqrt{3}}\right) + (1)f\left(+\frac{1}{\sqrt{3}}\right) \tag{6.103}$$

The formula provides an exact integral solution if the function to be integrated is a polynomial of third order or less.

If the function to be integrated is a higher order polynomial or other complicated functions, the same procedure as shown in Eqs. (6.98) - (6.103) can still be applied to derive the Gauss integration formulas to produce more accurate integral solutions. For example, a more accurate integral solution can be derived by using 3 terms in the formula as

$$I = \int_{-1}^{1} f(\xi)\,d\xi$$

$$= W_1 f(\xi_1) + W_2 f(\xi_2) + W_3 f(\xi_3) \tag{6.104}$$

The six unknowns in Eq. (6.104) can be determined to yield

$$W_1 = W_3 = 5/9 \quad ; \quad W_2 = 8/9 \tag{6.105a}$$

$$\xi_1 = -\sqrt{0.6} = -0.7745966692 \quad ; \quad \xi_2 = 0 \tag{6.105b}$$

$$\xi_3 = +\sqrt{0.6} = +0.7745966692 \tag{6.105c}$$

So that the integral formula for the three-point Gauss integration is

$$I = \left(\frac{5}{9}\right)f\left(-\sqrt{0.6}\right) + \left(\frac{8}{9}\right)f(0) + \left(\frac{5}{9}\right)f\left(+\sqrt{0.6}\right) \tag{6.106}$$

which provides an exact integral solution if the function to be integrated is a polynomial of fifth order or less. The same procedure above can be further applied to derive the Gauss integration formulas with the weights W_i and locations ξ_i for n Gauss points as shown in Table 6.3. The formulas are known as the *Gauss-Legendre formulas* for which the integral value is determined from

$$I = \int_{-1}^{1} f(\xi)\,d\xi \cong \sum_{i=1}^{n} W_i f(\xi_i) \tag{6.107}$$

From the explanation above, the Gauss integration formulas can produce an exact integral solution if the function to be integrated is a polynomial of order $2n-1$ or less.

For n Gauss points, the weights W_i and locations ξ_i as shown in Table 6.3 are determined by first assuming the function $f(\xi)$ in the form

$$f(\xi) = 1, \xi, \xi^2, \xi^3, ..., \xi^{2n-1} \tag{6.108}$$

Then, the same procedure as explained by Eqs. (6.99)-(6.100) for the two-point integration is applied to yield $2n$ equations with $2n$ unknowns as follows,

$$W_1 + W_2 + ... + W_n = 2$$

$$W_1 \xi_1 + W_2 \xi_2 + ... + W_n \xi_n = 0$$

$$W_1 \xi_1^2 + W_2 \xi_2^2 + \ldots + W_n \xi_n^2 = \frac{2}{3}$$

$$W_1 \xi_1^3 + W_2 \xi_2^3 + \ldots + W_n \xi_n^3 = 0 \qquad (6.109)$$

$$\vdots \qquad \vdots$$

$$W_1 \xi_1^{2n-2} + \ldots + W_n \xi_n^{2n-2} = \frac{2}{2n-1}$$

$$W_1 \xi_1^{2n-1} + \ldots + W_n \xi_n^{2n-1} = 0$$

The values of the weights W_i and locations ξ_i as shown in Table 6.3 are obtained by solving Eq. (6.109).

Table 6.3 Weights and locations of the Gauss-Legendre formulas.

Gauss points n	Locations $\pm \xi_i$	Weights W_i
1	0.0000000000	2.0000000000
2	0.5773502692	1.0000000000
3	0.0000000000	0.8888888889
	0.7745966692	0.5555555556
4	0.3399810436	0.6521451549
	0.8611363116	0.3478548451
5	0.0000000000	0.5688888889
	0.5384693101	0.4786286705
	0.9061798459	0.2369268850
6	0.2386191861	0.4679139346
	0.6612093865	0.3607615730
	0.9324695142	0.1713244924

The Gauss-Legendre formulas in Eq. (6.107) were developed for integrating a function $f(\xi)$ from the limits -1 to +1 along the ξ-coordinate. Since a given integrand is normally in form of the function $f(x)$ that needs to be integrated along the x-coordinate from the lower limit a to upper limit b, the coordinate transformation from x- to ξ-coordinate must be first performed. Transformation of the function $f(x)$ from x- to ξ-coordinate can be done easily by considering Fig. 6.14.

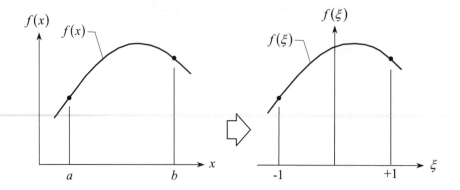

Figure 6.14 Transformation of the function to be integrated from x- to ξ-coordinate
before applying Gauss-Legendre integration formulas.

Coordinate transformation from x to ξ may be done by using linear relation in the form

$$x = c_0 + c_1 \xi \tag{6.110}$$

where c_0 and c_1 are constants that can be determined from the integration limits as follows

at $\qquad x = a \qquad ; \qquad a = c_0 + c_1(-1) \tag{6.111a}$

at $\qquad x = b \qquad ; \qquad b = c_0 + c_1(+1) \tag{6.111b}$

which give

$$c_0 = \frac{a+b}{2} \qquad \text{and} \qquad c_1 = \frac{b-a}{2} \tag{6.112}$$

Then, the relation between the two coordinates is

$$x = \frac{a+b}{2} + \frac{b-a}{2}\xi \tag{6.113}$$

And $\qquad\qquad\qquad dx = \frac{b-a}{2}d\xi \tag{6.114}$

In summary, the Gauss integration starts from the integral of a function $f(x)$ along the x-coordinate with the limits of a and b as

$$I = \int_a^b f(x)dx \tag{6.115}$$

The integral is first transformed from x- to ξ-coordinate by using Eqs. (6.113) and (6.114) so that the integration limits change from a and b to -1 and +1 as

$$I = \int_{-1}^{1} f\left(\frac{a+b}{2} + \frac{b-a}{2}\xi\right)\left(\frac{b-a}{2}\right)d\xi$$

$$= \left(\frac{b-a}{2}\right)\int_{-1}^{1} f\left(\frac{a+b}{2} + \frac{b-a}{2}\xi\right)d\xi \tag{6.116}$$

Then, the Gauss-Legendre formula according to Eq. (6.107) is applied to yield

$$I = \left(\frac{b-a}{2}\right)\sum_{i=1}^{n} W_i\, f\left(\frac{a+b}{2} + \frac{b-a}{2}\xi_i\right) \tag{6.117}$$

The entire process of the Guass integration method can be understood clearly by solving the integral in Example 6.1 again as shown in the following example.

Example 6.11 Use the two-point Gauss integration method to integrate

$$I = \int_0^2 f(x)\,dx = \int_0^2 \left(2x^3 - 5x^2 + 3x + 1\right)dx \tag{6.12}$$

Then, develop a computer program for integrating a given function by using one- to six-point Gauss formulas. Compare the computed integrals obtained from each case with the exact solution of 2.666667.

By starting from the coordinate transformation from x to ξ according to Eq. (6.113)

$$x = \frac{a+b}{2} + \frac{b-a}{2}\xi \tag{6.113}$$

Since the integration limits are $a = 0$ and $b = 2$, Eq. (6.113) reduces to

$$x = 1 + \xi \tag{6.118}$$

Then, the integral Eq. (6.12) becomes

$$I = \int_0^2 \left(2x^3 - 5x^2 + 3x + 1\right)dx$$

$$= \int_{-1}^1 \left[2(1+\xi)^3 - 5(1+\xi)^2 + 3(1+\xi) + 1\right]\left(\frac{2-0}{2}\right)d\xi$$

By using the two-point Gauss formula with $n = 2$, Eq. (6.117) is

$$I = \left(\frac{2-0}{2}\right)\sum_{i=1}^{2} W_i\left[2(1+\xi_i)^3 - 5(1+\xi_i)^2 + 3(1+\xi_i) + 1\right]$$

$$= W_1\left[2(1+\xi_1)^3 - 5(1+\xi_1)^2 + 3(1+\xi_1) + 1\right]$$

$$+ W_2\left[2(1+\xi_2)^3 - 5(1+\xi_2)^2 + 3(1+\xi_2) + 1\right]$$

But from Table 6.3, $W_1 = W_2 = 1$ and $\xi_1 = -\sqrt{1/3}$, $\xi_2 = +\sqrt{1/3}$, then

$$I = (1)[0.1509982 - 0.8931639 + 1.2679492 + 1]$$

$$+ (1)[7.8490018 - 12.440169 + 4.7320508 + 1]$$

$$I = 2.666667$$

which is equal to the exact solution because the two-point Gauss formula can provide exact integral if the integrand is a polynomial of third order or less.

Figure 6.15 shows a computer program for the Gauss-Legendre integration method by using one to six integration points. The program was developed so that the computational procedure is easy to understand. The program starts from declaring the values of the Gauss locations ξ and their weights W. The program reads the integration limits a and b from user input. The function $f(x)$ to be integrated is declared in the program. The function can be replaced by the new function desired for integration.

Fortran

```
      PROGRAM  GAUSINT
C.....A GAUSS-LEGENDRE INTEGRATION PROGRAM FOR
C.....ESTIMATING INTEGRATION OF FUNCTION F(X)
C.....USING 1 THROUGH 6 GAUSS POINTS
      DIMENSION  XI(21), W(21)
      DATA  XI /  0.0000000,-.5773503,+.5773503,
     *  -.7745967,0.0000000,+.7745967,-.8611363,
     *  -.3399810,+.3399810,+.8611363,-.9061798,
     *  -.5384693,0.0000000,+.5384693,+.9061798,
     *  -.9324695,-.6612094,-.2386192,+.2386192,
     *  +.6612094,+.9324695 /
      DATA  W /  2.0000000,1.0000000,1.0000000,
     *  0.5555556,0.8888889,0.5555556,0.3478549,
     *  0.6521452,0.6521452,0.3478549,0.2369269,
     *  0.4786287,0.5688889,0.4786287,0.2369269,
     *  0.1713245,0.3607616,0.4679139,0.4679139,
     *  0.3607616,0.1713245 /
C.....READ LIMITS OF INTEGRATION:
      READ(5,*)  A, B
      A0 = (A + B)/2.
      A1 = (B - A)/2.
C.....PERFORM 1 THRU 6 GUASS POINTS COMPUTATION:
      IC = 1
      DO 100  NG=1,6
      SUM = 0.
      DO 200  ITERMS=1,NG
      X = A0 + A1*XI(IC)
      AI = FUNC(X)
      SUM = SUM + W(IC)*AI
      IC = IC + 1
  200 CONTINUE
      SUM = A1*SUM
      WRITE(6,300)  NG, SUM
  300 FORMAT(' RESULT OF INTEGRATION WITH', I2,
     *        ' GAUSS POINT(S) IS', E12.6)
  100 CONTINUE
      STOP
      END
C-------------------------------------------------
      FUNCTION FUNC(X)
      FUNC = 2.*X*X*X - 5.*X*X + 3.*X + 1.
      RETURN
      END
```

MATLAB

```
%   PROGRAM  GAUSINT
%   A GAUSS-LEGENDRE INTEGRATION PROGRAM FOR
%   ESTIMATING INTEGRATION OF FUNCTION F(X)
%   USING 1 THROUGH 6 GAUSS POINTS
%-------------------------------------------------
func = inline('2.*x^3 - 5.*x^2 + 3.*x + 1.','x');
%-------------------------------------------------
xi = [ 0.0000000; -0.5773503;  0.5773503;
      -0.7745967;  0.0000000;  0.7745967;
      -0.8611363; -0.3399810;  0.3399810;
       0.8611363; -0.9061798; -0.5384693;
       0.0000000;  0.5384693;  0.9061798;
      -0.9324695; -0.6612094; -0.2386192;
       0.2386192;  0.6612094;  0.9324695];
w  = [ 2.0000000;  1.0000000;  1.0000000;
       0.5555556;  0.8888889;  0.5555556;
       0.3478549;  0.6521452;  0.6521452;
       0.3478549;  0.2369269;  0.4786287;
       0.5688889;  0.4786287;  0.2369269;
       0.1713245;  0.3607616;  0.4679139;
       0.4679139;  0.3607616;  0.1713245];
%   READ LIMITS OF INTEGRATION:
a = input('Enter value of a: ');
b = input('Enter value of b: ');
a0 = (a + b)/2.;
a1 = (b - a)/2.;
%   PERFORM 1 THRU 6 GUASS POINTS COMPUTATION:
ic = 1;
for ng = 1:6
    sum = 0.;
    for iterms = 1:ng
        x  = a0 + a1*xi(ic);
        AI = func(x);
        sum = sum + w(ic)*AI;
        ic = ic + 1;
    end
    sum = a1*sum;
    fprintf('\nRESULT OF INTEGRATION WITH %2d',ng );
    fprintf('  GAUSS POINT(S) IS %12.6e'     ,sum);
end
```

Figure 6.15 Computer program for Gauss integration method.

Figure 6.16 shows the computed integral values of Example 6.11 obtained from using the computer program in Fig. 6.15. The exact integral solution is 2.666667. The figure shows that the computed integral values are exact when two- to six-point Gauss integration formulas are used since the function to be integrated is a third-order polynomial. It should be noted that for a general function $f(x)$ that may be complicated, more accurate solutions are obtained by using the increased number of Gauss integration points.

```
RESULT OF INTEGRATION WITH 1 GAUSS POINT(S) IS .200000E+01
RESULT OF INTEGRATION WITH 2 GAUSS POINT(S) IS .266667E+01
RESULT OF INTEGRATION WITH 3 GAUSS POINT(S) IS .266667E+01
RESULT OF INTEGRATION WITH 4 GAUSS POINT(S) IS .266667E+01
RESULT OF INTEGRATION WITH 5 GAUSS POINT(S) IS .266667E+01
RESULT OF INTEGRATION WITH 6 GAUSS POINT(S) IS .266667E+01
```

Figure 6.16 Computed integral values of Example 6.11 from the Gauss integration computer program in Fig. 6.15.

6.9 Multiple Integration

Multiple integrations that are frequently used in science and engineering are the double integration over the area A

$$I \;=\; \iint_A f(x, y)\, dA \;\;=\; \iint_A f(x, y)\, dx\, dy \tag{6.119}$$

and the triple integration over the volume V

$$I \;=\; \iiint_V f(x, y)\, dV \;\;=\; \iiint_V f(x, y)\, dx\, dy\, dz \tag{6.120}$$

Multiple integrations can be done straightforwardly by performing the single integration in each coordinate direction, one at a time.

For example, the double integration of a function $f(x, y)$ in Eq. (6.119) for the intervals $a < x < b$ and $c < y < d$ can be expressed in the form

$$I \;=\; \iint_A f(x, y)\, dx\, dy \;\;=\; \int_a^b \left(\int_c^d f(x, y)\, dy \right) dx \tag{6.121}$$

The function $f(x, y)$ is first integrated along the y-coordinate direction by keeping x constant. The result is integrated again along the x-coordinate direction to obtain the final integral solution.

The double integral of the function $f(x, y)$ in Eq. (6.121) can also be performed by first integrating the function along the x-coordinate direction in the inner bracket as shown below

$$I \;=\; \int_c^d \left(\int_a^b f(x, y)\, dx \right) dy \tag{6.122}$$

A numerical method in the form of Eq. (6.95) can be applied to approximate the integral in the inner bracket along the x-direction for Eq. (6.122) as

$$I \;\cong\; \int_c^d \left(\sum_{i=1}^n W_{xi}\, f(x_i, y) \right) dy \tag{6.123}$$

For an example of using the trapezoidal rule ($n = 2$), the weights are $W_{x1} = W_{x2} = (b-a)/2$. The method is applied again to numerically integrate the result obtained along the y-coordinate direction to yield

$$I \cong \sum_{i=1}^{n} \sum_{j=1}^{n} W_{xi} W_{yj} f(x_i, y_j) \tag{6.124}$$

where $W_{y1} = W_{y2} = (d-c)/2$. Other numerical methods, such as the Simpson's rule and the Boole's rule, can be applied in the same fashion.

Example 6.12 Use the trapezoidal rule to determine the double integral

$$I = \int_0^1 \int_1^2 (1+x) y \, dx \, dy \tag{6.125}$$

Herein, $a=1$, $b=2$, $c=0$, $d=1$, $f(x,y)=(1+x)y$ and $W_{x1} = W_{x2} = (b-a)/2 = 0.5$, $W_{y1} = W_{y2} = (d-c)/2 = 0.5$, then, Eq. (6.124) yields

$$
\begin{aligned}
I &= W_{x1} W_{y1} f(x_1, y_1) + W_{x2} W_{y1} f(x_2, y_1) + W_{x1} W_{y2} f(x_1, y_2) + W_{x2} W_{y2} f(x_2, y_2) \\
&= W_{x1} W_{y1} f(a, c) + W_{x2} W_{y1} f(b, c) + W_{x1} W_{y2} f(a, d) + W_{x2} W_{y2} f(b, d) \\
&= (0.5)(0.5) f(1, 0) + (0.5)(0.5) f(2, 0) + (0.5)(0.5) f(1, 1) + (0.5)(0.5) f(2, 1) \\
&= (0.25)(0) + (0.25)(0) + (0.25)(2) + (0.25)(3) \\
&= 0 + 0 + 0.5 + 0.75
\end{aligned}
$$

$$I = 1.25 \tag{6.126}$$

The computed integral solution is exact because the given function $f(x, y)$ is linear in both x- and y-directions. The method won't provide exact solution if the function $f(x, y)$ is in a more complex form, such as $f(x,y) = (1+x^2) y^3$ or $f(x,y) = (1+\sin x) e^y$. However, accurate integral solution of these functions can be obtained by using composite trapezoidal rule with many segments.

In solving practical science and engineering problems, the Gauss integration method may be preferred because of its solution accuracy. To apply the Gauss integration method for the double integration, the integral limits from a to b in the x-direction are first changed to -1 and +1 in the ξ-direction. Similarly, the integral limits from c to d in the y-direction are first changed to -1 and +1 in the η-direction. The linear relations between the x-y and ξ-η coordinate systems similar to that explained in Eq. (6.110) and Example 6.11 can be used for both the x- and y-directions as

$$x = \frac{a+b}{2} + \frac{b-a}{2}\xi \quad ; \quad dx = \frac{b-a}{2}d\xi \tag{6.127a}$$

$$y = \frac{c+d}{2} + \frac{d-c}{2}\eta \quad ; \quad dy = \frac{d-c}{2}d\eta \tag{6.127b}$$

Then, the double integral in Eq. (6.122) becomes

$$I = \int_c^d \left(\int_a^b f(x, y) \, dx \right) dy$$

$$= \int_c^d \left(\int_{-1}^1 f\left(\frac{a+b}{2} + \frac{b-a}{2}\xi, y\right)\left(\frac{b-a}{2}\right)d\xi \right) dy$$

$$= \int_c^d \left(\left(\frac{b-a}{2}\right)\int_{-1}^1 f\left(\frac{a+b}{2} + \frac{b-a}{2}\xi, y\right)d\xi \right) dy$$

$$= \left(\frac{b-a}{2}\right)\left(\frac{d-c}{2}\right)\int_{-1}^1\int_{-1}^1 f\left(\frac{a+b}{2} + \frac{b-a}{2}\xi, \frac{c+d}{2} + \frac{d-c}{2}\eta\right)d\xi \, d\eta$$

$$= \left(\frac{b-a}{2}\right)\left(\frac{d-c}{2}\right)\sum_{i=1}^n\sum_{j=1}^n W_i W_j \, f\left(\frac{a+b}{2} + \frac{b-a}{2}\xi_i, \frac{c+d}{2} + \frac{d-c}{2}\eta_j\right) \quad (6.128)$$

For example, the two Gauss points ($n = 2$) in each ξ- and η-coordinate direction is used, the weights are $W_1 = W_2 = 1$ with the Gauss point locations of $\xi_1 = \eta_1 = -\sqrt{1/3}$ and $\xi_2 = \eta_2 = +\sqrt{1/3}$.

Example 6.13 Use the Gauss integration method to determine the double integral

$$I = \int_0^1\int_1^2 (1+x)y \, dx \, dy \quad (6.125)$$

Herein, $a = 1$, $b = 2$, $c = 0$, $d = 1$, then, Eqs. (6.127a-b) are

$$x = 1.5 + 0.5\xi \quad ; \quad dx = 0.5d\xi$$

$$y = 0.5 + 0.5\eta \quad ; \quad dy = 0.5d\eta$$

Then, Eq. (6.125) becomes

$$I = \int_{-1}^1\int_{-1}^1 [(1+1.5+0.5\xi)(0.5+0.5\eta)](0.5d\xi)(0.5d\eta)$$

$$= (0.25)\int_{-1}^1\int_{-1}^1 [(2.5+0.5\xi)(0.5+0.5\eta)]d\xi \, d\eta$$

$$= (0.0625)\int_{-1}^1\int_{-1}^1 (5+\xi)(1+\eta)d\xi \, d\eta$$

By using the two Gauss points ($n=2$) in each ξ- and η-coordinate direction,

$$I = (0.0625)\sum_{i=1}^2\sum_{j=1}^2 W_i W_j (5+\xi_i)(1+\eta_j)$$

$$= (0.0625)[W_1 W_1(5+\xi_1)(1+\eta_1) + W_1 W_2(5+\xi_1)(1+\eta_2)$$
$$+ W_2 W_1(5+\xi_2)(1+\eta_1) + W_2 W_2(5+\xi_2)(1+\eta_2)]$$

With the weights of $W_1 = W_2 = 1$, and the Gauss point locations of $\xi_1 = \eta_1 = -\sqrt{1/3}$, $\xi_2 = \eta_2 = +\sqrt{1/3}$, then

$$I = (0.0625)\left[(1)(1)\left(5-\sqrt{1/3}\right)\left(1-\sqrt{1/3}\right) + (1)(1)\left(5-\sqrt{1/3}\right)\left(1+\sqrt{1/3}\right)\right.$$

$$\left. + (1)(1)\left(5+\sqrt{1/3}\right)\left(1-\sqrt{1/3}\right) + (1)(1)\left(5+\sqrt{1/3}\right)\left(1+\sqrt{1/3}\right)\right]$$

$$= (0.0625)[1.869232+6.976068+2.357266+8.797435]$$

$$= (0.0625)(20)$$

$$I = 1.25$$

The Gauss integration method yields the exact integral solution because the given function $f(x, y)$ is linear in both x- and y-directions. If the given function $f(x, y)$ is complicated, many Gauss points must be used to produce a more accurate solution. Since the Gauss integration method provides high integral solution accuracy, the method is widely used for solving practical problems by embedding itself in many commercial software.

6.10 MATLAB Commands for Integration

One of the simple MATLAB commands for integration is `trapz`. The command employs the composite trapezoidal rule to estimate the integral value of a given function. The format for using the command is

$$z = \text{trapz}(x, y)$$

where   x   is the data along the x-coordinate
        y   is the function to be integrated
        z   is the integral value

Example 6.14 Determine the integral in Eq. (6.12) by using the MATLAB command `trapz`.

$$I = \int_0^2 f(x)\,dx = \int_0^2 \left(2x^3 - 5x^2 + 3x + 1\right)dx \tag{6.12}$$

Compare the computed value with the exact solution of 2.666667.

It is noted that the command `trapz` implicitly specifies the segment length h in the x input data. As an example of $h = 0.4$, the input data, function and output in x, y and z variables are as follows,

```
>> x = [0:0.4:2];
>> y = 2*x.^3-5*x.^2+3*x+1;
>> z = trapz(x,y)

z =

    2.7200
```

If the segment length h is reduced to 0.01, the output solution is closer to the exact solution.

```
>> x = [0:0.01:2];
>> y = 2*x.^3-5*x.^2+3*x+1;
>> z = trapz(x,y)

z =

    2.6667
```

Another useful MATLAB command for integration is `cumtrapz`. The command determines cumulative integral. The format of the command is

$$z = cumtrapz(x,y)$$

where   x   is the data along the *x*-coordinate
        y   is the function to be integrated
        z   is the cumulative integral value. For example, z(3) is the integral value for the interval from x(1) to x(3).

MATLAB also provides simple commands for double and triple integrations. The command format for double integration is,

$$I = dblquad(function,xmin,xmax,ymin,ymax,tol)$$

where   I   is the integral value
        function   is the function to be integrated input by using the command `inline`
        xmin   is the lower limit of x
        xmax   is the upper limit of x
        ymin   is the lower limit of y
        ymax   is the upper limit of y
        tol   is the specified tolerance, default value is 1×10^{-6}.

Example 6.15 Use MATLAB function `dblquad` to determine the integral

$$I = \int_0^1 \int_1^2 (1+x)y \, dx \, dy \qquad (6.125)$$

The command entered and the output integral value are as follows.

```
>> I = dblquad(inline('(1+x)*y','x','y'),1,2,0,1)

I =

    1.2500
```

The MATLAB command `triplequad` is used for determining the triple integration. The command format is similar to the `dblquad` command except the additional limits in the *z*-coordinate,

$$I = triplequad(function,xmin,xmax,ymin,ymax,zmin,zmax,tol)$$

Example 6.16 Use the MATLAB command `triplequad` to determine the integral

$$\int_{-2}^{2}\int_{0}^{2}\int_{-3}^{1}\left(x^{3}-3yz\right)dx\,dy\,dz \tag{6.129}$$

It is noted that the exact solution is -160. The command entered and the output integral value from MATLAB are as follows.

```
>> I = triplequad(inline('x.^3-3*y*z','x','y','z'),-3,1,0,2,-2,2)

I   =

    -160.0000
```

6.11 Differentiation

Understanding physical meanings of derivatives and procedures for determining them are essential to study higher-level numerical methods in the following chapters. The meaning for the derivative of a function $f(x)$ is explained by Fig. 6.17.

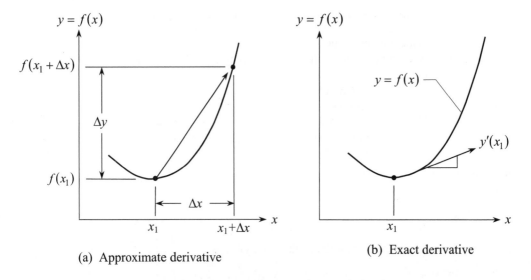

(a) Approximate derivative   (b) Exact derivative

Figure 6.17 Approximate and exact derivatives.

Figure 6.17(a) shows the distribution of a function y that varies with the independent variable x. An approximate derivative of the function y with respect to x is

$$\frac{\Delta y}{\Delta x} = \frac{f(x_1+\Delta x) - f(x_1)}{\Delta x} \tag{6.130}$$

As Δx approaches zero, the exact derivative at x_1 is

$$\frac{dy}{dx} = \lim_{\Delta x \to 0} \frac{f(x_1 + \Delta x) - f(x_1)}{\Delta x} \tag{6.131}$$

The exact derivative dy/dx is the slope at x_1 normally denoted by y' or $f'(x)$ as shown in Fig. 6.17(b).

Many functions $f(x)$ studied in high school or first year college are simple and their derivatives can be determined exactly. For example,

$$y = f(x) = x^n \tag{6.132}$$

then, its exact derivative is

$$\frac{dy}{dx} = \frac{df(x)}{dx} = nx^{n-1} \tag{6.133}$$

Most of the functions in practice, however, are complicated and their exact derivatives can not be determined easily. Thus, their approximate derivatives are determined instead by starting from the Taylor's series as shown in Eq. (2.29),

$$f(x_{i+1}) = f(x_i) + hf'(x_i) + \frac{h^2}{2!}f''(x_i) + \dots \tag{6.134}$$

where h is the width between the locations x_i and x_{i+1}. From Eq. (6.134), the first derivative of the function $f(x)$ at x_i is

$$f'(x_i) = \frac{f(x_{i+1}) - f(x_i)}{h} - \frac{h}{2!}f''(x_i) + \dots \tag{6.135}$$

Or,

$$f'(x_i) = \frac{f(x_{i+1}) - f(x_i)}{h} + O(h) \tag{6.136}$$

where $O(h)$ represents the error of order h if the first term on the right-hand side of Eq. (6.136) is used to approximate the derivative. The approximation is sometimes called the *first forward divided difference* because the two values of the function $f(x)$ at x_i and x_{i+1} are used to determine the derivative as shown in Fig. 6.18(a).

Similar to Eq. (6.134), the Taylor's series can be used to determine the value of the function $f(x)$ at location x_{i-1} as

$$f(x_{i-1}) = f(x_i) - hf'(x_i) + \frac{h^2}{2!}f''(x_i) - \dots \tag{6.137}$$

So that the first-order derivative at x_i is

$$f'(x_i) = \frac{f(x_i) - f(x_{i-1})}{h} + \frac{h}{2!}f''(x_i) - \dots \tag{6.138}$$

Or,

$$f'(x_i) = \frac{f(x_i) - f(x_{i-1})}{h} + O(h) \tag{6.139}$$

Approximate derivative by using the first term on the right-hand side of Eq. (6.139) is called the *first backward divided difference* because the two values of the function $f(x)$ at x_i and x_{i-1} are used to determine the derivative as shown in Fig. 6.18(a).

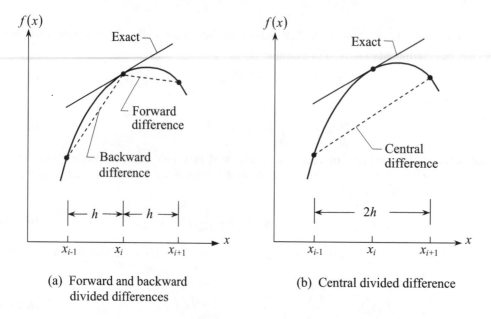

(a) Forward and backward
     divided differences

(b) Central divided difference

Figure 6.18 Comparisons between the approximate derivatives from the forward, backward
and central divided-differences with the exact derivative.

By subtracting Eq. (6.137) from (6.134)

$$f(x_{i+1}) - f(x_{i-1}) \;=\; 2h\,f'(x_i) + \frac{2h^3}{3!}f'''(x_i) + \dots \tag{6.140}$$

the first derivative at x_i is

$$f'(x_i) \;=\; \frac{f(x_{i+1}) - f(x_{i-1})}{2h} - \frac{h^2}{3!}f'''(x_i) - \dots \tag{6.141}$$

or,

$$f'(x_i) \;=\; \frac{f(x_{i+1}) - f(x_{i-1})}{2h} + O(h^2) \tag{6.142}$$

The first derivative is estimated by using values of the function $f(x)$ at x_{i+1} and x_{i-1}. The approximate derivative by using the first term on the right-hand-side of Eq. (6.142) as shown in Fig. 6.18(b) is called the *central divided difference* which has the error of order h^2.

Higher order derivatives of the function $f(x)$ can be derived in the same manner. For example, derivation of the second derivative may start from the Taylor's series for determining the function $f(x)$ at x_{i+2} from the values at x_i as follows.

$$f(x_{i+2}) = f(x_i) + (2h)f'(x_i) + \frac{(2h)^2}{2!}f''(x_i) + \dots \tag{6.143}$$

By multiplying the factor of two onto Eq. (6.134) and subtracting it from Eq. (6.143),

$$f(x_{i+2}) - 2f(x_{i+1}) = -f(x_i) + h^2 f''(x_i) + \dots \tag{6.144}$$

the second derivative of the function $f(x)$ at x_i is obtained as

$$f''(x_i) = \frac{f(x_{i+2}) - 2f(x_{i+1}) + f(x_i)}{h^2} + O(h) \tag{6.145}$$

The first term on the right-hand side of Eq. (6.145) is called the *second forward divided difference* which can be used to approximate the second derivative of the function. The approximate value has the error of order h. Tables 6.4, 6.5 and 6.6 summarize the first- to fourth-order forward, backward and central divided differences of the function $f(x)$ evaluated at x_i, respectively.

Table 6.4 Forward divided differences with error of order $O(h)$.

$$f'(x_i) = [f(x_{i+1}) - f(x_i)]/h$$

$$f''(x_i) = [f(x_{i+2}) - 2f(x_{i+1}) + f(x_i)]/h^2$$

$$f'''(x_i) = [f(x_{i+3}) - 3f(x_{i+2}) + 3f(x_{i+1}) - f(x_i)]/h^3$$

$$f''''(x_i) = [f(x_{i+4}) - 4f(x_{i+3}) + 6f(x_{i+2}) - 4f(x_{i+1}) + f(x_i)]/h^4$$

Table 6.5 Backward divided differences with error of order $O(h)$.

$$f'(x_i) = [f(x_i) - f(x_{i-1})]/h$$

$$f''(x_i) = [f(x_i) - 2f(x_{i-1}) + f(x_{i-2})]/h^2$$

$$f'''(x_i) = [f(x_i) - 3f(x_{i-1}) + 3f(x_{i-2}) - f(x_{i-3})]/h^3$$

$$f''''(x_i) = [f(x_i) - 4f(x_{i-1}) + 6f(x_{i-2}) - 4f(x_{i-3}) + f(x_{i-4})]/h^4$$

Table 6.6 Central divided differences with error of order $O(h^2)$.

$$f'(x_i) = [f(x_{i+1}) - f(x_{i-1})]/2h$$

$$f''(x_i) = [f(x_{i+1}) - 2f(x_i) + f(x_{i-1})]/h^2$$

$$f'''(x_i) = [f(x_{i+2}) - 2f(x_{i+1}) + 2f(x_{i-1}) - f(x_{i-2})]/2h^3$$

$$f''''(x_i) = [f(x_{i+2}) - 4f(x_{i+1}) + 6f(x_i) - 4f(x_{i-1}) + f(x_{i-2})]/h^4$$

The approximate derivatives presented in Tables 6.4 - 6.6 have errors of order h. For example, the first-order derivative of the function $f(x)$ as shown in Eq. (6.136) and in Table 6.4 has the error of order h. A more accurate derivative the function $f(x)$ can be derived by starting from Eq. (6.135)

$$f'(x_i) \; = \; \frac{f(x_{i+1}) - f(x_i)}{h} - \frac{h}{2} f''(x_i) + O(h^2) \tag{6.146}$$

and substituting the second-order derivative from Eq. (6.145) into it to yield

$$f'(x_i) \; = \; \frac{-f(x_{i+2}) + 4f(x_{i+1}) - 3f(x_i)}{2h} + O(h^2) \tag{6.147}$$

The first-order derivative of the function $f(x)$ as obtained in Eq. (6.147) has the error of order h^2. The same process as explained above can be applied to derive the high-order derivatives of the function $f(x)$. Their results by using the forward, backward and central divided differences are presented in Tables 6.7, 6.8 and 6.9, respectively.

Table 6.7 Forward divided differences with error of order $O(h^2)$.

$$f'(x_i) \; = \; [-f(x_{i+2}) + 4f(x_{i+1}) - 3f(x_i)]/2h$$

$$f''(x_i) \; = \; [-f(x_{i+3}) + 4f(x_{i+2}) - 5f(x_{i+1}) + 2f(x_i)]/h^2$$

$$f'''(x_i) \; = \; [-3f(x_{i+4}) + 14f(x_{i+3}) - 24f(x_{i+2}) + 18f(x_{i+1}) - 5f(x_i)]/2h^3$$

$$f''''(x_i) \; = \; [-2f(x_{i+5}) + 11f(x_{i+4}) - 24f(x_{i+3}) + 26f(x_{i+2}) - 14f(x_{i+1}) + 3f(x_i)]/h^4$$

Table 6.8 Backward divided differences with error of order $O(h^2)$.

$$f'(x_i) \; = \; [3f(x_i) - 4f(x_{i-1}) + f(x_{i-2})]/2h$$

$$f''(x_i) \; = \; [2f(x_i) - 5f(x_{i-1}) + 4f(x_{i-2}) - f(x_{i-3})]/h^2$$

$$f'''(x_i) \; = \; [5f(x_i) - 18f(x_{i-1}) + 24f(x_{i-2}) - 14f(x_{i-3}) + 3f(x_{i-4})]/2h^3$$

$$f''''(x_i) \; = \; [3f(x_i) - 14f(x_{i-1}) + 26f(x_{i-2}) - 24f(x_{i-3}) + 11f(x_{i-4}) - 2f(x_{i-5})]/h^4$$

Table 6.9 Central divided differences with error of order $O(h^4)$.

$$f'(x_i) \; = \; [-f(x_{i+2}) + 8f(x_{i+1}) - 8f(x_{i-1}) + f(x_{i-2})]/12h$$

$$f''(x_i) \; = \; [-f(x_{i+2}) + 16f(x_{i+1}) - 30f(x_i) + 16f(x_{i-1}) - f(x_{i-2})]/12h^2$$

$$f'''(x_i) \; = \; [-f(x_{i+3}) + 8f(x_{i+2}) - 13f(x_{i+1}) + 13f(x_{i-1}) - 8f(x_{i-2}) + f(x_{i-3})]/8h^3$$

$$f''''(x_i) \; = \; [-f(x_{i+3}) + 12f(x_{i+2}) - 39f(x_{i+1}) + 56f(x_i) - 39f(x_{i-1})$$
$$+ 12f(x_{i-2}) - f(x_{i-3})]/6h^4$$

Example 6.17 Determine the first-order derivative of the function in Fig. 6.3

$$f(x) = 2x^3 - 5x^2 + 3x + 1 \qquad (6.148)$$

at $x = 1.0$ with $h = 0.1$ by using the forward divided difference $O(h)$, backward divided difference $O(h)$ and central divided difference $O(h^2)$. Compare the approximate derivatives with the exact solution.

It is noted that the exact derivative of the function $f(x)$ in Eq. (6.148) is

$$f'(x) = 6x^2 - 10x + 3 \qquad (6.149)$$

Thus, the exact derivative value at $x = 1.0$ is

$$f'(x=1.0) = 6(1.0)^2 - 10(1.0) + 3 = -1.0 \qquad (6.150)$$

By using $h = 0.1$, then

$$x_{i-1} = 0.9 \quad ; \quad f(x_{i-1}) = 1.108$$

$$x_i = 1.0 \quad ; \quad f(x_i) = 1.000$$

$$x_{i+1} = 1.1 \quad ; \quad f(x_{i+1}) = 0.912$$

The above values are used to determine the approximate derivatives and their true errors according to those shown in Tables 6.4 to 6.6 as follows.

Derivative value and its error using the forward divided difference,

$$f'(1.0) = (0.912 - 1.000)/0.1 = -0.88 \quad ; \quad \varepsilon_t = 12\% \qquad (6.151)$$

Derivative value and its error using the backward divided difference,

$$f'(1.0) = (1.000 - 1.108)/0.1 = -1.08 \quad ; \quad \varepsilon_t = 8\% \qquad (6.152)$$

Derivative value and its error using the central divided difference,

$$f'(1.0) = (0.912 - 1.108)/0.2 = -0.98 \quad ; \quad \varepsilon_t = 2\% \qquad (6.153)$$

The true errors of the approximate derivatives obtained from the forward, backward and central divided differences in Eqs. (6.151) - (6.153) show that the central divided difference provides higher solution accuracy. Such result agrees with the fact that the central divided difference gives the error of order h^2, while both the forward and backward divided differences yield the error of order h.

Figure 6.19 presents a computer program for determining derivatives of a function at various x-locations by using the central divided difference. The program requires the input data of the beginning and ending locations including the number of locations for determining the derivatives. Typical results obtained from the program by using the function given in Example 6.17 are shown in Table 6.10.

Fortran

```
      PROGRAM NUMDIF
C.....A NUMERICAL DIFFERENTIATION PROGRAM
      DIMENSION FX(100), DIFF(100)
C.....READ END LOCATIONS AND NO. OF POINTS:
      READ(5,*)  A, B, N
      H = (B - A)/(N - 1)
      X = A
C.....COMPUTE FUNCTION VALUES AT POINTS:
      DO 100  I=1,N
      FX(I) = FUNC(X)
      X = X + H
  100 CONTINUE
C.....COMPUTE DERIVATIVES AT POINTS:
      DIFF(1) = (FX(2) - FX(1))/H
      DO 200  I=2,N-1
      DIFF(I) = (FX(I+1) - FX(I-1))/(2.*H)
  200 CONTINUE
      DIFF(N) = (FX(N) - FX(N-1))/H
      WRITE(6,300)
  300 FORMAT(5X,'X',11X,'FX',10X,'DERIVATIVE',/)
      X = A
      DO 400  I=1,N
      WRITE(6,500)  X, FX(I), DIFF(I)
  500 FORMAT(2X, F5.2, 4X, F10.3, 5X, F10.3)
      X = X + H
  400 CONTINUE
      STOP
      END
C------------------------------------------------
      FUNCTION FUNC(X)
      FUNC = 2.*X*X*X - 5.*X*X + 3.*X + 1.
      RETURN
      END
```

MATLAB

```
%  PROGRAM NUMDIF
%  A NUMERICAL DIFFERENTIATION PROGRAM
%-----------------------------------------------------
func = inline('2.*x^3 - 5.*x^2 + 3.*x + 1.', 'x');
%-----------------------------------------------------
%  READ END LOCATIONS AND NO. OF POINTS:
a = input('Enter end location (a): ');
b = input('Enter end location (b): ');
n = input('Enter number of point:  ');
h = (b - a)/(n - 1);
x = a;
%  COMPUTE FUNCTION VALUES AT POINTS:
for i = 1:n
    fx(i) = func(x);
    x = x + h;
end
%  COMPUTE DERIVATIVES AT POINTS:
diff(1) = (fx(2) - fx(1))/h;
for i = 2:n-1
    diff(i) = (fx(i+1) - fx(i-1))/(2.*h);
end
diff(n) = (fx(n) - fx(n-1))/h;
fprintf('\n     X               FX          DERIVATIVE');
x = a;
for i = 1:n
    fprintf('\n  %5.2f       %10.3f %10.3f', x, fx(i), ...
diff(i))
    x = x + h;
end
```

Figure 6.19 Computer program to determine derivatives of the function in Example 6.17 at various *x*-locations by using the central divided difference.

Table 6.10 Computed derivatives of the function in Example 6.17 at various *x*-locations by using the central divided difference.

X	FX	DERIVATIVE
0.50	1.500	−0.680
0.60	1.432	−0.820
0.70	1.336	−1.040
0.80	1.224	−1.140
0.90	1.108	−1.120
1.00	1.000	−0.980
1.10	0.912	−0.720
1.20	0.856	−0.340
1.30	0.844	0.160
1.40	0.888	0.780
1.50	1.000	1.120

6.12 MATLAB Commands for Differentiation

MATLAB has many useful commands to perform differentiation of a given function. One of the commands is `gradient` which is used to determine the gradient of the function $f(x, y)$ in the form of

$$\nabla f = \frac{\partial f}{\partial x}\hat{i} + \frac{\partial f}{\partial y}\hat{j}$$

If the given function f varies with only one independent variable x, then, the command determines the value of $\dfrac{df}{dx}$.

The format for using this command when the function f varies with one independent variable is

$$FX = \text{gradient(function,h)}$$

where   FX   is the computed gradients
       function   is the given function
       h   is the step size

Example 6.18 Determine the derivatives of the function given in Example 6.17 at different x by using the MATLAB command `gradient`.

$$f(x) = 2x^3 - 5x^2 + 3x + 1 \tag{6.148}$$

The derivatives can be obtained by typing

```
>> f = inline('2*x.^3-5*x.^2+3*x+1','x');
>> x = 0.5:0.1:1.5;
>> y = f(x);
>> FX = gradient(y,0.1)
FX =

Columns 1 through 7

-0.6800   -0.8200   -1.0400   -1.1400   -1.1200   -0.9800   -0.7200

Columns 8 through 11

-0.3400    0.1600    0.7800    1.1200
```

The computed derivatives are identical to those presented in Table 6.10.

6.13 Closure

Numerical methods for integration and differentiation of a given function are presented in this chapter. Fundamentals of these numerical methods are explained in details with examples and computer programs. The simplest integration method is based on the trapezoidal rule. The trapezoidal rule approximates the integral from the area under a linear function. A more accurate integral solution can be obtained by using the composite trapezoidal rule. The composite trapezoidal rule divides the integration range into a number of segments so that the single trapezoidal rule is applied to each segment. If the integral is approximate by the area under a quadratic function, the method is called the Simpson's rule. Improved integral solution accuracy can be obtained by determining the area under higher order polynomials. These methods lead to a family of the methods so called the Newton-Cotes formulas. Other integration methods that can provide high solution accuracy, such as the Romberg and Gauss integration methods, are also presented. The Gauss integration method is popular and widely used in many commercial software for solving practical problems. Numerical integration methods for two- and three-dimensional domains are also presented and explained. These methods follow the procedure used for one-dimensional domain by performing the integration in each coordinate direction, one at a time.

Numerical methods for determining derivatives of a given function at various locations are presented at the end of the chapter. The approximate derivatives can be determined by using the forward, backward and central divided differences. Understanding the methods for integration and differentiation is essential to further study the higher-level numerical methods in the following chapters.

Exercises

1.   Determine the integral

$$I = \int_{2}^{8} \left(4x^5 - 3x^4 + x^3 - 6x + 2\right)dx$$

by using the trapezoidal rule and composite trapezoidal rule with $n = 2, 4$ and 6. Plot distribution of the integrand between the given lower and upper limits. Compare the computed values with the exact solution and determine the true percentage errors.

2.   Determine the integral

$$I = \int_{0}^{0.5} \frac{dx}{1+x^4}$$

by using the trapezoidal rule and composite trapezoidal rule with $n = 2$ and 5. Compare the computed values with the exact solution and determine their true percentage errors. Check the computed values with the solutions obtained by modifying the computer program in Fig. 6.6.

3.   Determine the integral

$$I = \int_{0}^{4} xe^{2x} dx$$

by using the composite trapezoidal rule with $n = 2, 4$ and 8. Compare the computed values with the exact solution and determine the true percentage errors.

4.   Determine the integral

$$I = \int_{-\pi}^{\pi} \frac{dx}{1+\sin^2 x}$$

by using the trapezoidal rule and composite trapezoidal rule with $n = 2, 4$ and 6. Compare the computed values with the exact solution and determine the true percentage errors.

5.   Determine the integral

$$I = \int_{0}^{1} \frac{x^4 e^x}{\left(e^x - 1\right)^2} dx$$

by using the trapezoidal rule and composite trapezoidal rule with $n = 2, 4$ and 5. Compare the computed values with the solutions obtained by modifying the computer program in Fig. 6.6.

6.   Determine the elliptic integral of the first kind as shown in Eq. (6.5)

$$K(\theta) = \int_{0}^{\pi/2} \frac{dx}{\sqrt{1 - \sin^2 \frac{\theta}{2} \sin^2 x}}$$

when $\theta = \pi/6$ by using the trapezoidal rule and composite trapezoidal rule with $n = 2, 3$ and 6. Compare the computed values with the solutions obtained by modifying the computer program in Fig. 6.6. Determine the number of segments n required to provide the solution accuracy up to 4 significant figures.

7.   Derive the integral expression by using the Simpson's 1/3 rule as shown in Eq. (6.43). The derivation starts by integrating the second-order Lagrange polynomial in Eq. (6.42). Show the derivation in details.

8.   Show that the solution error produced by using the Simpson's 1/3 rule for integrating a function f from the limit a to b is

$$E_t = -\frac{(b-a)^5}{2,880} f^{(4)}(\xi)$$

Then, explain physical meaning of the error expression above.

9.   Determine the integral

$$I = \int_{-1}^{2} \left(x^7 + 2x^3 - 1\right) dx$$

by using the Simpson's 1/3 rule when $n = 2, 4$ and 6. Compare the computed values with the exact solution and determine the true percentage errors. Then, solve the problem again but by using by using the Simpson's 3/8 rule with $n = 3, 6$.

10.   Determine the integral

$$I = \int_0^{\pi/4} \frac{x}{\cos^2 x} \, dx$$

by using the Simpson's 1/3 rule when $n = 2, 4$. Show the computational procedure in details. Solve the problem again but by using the Simpson's 3/8 rule with $n = 3$ and 6.

11.   Determine the integral

$$I = \int_0^{\pi/2} \frac{\sin x}{\sqrt{1 - 0.25\sin^2 x}} \, dx$$

by using the Simpson's 1/3 rule when $n = 2, 4$ and 6. Show the computational procedure in details. Check the computed values with those obtained by modifying the computer program in Fig. 6.9.

12.   Develop a computer program to integrate a function by using the Simpson's 3/8 rule. Verify the program by solving Example 6.8. Then, use the program to determine the integral

$$I = \int_0^1 \frac{e^x}{1 + e^x} \, dx$$

with $n = 3, 30$ and 300, respectively. Provide comments on the accuracy of the computed values.

13.   Show that the solution error that occurs by using the composite Simpson's 1/3 rule with n segments to integrate a function f from the lower limit a to upper limit b is

$$E_a = -\frac{(b-a)^5}{180n^4} \bar{f}^{(4)}$$

where $\bar{f}^{(4)}$ is the average fourth-order derivative of all segments. Then, set up an example to demonstrate the validity of the error expression above.

14.   Derive Eqs. (6.56) and (6.60) for determining the integral of a given function and the associated error, respectively, using the Simpson's 3/8 rule. Show detailed derivation and explain physical meaning of the terms in these two equations.

15.   In practice, the integral values are determined from a set of experimental data because their continuous functions may not be available. Modify the computer programs of the composite trapezoidal and Simpson's rules as shown in Figs. 6.6 and 6.9 to determine the integral values of the data below from the limits $a = 0$ to $b = 2$.

x	0	0.2	0.4	0.6	0.8	1.0
$f(x)$	-1.000	-0.992	-0.936	-0.784	-0.488	0.000
x	1.0	1.2	1.4	1.6	1.8	2.0
$f(x)$	0.000	0.728	1.744	3.096	4.832	7.000

The modified computer programs should be validated by solving the integral below that has exact solution.

$$I = \int_0^2 f(x)\,dx = \int_0^2 \left(x^3 - 1\right)dx$$

16. Derive the integral expression and associated error that occur by using the Boole's rule as shown in Eq. (6.68). Show the derivation in details and set up an example such that the Boole's rule can not provide exact integral solution.

17. Prove the Romberg integration formula in Eq. (6.87) by showing the derivation in details. Explain physical meanings of the integral expressions when $k = 1, 2, 3$ and 4, respectively.

18. Study the computer program for the Romberg integration in Fig. 6.11. Draw a flow chart and explain the computational procedure that occurs in the program. Then, use the program to solve Example 6.10 and discuss on the accuracy of the computed integral value.

19. Apply the Romberg integration method to determine the integral in Eq. (6.4)

$$I = \int_0^{0.8} e^{-x^2}\,dx$$

Show the solution procedure so that the computed integral value has the relative error less than 0.0001%.

20. Apply the Romberg integration method for 4 rounds to determine the integral

$$I = \int_{\pi/4}^{\pi} \left(e^{2x} + \cos x\right)dx$$

Compare the solution obtained from each round with the exact solution.

21. Apply the Romberg integration method for 3 rounds to determine the integral

$$I = \int_0^{\pi/4} e^{3x} \sin 2x \, dx$$

Show the solution procedure in details. Compare the computed integral values with those obtained by using the computer program in Fig. 6.11.

22. Apply the Romberg integration method for 4 rounds to determine the integral

$$I = \int_1^{1/2} e^{-x}\,dx$$

Show the computational procedure in details.

23.   Determine the elliptic integral of the first kind in problem 6 again but by using the Romberg integration method. Apply the method until computed solution has the relative error less than 0.0001%. Verify the computed solution with that obtained from modifying the computer program in Fig. 6.11.

24.   Derive the weights and locations of the 3-point Gauss integration method as shown in Eq. (6.105a-c). Explain necessary conditions required to derive these values.

25.   Determine the integral in Example 6.11 again but by using the 3-point Gauss integration method. Show the solution procedure in details. Is it possible that the solution accuracy decreases with the increased number of Gauss point? If the answer is yes, explain the source of the error.

26.   Modify the computer program in Fig. 6.15 for determining the integral

$$I \;=\; \int_0^{\pi/2} \sin x \, dx$$

It is noted that the above integral was determined by using the Romberg integration method in Example 6.10. Provide comments on the solution accuracy obtained from using the two methods.

27.   Apply the Gauss integration method to determine

$$I \;=\; \int_3^4 \frac{x}{\sqrt{x^2 + 4}} \, dx$$

by using the number of Gauss points of $n = 2, 3$ and 4. Show the computational procedure in details.

28.   Apply the Gauss integration method to determine

$$I \;=\; \int_0^1 x^2 e^{-x} \, dx$$

by using the number of Gauss points of $n = 2, 3$ and 4. Show the computational procedure in details. Compare the computed integral values with those obtained by modifying the computer program in Fig. 6.15.

29.   Apply the Gauss integration method to determine

$$I \;=\; \int_0^1 \left(1 + x^2\right)^{3/2} dx$$

by using the number of Gauss points of $n = 2, 3, 4, 5$ and 6. Compare the computed integral values with the exact solution of 1.567951962.

30. Apply the Gauss integration method to determine

$$I = \int_0^5 \frac{dx}{1+(x+\pi)^2}$$

by using the number of Gauss points of $n = 2, 3, 4, 5$ and 6. Compare the computed integral values with the exact solution of

$$I = \tan^{-1}(5+\pi) - \tan^{-1}(\pi)$$

Then, give comment on the possibility that the Gauss integration method can provide exact integral solution. If it is possible, how many Gauss points are needed?

31. The computer program for the Gauss integration method was developed so that it is easy to understand. The program does not take the advantage that the weights W_i and locations ξ_i are symmetric. The data statements for the weights and locations can be reduced by using their symmetrical property. Improve the program by using the symmetrical property of these data in order to reduce the memory requirement. Then validate the program by solving the integrals in Examples 6.10 and 6.11.

32. Use the `trapz` command in MATLAB to determine the integral

$$I = \int_0^2 \sqrt{2x - x^2}\, dx$$

Select appropriate segment length so that the true error is less than 0.0001. Check the accuracy of the computed integral value by comparing with the exact solution determined from

$$\int \sqrt{2x - x^2}\, dx = \frac{x-1}{2}\sqrt{2x - x^2} + \frac{1}{2}\arcsin(x-1)$$

33. Use the `trapz` command in MATLAB to determine the integral

$$I = \int_1^3 \frac{1}{1+2\sin x}\, dx$$

Select appropriate segment length so that the true error is less than 0.0001. Derive the exact integral solution to measure the accuracy of the computed integral value.

$$\int \frac{1}{1+2\sin x}\, dx = \frac{1}{\sqrt{3}}\ln\left[\frac{\tan(x/2)+2-\sqrt{3}}{\tan(x/2)+2+\sqrt{3}}\right]$$

34. Use the `trapz` command in MATLAB to determine the integral

$$I = \int_0^3 \frac{1}{\sqrt{2x^2+3x+1}}\, dx$$

Select appropriate segment length so that the true error is less than 0.0001. Check the accuracy of the computed integral value by comparing with the exact solution determined from

$$\int \frac{1}{\sqrt{2x^2+3x+1}}\,dx = \frac{1}{\sqrt{2}}\ln\left|\,2\sqrt{4x^2+6x+2}\,+4x+3\,\right|$$

35.   Use the single trapezoidal rule to determine the integral

$$I = \int_0^3\int_1^4 \left(x^2y+8y\right)\sin x\,dx\,dy$$

Repeat the problem but by using the Simpson's 1/3 rule. Compare the accuracy of the computed integral values with the exact solution.

36.   Modify the computer program that uses the composite trapezoidal rule to determine the double integral

$$I = \int_0^\pi\int_1^3 \frac{\cos x\left(2y^2+\sin x\right)}{\sqrt{1+y}}\,dx\,dy$$

Use the number of segments of 5, 10 and 20 in each x- and y-direction to verify the convergence of the computed integral values.

37.   Apply the Gauss integration to determine the double integral

$$I = \int_2^3\int_1^2 e^x e^y\,dx\,dy$$

Use $n=2$ in each direction. Repeat the problem but by using $n=3$ to verify the convergence of the computed integral values.

38.   Determine the integral in Problem 36 again but by using the dblquad command in MATLAB. Use the required tolerance of 1×10^{-5} and compare the solution with that obtained from Problem 36.

39.   Use the dblquad command in MATLAB to determine the double integral

$$I = \int_{-2}^3\int_2^4 2x^2y+3xy^2\,dx\,dy$$

Specify an appropriate tolerance and compare the computed integral value with the exact solution of $910/3$.

40.   Use the dblquad command in MATLAB to determine the double integral

$$I = \int_0^{0.5}\int_0^\pi x\cos(xy)\cos^2(\pi x)\,dx\,dy$$

Specify an appropriate tolerance and compare the computed integral value with the exact solution of $1/(3\pi)$.

41. Use the single trapezoidal rule to determine the triple integral

$$I = \int_2^3 \int_0^1 \int_1^2 (1+x)yz^2 \, dx \, dy \, dz$$

Then, repeat the problem but by using the Gauss integration method with $n = 2$ in each x-, y- and z-direction. Compare the computed integral value with the exact solution.

42. Use the `triplequad` command in MATLAB to determine the triple integral

$$I = \int_0^{\pi/2} \int_0^1 \int_0^2 zr^2 \sin\theta \, dz \, dr \, d\theta$$

Specify an appropriate tolerance and compare the computed integral value with the exact solution of $2/3$.

43. Use the `triplequad` command in MATLAB to determine the triple integral

$$I = \int_0^{2\pi} \int_0^{\pi} \int_0^5 (\rho^4 \sin\phi) d\rho \, d\phi \, d\theta$$

Specify an appropriate tolerance and compare the computed integral value with the exact solution of $2{,}500\,\pi$.

44. Show the derivation for the expressions of the first and second derivatives as given in Tables 6.4 - 6.6 in details.

45. Show the derivation for the expression of the third derivative in Table 6.6 by using the central divided difference.

46. Determine the first derivative of the function

$$f(x) = e^x$$

at $x = 2$ with $h = 0.25$ by using the forward divided difference $O(h)$, backward divided difference $O(h)$ and central divided difference $O(h^2)$. Show detailed derivation and determine the true errors of the computed derivatives with the exact solution.

47. Determine the first and second derivatives of the function

$$f(x) = \tan^{-1}(x^2 - x + 1)$$

at $x=1$ with $h=0.1$ by using the forward divided difference $O(h)$, backward divided difference $O(h)$ and central divided difference $O(h^2)$. Then, repeat the problem again but by using $h=0.05$.

48.   Show the derivation for the expressions of the first and second derivatives as given in Tables 6.7 - 6.9 in details.

49.   Determine the first to the fourth derivatives of the function

$$f(x) \;=\; e^{x/3} + x^2$$

at $x=-2.5$ with $h=0.1$ by using the forward divided difference $O(h^2)$, backward divided difference $O(h^2)$ and central divided-difference $O(h^4)$. Show detailed derivation and determine the true errors of the computed derivatives with the exact solution.

50.   Modify the computer program for determining the first derivative as shown in Fig. 6.19 by using the more accurate derivative expressions in Tables 6.7 - 6.9. Then, use the program to solve Example 6.17 and compare the computed solutions with those shown in the example.

Chapter
7

Ordinary Differential Equations

7.1 Introduction

Ordinary differential equations occur in solving many scientific and engineering problems. For example, in the determination of the shuttle velocity v that varies with time t during descending through the earth atmosphere, the governing differential Eq. (1.5) representing the Newton's second law is

$$\frac{dv}{dt} = g - \frac{c}{m}v \tag{7.1}$$

where g is the gravitational acceleration constant, c is the air drag coefficient and m is the mass of the shuttle. Equation (7.1) is called an ordinary differential equation (ODE) because the dependent variable v varies only with the independent variable t. The ordinary differential equation differs from the partial differential equation (PDE) which is presented in the following chapter. A dependent variable in the partial differential equation varies with two or more independent variables. Solutions to the partial differential equations thus are more complex and difficult to derive.

The ordinary differential equation can be classified into several types. Equation (7.1) is called the first-order ordinary differential equation because the highest derivative term is of the first-order. Some ordinary differential equations contain the second-order derivative terms. For example, the equilibrium equation of a swinging pendulum as shown in Fig. 7.1 is in the form of the second-order ordinary differential equation. The equation is derived from the equilibrium condition at any instant during swinging by using the Newton's second law.

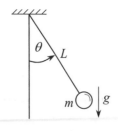

The equilibrium condition of the forces acting on the mass in the tangential direction leads to the second-order differential equation in the form,

$$\frac{d^2\theta}{dt^2} + \frac{g}{L}\sin\theta = 0 \tag{7.2}$$

where θ is the swinging angle that varies with time t and L is the cord length.

Figure 7.1 Swinging pendulum.

It is noted that the second-order differential equation as shown in Eq. (7.2) can be separated into two first-order differential equations by introducing a new variable. For example, if

$$\frac{d\theta}{dt} = \beta \tag{7.3a}$$

then, Eq. (7.2) becomes

$$\frac{d\beta}{dt} = -\frac{g}{L}\sin\theta \tag{7.3b}$$

Because a second-order differential equation can be separated into two first-order differential equations, the numerical methods for solving the first-order differential equation are presented in this chapter. The same methods can be applied to solve the second-order and other higher-order ordinary differential equations.

Equations (7.1) and (7.2) as shown above are linear and nonlinear ordinary differential equations, respectively. A linear differential equation can be identified by considering the differential equation in the general form of

$$a\frac{d^2y}{dx^2} + b\frac{dy}{dx} + cy = d \tag{7.4}$$

If the coefficients a, b, c, d are constant or function of the independent variable x, the equation is *linear*. Equation (7.4) becomes *nonlinear* if the coefficients are function of the dependent variable y, such as $a = y^2$ or $b = 2y$. Knowing that a differential equation is either linear or nonlinear is useful for finding its solution. Exact solutions for most nonlinear differential equations cannot be derived. Numerical methods are the only way to find their approximate solutions. Since most of the differential equations that occur in practical problems are nonlinear, the numerical methods thus help analysts to obtain solutions that could not be found in the past.

Equation (7.2) that governs the swinging phenomenon of the pendulum is a nonlinear differential equation. The equation is nonlinear because the $\sin\theta$ function can be expressed in form of infinite series of the dependent variable θ as

$$\sin\theta = \theta - \frac{\theta^3}{3!} + \frac{\theta^5}{5!} - \frac{\theta^7}{7!} + \dots \tag{7.5}$$

Thus, the exact solution to the differential Eq. (7.2) is difficult or even impossible to derive. Exact solution to Eq. (7.2) can be obtained easier if the differential equation becomes linear. The differential equation becomes linear if

$$\sin \theta = \theta \tag{7.6}$$

which is valid for small swinging angle θ. By substituting Eq. (7.6) into Eq. (7.2), the linear differential equation is

$$\frac{d^2\theta}{dt^2} + \frac{g}{L}\theta = 0 \tag{7.7}$$

Its exact solution for the swinging angle θ that varies with time t is

$$\theta(t) = \theta_0 \cos\sqrt{\frac{g}{L}}\, t \tag{7.8}$$

The exact solution above is obtained by using the initial angle θ_0 and zero velocity at time $t = 0$ as

$$\theta(t = 0) = \theta_0 \qquad \text{and} \qquad \frac{d\theta}{dt}(t = 0) = 0 \tag{7.9}$$

The example shows that two initial conditions are required to solve a second-order differential equation. Similarly, an initial condition is needed for solving a first-order differential equation such as Eq. (7.1). Solutions to the ordinary differential equation thus depend on the initial conditions of the problems as shown by the following example.

A first-order ordinary differential equation can be written in a general form as,

$$\frac{dy}{dx} = f(x, y) \tag{7.10}$$

As an example, if

$$f(x, y) = 3x^2 - 2x + 1 \tag{7.10a}$$

then,

$$\frac{dy}{dx} = 3x^2 - 2x + 1 \tag{7.10b}$$

The general solution is obtained after performing integration,

$$y = x^3 - x^2 + x + C \tag{7.10c}$$

where C is the integrating constant. The integration constant C is determined from the initial condition of the problem. As an example, if the initial condition is given by $y(x = 0) = 2$, then $C = 2$. Thus, the solution to the differential equation with the given initial condition is

$$y = x^3 - x^2 + x + 2 \tag{7.10d}$$

Equation (7.10d) shows that the final solution of a differential equation depends on the initial condition. The same solution behavior occurs for the swinging pendulum problem where its final solution depends on the initial angle θ_0.

In this chapter, popular methods for solving the ordinary differential equations are presented. These methods are: (1) the Euler's method, (2) the Heun's method, (3) the modified Euler's method, (4) the Runge-Kutta method, (5) the methods for solving a system of first-order differential equation, and (6) the multistep methods. Corresponding computer programs are also presented to aid understanding of these methods for obtaining solutions.

7.2 Euler's Method

The Euler's method is the simplest method for solving the ordinary differential equation in the form,

$$\frac{dy}{dx} = f(x, y) \tag{7.10}$$

The concept of the method is explained by considering Fig. 7.2.

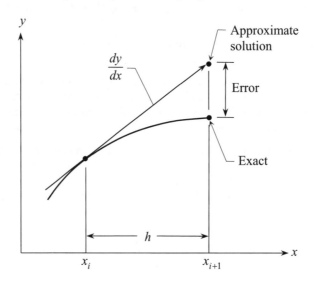

Figure 7.2 The Euler's method.

From Fig. 7.2, the approximate solution y_{i+1} at x_{i+1} is determined from the solution y_i at x_i by estimating the slope from

$$\frac{dy}{dx} = \frac{y_{i+1} - y_i}{x_{i+1} - x_i} = \frac{y_{i+1} - y_i}{h} \tag{7.11}$$

where $h = x_{i+1} - x_i$ is the step size used in the computation. By substituting the slope from Eq. (7.11) into Eq. (7.10),

$$\frac{y_{i+1} - y_i}{h} = f(x_i, y_i)$$

i.e., $$y_{i+1} = y_i + f(x_i, y_i)h \tag{7.12}$$

Equation (7.12) shows that the approximate solution y_{i+1} is determined from the solution y_i at x_i by using the step size h. As shown in the figure, the solution error varies directly with the step size h. Higher solution accuracy is obtained by using smaller step size as demonstrated in the following example.

Example 7.1 Use the Euler's method to solve the first-order differential equation,

$$\frac{dy}{dx} = y \cos x \qquad (7.13)$$

with the initial condition of $y(0)=1$. Determine the approximate solutions by using the three different step sizes of $h = 1.00, 0.50$ and 0.25. Compare the Euler's solutions with the exact solution.

The exact solution to the differential Eq. (7.13) can be derived by separating the two variables onto opposite sides of the equation and performing integration,

$$\int \frac{dy}{y} = \int \cos x \, dx \qquad (7.14)$$

The integration yields,

$$\ln y = \sin x + A \qquad (7.15)$$

where A is the integrating constant which can be determined from the initial condition of $y(0)=1$,

$$\ln 1 = 0 + A$$

to give,

$$A = \ln 1 = 0$$

Then, Eq. (7.15) becomes

$$\ln y = \sin x$$

leading to the exact solution of

$$y = y(x) = e^{\sin x} \qquad (7.16)$$

To determine the Euler's solution, Eq. (7.12) is employed together with the given function $f(x, y)$,

$$y_{i+1} = y_i + (y_i \cos x_i) h \qquad (7.17)$$

By starting from the initial condition of $y(x = 0) = 1$, i.e., $x_0 = 0$ and $y_0 = 1$, then Eq. (7.17) gives

$$y_1 = 1 + [(1)(1)](1) = 2 \qquad (7.18a)$$

so that the exact error is

$$E_t = e^{\sin (1)} - 2 = 0.31978 \qquad (7.18b)$$

The approximate solution obtained from Eq. (7.18a) is used to determine the solution for the second round. The Euler's Eq. (7.17) is used again with $x_1 = 1$ and $y_1 = 2$,

$$y_2 = 2 + [(2)(0.54030)](1) = 3.08060 \qquad (7.19a)$$

for which the exact error is

$$E_t = e^{\sin (2)} - 3.08060 = -0.59802 \qquad (7.19b)$$

The process is repeated to determine the approximate solution at the third round with $x_2 = 2$ and $y_2 = 3.08060$ to give

$$y_3 = 3.08060 + [(3.08060)(-0.41615)](1) = 1.79862 \qquad (7.20a)$$

and the exact error is,

$$E_t = e^{\sin(3)} - 1.79862 = -0.64706 \qquad (7.20b)$$

After three rounds of computation, the approximate solutions are obtained and compared with the exact solution as shown in Fig. 7.3.

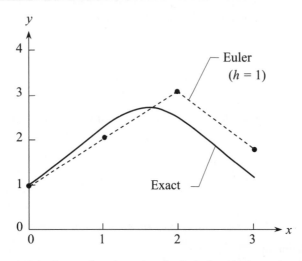

Figure 7.3 Comparison between the Euler's solution using $h = 1$ with the exact solution in Example 7.1.

Accuracy of the approximate solutions obtained from the Euler's method increases by reducing the step size h. Table 7.1 shows the approximate solutions obtained from using the three different step sizes of $h = 1.00$, 0.50 and 0.25 as compared to the exact solution. These approximate solutions are compared with the exact solution as plotted in Fig. 7.4.

Table 7.1   Comparison of the Euler's solutions by using different step sizes with the exact solution.

x	Exact solution	Euler's solutions		
		$h=1.00$	$h=0.50$	$h=0.25$
0.00	1.00000	1.00000	1.00000	1.00000
0.25	1.28070			1.25000
0.50	1.61515		1.50000	1.55279
0.75	1.97712			1.89346
1.00	2.31978	2.00000	2.15819	2.23982
1.25	2.58309			2.54236
1.50	2.71148		2.74122	2.74278
1.75	2.67510			2.79128
2.00	2.48258	3.08060	2.83818	2.66690
2.25	2.17727			2.38944
2.50	1.81934		2.24763	2.01419
2.75	1.46472			1.61078
3.00	1.15156	1.79862	1.34729	1.23857

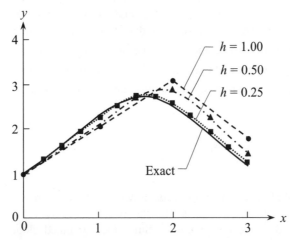

Figure 7.4 Comparison of the Euler's solutions using three different step sizes
with the exact solution in Example 7.1.

Even though the Euler's method is simple, developing a computer program is necessa-ry for obtaining solutions for a large number of steps. This is because the computational procedure is repeated by using the values obtained from the earlier steps. Miscalculating a value thus affects the solutions at the later steps. Figure 7.5 shows a computer program of the Euler's method for obtaining the solutions in Example 7.1.

Fortran

```
      PROGRAM  EULER
C.....A PROGRAM FOR SOLVING ORDINARY DIFFERENTIAL
C.....EQUATION USING THE EULER'S METHOD
C.....READ INITIAL CONDITIONS, NUMBER OF STEPS,
C.....AND STEP SIZE:
      READ(5,*)  X, Y, N, H
      WRITE(6,100)  H
 100 FORMAT(' SOLUTION WITH STEP SIZE =', E10.4,
     *   ' IS:', /, 10X, 'X', 15X, 'Y')
      WRITE(6,200)  X, Y
 200 FORMAT(2E16.6)
      DO 300  I=1,N
      SLOPE = FUNC(X,Y)
      Y = Y + SLOPE*H
      X = X + H
      WRITE(6,200)  X, Y
 300 CONTINUE
      STOP
      END
C-----------------------------------------------
      FUNCTION  FUNC(X,Y)
      FUNC = Y*COS(X)
      RETURN
      END
```

MATLAB

```
%  PROGRAM  EULER;
%  A PROGRAM FOR SOLVING ORDINARY DIFFERENTIAL.
%  EQUATION USING THE EULER'S METHOD
%  READ INITIAL CONDITIONS, NUMBER OF STEPS,
%  AND STEP SIZE:
%-----------------------------------------------
func = inline('y*cos(x)','x','y');
%-----------------------------------------------
x = input('\nEnter value of x:   ');
y = input( 'Enter value of y:   ');
n = input( 'Enter value of n:   ');
h = input( 'Enter value of h:   ');
fprintf('\nSOLUTION WITH STEP SIZE = %10.4e IS:',h);
fprintf('\n          X                Y');
fprintf('\n%16.6e%16.6e',x,y);
for i = 1:n
    slope = func(x,y);
    y = y + slope*h;
    x = x + h;
    fprintf('\n%16.6e%16.6e',x,y);
end
```

Figure 7.5 Computer program for solving the first-order ordinary differential equation
by using the Euler's method in Example 7.1.

Solutions obtained from the Euler's method as shown in Table 7.1 and Fig. 7.4 indicate that their errors vary with the step size *h*. Such errors consist of: (1) the local error from using the Euler's formula in Eq. (7.12) and (2) the error that is accumulated as the Euler's process repeats. Combination of the two errors is known as the global error. Magnitude of the global error varies in the same order with the step size h, or $O(h)$. If the step size is reduced by a half, the global error also reduces by a half. The Euler's method is thus sometimes called the first-order method.

In the next section, another method that can provide solution with the second-order of accuracy, $O(h^2)$, is presented. The method reduces the solution error into a quarter if the step size h is cut by a half.

7.3 Heun's Method

The Heun's method is modified from the Euler's method to increase the solution accuracy. Figure 7.2 of the Euler's method suggests that the solution accuracy increases if a proper slope $(y' = dy/dx)$ is employed in the computation. Since the computed slope at x_i from the Euler's method is

$$y_i' = f(x_i, y_i) \tag{7.21}$$

Such slope is used for determining the solution at x_{i+1} from

$$y_{i+1}^0 = y_i + f(x_i, y_i)h \tag{7.22}$$

The approximate solution obtained from Eq. (7.22) can be employed to determine the slope at the same location x_{i+1} as shown in Fig. 7.6(a) as

$$y_{i+1}' = f\left(x_{i+1}, y_{i+1}^0\right) \tag{7.23}$$

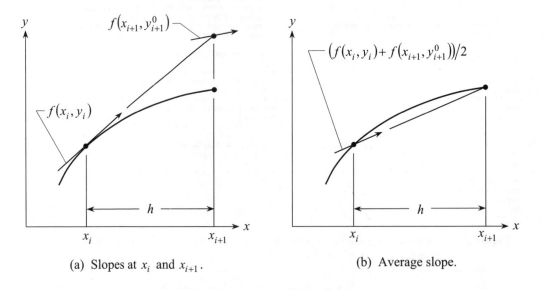

(a) Slopes at x_i and x_{i+1}.   (b) Average slope.

Figure 7.6 The Heun's method.

The new slope at x_{i+1} is then averaged with the slope at x_i. The average slope is then used at x_i for determining the solution at x_{i+1} as shown in Fig. 7.6(b). The process should provide improved solution at x_{i+1} because a more accurate slope is employed in the computation.

From Eqs. (7.21) and (7.23), the average slope is

$$\bar{y}' = \frac{y_i' + y_{i+1}'}{2} = \frac{f(x_i, y_i) + f\left(x_{i+1}, y_{i+1}^0\right)}{2} \tag{7.24}$$

Thus, the approximate solution at x_{i+1} is

$$y_{i+1} = y_i + \frac{f(x_i, y_i) + f\left(x_{i+1}, y_{i+1}^0\right)}{2}h \tag{7.25}$$

In summary, the Heun's method consists of two computational steps. The first step is to determine the slope from the predicted solution y_{i+1}^0 at x_{i+1}. The process of this first step is known as the predictor. The predicted slope y_{i+1}^0 is averaged with the slope at x_i in the second step. The averaged slope is then used at x_i to determine the solution at x_{i+1}. The second step is called the corrector that can provide a more accurate solution as compared to the original Euler's method. The Heun's method is thus sometimes called the predictor-corrector method because of the two processes above are employed. These two processes are summarized as follows:

Predictor:
$$y_{i+1}^0 = y_i + f(x_i, y_i)h \tag{7.26a}$$

Corrector:
$$y_{i+1} = y_i + \frac{f(x_i, y_i) + f\left(x_{i+1}, y_{i+1}^0\right)}{2}h \tag{7.26b}$$

Example 7.2 Use the Heun's method to solve the same first-order ordinary differential equation

$$\frac{dy}{dx} = y \cos x \tag{7.13}$$

with the initial condition of $y(0) = 1$. The problem was solved earlier by the Euler's method in Example 7.1 by employing the step size of $h = 1.00$. Compare the solution with the Euler's and exact solutions. Then, develop a computer program to obtain a more accurate solution by using a smaller step size of $h = 0.25$.

For the first round at $x_0 = 0$ and $y_0 = 1$ with $h = 1$, the predictor from Eq. (7.26a) is

$$y_1^0 = 1 + [(1)(1)](1) = 2 \tag{7.27a}$$

Then, the slope at the end of the step is

$$f\left(x_1, y_1^0\right) = (2)\cos(1) = 1.08060$$

Thus, the corrector which is the solution at the end of the step can be determined from Eq. (7.26b)

$$y_1 = 1 + \frac{1 + 1.08060}{2}(1) = 2.04030 \tag{7.27b}$$

so that the exact error is

$$E_t = e^{\sin(1)} - 2.04030 = 0.27948 \tag{7.27c}$$

For the second round $x_1 = 1$, $y_1 = 2.04030$, the predictor from Eq. (7.26a) is

$$y_2^0 \;=\; 2.04030 + [2.04030\cos(1)](1) \;=\; 3.14268 \qquad (7.28a)$$

The slope at the end of the step, i.e. at $x_2 = 2$, is

$$f(x_2, y_2^0) \;=\; (3.14268)\cos(2) \;=\; -1.30782$$

Then, with the corrector from Eq. (7.26b),

$$y_2 \;=\; 2.04030 + \frac{1.10238 - 1.30782}{2}(1) \;=\; 1.93758 \qquad (7.28b)$$

which gives the exact error of

$$E_t \;=\; e^{\sin(2)} - 1.93758 \;=\; 0.54500 \qquad (7.28c)$$

For the third round starting from $x_2 = 2$, $y_2 = 1.93758$, the predictor, the slope, the corrector and the exact error are,

$$y_3^0 \;=\; 1.93758 + [1.93758\cos(2)](1) \;=\; 1.13126 \qquad (7.29a)$$

$$f(x_3, y_3^0) \;=\; (1.13126)\cos(3) \;=\; -1.11994$$

$$y_3 \;=\; 1.93758 + \frac{-0.80632 - 1.11994}{2}(1) \;=\; 0.97445 \qquad (7.29b)$$

$$E_t \;=\; e^{\sin(3)} - 0.97445 \;=\; 0.17711 \qquad (7.29c)$$

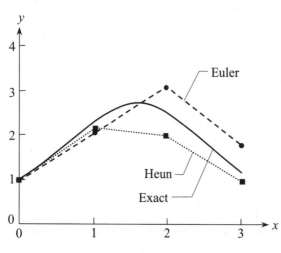

Figure 7.7 Comparison of the Heun's solution using $h = 1$ with the Euler's and exact solutions in Example 7.2.

The Heun's solution is plotted to compare with the Euler's and exact solutions as shown in Fig. 7.7. Figure 7.8 presents a computer program of the Heun's method for solving the differential equation in Example 7.2. The program is similar to that of the Euler's method in Fig. 7.5 except few additional commands are included for determining the predictor and corrector. Table 7.2 shows the Heun's solution which is more accurate than the Euler's solution. The Heun's method is the second-order method that provides second-order accuracy of solution, $O(h^2)$. It is noted that the Euler's method as explained earlier can be modified to yield the second-order solution accuracy similar to the Heun's method as explained in the following section.

Fortran

```
      PROGRAM  HEUN
C.....A PROGRAM FOR SOLVING ORDINARY DIFFERENTIAL
C.....EQUATION USING THE HEUN'S METHOD
C.....READ INITIAL CONDITIONS, NUMBER OF STEPS,
C.....AND STEP SIZE:
      READ(5,*)  X, Y, N, H
      WRITE(6,100)  H
 100  FORMAT(' SOLUTION WITH STEP SIZE =', E10.4,
     *   ' IS:', /, 10X, 'X', 15X, 'Y')
      WRITE(6,200)  X, Y
 200  FORMAT(2E16.6)
      DO 300  I=1,N
      S0 = FUNC(X,Y)
      Y1 = Y + S0*H
      X  = X + H
      S1 = FUNC(X,Y1)
      SA = (S0 + S1)/2.
      Y  = Y + SA*H
      WRITE(6,200)  X, Y
 300  CONTINUE
      STOP
      END
C------------------------------------------------
      FUNCTION  FUNC(X,Y)
      FUNC = Y*COS(X)
      RETURN
      END
```

MATLAB

```
%   PROGRAM  HEUN
%   A PROGRAM FOR SOLVING ORDINARY DIFFERENTIAL
%   EQUATION USING THE HEUN'S METHOD
%   READ INITIAL CONDITIONS, NUMBER OF STEPS,
%   AND STEP SIZE:
%------------------------------------------------
func = inline('y*cos(x)','x','y');
%------------------------------------------------
x = input('\nEnter value of x:  ');
y = input( 'Enter value of y:  ');
n = input( 'Enter value of n:  ');
h = input( 'Enter value of h:  ');
fprintf('\nSOLUTION WITH STEP SIZE = %10.4e IS:',h);
fprintf('\n            X                Y');
fprintf('\n%16.6e%16.6e',x,y);
for i = 1:n
    s0 = func(x,y);
    y1 = y + s0*h;
    x  = x + h;
    s1 = func(x,y1);
    sa = (s0 + s1)/2.;
    y  = y + sa*h;
    fprintf('\n%16.6e%16.6e',x,y);
end
```

Figure 7.8 Computer program for solving the first-order ordinary differential equation by using the Heun's method in Example 7.2.

7.4 Modified Euler's Method

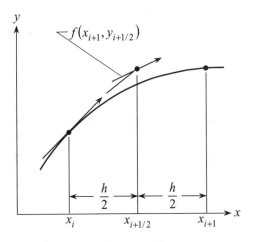

(a) Slope at mid-step.

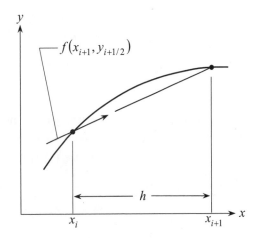

(b) Use of the average slope.

Figure 7.9 The modified Euler's method.

The idea of the modified Euler's method is similar to the Heun's method in the sense that accurate slope at the beginning of the step leads to an improved solution accuracy at the end of the step. The modified Euler's method determines the value and its slope at the mid-step as shown in Fig. 7.9(a). The computed slope is then used at the beginning of the step to predict the solution at the end of the step as illustrated in Fig. 7.9(b).

The solution at mid-step is first determined according to the Euler's method,

$$y_{i+1/2} = y_i + f(x_i, y_i)\frac{h}{2} \tag{7.30}$$

The slope at mid-step is then determined,

$$y'_{i+1/2} = f(x_{i+1/2}, y_{i+1/2}) \tag{7.31}$$

The computed slope is used at the beginning of the step to determine the solution at the end of the step from

$$y_{i+1} = y_i + f(x_{i+1/2}, y_{i+1/2})h \tag{7.32}$$

Example 7.3 Use the modified Euler's method to solve the ordinary differential equation,

$$\frac{dy}{dx} = y \cos x \tag{7.13}$$

with the initial condition of $y(0) = 1$. Use the step size of $h = 1.00$ in the computation. Then, develop a corresponding computer program to solve the problem again but by decreasing the step size to $h = 0.25$. It is noted that the problem is identical to Examples 7.1 and 7.2 that were previously solved by using the Euler's and Heun's method, respectively.

For the first round with $h = 1.00$ at $x_0 = 0$ and $y_0 = 1$, the solution at mid-step is determined from Eq. (7.30),

$$y_{1/2} = 1 + [(1)(1)](0.5) \qquad = 1.5 \tag{7.33a}$$

The slope at mid-step can then be determined by using Eq. (7.31),

$$y'_{1/2} = 1.5\cos(0.5) \qquad = 1.31637$$

The computed slope is used at the beginning of the step to determine the solution at the end of the step according to Eq. (7.32) as

$$y_1 = 1 + (1.31637)(1) \qquad = 2.31637 \tag{7.33b}$$

which has the exact error of

$$E_t = e^{\sin(1)} - 2.31637 \qquad = 0.00341 \tag{7.33c}$$

For the second round, $x_1 = 1$ and $y_1 = 2.31637$, the solution and its slope at mid-step, the computed solution at the end of the step and the exact error are

$$y_{1+1/2} = 2.31637 + [2.31637\cos(1)](0.5) \qquad = 2.94214 \tag{7.34a}$$

$$y'_{1+1/2} = 2.94214\cos(1.5) \qquad = 0.20812$$

$$y_2 \quad = \quad 2.31637 + (0.20812)(1) \qquad\qquad = \quad 2.52449 \qquad\qquad (7.34b)$$

$$E_t \quad = \quad e^{\sin(2)} - 2.52449 \qquad\qquad = \quad -0.04191 \qquad\qquad (7.34c)$$

Similarly, for the third round at $x_2 = 2$ with $y_2 = 2.52449$, the solution and its slope at mid-step, the computed solution at the end of the step and the exact error are

$$y_{2+1/2} \quad = \quad 2.52449 + [2.52449 \cos(2)](0.5) \quad = \quad 1.99921 \qquad (7.35a)$$

$$y'_{2+1/2} \quad = \quad 1.99921 \cos(2.5) \qquad\qquad = \quad -1.60165$$

$$y_3 \quad = \quad 2.52449 + (-1.60165)(1) \qquad = \quad 0.92284 \qquad (7.35b)$$

$$E_t \quad = \quad e^{\sin(3)} - 0.92284 \qquad\qquad = \quad 0.22872 \qquad (7.35c)$$

The solution obtained from the modified Euler's method using the step size of $h = 1$ is compared with the Euler's, Heun's and exact solutions as plotted in Fig. 7.10.

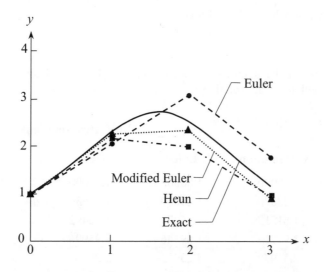

Figure 7.10 Comparison of the solutions obtained from the Euler's, Heun's and modified Euler's methods with the exact solution in Example 7.3.

Figure 7.11 shows the corresponding computer program of the modified Euler's method for solving the differential equation in Example 7.3. The computed solution using $h = 0.25$ is compared with the Euler's, Heun's and exact solutions in Table 7.2. The modified Euler's solution in the Table shows that the method provides the second-order solution accuracy, $O(h^2)$, in the same way as the Heun's method.

Fortran

```
      PROGRAM  MEULER
C.....A PROGRAM FOR SOLVING ORDINARY DIFFERENTIAL
C.....EQUATION USING THE MODIFIED EULER'S METHOD
C.....READ INITIAL CONDITIONS, NUMBER OF STEPS,
C.....AND STEP SIZE:
      READ(5,*)  X, Y, N, H
      WRITE(6,100)  H
  100 FORMAT(' SOLUTION WITH STEP SIZE =', E10.4,
     *  ' IS:', /, 10X, 'X', 15X, 'Y')
      WRITE(6,200)  X, Y
  200 FORMAT(2E16.6)
      DO 300  I=1,N
      S0 = FUNC(X,Y)
      Y1 = Y + S0*H/2.
      X1 = X + H/2.
      SA = FUNC(X1,Y1)
      Y  = Y + SA*H
      X  = X + H
      WRITE(6,200)  X, Y
  300 CONTINUE
      STOP
      END
C-----------------------------------------------------
      FUNCTION  FUNC(X,Y)
      FUNC = Y*COS(X)
      RETURN
      END
```

MATLAB

```
%    PROGRAM  MEULER
%    A PROGRAM FOR SOLVING ORDINARY DIFFERENTIAL
%    EQUATION USING THE MODIFIED EULER'S METHOD
%    READ INITIAL CONDITIONS, NUMBER OF STEPS,
%    AND STEP SIZE:
%-------------------------------------------------
func = inline('y*cos(x)','x','y');
%-------------------------------------------------
x = input('\nEnter value of x:  ');
y = input( 'Enter value of y:  ');
n = input( 'Enter value of n:  ');
h = input( 'Enter value of h:  ');
fprintf('\nSOLUTION WITH STEP SIZE = %10.4e IS:',h);
fprintf('\n          X                Y');
fprintf('\n%16.6e%16.6e',x,y);
for i = 1:n
    s0 = func(x,y);
    y1 = y + s0*h/2.;
    x1 = x + h/2.;
    sa = func(x1,y1);
    y  = y + sa*h;
    x  = x + h;
    fprintf('\n%16.6e%16.6e',x,y);
end
```

Figure 7.11 Computer program for solving the first-order ordinary differential equation by using the modified Euler's method in Example 7.3.

Table 7.2 Comparison of the solutions obtained from the Euler's, Heun's and modified Euler's methods with the exact solution.

		Numerical solutions		
x	**Exact solution**	**Euler**	**Heun**	**Modified Euler**
0.00	1.00000	1.00000	1.00000	1.00000
0.25	1.28070	1.25000	1.27639	1.27906
0.50	1.61515	1.55279	1.60492	1.61263
0.75	1.97712	1.89346	1.95996	1.97545
1.00	2.31978	2.23982	2.29581	2.32096
1.25	2.58309	2.54236	2.55358	2.58805
1.50	2.71148	2.74278	2.67858	2.71888
1.75	2.67510	2.79128	2.64153	2.68173
2.00	2.48258	2.66690	2.45139	2.48539
2.25	2.17727	2.38944	2.15141	2.17541
2.50	1.81934	2.01419	1.80087	1.81444
2.75	1.46472	1.61078	1.45413	1.45952
3.00	1.15156	1.23857	1.14776	1.14820

7.5 Runge-Kutta Method

The Runge-Kutta method is popular and has been used to solve scientific and engineering problems that require high solution accuracy. The key idea of the method is to obtain a proper slope at x_i in order to determine accurate solution y_{i+1} at x_{i+1}. The predicted solution is first written in the form similar to the three methods previously explained,

$$y_{i+1} = y_i + \phi(x_i, y_i, h)\, h \tag{7.36}$$

where $\phi(x_i, y_i, h)$ is called the increment function. The increment function representing the average slope over the step h is expressed in a general form as

$$\phi = a_1 k_1 + a_2 k_2 + a_3 k_3 + \ldots + a_n k_n \tag{7.37}$$

where $a_i, i = 1, 2, 3, \ldots, n$ are constants and

$$k_1 = f(x_i, y_i) \tag{7.38a}$$

$$k_2 = f(x_i + p_1 h, y_i + q_{11} k_1 h) \tag{7.38b}$$

$$k_3 = f(x_i + p_2 h, y_i + q_{21} k_1 h + q_{22} k_2 h) \tag{7.38c}$$

$$\vdots \qquad\qquad \vdots$$

$$k_n = f\big(x_i + p_{n-1} h, y_i + q_{n-1,1} k_1 h + q_{n-1,2} k_2 h$$
$$+ \ldots + q_{n-1,n-1} k_{n-1} h\big) \tag{7.38n}$$

The subscript n in Eq. (7.37) denotes the order of the Runge-Kutta method. For example, the method is called the first-order Runge-Kutta method if $n = 1$. By considering Eqs. (7.36) - (7.38a), the first-order Runge-Kutta method is equivalent to the Euler method. If $n = 2$, Eqs. (7.36) - (7.38b) yield the second-order Runge-Kutta method. The parameters $k_i, i = 1, 2, 3, \ldots, n$ in Eq. (7.38) depend on the given function on the right-hand side of the ordinary differential equation. The coefficients p and q are constants which can be determined and will be shown later. Determination of these coefficients for the fourth-order Runge-Kutta method is also presented in details in Appendix C. It is noted that the value for the parameter k_1 must be known prior to determining the parameter k_2. Similarly, the value of the parameter k_2 must be known earlier in order to determine the parameter k_3. To understand the Runge-Kutta method clearly, the following section explains the second-order Runge-Kutta method ($n = 2$) in details.

7.5.1 Second-order

For the second-order Runge-Kutta method ($n = 2$), the general form of Eqs. (7.36) - (7.38) become

$$y_{i+1} = y_i + \big(a_1 k_1 + a_2 k_2\big) h \tag{7.39}$$

where
$$k_1 = f(x_i, y_i) \tag{7.40}$$

and
$$k_2 = f(x_i + p_1 h, y_i + q_{11} k_1 h) \tag{7.41}$$

There are four unknowns of a_1, a_2, p_1 and q_{11} in Eqs. (7.39) - (7.41). These four unknowns are determined such that the second-order Runge-Kutta method provides the same solution accuracy as that obtained from the Taylor series expansion with three terms,

$$y_{i+1} = y_i + f(x_i, y_i)h + f'(x_i, y_i)\frac{h^2}{2} + \dots \tag{7.42}$$

From chain-rule, the first-order derivative term in Eq. (7.42) can be expressed as

$$f'(x_i, y_i) = \frac{\partial f}{\partial x} + \frac{\partial f}{\partial y}\frac{dy}{dx} = \frac{\partial f}{\partial x} + f(x_i, y_i)\frac{\partial f}{\partial y} \tag{7.43}$$

By substituting Eq. (7.43) into Eq. (7.42),

$$y_{i+1} = y_i + f(x_i, y_i)h + \left(\frac{\partial f}{\partial x} + f(x_i, y_i)\frac{\partial f}{\partial y}\right)\frac{h^2}{2} + \dots \tag{7.44}$$

Thus, in order to determine the four unknowns, the second-order Runge-Kutta Eq. (7.39) must be written in the form of the Taylor series in Eq. (7.44). To do that, the Taylor series expansion with two variables

$$g(x+r, y+s) = g(x, y) + r\frac{\partial g}{\partial x} + s\frac{\partial g}{\partial y} \tag{7.45}$$

is first applied to Eq. (7.41) leading to

$$k_2 = f(x_i, y_i) + p_1 h\frac{\partial f}{\partial x} + q_{11}k_1 h\frac{\partial f}{\partial y} + \dots \tag{7.46}$$

By substituting k_1 from Eq. (7.40) into Eq. (7.46), then further substituting Eq. (7.46) into Eq. (7.39) and rearranging terms, Eq. (7.39) finally becomes

$$y_{i+1} = y_i + \left[a_1 f(x_i, y_i) + a_2 f(x_i, y_i)\right] h$$

$$+ \left[a_2 p_1 \frac{\partial f}{\partial x} + a_2 q_{11}f(x_i, y_i)\frac{\partial f}{\partial y}\right]h^2 + \dots \tag{7.47}$$

By comparing the Runge-Kutta Eq. (7.47) with the Taylor series Eq. (7.44), the following three conditions are obtained,

$$a_1 + a_2 = 1 \tag{7.48a}$$

$$a_2 p_1 = 1/2 \tag{7.48b}$$

$$a_2 q_{11} = 1/2 \tag{7.48c}$$

Since there are four unknowns with the only three available conditions, an unknown must be given in order to determine the rest of the unknowns.

If $a_2 = 1/2$ is selected, then Eqs. (7.48a-c) yield

$$a_1 = 1/2 \quad ; \quad p_1 = 1 \quad ; \quad q_{11} = 1 \tag{7.49}$$

Then, the second-order Runge-Kutta Eq. (7.39) is

$$y_{i+1} = y_i + \left(\frac{1}{2} k_1 + \frac{1}{2} k_2 \right) h \tag{7.50}$$

and Eqs. (7.40) - (7.41) are

$$k_1 = f(x_i, y_i) \qquad = \quad \text{slope at the beginning of step } h$$

$$k_2 = f(x_i + h, y_i + hk_1) \qquad = \quad \text{slope at the end of step } h$$

It is noted that Eq. (7.50) is identical to Eq. (7.26b) which is the Huen's method presented in Section 7.3.

If $a_2 = 1$ is selected, then Eqs. (7.48a-c) yield

$$a_1 = 0 \quad ; \quad p_1 = 1/2 \quad ; \quad q_{11} = 1/2 \tag{7.51}$$

Then, the second-order Runge-Kutta Eq. (7.39) is

$$y_{i+1} = y_i + (0 + k_2) h \tag{7.52}$$

and Eqs. (7.40) – (7.41) are

$$k_1 = f(x_i, y_i) \qquad = \quad \text{slope at the beginning of step } h$$

$$k_2 = f\left(x_i + \frac{1}{2}h, y_i + \frac{1}{2}hk_1 \right) \qquad = \quad \text{slope at the end of step } h$$

It is also noted that Eq. (7.52) is identical to Eq. (7.32) which is the modified Euler's method presented in Section 7.4.

Equations (7.50) and (7.52) for the second-order Runge-Kutta method provide a solution with the second-order of accuracy, $O(h^2)$. The second-order Runge-Kutta method reduces the solution error to a quarter if the step size h is cut by a half.

7.5.2 Third-order

The third-order Runge-Kutta method ($n = 3$) can be obtained from the general expressions as shown in Eqs. (7.36) - (7.3.8). The widely used third-order Runge-Kutta method is in the following form,

$$y_{i+1} = y_i + \left[\frac{1}{6} (k_1 + 4 k_2 + k_3) \right] h \tag{7.53}$$

where

$$k_1 = f(x_i, y_i) \tag{7.54a}$$

$$k_2 = f\left(x_i + \frac{1}{2}h, \ y_i + \frac{1}{2}hk_1\right) \tag{7.54b}$$

$$k_3 = f(x_i + h, \ y_i - hk_1 + 2hk_2) \tag{7.54c}$$

Solution error obtained from Eq. (7.53) decreases with the step size h as the third order, $O(h^3)$. The computational procedure of the third-order Runge-Kutta method is simple as shown in the following example.

Example 7.4 Develop a computer program by employing the third-order Runge-Kutta method to solve the differential equation

$$\frac{dy}{dx} = y\cos x \tag{7.13}$$

with the initial condition of $y(0)=1$. Use the step size of $h = 0.25$ in the computation and compare the solution with the exact and Heun's solutions.

Starting from the initial condition of $y_0 = 1$ at $x_0 = 0$, Eqs. (7.53) - (7.54) give

$$k_1 = (1)\cos(0) \qquad\qquad\qquad = 1$$

$$k_2 = \left[1 + \left(\frac{1}{2}\right)(0.25)(1)\right]\cos\left(\frac{0.25}{2}\right) \qquad = 1.11622$$

$$k_3 = [1 - (0.25)(1) + 2(0.25)(1.11622)]\cos(0.25) \qquad = 1.26745$$

$$y_1 = 1 + \left[\frac{1}{6}(1 + 4(1.11622) + 1.26745)\right](0.25) \qquad = 1.28051$$

The solution of $y_1 = 1.28051$ obtained from the first step at $x_1 = 0.25$ is used to determine the solution at the end of the second step as

$$k_1 = (1.28051)\cos(0.25) \qquad\qquad\qquad = 1.24071$$

$$k_2 = \left[1.28051 + \left(\frac{1}{2}\right)(0.25)(1.24071)\right]\cos\left(0.25 + \frac{0.25}{2}\right) \qquad = 1.33584$$

$$k_3 = [1.28051 - (0.25)(1.24071) + 2(0.25)(1.33584)]\cos(0.25 + 0.25) \qquad = 1.43771$$

$$y_2 = 1.28051 + \left[\frac{1}{6}(1.24071 + 4(1.33584) + 1.43771)\right](0.25) \qquad = 1.61475$$

The same process is repeated to determine the solutions at the later steps. Figure 7.12 shows the computer programs of the third-order Runge-Kutta method for solving the differential equation in this example. The computed solutions at different time steps are compared with the exact and Heun's solutions as shown in Table 7.3.

Fortran

```
      PROGRAM  RK3
C.....A PROGRAM FOR SOLVING ORDINARY DIFFERENTIAL
C.....EQUATION BY THIRD-ORDER RUNGE-KUTTA METHOD
C.....READ INITIAL CONDITIONS, NUMBER OF STEPS,
C.....AND STEP SIZE:
      READ(5,*)  X, Y, N, H
      WRITE(6,100)  H
100   FORMAT(' SOLUTION WITH STEP SIZE =', E10.4,
     *  ' IS:', /, 10X, 'X', 15X, 'Y')
      WRITE(6,200)  X, Y
200   FORMAT(2E16.6)
      DO 300  I=1,N
      AK1 = FUNC(X,Y)
      XX  = X + H/2.
      YY  = Y + H*AK1/2.
      AK2 = FUNC(XX,YY)
      XX  = X + H
      YY  = Y - H*AK1 + 2.*H*AK2
      AK3 = FUNC(XX,YY)
      Y   = Y + (AK1 + 4.*AK2 + AK3)*H/6.
      X   = X + H
      WRITE(6,200)  X, Y
300   CONTINUE
      STOP
      END
C------------------------------------------------
      FUNCTION  FUNC(X,Y)
      FUNC = Y*COS(X)
      RETURN
      END
```

MATLAB

```
%   PROGRAM  RK3
%   A PROGRAM FOR SOLVING ORDINARY DIFFERENTIAL
%   EQUATION BY THIRD-ORDER RUNGE-KUTTA METHOD
%   READ INITIAL CONDITIONS, NUMBER OF STEPS,
%   AND STEP SIZE:
%------------------------------------------------
func = inline('y*cos(x)','x','y');
%------------------------------------------------
x = input('\nEnter value of x:  ');
y = input( 'Enter value of y:  ');
n = input( 'Enter value of n:  ');
h = input( 'Enter value of h:  ');
fprintf('\nSOLUTION WITH STEP SIZE = %10.4e IS:',h);
fprintf('\n          X                   Y');
fprintf('\n%16.6e%16.6e',x,y);
for i = 1:n
    ak1 = func(x,y);
    xx  = x + h/2.;
    yy  = y + h*ak1/2.;
    ak2 = func(xx,yy);
    xx  = x + h;
    yy  = y - h*ak1 + 2.*h*ak2;
    ak3 = func(xx,yy);
    y   = y + (ak1 + 4.*ak2 + ak3)*h/6.;
    x   = x + h;
    fprintf('\n%16.6e%16.6e',x,y);
end
```

Figure 7.12 Computer program for solving the first-order ordinary differential equation by using the third-order Runge-Kutta method in Example 7.4.

7.5.3 Fourth-order

The fourth-order Runge-Kutta method ($n = 4$) is widely used for solving many scientific and engineering problems by embedding itself in commercial software. The method provides solution with the fourth-order of accuracy, $O(h^4)$. The most popular form of the fourth-order Runge-Kutta method which was derived from the general form of Runge-Kutta Eqs. (7.36) - (7.38) is,

$$y_{i+1} = y_i + \left[\frac{1}{6}(k_1 + 2 k_2 + 2k_3 + k_4)\right] h \tag{7.55}$$

where

$$k_1 = f(x_i, y_i) \tag{7.56a}$$

$$k_2 = f\left(x_i + \frac{1}{2}h, y_i + \frac{1}{2}hk_1\right) \tag{7.56b}$$

$$k_3 = f\left(x_i + \frac{1}{2}h, y_i + \frac{1}{2}hk_2\right) \tag{7.56c}$$

$$k_4 = f(x_i + h, y_i + hk_3) \tag{7.56d}$$

Details for the derivation of Eqs. (7.55) - (7.56a-d) are presented in Appendix C.

Example 7.5 Use the fourth-order Runge-Kutta method to solve the differential equation

$$\frac{dy}{dx} = y \cos x \qquad (7.13)$$

with the initial condition of $y(0)=1$ by developing a computer program. Then, employ the program to determine the solution by using the step size of $h = 0.25$. Compare the solution with the second-, third-order Runge-Kutta and exact solutions.

Details of the computational procedure for the first two steps are shown below. With the initial condition of $y_0 = 1$ at $x_0 = 0$, Eqs. (7.55) - (7.56) for the first step are

$$k_1 = (1)\cos(0) \qquad\qquad = 1$$

$$k_2 = \left[1+\left(\frac{1}{2}\right)(0.25)(1)\right]\cos\left(\frac{0.25}{2}\right) \qquad\qquad = 1.11622$$

$$k_3 = \left[1+\left(\frac{1}{2}\right)(0.25)(1.11622)\right]\cos\left(\frac{0.25}{2}\right) \qquad\qquad = 1.13064$$

$$k_4 = [1+(0.25)(1.13064)]\cos(0.25) \qquad\qquad = 1.24278$$

$$y_1 = 1 + \left[\frac{1}{6}(1+2(1.11622)+2(1.13064)+1.24278)\right](0.25) \qquad = 1.28069$$

For the second step with the computed $y_1 = 1.28069$ at $x_1 = 0.25$, Eqs. (7.55) - (7.56) are

$$k_1 = (1.28069)\cos(0.25) \qquad\qquad = 1.24087$$

$$k_2 = \left[1.28069+\left(\frac{1}{2}\right)(0.25)(1.24087)\right]\cos\left(0.25+\frac{0.25}{2}\right) \qquad = 1.33602$$

$$k_3 = \left[1.28069+\left(\frac{1}{2}\right)(0.25)(1.33602)\right]\cos\left(0.25+\frac{0.25}{2}\right) \qquad = 1.34709$$

$$k_4 = [1.28069+(0.25)(1.34709)]\cos(0.25+0.25) \qquad\qquad = 1.41945$$

$$y_2 = 1.28069 + \left[\frac{1}{6}(1.24087+2(1.33602)+2(1.34709)+1.41945)\right](0.25) = 1.61513$$

The process is repeated to determine solution for the other steps. Miscalculating a parameter in any step will cause error in the later steps. Thus, solving the problem by developing a computer program is essential. Figure 7.13 shows a fourth-order Runge-Kutta computer program for solving this problem. User can employ the program to solve other differential equations by simply modifying the function declared in the program.

The computed solutions at different x are compared with the exact and the second- and third-order Runge-Kutta solutions in Table 7.3. The table shows that the fourth-order Runge-Kutta solutions are more accurate than those obtained from the second- and third-order Runge-Kutta methods. For example, at $x = 1.00$, the fourth-order Runge-Kutta solution is 2.31974. Such solution has only 0.002% error as compared to the exact solution of 2.31978. The solutions obtained from the Euler's, second- and third-order Runge-Kutta methods contain the errors of 3.45%, 1.03% and 0.02%, respectively. High solution accuracy of the fourth-order Runge-Kutta method has led to the method's popularity for solving many practical applications as well as using in academic research.

Fortran

```
      PROGRAM  RK4
C.....A PROGRAM FOR SOLVING ORDINARY DIFFERENTIAL
C.....EQUATION BY FOURTH-ORDER RUNGE-KUTTA METHOD
C.....READ INITIAL CONDITIONS, NUMBER OF STEPS,
C.....AND STEP SIZE:
      READ(5,*)  X, Y, N, H
      WRITE(6,100)  H
 100  FORMAT(' SOLUTION WITH STEP SIZE =', E10.4,
     *  ' IS:', /, 10X, 'X', 15X, 'Y')
      WRITE(6,200)  X, Y
 200  FORMAT(2E16.6)
      DO 300  I=1,N
      AK1 = FUNC(X,Y)
      XX  = X + H/2.
      YY  = Y + H*AK1/2.
      AK2 = FUNC(XX,YY)
      YY  = Y + H*AK2/2.
      AK3 = FUNC(XX,YY)
      XX  = X + H
      YY  = Y + H*AK3
      AK4 = FUNC(XX,YY)
      Y   = Y + (AK1 + 2.*AK2 + 2.*AK3 + AK4)*H/6.
      X   = X + H
      WRITE(6,200)  X, Y
 300  CONTINUE
      STOP
      END
C-------------------------------------------------
      FUNCTION  FUNC(X,Y)
      FUNC = Y*COS(X)
      RETURN
      END
```

MATLAB

```
%  PROGRAM  RK4
%  A PROGRAM FOR SOLVING ORDINARY DIFFERENTIAL
%  EQUATION BY FOURTH-ORDER RUNGE-KUTTA METHOD
%  READ INITIAL CONDITIONS, NUMBER OF STEPS,
%  AND STEP SIZE:
%---------------------------------------------
func = inline('y*cos(x)','x','y');
%---------------------------------------------
x = input('\nEnter value of x:  ');
y = input( 'Enter value of y:  ');
n = input( 'Enter value of n:  ');
h = input( 'Enter value of h:  ');
fprintf('\nSOLUTION WITH STEP SIZE = %10.4e IS:',h);
fprintf('\n          X                Y');
fprintf('\n%16.6e%16.6e',x,y);
for i = 1:n
    ak1 = func(x,y);
    xx  = x + h/2.;
    yy  = y + h*ak1/2.;
    ak2 = func(xx,yy);
    yy  = y + h*ak2/2.;
    ak3 = func(xx,yy);
    xx  = x + h;
    yy  = y + h*ak3;
    ak4 = func(xx,yy);
    y   = y + (ak1 + 2.*ak2 + 2.*ak3 + ak4)*h/6.;
    x   = x + h;
    fprintf('\n%16.6e%16.6e',x,y);
end
```

Figure 7.13 Computer program for solving the first-order ordinary differential equation by using the fourth-order Runge-Kutta method in Example 7.5.

Table 7.3 Comparison of the second-, third- and fourth-order Runge-Kutta solutions obtained from solving the differential Eq. (7.13) with the exact solution.

		Order of Runge-Kutta method		
x	Exact	Second	Third	Fourth
0.00	1.00000	1.00000	1.00000	1.00000
0.25	1.28070	1.27639	1.28051	1.28069
0.50	1.61515	1.60492	1.61475	1.61513
0.75	1.97712	1.95996	1.97657	1.97709
1.00	2.31978	2.29581	2.31923	2.31974
1.25	2.58309	2.55358	2.58273	2.58304
1.50	2.71148	2.67858	2.71149	2.71144
1.75	2.67510	2.64153	2.67555	2.67505
2.00	2.48258	2.45139	2.48342	2.48254
2.25	2.17727	2.15141	2.17832	2.17723
2.50	1.81934	1.80087	1.82034	1.81931
2.75	1.46472	1.45413	1.46547	1.46469
3.00	1.15156	1.14775	1.15201	1.15155

7.6 System of Equations

Practical scientific and engineering problems often require solving many first-order differential equations which are coupled. For example, the swinging pendulum problem explained in Section 7.1 consists of two first-order differential Eqs. (7.57a-b). The two differential equations are coupled and must be solved simultaneously for their solutions. In general, a system of n first-order differential equations can be written in the form,

$$\frac{dy_1}{dx} = f_1(x, y_1, y_2, ..., y_n) \tag{7.57a}$$

$$\frac{dy_2}{dx} = f_2(x, y_1, y_2, ..., y_n) \tag{7.57b}$$

$$\vdots \qquad\qquad \vdots$$

$$\frac{dy_n}{dx} = f_n(x, y_1, y_2, ..., y_n) \tag{7.57n}$$

The methods, such as the Euler's and Runge-Kutta methods explained earlier in Sections 7.2 - 7.5, can be modified to solve the above set of coupled first-order differential equations. The example below presents the use of the presented methods to solve a second-order differential equation which is separated into two coupled first-order differential equations.

Example 7.6 Employ the Euler's method to solve the second-order differential equation

$$\frac{d^2 y}{dx^2} + 2\frac{dy}{dx} + 4y = 0 \tag{7.58}$$

with the initial conditions of $y(0) = 2$ and $dy/dx(0) = 0$. Determine the solution from $x = 0$ to 3 by using the step sizes of $h = 0.1$ and 0.01.

The exact solution to the differential Eq. (7.58) with the given initial conditions is

$$y(x) = 2e^{-x}\left[\cos\left(\sqrt{3}\,x\right) + \frac{1}{\sqrt{3}}\sin\left(\sqrt{3}\,x\right)\right] \tag{7.59}$$

The Euler's method can be modified to solve the second-order differential Eq. (7.58). The differential equation is first separated into two first-order differential equations. For example, by assigning

$$\frac{dy}{dx} = z \tag{7.60a}$$

then, Eq. (7.58) becomes

$$\frac{dz}{dx} = -2z - 4y \tag{7.60b}$$

with the initial conditions of $y(x = 0) = 2$ and $z(x = 0) = 0$.

The Euler's Eq. (7.12) is then applied to Eqs. (7.60a-b) as follows

$$y_{i+1} = y_i + f_1(x_i, y_i, z_i)\,h \qquad\qquad = y_i + z_i h \tag{7.61a}$$

$$z_{i+1} = z_i + f_2(x_i, y_i, z_i)\,h \qquad\qquad = z_i + (-2z_i - 4y_i)h \tag{7.61b}$$

With the step size of $h = 0.1$, the first step yields

$$y_1 = 2 + (0)(0.1) = 2$$

$$z_1 = 0 + [-2(0) - 4(2)](0.1) = -0.8$$

The computed solutions are then used in the second step to yield

$$y_2 = 2 + (-0.8)(0.1) = 1.92$$

$$z_2 = -0.8 + [-2(-0.8) - 4(2)](0.1) = -1.44$$

The process is repeated until the last step at $x = 3$ is reached. Figure 7.14 shows the corresponding computer program that follows the above procedure for solving this example. The computed solutions by using the step sizes of $h = 0.1$ and 0.01 are compared with the exact solution as shown in Table 7.4.

Fortran

```
      PROGRAM  SYSEUL
C.....A PROGRAM FOR SOLVING A SET OF TWO ORDINARY
C.....FIRST-ORDER DIFFERENTIAL EQUATIONS USING
C.....THE EULER'S METHOD
C.....READ INITIAL CONDITIONS, NUMBER OF STEPS,
C.....AND STEP SIZE:
      READ(5,*)  X, Y, Z, N, H
      WRITE(6,100)  H
100   FORMAT(' SOLUTION WITH STEP SIZE =', E10.4,
     *  ' IS:', /, 10X, 'X', 15X, 'Y')
      WRITE(6,200)  X, Y, Z
200   FORMAT(3E16.6)
      DO 300  I=1,N
      F1 = FUNC1(X,Y,Z)
      F2 = FUNC2(X,Y,Z)
      Y  = Y + F1*H
      Z  = Z + F2*H
      X  = X + H
      WRITE(6,200)  X, Y, Z
300   CONTINUE
      STOP
      END
C----------------------------------------------------
      FUNCTION   FUNC1(X,Y,Z)
      FUNC1 = Z
      RETURN
      END
C----------------------------------------------------
      FUNCTION   FUNC2(X,Y,Z)
      FUNC2 = -2.*Z - 4.*Y
      RETURN
      END
```

MATLAB

```
%    PROGRAM  SYSEUL
%    A PROGRAM FOR SOLVING A SET OF TWO ORDINARY
%    FIRST-ORDER DIFFERENTIAL EQUATIONS USING
%    THE EULER'S METHOD
%    READ INITIAL CONDITIONS, NUMBER OF STEPS,
%    AND STEP SIZE:
%-------------------------------------------------
func1 = inline('z','x','y','z');
%-------------------------------------------------
func2 = inline('-2.*z - 4.*y','x','y','z');
%-------------------------------------------------
x = input('\nEnter value of x:  ');
y = input( 'Enter value of y:  ');
z = input( 'Enter value of z:  ');
n = input( 'Enter value of n:  ');
h = input( 'Enter value of h:  ');
fprintf('\nSOLUTION WITH STEP SIZE = %10.4e IS:',h);
fprintf('\n       X              Y              Z');
fprintf('\n%16.6e%16.6e%16.6e',x,y,z);
for i = 1:n
    f1 = func1(x,y,z);
    f2 = func2(x,y,z);
    y  = y + f1*h;
    z  = z + f2*h;
    x  = x + h;
    fprintf('\n%16.6e%16.6e%16.6e',x,y,z);
end
```

Figure 7.14 Computer program for solving the second-order ordinary differential equation by using the Euler's method in Example 7.6.

Example 7.7 Solve the problem in Example 7.6 again but by developing a computer program that employs the fourth-order Runge-Kutta method. Use the step size of $h = 0.1$ and compare the solution obtained with the exact and the Euler's solutions.

The second-order differential equation in Eq. (7.58) is firstly separated into the two first-order differential equations as shown in Eqs. (7.60a-b). The fourth-order Runge-Kutta method as shown in Eqs. (7.55) - (7.56) is then applied to each differential equation as follows,

$$y_{i+1} = y_i + \left[\frac{1}{6}\left(k_{1y} + 2\,k_{2y} + 2k_{3y} + k_{4y}\right)\right] h \qquad (7.62a)$$

$$z_{i+1} = z_i + \left[\frac{1}{6}\left(k_{1z} + 2\,k_{2z} + 2k_{3z} + k_{4z}\right)\right] h \qquad (7.62b)$$

where

$$k_{1y} = f_1(x_i, y_i, z_i) \qquad (7.63a)$$

$$k_{2y} = f_1\left(x_i + \frac{1}{2}h,\ y_i + \frac{1}{2}hk_{1y},\ z_i + \frac{1}{2}hk_{1z}\right) \qquad (7.63b)$$

$$k_{3y} = f_1\left(x_i + \frac{1}{2}h,\ y_i + \frac{1}{2}hk_{2y},\ z_i + \frac{1}{2}hk_{2z}\right) \qquad (7.63c)$$

$$k_{4y} = f_1\left(x_i + h,\ y_i + hk_{3y},\ z_i + hk_{3z}\right) \qquad (7.63d)$$

$$k_{1z} = f_2(x_i, y_i, z_i) \qquad (7.64a)$$

$$k_{2z} = f_2\left(x_i + \frac{1}{2}h,\ y_i + \frac{1}{2}hk_{1y},\ z_i + \frac{1}{2}hk_{1z}\right) \qquad (7.64b)$$

$$k_{3z} = f_2\left(x_i + \frac{1}{2}h,\ y_i + \frac{1}{2}hk_{2y},\ z_i + \frac{1}{2}hk_{2z}\right) \qquad (7.64c)$$

$$k_{4z} = f_2\left(x_i + h,\ y_i + hk_{3y},\ z_i + hk_{3z}\right) \qquad (7.64d)$$

The functions f_1 and f_2 are

$$f_1(x_i, y_i, z_i) = z_i \qquad \text{and} \qquad f_2(x_i, y_i, z_i) = -2z_i - 4y_i$$

For example, at the first step, $x_0 = 0$, $y_0 = 2$, $z_0 = 0$ where $h = 0.1$, Eqs. (7.63) - (7.64) are

$$
\begin{aligned}
k_{1y} &= f_1(0, 2, 0) & &= & 0 \\
k_{1z} &= f_2(0, 2, 0) & &= & -8 \\
k_{2y} &= f_1(0.05, 2, -0.4) & &= & -0.4 \\
k_{2z} &= f_2(0.05, 2, -0.4) & &= & -7.2 \\
k_{3y} &= f_1(0.05, 1.98, -0.36) & &= & -0.36 \\
k_{3z} &= f_2(0.05, 1.98, -0.36) & &= & -7.2 \\
k_{4y} &= f_1(0.10, 1.964, -0.72) & &= & -0.72 \\
k_{4z} &= f_2(0.10, 1.964, -0.72) & &= & -6.416
\end{aligned}
$$

Then, the new solutions of y and z in Eqs. (7.62a-b) are

$$y_1 = 2 + \left[\frac{1}{6}\left(0 + 2\,(-0.4) + 2(-0.36) + (-0.72)\right)\right](0.1) \quad = \quad 1.962667$$

$$z_1 = 0 + \left[\frac{1}{6}\left(-8 + 2\,(-7.2) + 2(-7.2) + (-6.416)\right)\right](0.1) \quad = \quad -0.720267$$

These solutions are used in the computation of the second step. The process is repeated until the last step is reached. Figure 7.5 shows the corresponding computer program for solving this example by using the fourth-order Runge-Kutta method. The computed solution is compared with the exact and Euler's solutions in Table 7.4.

Fortran

```
      PROGRAM SYSRK4
C.....A PROGRAM FOR SOLVING A SET OF TWO ORDINARY
C.....FIRST-ORDER DIFFERENTIAL EQUATIONS USING
C.....THE FOURTH-ORDER RUNGE-KUTTA METHOD
C.....READ INITIAL CONDITIONS, NUMBER OF STEPS,
C.....AND STEP SIZE:
      READ(5,*)  X, Y, Z, N, H
      WRITE(6,100)   H
 100 FORMAT(' SOLUTION WITH STEP SIZE =', E10.4,
     *    ' IS:', /, 10X, 'X', 15X, 'Y')
      WRITE(6,200)   X, Y, Z
 200 FORMAT(3E16.6)
      DO 300  I=1,N
      AK1Y = FUNC1(X,Y,Z)
      AK1Z = FUNC2(X,Y,Z)
      XX = X + H/2.
      YY = Y + H*AK1Y/2.
      ZZ = Z + H*AK1Z/2.
      AK2Y = FUNC1(XX,YY,ZZ)
      AK2Z = FUNC2(XX,YY,ZZ)
      YY = Y + H*AK2Y/2.
      ZZ = Z + H*AK2Z/2.
      AK3Y = FUNC1(XX,YY,ZZ)
      AK3Z = FUNC2(XX,YY,ZZ)
      XX = X + H
      YY = Y + H*AK3Y
      ZZ = Z + H*AK3Z
      AK4Y = FUNC1(XX,YY,ZZ)
      AK4Z = FUNC2(XX,YY,ZZ)
      Y = Y + (AK1Y+2.*AK2Y+2.*AK3Y+AK4Y)*H/6.
      Z = Z + (AK1Z+2.*AK2Z+2.*AK3Z+AK4Z)*H/6.
      X = X + H
      WRITE(6,200)   X, Y, Z
 300 CONTINUE
      STOP
      END
C-------------------------------------------------
      FUNCTION  FUNC1(X,Y,Z)
      FUNC1 = Z
      RETURN
      END
C-------------------------------------------------
      FUNCTION  FUNC2(X,Y,Z)
      FUNC2 = -2.*Z - 4.*Y
      RETURN
      END
```

MATLAB

```
%  PROGRAM  SYSRK4
%  A PROGRAM FOR SOLVING A SET OF TWO ORDINARY
%  FIRST-ORDER DIFFERENTIAL EQUATIONS USING
%  THE FOURTH-ORDER RUNGE-KUTTA METHOD
%  READ INITIAL CONDITIONS, NUMBER OF STEPS,
%  AND STEP SIZE:
%------------------------------------------------
func1 = inline('z','x','y','z');
%------------------------------------------------
func2 = inline('-2.*z - 4.*y','x','y','z');
%------------------------------------------------
x = input('\nEnter value of x:  ');
y = input( 'Enter value of y:  ');
z = input( 'Enter value of z:  ');
n = input( 'Enter value of n:  ');
h = input( 'Enter value of h:  ');
fprintf('\nSOLUTION WITH STEP SIZE = %10.4e IS:',h);
fprintf('\n       X             Y            Z');
fprintf('\n%16.6e%16.6e%16.6e',x,y,z);
for i = 1:n
    ak1y = func1(x,y,z);
    ak1z = func2(x,y,z);
    xx   = x + h/2.;
    yy   = y + h*ak1y/2.;
    zz   = z + h*ak1z/2.;
    ak2y = func1(xx,yy,zz);
    ak2z = func2(xx,yy,zz);
    yy   = y + h*ak2y/2.;
    zz   = z + h*ak2z/2.;
    ak3y = func1(xx,yy,zz);
    ak3z = func2(xx,yy,zz);
    xx   = x + h;
    yy   = y + h*ak3y;
    zz   = z + h*ak3z;
    ak4y = func1(xx,yy,zz);
    ak4z = func2(xx,yy,zz);
    y    = y + (ak1y + 2.*ak2y + 2.*ak3y + ak4y)*h/6.;
    z    = z + (ak1z + 2.*ak2z + 2.*ak3z + ak4z)*h/6.;
    x    = x + h;
    fprintf('\n%16.6e%16.6e%16.6e',x,y,z);
end
```

Figure 7.15 Computer program for solving the second-order ordinary differential equation by using the fourth-order Runge-Kutta method in Example 7.7.

Table 7.4 highlights the solution accuracy obtained from the fourth-order Runge-Kutta method. The table compares the Runge-Kutta solution with the Euler's solutions that use different step sizes. For example, the exact solution at $x = 1.0$ is 0.301149. By using the step size of $h = 0.1$, the Runge-Kutta solution has the error only 0.004% while the Euler's solution produces the error 38%. The Euler's method still yields the error 3.6% even though the step size is reduced to one-tenth with $h = 0.01$. Since the fourth-order Runge-Kutta method can provide high solution accuracy and a corresponding computer program can be easily developed, the method is widely used for solving ordinary differential equations.

Table 7.4 Comparison of the Euler's and fourth-order Runge-Kutta solutions obtained from solving examples 7.6 and 7.7 with the exact solution.

		Euler		Runge-Kutta
x	**Exact**	*h* = 0.1	*h* = 0.01	*h* = 0.1
0.00	2.000000	2.000000	2.000000	2.000000
0.50	1.319400	1.359360	1.322049	1.319407
1.00	0.301149	0.185380	0.290313	0.301136
1.50	-0.248709	-0.429154	-0.264388	-0.248732
2.00	-0.306246	-0.400115	-0.314619	-0.306259
2.50	-0.149181	-0.121281	-0.147723	-0.149181
3.00	-0.004579	0.076168	0.001437	-0.004571

Example 7.8 Use the Euler's and fourth-order Runge-Kutta methods to solve the swinging pendulum angle $\theta(t)$ from the second-order differential equation

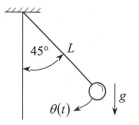

Figure 7.16 Swinging pendulum.

$$\frac{d^2\theta}{dt^2} + \frac{g}{L}\sin\theta = 0 \qquad (7.2)$$

Use the gravitational accelerating constant of $g = 9.8$ m/sec^2 and the chord length of $L = 0.5$ m. The pendulum is released from the angle of $\theta_0 = \pi/4$ radian.

As mentioned earlier in section 7.1, the governing differential Eq. (7.2) is nonlinear. Derivation of the exact solution is lengthy and difficult. For small swinging angle, the governing differential equation becomes linear and the exact solution can be derived easily as shown by Eqs. (7.5) - (7.9). The exact solution for small swinging angle is

$$\theta(t) = \theta_0 \cos\sqrt{\frac{g}{L}}\,t \qquad (7.8)$$

By substituting the given values θ_0, g and L, the solution is

$$\theta(t) = \frac{\pi}{4}\cos\sqrt{\frac{9.8}{0.5}}\,t \qquad (7.65)$$

as shown in Table 7.5.

Approximate solution to the nonlinear differential Eq. (7.2) can be obtained conveniently by applying the Euler's or Rung-Kutta method. The procedure starts from separating the governing second-order differential equation into two first-order differential equations as follows,

$$\frac{d\theta}{dt} = \beta \qquad (7.3a)$$

$$\frac{d\beta}{dt} = -\frac{g}{L}\sin\theta \qquad (7.3b)$$

The Euler's or the Runge-Kutta method can then be applied to solve these two equations simultaneously by using the same procedure as explained in Examples 7.6 and 7.7. Corresponding computer programs similar to those shown in Figs. 7.14 and 7.15 can also developed for their solutions. Table 7.5 shows the comparison of the solutions obtained from the two methods by using different step sizes. The table also shows a relatively large difference between the solutions arisen from solving the linear and nonlinear differential equations. The solution difference suggests the importance for obtaining the solution of the original nonlinear equation. This is because nonlinear differential equations always occur in practical scientific and engineering problems.

Table 7.5 Comparison of the solutions for the swinging pendulum in Example 7.8 at different times.

Time *t* (sec)	Linear solution (*θ*, radian)	Nonlinear solution (*θ*, radian)		
		Euler *h* = 0.01	Euler *h* = 0.001	Runge-Kutta *h* = 0.01
0.0	0.785398	0.785398	0.785398	0.785398
0.2	0.497118	0.531681	0.522414	0.521418
0.4	-0.156096	-0.106033	-0.104951	-0.104747
0.6	-0.694720	-0.691733	-0.659939	-0.656418
0.8	-0.723351	-0.820165	-0.764768	-0.758862
1.0	-0.220971	-0.398713	-0.353641	-0.349142

7.7 MATLAB Commands

MATLAB has several commands for solving ordinary differential equations. These commands include `ode45`, `ode23`, `ode113`, etc. The format for using these commands is

$$[x,y] = solver(odefunc, span, y0)$$

where   `solver`   is the command used, such as `ode45`, `ode23` or `ode113`.
            `odefunc`   is the function on to be solved
            `span`   is the interval and step size
            `y0`   is the initial condition

Example 7.9 Employ the MATLAB command `ode45` to solve the differential equation

$$\frac{dy}{dx} = y \cos x \tag{7.13}$$

with the initial condition of $y(0) = 1$. Use the step size of 0.25 for the time interval of $x = 0$ t0 $x = 3$.

The MATLAB commands for solving the differential Eq. (7.13) and its solution are as follows

```
>> f = inline('y*cos(x)','x','y');
>> [x,y] = ode45(f,[0:0.25:3],1);
>> y

y =

    1.0000
    1.2807
    1.6151
    1.9771
    2.3198
    2.5831
    2.7115
    2.6751
    2.4826
    2.1773
    1.8193
    1.4647
    1.1516
```

The solutions above are quite accurate as compared to the exact solution. Details of the MATLAB commands and their solution accuracy are provided in Table 7.6.

Table 7.6   Details of the commands `ode23`, `ode45` and `ode113` for solving a first-order ordinary differential equation.

Command	Accuracy	Details
ode23	Low	Use the second- and third-order Runge-Kutta methods.
ode45	Medium	Use the fourth- and fifth-order Runge-Kutta methods. The command is popular and widely used.
ode113	Medium to high	Use the multistep methods of Adams-Bashforth and Adams-Moulton formula explained in section 7.8.

Example 7.10 Employ a MATLAB command `ode45` to solve the set of two first-order differential equations in Example 7.6

$$\frac{dy_1}{dx} = y_2$$

$$\frac{dy_2}{dx} = -2y_2 - 4y_1$$

with the initial conditions of $y_1(x=0) = 2$ and $y_2(x=0) = 0$.

An m-file under the name `sysode.m` corresponding to the given differential equations is firstly established as follows

```
function dy = sysode(x,y)
dy = zeros(2,1);
dy(1) = y(2);
dy(2) = -2*y(2) - 4*y(1);
```

Then, the MATLAB command `ode45` is employed by using the format

```
>> [x,y] = ode45(@sysode,[0:0.5:3],[2 0])
```

The response from MATLAB is as follows

```
x =

         0
    0.5000
    1.0000
    1.5000
    2.0000
    2.5000
    3.0000

y =

    2.0000         0
    1.3194   -2.1339
    0.3012   -1.6772
   -0.2487   -0.5331
   -0.3063    0.1981
   -0.1492    0.3518
   -0.0046    0.2036
```

The y-solution above consists of two columns. The first column (y_1) is the solution to the problem while the second column contains the values of y_2, i.e., derivatives of y_1. The solution (y_1) to the problem is accurate as compared to the exact solution in Table 7.4.

7.8 Multistep Methods

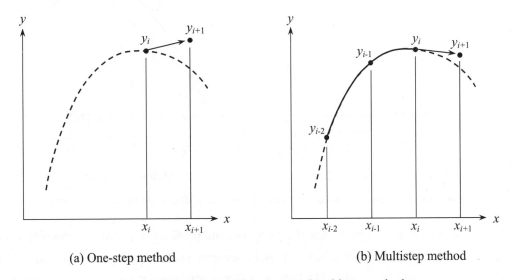

(a) One-step method   (b) Multistep method

Figure 7.17 Concept of the one- and multistep methods.

All methods presented in Sections 7.2 - 7.6 are for determining a new solution y_{i+1} at x_{i+1} from the computed solution y_i at x_i as depicted in Fig. 7.17(a). These methods are called *one-step method* because only the computed solution from the previous step is used to determine a new solution at the next step. Since many solutions have been computed, they can be used to determine a new solution as described by Fig. 7.17(b). The new solution should be more accurate because it was determined from many previous computed solutions. Such concept has led the so called *multistep method* as will be presented in this section.

7.8.1 Non-self-starting Heun's method

One of the simple multistep methods is based on the Heun's method studied in section 7.3. The Heun's method consists of two steps for determining the predictor and corrector as shown in Eqs. (7.26a-b). The predictor y_{i+1}^0 is determined from

$$y_{i+1}^0 = y_i + f(x_i, y_i)h \qquad (7.26a)$$

where $f(x_i, y_i)$ is the slope at x_i as shown in Eq. (7.18a). The computed y_{i+1}^0 from Eq. (7.26a) is used to determine the slope at x_{i+1}. Then, the corrector is determined from

$$y_{i+1} = y_i + \frac{f(x_i, y_i) + f\left(x_{i+1}, y_{i+1}^0\right)}{2}h \qquad (7.26b)$$

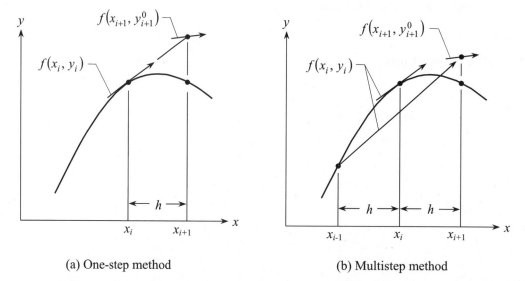

(a) One-step method   (b) Multistep method

Figure 7.18 Concept of the one- and multistep non-self-starting Heun's methods.

The above procedure suggests that if a more accurate predictor y_{i+1}^0 can be determined, the computed solution accuracy is increased. A more accurate predictor y_{i+1}^0 can be obtained by following the concept of Fig. 7.18(b). The slope at x_i for determining the solution y_{i+1} is computed from the solution at x_{i-1} with the step size of $2h$ as

$$y_{i+1}^0 = y_{i-1} + f(x_i, y_i) 2h \tag{7.66}$$

Improved solution accuracy obtained from the predictor in Eq. (7.66) will be demonstrated in Example 7.11. It is noted that, at the very beginning of the computational process for determining y_1, only y_0 is available while y_{-1} is not known. Such incomplete information at the starting point of the computation, has led to the name of the *non-self-starting Heun's method*.

In conclusion, the non-self-starting Heun's method is the predictor-corrector method that consists of two steps:

Predictor: $$y_{i+1}^0 = y_{i-1} + f(x_i, y_i) 2h \tag{7.67a}$$

Corrector: $$y_{i+1} = y_i + \frac{f(x_i, y_i) + f(x_{i+1}, y_{i+1}^0)}{2} h \tag{7.67b}$$

Example 7.11 Employ the non-self-starting Heun's method to solve the differential equation

$$\frac{dy}{dx} = y \cos x \tag{7.13}$$

with the initial condition of $y(0) = 1$. Use the step size of $h = 1.00$ and compare the computed solution with those obtained from the one-step Heun's method in Example 7.2.

It is noted that the non-self-starting Heun's method requires the value y_{-1} at $x = -h = -1$ to determine the solution at the first step. The value y_{-1} may be determined from the fourth-order Runge-Kutta method as described in section 7.5.3 or by employing the computer program in Fig. 7.13. The method yields $y_{-1} = 0.43036$. The predictor in Eq. (7.67a) for the first step of the computation is

$$y_1^0 = 0.43036 + [(1)(1)](2)(1) = 2.43036 \tag{7.68a}$$

Then, the slope at the end of the step is

$$f(x_1, y_1^0) = (2.43036) \cos(1) = 1.31313$$

and the corrector in Eq. (7.67b) gives the solution

$$y_1 = 1 + \frac{1 + 1.31313}{2}(1) = 2.15657 \tag{7.68b}$$

The true error is

$$E_t = e^{\sin(1)} - 2.15657 = 0.16321 \tag{7.68c}$$

which is less than the error of 0.27948 obtained from the one-step Heun's method as shown in Eq. (7.27c).

Similarly, the second step of the computation starts from $x_1 = 1$ and $y_1 = 2.15657$. The predictor, slope at end of the step, corrector and true error are

$$y_2^0 = 1 + [2.15657 \cos(1)](2)(1) = 3.33040 \tag{7.69a}$$

$$f(x_2, y_2^0) = (3.33040) \cos(2) = -1.38594$$

$$y_2 \; = \; 2.15657 + \frac{1.16520 - 1.38594}{2}(1) \quad = \quad 2.04620 \qquad (7.69\text{b})$$

$$E_t \; = \; e^{\sin(2)} - \; 2.04620 \qquad\qquad\quad = \quad 0.43638 \qquad (7.69\text{c})$$

Again, the solution error produced by the non-self-starting Heun's method is less than that from the one-step Heun's method as shown in Eq. (7.28c).

7.8.2 Adams-Bashforth method

One of the multistep methods widely used and implemented in many commercial software is based on the Adams-Bashforth formulas. The formulas are sometimes called the Adams *open* formulas. The formulas were derived from the Taylor series expansion as shown in Eq. (6.134)

$$y_{i+1} \; = \; y_i \; + \; f_i h \; + \; \frac{f_i'}{2!} h^2 \; + \; \frac{f_i''}{3!} h^3 \; + \; \cdots$$

or $$y_{i+1} \; = \; y_i \; + \; h \left[f_i \; + \; \frac{h}{2} f_i' \; + \; \frac{h^2}{6} f_i'' \; + \; \cdots \right] \qquad (7.70)$$

For example, the first backward divided-difference as shown in Eq. (6.138) is

$$f_i' \; = \; \frac{f_i - f_{i-1}}{h} \; + \; \frac{h}{2} f_i'' \; + \; O(h^2) \qquad (7.71)$$

By substituting Eq. (7.71) into Eq. (7.70)

$$y_{i+1} \; = \; y_i \; + \; h \left[f_i \; + \; \frac{h}{2} \left(\frac{f_i - f_{i-1}}{h} + \frac{h}{2} f_i'' \; + \; O(h^2) \right) \; + \; \frac{h^2}{6} f_i'' \; + \; \cdots \right]$$

and arranging terms to get

$$y_{i+1} \; = \; y_i \; + \; h \left(\frac{3}{2} f_i - \frac{1}{2} f_{i-1} \right) \; + \; \frac{5}{12} h^3 f_i'' \; + \; O(h^4)$$

or $$y_{i+1} \; = \; y_i \; + \; h \left(\frac{3}{2} f_i - \frac{1}{2} f_{i-1} \right) \; + \; O(h^3) \qquad (7.72)$$

Equation (7.72) is called the second-order Adams open formula. It is called the open formula because the new solution y_{i+1} can be determined directly from the known functions f_i and f_{i-1} previously computed in the earlier steps. It should be noted that for the first step at $x = 0$, y_i and f_i are known and y_{i-1} may be determined by using the Runge-Kutta method as explained in Example 7.11. Application of the Adams open formula is presented in Example 7.12.

Similarly, the third-order Adams open formula can be derived from the second backward divided-difference from Table 6.5

$$f_i'' \; = \; \frac{f_i - 2 f_{i-1} + f_{i-2}}{h^2} \; + \; O(h) \qquad (7.73)$$

By substituting Eqs. (7.71) and (7.73) into Eq. (7.70) and arranging terms, the new solution y_{i+1} is

$$y_{i+1} = y_i + h\left(\frac{23}{12}f_i - \frac{16}{12}f_{i-1} + \frac{5}{12}f_{i-2}\right) + O(h^4) \tag{7.74}$$

which is called the third-order open Adams formula. High-order open Adams-Bashforth formulas can be derived by using the same procedure. The new solution y_{i+1} can be written in a more general form as

$$y_{i+1} = y_i + h\sum_{k=0}^{n-1}\beta_k f_{i-k} + O(h^{n+1}) \tag{7.75}$$

where n is the order of the Adams-Bashforth formula and the coefficients β_k are shown in Table 7.7.

Table 7.7 Coefficients β_k in the Adams-Bashforth formula.

Order	β_0	β_1	β_2	β_3	β_4	β_5
1	1					
2	$\frac{3}{2}$	$-\frac{1}{2}$				
3	$\frac{23}{12}$	$-\frac{16}{12}$	$\frac{5}{12}$			
4	$\frac{55}{24}$	$-\frac{59}{24}$	$\frac{37}{24}$	$-\frac{9}{24}$		
5	$\frac{1,901}{720}$	$-\frac{2,774}{720}$	$\frac{2,616}{720}$	$-\frac{1,274}{720}$	$\frac{251}{720}$	
6	$\frac{4,277}{1,440}$	$-\frac{7,923}{1,440}$	$\frac{9,982}{1,440}$	$-\frac{7,298}{1,440}$	$\frac{2,877}{1,440}$	$-\frac{475}{1,440}$

Example 7.12 Develop a computer program by using the fourth-order Adams-Bashforth formula to solve the differential equation

$$\frac{dy}{dx} = y\cos x \tag{7.13}$$

with the initial condition of $y(0)=1$. Use the step size of $h = 0.25$ and compare the solution obtained with the exact solution.

From Eq. (7.75) and Table 7.7, the fourth-order Adams-Bashforth formula is

$$y_{i+1} = y_i + h\left(\frac{55}{24}f_i - \frac{59}{24}f_{i-1} + \frac{37}{24}f_{i-2} - \frac{9}{24}f_{i-3}\right) \tag{7.76}$$

For the first step $i = 0$ of the computation at $x_0 = 0$, the given initial condition is $y_0 = 1$. However, Eq. (7.76) in the formula needs the values y_{-1}, y_{-2}, y_{-3} in order to determine f_{-1}, f_{-2} and f_{-3}, respectively. The fourth-order Runge-Kutta computer program as shown in Fig. 7.13 can be used to yield $y_{-1} = 0.78083$, $y_{-2} = 0.61914$ and $y_{-3} = 0.50579$. Thus, the values of the functions in Eq. (7.76) when $i = 0$ are

$$f_0 \;=\; (1)\cos(0) \qquad\qquad = \; 1.00000$$

$$f_{-1} \;=\; (0.78083)\cos(-0.25) \qquad = \; 0.75656$$

$$f_{-2} \;=\; (0.61914)\cos(-0.50) \qquad = \; 0.54335$$

$$f_{-3} \;=\; (0.50579)\cos(-0.75) \qquad = \; 0.37008$$

Thus, the solution y_1 at the first step by using Eq. (7.76) is

$$y_1 \;=\; 1 \,+\, 0.25\left(\frac{55}{24}(1.00000) - \frac{59}{24}(0.75656) + \frac{37}{24}(0.54335) - \frac{9}{24}(0.37008)\right)$$

$$=\; 1.28267$$

The solution obtained from the first step is used to determine the solution at the second step. The same process is repeated for determining the solutions at the later steps. The process is terminated when the specified number of steps is met or the last step is reached. A corresponding computer program that uses the fourth-order Adams-Bashforth formula is shown in Fig. 7.19. The values y_{-1}, y_{-2}, y_{-3} required by the program can be determined by using the fourth-order Runge-Kutta computer program as shown in Fig. 7.13. The computed solution obtained from the fourth-order Adams-Bashforth computer program is compared with the exact solution as shown in Table 7.9.

Fortran

```
      PROGRAM  ADAMBAS
C.....A PROGRAM FOR SOLVING ORDINARY DIFFERENTIAL
C.....EQ. BY FOURTH-ORDER ADAMS-BASHFORTH METHOD
C.....READ INITIAL CONDITIONS OF  X0, Y0, Y-1,
C.....Y-2, Y-3, NUMBER OF STEPS AND STEP SIZE:
      READ(5,*)  X, Y, YM1, YM2, YM3, N, H
      WRITE(6,100)  H
  100 FORMAT(' SOLUTION WITH STEP SIZE =', E10.4,
     *  ' IS:', /, 10X, 'X', 15X, 'Y')
      WRITE(6,200)  X, Y
  200 FORMAT(2E16.6)
      DO 300  I=1,N
      F0  = FUNC(X,Y)
      XM1 = X - H
      F1  = FUNC(XM1,YM1)
      XM2 = X - 2.*H
      F2  = FUNC(XM2,YM2)
      XM3 = X - 3.*H
      F3  = FUNC(XM3,YM3)
      YM3 = YM2
      YM2 = YM1
      YM1 = Y
      Y = Y + (55*F0 - 59*F1 + 37*F2 - 9*F3)*H/24.
      X   = X + H
      WRITE(6,200)  X, Y
  300 CONTINUE
      STOP
      END
C-------------------------------------------------
      FUNCTION  FUNC(X,Y)
      FUNC = Y*COS(X)
      RETURN
      END
```

MATLAB

```
%  PROGRAM  ADAMBAS
%  A PROGRAM FOR SOLVING ORDINARY DIFFERENTIAL
%  EQ. BY FOURTH-ORDER ADAMS-BASHFORTH METHOD
%  READ INITIAL CONDITIONS OF  X0, Y0, Y-1,
%  Y-2, Y-3, NUMBER OF STEPS AND STEP SIZE:
%-------------------------------------------------
func = inline('y*cos(x)','x','y');
%-------------------------------------------------
x   = input('\nEnter value of x:   ');
y   = input( 'Enter value of y:   ');
ym1 = input( 'Enter value of ym1: ');
ym2 = input( 'Enter value of ym2: ');
ym3 = input( 'Enter value of ym3: ');
n   = input( 'Enter value of n:   ');
h   = input( 'Enter value of h:   ');
fprintf('\nSOLUTION WITH STEP SIZE = %10.4e IS:',h);
fprintf('\n          X               Y');
fprintf('\n%16.6e%16.6e',x,y);
for i = 1:n
    f0  = func(x,y);
    xm1 = x - h;
    f1  = func(xm1,ym1);
    xm2 = x - 2.*h;
    f2  = func(xm2,ym2);
    xm3 = x - 3.*h;
    f3  = func(xm3,ym3);
    ym3 = ym2;
    ym2 = ym1;
    ym1 = y;
    y   = y + (55*f0 - 59*f1 + 37*f2 - 9*f3)*h/24.;
    x   = x + h;
    fprintf('\n%16.6e%16.6e',x,y);
end
```

Figure 7.19 Computer program for solving the first-order ordinary differential equation by using the fourth-order open Adams-Bashforth formula in Example 7.12.

7.8.3 Adams-Moulton method

The Adams-Moulton method is similar to the Adams-Bashforth method but can provide higher solution accuracy for general problems. The Adams-Moulton formulas are derived by using the backward Taylor series expansion in the form

$$y_i = y_{i+1} - f_{i+1}h + \frac{f'_{i+1}}{2!}h^2 - \frac{f''_{i+1}}{3!}h^3 + \cdots$$

or

$$y_{i+1} = y_i + h\left[f_{i+1} - \frac{h}{2!}f'_{i+1} + \frac{h^2}{3!}f''_{i+1} - \cdots\right] \tag{7.77}$$

The first-order derivative term in Eq. (7.77) is approximate by using the backward divided-difference as

$$f'_{i+1} = \frac{f_{i+1} - f_i}{h} + \frac{h}{2}f''_{i+1} + O(h^2) \tag{7.78}$$

By substituting Eq. (7.78) into Eq. (7.77) and arranging terms to yield

$$y_{i+1} = y_i + h\left(\frac{1}{2}f_{i+1} + \frac{1}{2}f_i\right) + O(h^3) \tag{7.79}$$

Equation (7.79) is called the second-order Adams closed formula. It is called the closed formula because the function f_{i+1} on the right-hand side of Eq. (7.79) depends on the value y_{i+1} which is also unknown.

Other high-order Adams closed formulas can be derived by using the same procedure as explained above. Details of the derivation are omitted herein and are left as exercise. The Adams-Moulton formulas can be written in a general for as follow

$$y_{i+1} = y_i + h\sum_{k=0}^{n-1}\beta_k f_{i-k+1} + O(h^{n+1}) \tag{7.80}$$

where n is the order of the formula and β_k are the coefficients as shown in Table 7.8.

Table 7.8 Coefficients β_k in the Adams-Moulton formulas.

Order	β_0	β_1	β_2	β_3	β_4	β_5
1	1					
2	$\frac{1}{2}$	$\frac{1}{2}$				
3	$\frac{5}{12}$	$\frac{8}{12}$	$-\frac{1}{12}$			
4	$\frac{9}{24}$	$\frac{19}{24}$	$-\frac{5}{24}$	$\frac{1}{24}$		
5	$\frac{251}{720}$	$\frac{646}{720}$	$-\frac{264}{720}$	$\frac{106}{720}$	$-\frac{19}{720}$	
6	$\frac{475}{1,440}$	$\frac{1,427}{1,440}$	$-\frac{798}{1,440}$	$\frac{482}{1,440}$	$-\frac{173}{1,440}$	$\frac{27}{1,440}$

Example 7.13 Apply the fourth-order Adams-Moulton formula to solve the differential equation

$$\frac{dy}{dx} = y \cos x \tag{7.13}$$

with the initial condition of $y(0) = 1$. Use the step size of $h = 0.25$ and compare the solution obtained with the exact solution and the solution from the fourth-order Adams-Bashforth formula in Example 7.12.

From Eq. (7.80) and Table 7.8, the fourth-order Adams-Moulton formula is

$$y_{i+1} = y_i + h\left(\frac{9}{24}f_{i+1} + \frac{19}{24}f_i - \frac{5}{24}f_{i-1} + \frac{1}{24}f_{i-2}\right) \tag{7.81}$$

Similar to the Adams-Bashforth method in Example 7.12, the initial condition at the first step $x_0 = 0$ is $y_0 = 1$. The fourth-order Adams-Moulton formula in Eq. (7.81) needs f_{-1} and f_{-2} which are determined from y_{-1} and y_{-2}, respectively. These values y_{-1} and y_{-2} can be determined by using the fourth-order Runge-Kutta computer program in Fig. 7.13 to yield $y_{-1} = 0.78083$ and $y_{-2} = 0.61914$. The given y_0 and the computed y_{-1} and y_{-2} lead to the three function values of $f_0 = 1.00000$, $f_{-1} = 0.75656$, $f_{-2} = 0.54335$. Thus, Eq. (7.81) for the first step is

$$y_1 = 1 + 0.25\left[\frac{9}{24}f_1 + \frac{19}{24}(1.00000) - \frac{5}{24}(0.75656) + \frac{1}{24}(0.54335)\right] \tag{7.82}$$

It is noted that, by using the given function

$$f_1 = y_1 \cos(x_1) = y_1 \cos(0.25) = 0.96891y_1$$

thus, the solution of y_1 from Eq. (7.82) is

$$y_1 = 1.28049$$

which is more accurate than that from the Adams-Bashforth formula as compared to the exact solution. Table 7.9 compares the Adams-Bashforth and Adams-Moulton solutions with the exact solution. The table also shows the true errors produced by the two methods.

Table 7.9 Comparison of the Adams-Bashforth and Adams-Moulton solutions with the exact solution for Example 7.13.

		Adams-Bashforth		Adams-Moulton	
x	Exact solution	Solution	Error	Solution	Error
0.00	1.00000	1.00000	0.00000	1.00000	0.00000
0.25	1.28070	1.28267	-0.00197	1.28049	0.00021
0.50	1.61515	1.62075	-0.00056	1.61468	0.00047
0.75	1.97712	1.98633	-0.00921	1.97651	0.00061
1.00	2.31978	2.33009	-0.01031	2.31934	0.00044
1.25	2.58309	2.58982	-0.00673	2.58316	-0.00007
1.50	2.71148	2.71076	0.00072	2.71215	-0.00067
1.75	2.67510	2.66770	0.00740	2.67603	-0.00093
2.00	2.48258	2.47414	0.00844	2.48324	-0.00066
2.25	2.17727	2.17385	0.00342	2.17733	-0.00006
2.50	1.81934	1.82303	-0.00369	1.81887	0.00047
2.75	1.46472	1.47328	-0.00856	1.46408	0.00064
3.00	1.15156	1.16093	-0.00937	1.15106	0.00050

7.9 Closure

Popular methods for solving ordinary differential equations are presented in this chapter. These methods are the Euler's, Heun's, modified Euler's and Runge-Kutta methods. The methods can be applied to solve higher-order differential equations or a set of first-order differential equations. Among these methods, the Euler's method is considered as the simplest one. The method is easy to understand and can be applied to solve both linear and nonlinear ordinary differential equations.

The drawback of the Euler's method is that it does not produce solution with high accuracy, especially when using with a large step size. The Heun's and modified Euler's methods provide solutions with higher accuracy by determining proper slope within the step that is used to determine the new solution. Such idea for determining more accurate slope within the step is used as a basis in the development of the Runge-Kutta method. The Runge-Kutta method provides proper slope within the time step to further produce a more accurate solution. The proper slope is determined from matching the coefficients in the Runge-Kutta equation with the Taylor series expansion. The fourth Runge-Kutta method is widely used and implemented in many commercial software for solving ordinary differential equations.

The methods mentioned above are called the one-step method because the only solution computed at the previous step is used to determine the new solution. If the computed solutions at the earlier steps are used to determine the new solution, the method is called the multistep method. The Adams-Bashforth and Adams-Moulton methods are the multistep method that can yield more accurate solution than the solution obtained from the one-step method.

Concepts and theoretical formulation of the one-step and multistep methods were presented in details in this chapter. Examples were presented so that detailed computational procedure can be clearly understood. Corresponding computer programs were also developed and their usage were demonstrated. These computer programs can be modified to solve other differential equations encountered in research and applications.

Exercises

1. Use the Euler method to solve the ordinary differential equation

$$\frac{dy}{dx} = 1 + \frac{y}{x}$$

for $1 \le x \le 2$ with the initial condition of $y(1) = 2$. Use the step size of $h = 0.25$ in the computation. Plot to compare the computed solution with the exact solution of $y(x) = x \ln x + 2x$.

2. Solve the ordinary differential equation in Problem 1 again but by developing a computer program. Use three different step sizes of $h = 0.25, 0.1$ and 0.01 in the computation. Compare the solutions obtained from using the three step sizes with the exact solution.

3.   Use the Euler method to solve the ordinary differential equation

$$\frac{dy}{dx} = 1 + \frac{y}{x} + \left(\frac{y}{x}\right)^2$$

for $1 \le x \le 3$ with the initial condition of $y(1) = 0$. Use the step size of $h = 0.2$ in the computation. Plot to compare the computed solution with the exact solution of $y(x) = x \tan(\ln x)$.

4.   Solve the ordinary differential equation in Problem 3 again but by developing a computer program. Use three different step sizes of h = 0.2, 0.1 and 0.01 in the computation. Compare the solutions obtained from using the three step sizes with the exact solution.

5.   Use the Euler method to solve the ordinary differential equation

$$\frac{dy}{dx} = -y + x + 1$$

for $0 \le x \le 5$ with the initial condition of $y(0) = 1$ by developing a computer program. Employ three different step sizes of $h = 0.5, 0.1, 0.01$ in the computation. Compare the computed solutions obtained from using the three step sizes with the exact solution of $y(x) = e^{-x} + x$.

6.   Use the Heun's and the modified Euler methods to solve the ordinary differential equation

$$\frac{dy}{dx} = 4x - \frac{2y}{x}$$

for $1 \le x \le 2$ with the initial condition of $y(1) = 1$. Use the step size of $h = 0.25$ in the computation. Show detailed computations and plot to compare the solution obtained from each case with the exact solution of $y(x) = x^2$.

7.   Solve the ordinary differential equation in Problem 6 again but by developing a computer program. Use three different step sizes of $h = 0.25, 0.1$ and 0.01 in the computation. In each case, show the comparison between the computed and exact solutions together with the true error.

8.   Show that the solution error produced by using the Euler method is of order h, i.e., $O(h)$. Then, show such solution error by using an example with the step sizes of $h, h/2$ and $h/4$. Plot the solution error versus the step size h on the logarithmic scale.

9.   Show that the solution error produced by using the Heun's method is second-order of h, i.e., $O(h^2)$. Then, show such solution error by using an example with the step sizes of $h, h/2$ and $h/4$. Plot the solution error versus the step size h on the logarithmic scale.

10. Use the Heun's and the modified Euler methods to solve the ordinary differential equation

$$\frac{dy}{dx} - \cos 2x - \sin 3x = 0$$

for $0 \le x \le 1$ with the initial condition of $y(0) = 1$. Use the step size of $h = 0.25$ in the computation. Show detailed computational procedures. Plot to compare the solutions obtained from the two methods with the exact solution of

$$y(x) = \frac{1}{2}\sin 2x - \frac{1}{3}\cos 3x + \frac{4}{3}$$

11. Solve the ordinary differential equation in Problem 10 again by developing a computer program. Then, use the step sizes of $h = 0.25$, 0.1 and 0.01 in the computation. Set up a table to compare the computed solution at different steps with the exact solution.

12. Use the Heun's and the modified Euler methods to solve the ordinary differential equation

$$\frac{dy}{dx} - (1 - y)(4 - y) = 0$$

for $0 \le x \le 1$ with the initial condition of $y(0) = 0$. Use the step size of $h = 0.25$ in the computation. Show detailed computational procedures. Plot to compare the solutions obtained from the two methods with the exact solution of

$$y(x) = \frac{4\left(e^{3x} - 1\right)}{\left(4e^{3x} - 1\right)}$$

13. Solve the ordinary differential equation in Problem 12 again by developing a computer program. Then, obtain the solutions by using the step sizes of $h = 0.25$, 0.1 and 0.01. Plot to compare the solutions at different steps with the exact solution. Also, set up a table to show the true errors produced by the two methods.

14. Use the Heun's and the modified Euler methods to solve the ordinary differential equation

$$\frac{dy}{dx} + 2xy^2 = 0$$

for $0 \le x \le 20$ with the initial condition of $y(0) = 1$. Develop a computer program to solve for the solutions by using the step size of $h = 0.1$. Compare the solutions at every $\Delta x = 1$ with the exact solution of $y(x) = 1/\left(1 + x^2\right)$.

15. Use the second-, third- and fourth-order Runge-Kutta methods to solve the ordinary differential equation

$$\frac{dy}{dx} = -2xy^2$$

for $1 \le x \le 2$ with the initial condition of $y(1)=1$. Employ the step size of $h = 0.25$ in the computation. Show detailed computational procedures. Set up a table to compare the solutions obtained from the three methods with the exact solution of $y(x) = 1/x^2$.

16. Repeat Problem 12 again but by developing computer programs. Determine the solutions by using the step sizes of $h = 0.25, 0.1$ and 0.01. Plot to compare the solutions obtained from using different step sizes with the exact solution. Also, set up a table to show the true errors produced by the three solutions from each method.

17. Use the fourth-order Runge-Kutta method to solve the ordinary differential equation

$$\frac{dy}{dx} - y = x^2$$

for $1 \le x \le 2$ with the initial condition of $y(1)=1$ by developing a computer program. Then, use the step sizes of $h = 0.1, 0.01, 0.001$ and 0.0001 to obtain solutions. Set up a table to compare the solutions obtained with the exact solution of $y(x) = 6e^{x-1} - x^2 - 2x - 2$.

18. Use the fourth-order Runge-Kutta method to solve the ordinary differential equation

$$\frac{dy}{dx} + y^2 + \frac{y}{x} = \frac{1}{x^2}$$

for $1 \le x \le 2$ with the initial condition of $y(1)=-1$. Employ the step size of $h = 0.25$ in the computation. Plot to compare the solution with the exact solution of $y(x) = -1/x$.

19. Solve the ordinary differential equation in Problem 18 again by using the computer programs developed for the Euler, Heun's, modified Euler and fourth-order Rung-Kutta methods. Use the step size of $h = 0.05$ in the computation for all cases. Set up a table to show the computed solutions and the exact solution with the true errors. Provide comments on the solution accuracy and computational time for each method.

20. Use the second-, third- and fourth-order Runge-Kutta methods to solve the ordinary differential equation

$$\frac{dy}{dx} - \frac{2}{x}y = x^2 e^x$$

for $1 \le x \le 2$ with the initial condition of $y(1)=0$ by using computer programs. Employ the step size of $h = 0.1$ in the computation for all cases. Set up a table to compare the solutions with the exact solution of $y(x) = x^2 (e^x - e)$.

21. Employ the Euler and the fourth-order Runge-Kutta methods to solve the ordinary differential equation

$$\frac{dy}{dx} = \frac{e^x}{y}$$

for $0 \le x \le 2$ with the initial condition of $y(0)=1$. Use the step size of $h = 0.5$ in the computation. Show detailed computational procedures. Plot to compare the solutions obtained from the two methods with the exact solution of $y(x) = \sqrt{2e^x - 1}$.

22. Use the Euler and the fourth-order Runge-Kutta methods to solve the ordinary differential equation

$$x\frac{dy}{dx} - 4y = x^5 e^x$$

for $1 \le x \le 2$ with the initial condition of $y(1) = 0$. Select an appropriate step size h to obtain the solutions. Set up a table to compare the solutions obtained from using the two methods with the exact solution of $y(x) = x^4 \left(e^x - e\right)$.

23. Employ the MATLAB commands `ode23` and `ode45` to solve the ordinary differential equation

$$\frac{dy}{dt} = -\left(1 + x + x^2\right) - (2x+1)y - y^2$$

for $0 \le x \le 3$ with the initial condition of $y(0) = -1/2$. Then, further use MATLAB to plot the solutions for comparing with the exact solution of $y(x) = -x - \left(e^x + 1\right)^{-1}$.

24. Use the MATLAB commands `ode23` and `ode45` to solve the ordinary differential equation

$$\frac{dy}{dx} = 3x - \frac{y}{x}$$

for $1 \le x \le 6$ with the initial condition of $y(1) = 0$. Also, use MATLAB to plot the solutions for comparing with the exact solution of $y(x) = x^2 - x^{-1}$. Repeat the problem again but by using the initial condition of $y(1) = 1$ for which the exact solution is $y(x) = x^2$.

25. Employ the MATLAB commands `ode23` and `ode45` to solve the ordinary differential equation

$$\frac{dy}{dx} = 7x^2 - \frac{4y}{x}$$

for $1 \le x \le 6$ with the initial condition of $y(1) = 2$. Then, further use MATLAB to plot the solutions for comparing with the exact solution of $y(x) = x^3 + x^{-4}$. Repeat the problem again but by using the initial condition of $y(1) = 1$ for which the exact solution is $y(x) = x^3$.

26. Solve the second-order ordinary differential equation

$$\frac{d^2 y}{dx^2} - 2\frac{dy}{dx} + y = x\left(e^x - 1\right)$$

for $0 \leq x \leq 1$ with the initial conditions of $y(0) = y'(0) = 1$ by using the fourth-order Runge-Kutta method. Use the step size of $h = 0.25$ in the computation. Compare the solution with the exact solution of

$$y(x) = e^x \left(\frac{x^3}{6} - x + 3 \right) - x - 2$$

27.   Solve the second-order ordinary differential equation

$$x^2 \frac{d^2 y}{dx^2} - 2x \frac{dy}{dx} + 2y = x^3 \ln x$$

for $1 \leq x \leq 2$ with the initial conditions of $y(1) = 1$ and $y'(1) = 0$ by developing the computer programs for the Euler and the fourth-order Runge-Kutta methods. Then, use the programs to determine the solutions by using the step sizes of $h = 0.1$ and 0.01. Compare the solution accuracy with the exact solution of

$$y(x) = \frac{7}{4} x + \frac{1}{2} x^3 \ln x - \frac{3}{4} x^3$$

by determining the true errors.

28.   Solve the third-order ordinary differential equation

$$\frac{d^3 y}{dx^3} + 6y^4 = 0$$

for $1 \leq x \leq 1.5$ with the initial conditions of $y(1) = -1$, $y'(1) = -1$, and $y''(1) = 2$ by developing the computer programs for the Euler and the fourth-order Runge-Kutta methods. Then, use the programs to determine the solutions by using the step sizes of $h = 0.1$ and 0.05. Compare the solutions with the exact solution of $y(x) = (x - 2)^{-1}$ by plotting and determining the true errors.

29.   Solve a set of nonlinear ordinary differential equations

$$\frac{dy_1}{dx} = y_1 - y_2 - y_1 y_3$$

$$\frac{dy_2}{dx} = y_1 + y_2 - y_2 y_3$$

$$\frac{dy_3}{dx} = y_1^2 + y_2^2 - y_3$$

for $0 \leq x \leq 10$ with the initial conditions of $y_1(0) = 2$, $y_2(0) = 0$ and $y_3(0) = 1$ by developing the computer programs for the Euler and the fourth-order Runge-Kutta methods. Then, employ the programs with appropriate step size h to determine their solutions. Give comments on how to measure the accuracy of the obtained solutions.

30.   Show detailed derivation of the coefficients in Table 7.7 for the Adams-Bashforth formula.

31. Solve the ordinary differential equation in Example 7.12 again but by using a computer program developed for the sixth-order Adams-Bashforth formula. Compare the computed solution with the solutions obtained from the fourth-order Adams-Bashforth and the fourth-order Runge-Kuta methods. Provide comments on the solution accuracy obtained these methods.

32. Show detailed derivation of the coefficients in Table 7.8 for the Adams-Moulton formula.

33. Develop a computer program for the fourth-order Adams-Moulton formula. Then, use the program to solve the ordinary differential equation in Example 7.13 and compare the solution obtained with that shown in Table 7.9.

34. Solve the ordinary differential equation in Example 7.13 again but by using a computer program developed for the sixth-order Adams-Moulton formula. Compare the computed solution with the solutions from the fourth-order Adam-Bashforth and the fourth-order Adams-Moulton formulas as shown in Table 7.9.

35. Solve the ordinary differential equation in Problem 19 again but by using the non-self-starting Heun's method and the fourth-order Adams-Bashforth method. Compare the solutions obtained from the two methods with that from the fourth-order Runge-Kutta method. Give comments on the solution accuracy of these methods.

36. Solve the ordinary differential equation in Problem 17 again but by developing the computer programs for the fourth-order Adams-Bashforth and Adams-Moulton methods. Use the step size of $h = 0.05$ for obtaining the solutions. Compare the solutions with that from the fourth-order Runge-Kutta method.

37. A lumped mass of $m = 12$ kg with the initial temperature of $T_0 = 100°C$ is dropped into a water reservoir at the temperature of $T_a = 30°C$. The transient temperature response T of the lumped mass that varies with time t is determined from the ordinary differential equation

$$\frac{dT}{dt} + \frac{hA}{mc}(T - T_a) = 0$$

Where $h = 425$ J/sec-m^2-°C is the convection coefficient of the lumped mass surface, $A = 0.001$ m^2 is the lumped mass surface area, $c = 930$ J/kg-°C is the lumped mass specific heat. Employ the Euler and the fourth-order Runge-Kutta methods to determine the transient temperature response $T(t)$ of the lumped mass from 0 to 20 sec by selecting an appropriate time step. Set up a table to compare the solutions with the exact solution.

38. A lumped mass of $m = 35$ kg with the initial temperature of $T_0 = 2000$ K radiates heat from its surface of $A = 0.0015$ m^2 to a medium at the temperature of $T_a = 100$ K. The lumped transient temperature response T that varies with time t is determined from the ordinary differential equation

$$\frac{dT}{dt} + \frac{\varepsilon \sigma A}{mc}\left(T^4 - T_a^4\right) = 0$$

where $\varepsilon = 0.7$ is the surface emissivity, $\sigma = 5.67 \times 10^{-8}$ J/sec-m^2-K^4 is the Stefan-Boltzmann constant, and $c = 445$ J/kg-K is the specific heat of the lumped mass. Employ the Euler and the fourth-order Runge-Kutta methods to determine the transient temperature response $T(t)$ of the lumped mass from 0 to 10 sec. Select an appropriate time step to obtain solutions from the two methods. Provide comments on how to measure accuracy of the solutions obtained from the two methods.

39.   If the surface convection coefficient of the lumped mass varies linearly with the temperature, the governing equation for determining the transient temperature response T that varies with time t is in a form of nonlinear ordinary differential equation

$$\frac{dT}{dt} + (a + bT)T = 0$$

For the initial lumped mass temperature of $T_0 = 100$ where $a = 1$ and $b = 0.03$, employ the Euler and the fourth-order Runge-Kutta methods to determine the lumped mass temperature response $T(t)$ for $0 \le t \le 10$. Use the time steps of $h = 0.2, 0.1$ and 0.05 in the computation. Plot to compare the solutions obtained from both methods with the exact solution of

$$T(t) = aT_0 e^{-at} / \left[a + bT_0\left(1 - e^{-at}\right)\right]$$

40.   Solve Problems 39 again by using a computer program developed for the fourth-order Adams-Bashforth formula. Compare the solution with the exact solution and the solution obtained from the fourth-order Runge-Kutta method.

41.   Solve Problem 39 again by using a computer program developed for the fourth-order Adams-Moulton formula. Explain detailed computational procedures used in the program. Compare the solution obtained from the program with the exact solution.

Chapter

8

Partial Differential Equations

8.1 Introduction

Most of scientific and engineering problems are governed by partial differential equations that describe their physical phenomena. For example, in the analysis of deformation and stresses in a plate subjected to an in-plane loading, the governing differential equations representing the equilibrium of forces must be solved. Similarly, in the analysis of temperature distribution in a plate under a specified heating on its surface, the governing differential equation representing the conservation of energy at any point on the plate must also be solved. The differential equations that occur in scientific and engineering problems are in different forms. Thus, different types of numerical methods are needed to obtain accurate solutions. This chapter begins with the classifications of partial differential equations. Appropriate numerical methods for solving the different classes of partial differential equations will then be explained. Several examples will be presented to aid understanding of the methods as well as physical meanings of the problems and their solutions. Such understanding will help solving more difficult problems that are governed by complex partial differential equations.

8.1.1 Definitions

Differential equation is called a partial differential equation if its dependent variable varies with two or more independent variables. For example,

$$\frac{\partial^2 u}{\partial x^2} + \frac{\partial^2 u}{\partial y^2} = 0 \qquad (8.1)$$

where u is the dependent variable that varies with the independent variables x and y. Equation (8.1) is called a second-order differential equations because the highest partial derivative of u is two. It is noted that most of the differential equations that occur in scientific and engineering problems contain partial derivatives of the dependent variables up to the order of four. The independent variables are the three spatial coordinates x, y, z for general three-dimensional problems and the time t.

A partial differential equation is *linear* if the coefficients of all terms are constant or function of the independent variables only equations (8.2) and (8.3) below are linear differential equations because the coefficient of each term is either constant or function of x and y.

$$\frac{\partial^2 u}{\partial x^2} + \frac{\partial^2 u}{\partial y^2} + 7u = 12 \tag{8.2}$$

and

$$\left(x^2 - 3y\right)\frac{\partial^2 u}{\partial x^2} + \frac{\partial^2 u}{\partial x \partial y} + \frac{\partial u}{\partial y} = y^3 \tag{8.3}$$

But the partial differential equation,

$$\frac{\partial^2 u}{\partial x^2} + \left(\frac{\partial u}{\partial y}\right)^{0.5} = 0 \tag{8.4}$$

is *nonlinear* because the second term has its power order of a half, i.e., not an integer. Also, the partial differential Eq. (8.5) below is nonlinear,

$$u\frac{\partial u}{\partial x} + \left(\frac{\partial u}{\partial y}\right)^2 + \left(\frac{\partial u}{\partial x}\right)\left(\frac{\partial u}{\partial y}\right) = u^2 \tag{8.5}$$

because the coefficients of all terms are function of the dependent variable u. Nonlinear partial differential equations always occur in practical problems and require complex numerical methods for solving them.

Some simple linear partial differential equations can be solved for exact solutions by using advanced mathematical techniques. However, they cannot be solved if the shapes of the problems are irregular. Thus, numerical methods are needed to obtain approximate solutions instead. The popular numerical methods are the finite difference method, the finite element method and the finite volume method. The finite difference method is considered as the simplest one. The method is easy to understand and convenient to apply to problems with regular shapes. The finite difference method will be explained in details for solving different types of the partial differential equations in this chapter. For the problems with irregular shapes, the finite element method is more efficient and widely used. Details of the finite element method are presented in the next chapter.

8.1.2 Types of equations

Because the partial differential equations that arise in scientific and engineering problems are in many forms, the popular forms are considered herein. These popular forms can be written in general as,

$$a\frac{\partial^2 u}{\partial x^2} + b\frac{\partial^2 u}{\partial x \partial y} + c\frac{\partial^2 u}{\partial y^2} = f \tag{8.6}$$

where a, b, c may be constants or function of x and y. The function f on the right-hand side of the above equation may be constant of function of x, y, u, $\partial u/\partial x$ and $\partial u/\partial y$. Equation (8.6) can be classified into different types as follows.

(a) Equation (8.6) is called the *Elliptic equation* if $b^2 - 4ac < 0$. The Laplace's equation which has the form of

$$\frac{\partial^2 u}{\partial x^2} + \frac{\partial^2 u}{\partial y^2} = 0 \tag{8.7}$$

is an example of the elliptic equation. In general, the solutions of the elliptic equation are smooth. Steady-state heat transfer in a plate is an example problem that is governed by such differential equation. The dependent variable u in Eq. (8.7) represents the temperature distribution that varies with x-y coordinates of the plate as shown in Fig. 8.1. The finite difference method for solving the elliptic equation will be presented in section 8.2.

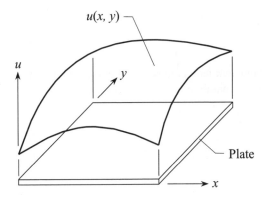

Figure 8.1 Temperature distribution on a plate governed
by the elliptic partial differential equation.

(b) The partial differential Eq. (8.6) is called the *Parabolic equation* if $b^2 - 4ac = 0$. An example problem that is governed by the parabolic equation is the transient heat conduction in a bar. The governing partial differential equation is,

$$k \frac{\partial^2 u}{\partial x^2} = \frac{\partial u}{\partial t} \tag{8.8}$$

where k is the thermal conductivity coefficient of the bar material. The transient temperature u varies with x-coordinate of the bar and time t. Typical transient temperature distributions along the bar at different times are shown in Fig. 8.2. Details of the finite difference method for solving such parabolic equation will be explained in section 8.3.

(c) The partial differential Eq. (8.6) is called the *Hyperbolic equation* if $b^2 - 4ac > 0$. An example problem is the oscillation behavior of a string that is governed by,

$$k^2 \frac{\partial^2 u}{\partial x^2} = \frac{\partial^2 u}{\partial t^2} \tag{8.9}$$

where k^2 represents tension in the string which is always positive. The deflection u varies with x-coordinate of the string and time t. Typical string deflection behaviors are shown in Fig. 8.3. The finite difference method for solving such hyperbolic equation will be presented in section 8.4.

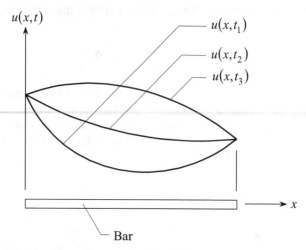

Figure 8.2 Transient temperature distributions in a bar which is governed by the parabolic differential equation.

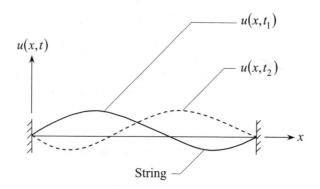

Figure 8.3 Behavior of string oscillation which is governed by the hyperbolic differential equation.

8.1.3 Boundary and initial conditions

Solution behaviors obtained from solving the partial differential Eqs. (8.7), (8.8) and (8.9) which are in the form of the elliptic, parabolic and hyperbolic equations, respectively, depend on the boundary and initial conditions. Two types of the boundary conditions frequently encountered are:

(a) *Dirichlet condition.* The condition specifies values of the dependent variable u at the boundaries. For example, temperature values are specified at both ends of the bar in Fig. 8.2.

(b) *Neumann condition.* The condition specifies values of the dependent variable gradient $\partial u/\partial x$. For example, if one end of the bar in Fig. 8.2 is insulated, then the slope of temperature $\partial u/\partial y = 0$ at that end.

The initial condition is used at the beginning of the solution process. For example, the temperature distribution along the bar in Fig. 8.2 may be specified by a function $u(x,0)=f(x)$ at time $t=0$.

Appropriate boundary and initial conditions are applied in the solving process of the elliptic, parabolic and hyperbolic equations as will be demonstrated in the following sections.

8.2 Elliptic Equation

8.2.1 Differential equation

An example of steady-state heat conduction in a plate is used herein to aid understanding for solving the elliptic equation by using the finite difference method. The problem is selected because heat transfer phenomenon and the temperature solution are easy to understand. Figure 8.4 shows a rectangular plate that lies in *x-y* coordinates under the steady-state conduction heat transfer. The figure also shows an infinitesimal element with the sizes of Δx and Δy. The plate has the thickness of t and is made from a material with the thermal conductivity coefficient of k.

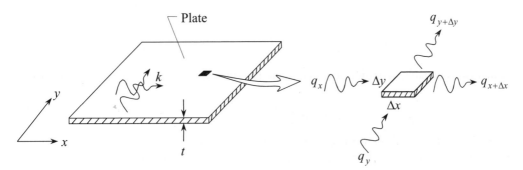

Figure 8.4 Conduction heat transfer in a plate.

To derive the governing differential equation, the conservation of energy is applied to the infinitesimal element such that,

$$\text{Heat flux in} - \text{Heat flux out} \;=\; 0 \tag{8.10}$$

i.e.,

$$\left(q_x + q_y\right) - \left(q_{x+\Delta x} + q_{y+\Delta y}\right) \;=\; 0 \tag{8.11}$$

where q_x and q_y are the heat fluxes in *x*- and *y*-directions, respectively. These heat fluxes depend on the temperature gradient according to the Fourier's law,

$$q_x \;=\; -k(t\,\Delta y)\frac{\partial T}{\partial x} \tag{8.12a}$$

$$q_y \;=\; -k(t\,\Delta x)\frac{\partial T}{\partial y} \tag{8.12b}$$

By substituting Eqs. (8.12a-b) into Eqs. (8.11) and applying the Taylor series expansion in the same fashion as shown in Eq. (2.29) onto $q_{x+\Delta x}$ and $q_{y+\Delta y}$,

$$- k(t\,\Delta y)\frac{\partial T}{\partial x} - k(t\,\Delta x)\frac{\partial T}{\partial y}$$

$$+ k(t\,\Delta y)\frac{\partial T}{\partial x} + \frac{\partial}{\partial x}\left[k(t\,\Delta y)\frac{\partial T}{\partial x}\right]\Delta x + \frac{1}{2}\frac{\partial^2}{\partial x^2}\left[k(t\,\Delta y)\frac{\partial T}{\partial x}\right](\Delta x)^2 + \dots$$

$$+ k(t\,\Delta x)\frac{\partial T}{\partial y} + \frac{\partial}{\partial y}\left[k(t\,\Delta x)\frac{\partial T}{\partial y}\right]\Delta y + \frac{1}{2}\frac{\partial^2}{\partial y^2}\left[k(t\,\Delta x)\frac{\partial T}{\partial y}\right](\Delta y)^2 + \dots = 0$$

Or,

$$\frac{\partial}{\partial x}\left(k\frac{\partial T}{\partial x}\right)t\,\Delta x\,\Delta y + \frac{1}{2}\frac{\partial^2}{\partial x^2}\left(k\frac{\partial T}{\partial x}\right)t\,(\Delta x)^2\,\Delta y + \dots$$

$$+ \frac{\partial}{\partial y}\left(k\frac{\partial T}{\partial y}\right)t\,\Delta x\,\Delta y + \frac{1}{2}\frac{\partial^2}{\partial y^2}\left(k\frac{\partial T}{\partial y}\right)t\,\Delta x\,(\Delta y)^2 + \dots = 0 \tag{8.13}$$

Equation (8.13) is then divided through by $t\,\Delta x\,\Delta y$. By letting $\Delta x \to 0$ and $\Delta y \to 0$, Eq. (8.13) becomes,

$$\frac{\partial}{\partial x}\left(k\frac{\partial T}{\partial x}\right) + \frac{\partial}{\partial y}\left(k\frac{\partial T}{\partial y}\right) = 0 \tag{8.14}$$

If the thermal conductivity coefficient k is constant, then the partial differential Eq. (8.14) reduces to

$$\frac{\partial^2 T}{\partial x^2} + \frac{\partial^2 T}{\partial y^2} = 0 \tag{8.15}$$

Equation (8.15) is in the form of the elliptic equation as shown in Eq. (8.7). This means the steady-state temperature distribution in a plate is determined by solving the elliptic equation which is in the form of the Laplace's equation.

If the plate is subjected to a specified heating on its surface, the corresponding partial differential equation can be derived in the same way. In this later case, the partial differential equation is in the form,

$$\frac{\partial^2 T}{\partial x^2} + \frac{\partial^2 T}{\partial y^2} = f(x, y) \tag{8.16}$$

where $f(x, y)$ denotes the specified heating function which may vary with x- and y-coordinates. The partial differential equation in the form of Eq. (8.16) is known as the Poisson's equation.

8.2.2 Computational procedures

To ease understanding on the application of the finite difference method for solving the elliptic equation, the problem of steady-state heat conduction in a rectangular plate is considered. The plate is first divided into a number of intervals in both x- and y-directions as shown in Fig. 8.5. The lengths of the intervals are Δx and Δy in x- and y-directions, respectively. The unknowns are at grid points where the horizontal and vertical lines intersect, e.g., at the location i, j, as shown in the figure.

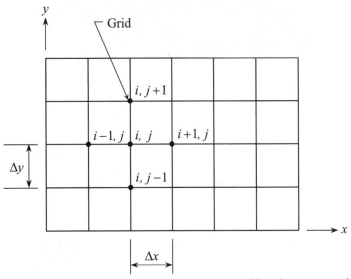

Figure 8.5 Dividing rectangular plate into intervals with unknowns at grid points by using the finite difference method.

The key step of the finite difference method is to transform the governing differential Eq. (8.15) into an algebraic equation. Herein, the central difference technique as shown in Table 6.6 is used to approximate the second-order derivative terms,

$$\frac{\partial^2 T}{\partial x^2} = \frac{T_{i+1,\,j} - 2T_{i,\,j} + T_{i-1,\,j}}{(\Delta x)^2} \tag{8.17a}$$

and

$$\frac{\partial^2 T}{\partial y^2} = \frac{T_{i,\,j+1} - 2T_{i,\,j} + T_{i,\,j-1}}{(\Delta y)^2} \tag{8.17b}$$

By substituting Eqs. (8.17a-b) into the Laplace's Eq. (8.15) and using $\Delta x = \Delta y$, the approximate differential equation is obtained,

$$T_{i+1,\,j} + T_{i-1,\,j} + T_{i,\,j+1} + T_{i,\,j-1} - 4T_{i,\,j} = 0 \tag{8.18}$$

Such expression can be written in a stencil form as,

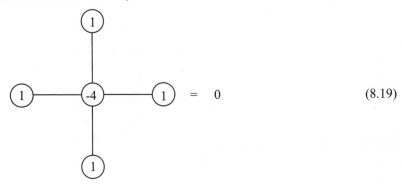

$$= 0 \tag{8.19}$$

The stencil form above is applied at the grid points where the temperatures are unknowns. For example, a rectangular plate with the size of 4×2 units as shown in Fig. 8.6 has specified temperatures along the four edges. If the plate is divided into 4 and 2 intervals in x- and y-directions, respectively, then the temperature unknowns are at the three grid points 2,2, 3,2 and 4,2.

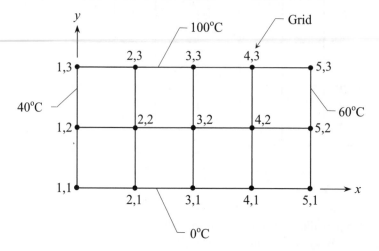

Figure 8.6 Application of the stencil form to establish a set of simultaneous equations for conduction heat transfer in a rectangular plate.

Application of the stencil form as shown in Eq. (8.19) at the grid point 2,2 gives the algebraic equation,

$$100 + 40 + 0 + T_{3,2} - 4T_{2,2} = 0$$

$$4T_{2,2} - T_{3,2} = 140 \qquad (8.20)$$

Similarly, at the grid point 3,2

$$100 + T_{2,2} + 0 + T_{4,2} - 4T_{3,2} = 0$$

$$-T_{2,2} + 4T_{3,2} - T_{4,2} = 100 \qquad (8.21)$$

and at the grid point 4,2

$$100 + T_{3,2} + 0 + 60 - 4T_{4,2} = 0$$

$$-T_{3,2} + 4T_{4,2} = 160 \qquad (8.22)$$

The three Eqs. (8.20-8.22) lead to a set of simultaneous equations,

$$4T_{2,2} - T_{3,2} = 140$$

$$-T_{2,2} + 4T_{3,2} - T_{4,2} = 100 \qquad (8.23)$$

$$- T_{3,2} + 4T_{4,2} = 160$$

which can be written in matrix form,

$$\begin{bmatrix} 4 & -1 & 0 \\ -1 & 4 & -1 \\ 0 & -1 & 4 \end{bmatrix} \begin{Bmatrix} T_{2,2} \\ T_{3,2} \\ T_{4,2} \end{Bmatrix} = \begin{Bmatrix} 140 \\ 100 \\ 160 \end{Bmatrix} \qquad (8.24)$$

It should be noted that the square matrix on the left-hand side of Eq. (8.24) has positive coefficients along the diagonal line. The magnitudes of these coefficients are relatively large as compared to the other coefficients on the off-diagonal line. With such matrix property, the Gauss-Seidel iteration method is effective for solving the set of simultaneous equations. It is also noted that the procedure for solving such problem is sometimes called as the Liebmann's method. The method requires less computational time and memory as compared to the other direct methods, such as the Gauss elimination or the LU decomposition method, especially when the model contains a large number of grid points. The corresponding computer program can also be developed easily as will be shown in the following example.

8.2.3 Example

In this section, an example for solving the elliptic equation by using the finite difference method is presented in details. A corresponding computer program is also developed to demonstrate the application of the method when the finite difference model contains many unknowns. The example has the exact solution of the temperature distribution, so that the finite difference solutions at grid points can be compared to measure the method efficiency.

Example 8.1 A rectangular plate with the size of 2×1 units has specified zero temperature along the left, lower and right edges as shown in Fig. 8.7. The plate is subjected to the specified temperature distribution that varies as a sine function along the upper edge as shown in the figure. Use the finite difference method to determine the plate temperature distribution by dividing the plate into 8 and 4 intervals in x- and y-directions, respectively.

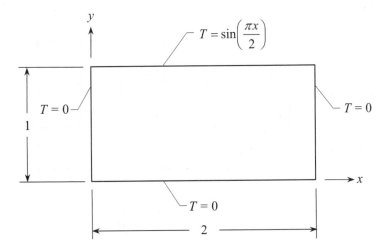

Figure 8.7 Plate with specified temperature along the four edges.

Compare the computed solutions with the exact temperature solution of,

$$T(x, y) = \sin\frac{\pi x}{2} \sinh\frac{\pi y}{2} \Big/ \sinh\frac{\pi}{2} \qquad (8.25)$$

The plate temperature distribution according to the exact solution in Eq. (8.25) is shown as a carpet plot in Fig. 8.8. The temperature distribution is in the form of sine function along the upper edge and decays to zero along the left, lower and right edges.

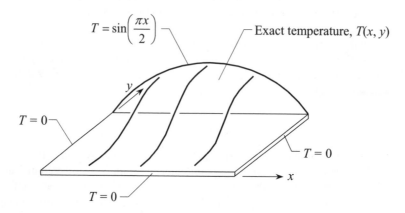

Figure 8.8 Exact temperature distribution over the plate according to Eq. (8.25).

The finite difference model with 8 and 4 intervals in x- and y-directions is shown in Fig. 8.9 with the grid point numbers. For this model, the lengths of the intervals are $\Delta x = \Delta y = 0.25$ unit. The model contains a total of 45 grid points for which 21 interior grid points are unknowns.

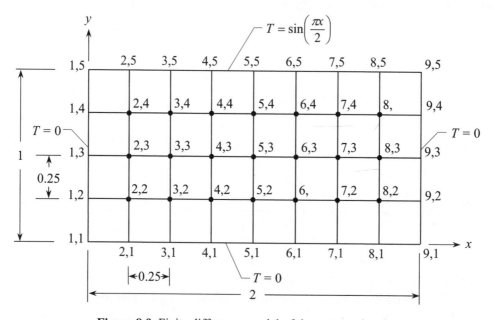

Figure 8.9 Finite difference model of the rectangular plate.

The stencil form as shown in Eq. (8.19) is applied to these 21 interior grid points together with the boundary conditions along the four edges as,

$$T(x, y = 0) = 0 \tag{8.26a}$$

$$T(x = 0, y) = 0 \tag{8.26b}$$

$$T(x = 2, y) = 0 \tag{8.26c}$$

and
$$T(x, y = 1) = \sin\frac{\pi x}{2} \tag{8.26d}$$

Such applications lead to a set of 21 algebraic equations. The algebraic equations are then solved by using the Gauss-Seidel iteration method for the temperature solutions at grid points from Eq. (8.18), i.e.,

$$T_{i, j} = \left(T_{i-1, j} + T_{i+1, j} + T_{i, j-1} + T_{i, j+1}\right)/4 \tag{8.27}$$

At the end of each iteration, the computed temperatures are compared with those obtained from the previous one. The iteration process is terminated if the temperature differences between the two successive iterations are less than the specified tolerance.

Because a typical finite difference model must contain a large number of grid points in order to obtain accurate solutions, a computer program should be developed for reducing the computational effort. Analysts can refine the model by using many small intervals in both x- and y-directions. For example, the plate can be divided into 200 and 100 intervals in x- and y-directions, respectively. In this latter case, the finite difference method will lead to a set of 19,701 equations which is impractical to solve them by hand.

Table 8.1 shows the solutions at grid points obtained from the finite difference method by using the computer program in Fig. 8.10. The computer program employs the Gauss-Seidel iteration technique to solve the set of simultaneous equations with the stopping tolerance criterion of 0.00001. The finite difference solutions are obtained after 24 iterations. These computed solutions at grid points are less than 1% difference from the exact solutions.

Table 8.1 Comparison of temperatures between the finite difference and exact solutions (values in bracket). Locations of numbers correspond to grid point locations in Fig. 8.9.

0.000	0.383	0.707	0.924	1.000	0.924	0.707	0.383	0.000
(0.000)	(0.383)	(0.707)	(0.924)	(1.000)	(0.924)	(0.707)	(0.383)	(0.000)
0.000	0.245	0.453	0.592	0.641	0.592	0.453	0.245	0.000
(0.000)	(0.244)	(0.452)	(0.590)	(0.639)	(0.590)	(0.452)	(0.244)	(0.000)
0.000	0.145	0.269	0.351	0.380	0.351	0.269	0.145	0.000
(0.000)	(0.144)	(0.267)	(0.349)	(0.377)	(0.349)	(0.267)	(0.144)	(0.000)
0.000	0.068	0.125	0.163	0.177	0.163	0.125	0.068	0.000
(0.000)	(0.067)	(0.124)	(0.162)	(0.175)	(0.162)	(0.124)	(0.067)	(0.000)
0.000	0.000	0.000	0.000	0.000	0.000	0.000	0.000	0.000
(0.000)	(0.000)	(0.000)	(0.000)	(0.000)	(0.000)	(0.000)	(0.000)	(0.000)

Fortran

```
          PROGRAM ELLIP
C.....A FINITE DIFFERENCE PROGRAM FOR SOLVING
C.....TEMPERATURE DISTRIBUTION IN A PLATE
          DIMENSION T(9,5)
C.....ASSIGN TOLERANCE AND MAX. NO. OF ITERATIONS:
          TOL = .00001
          MXITER = 100
C.....SET UP BOUNDARY CONDITIONS:
          DO 10  I=1,9
          T(I,1) = 0.
     10 CONTINUE
          DO 20  J=1,5
          T(1,J) = 0.
          T(9,J) = 0.
     20 CONTINUE
          X  = .25
          DX = X
          PI = 4.*ATAN(1.)
          DO 30  I=2,8
          T(I,5) = SIN(PI*X/2.)
          X = X + DX
     30 CONTINUE
C.....SET UP INITIAL TEMPERATURE VALUES:
          DO 40  I=2,8
          DO 40  J=2,4
          T(I,J) = J*T(I,5)/5
     40 CONTINUE
C.....SOLVE UNKNOWN TEMPERATURES AT GRID POINTS
C.....USING GAUSS-SEIDEL ITERATION TECHNIQUE:
          DO 100 ITER=1,MXITER
          IFLAG = 0
          DO 200  I=2,8
          DO 200  J=2,4
          TEMP = ( T(I-1,J) + T(I+1,J)
     1          + T(I,J+1) + T(I,J-1) )/4.
          DIFF = T(I,J) - TEMP
          IF(ABS(DIFF).GT.TOL)  IFLAG = 1
          T(I,J) = TEMP
    200 CONTINUE
          IF(IFLAG.EQ.0)  GO TO 300
    100 CONTINUE
          WRITE(6,110)
    110 FORMAT(' SOLUTION NOT CONVERGED WITHIN THE',
         * ' SPECIFIED NO. OF ITERATIONS & TOLERANCE')
          GO TO 1000
    300 CONTINUE
C.....PRINT OUT TEMPARTURES AT GRID POINTS IN THE
C.....FORMAT CORRESPONDING TO THE PROBLEM FIGURE:
          DO 500  J=5,1,-1
          WRITE(6,510)  (T(I,J), I=1,9)
    510 FORMAT(9F6.3)
    500 CONTINUE
   1000 CONTINUE
          STOP
          END
```

MATLAB

```
%  PROGRAM ELLIP
%  A FINITE DIFFERENCE PROGRAM FOR SOLVING
%  TEMPERATURE DISTRIBUTION IN A PLATE
%  ASSIGN TOLERANCE AND MAX. NO. OF ITERATIONS:
tol = 0.00001;
mxiter = 100.;
%  SET UP BOUNDARY CONDITIONS:
t = zeros(5,9);
x = 0.25;
dx = x;
for i = 2:8
    t(1,i) = sin(pi*x/2);
    x = x + dx;
end
%  SET UP INITIAL TEMPERATURE VALUES:
for i = 2:4
    for j = 2:8
        t(i,j) = j*t(i,5)/5.;
    end
end
%  SOLVE UNKNOWN TEMPERATURES AT GRID POINTS
%  USING GAUSS-SEIDEL ITERATION TECHNIQUE:
for iter = 1:mxiter
    iflag = 0.;
    for i = 2:4
        for j = 2:8
            temp = (t(i-1,j) + t(i+1,j) + t(i,j+1) ... +
t(i,j-1))/4.;
            diff = t(i,j) - temp;
            if abs(diff) > tol
                iflag = 1.;
            end
            t(i,j) = temp;
        end
    end
    if iflag == 0.
        continue
    end
end
if iflag == 1
    fprintf(' SOLUTION NOT CONVERGED WITHIN THE')
    fprintf(' SPECIFIED NO. OF ITERATIONS & TOLERANCE')
    break
end
%  PRINT OUT TEMPARTURES AT GRID POINTS IN THE
%  FORMAT CORRESPONDING TO THE PROBLEM FIGURE:
for i = 1:5
    fprintf('\n');
    for j = 1:9
        fprintf('%6.3f', t(i,j));
    end
end
```

Figure 8.10 Computer program for solving grid point temperatures on a rectangular plate using the finite difference method.

Example 8.2 From the problem statement and the plate temperature distribution in example 8.1, it can be seen that the problem has solution symmetry. Thus, only one half of the plate can be used in the analysis to obtain the same temperature solution. Using only one half of the plate can reduce the total number of unknowns and thus the computational time. Figure 8.11 shows only the left half of the plate that can be used in the analysis. The boundary conditions are identical to those of the original plate except the condition on the right boundary which represents zero conduction heat transfer.

The computational procedures follow those explained in example 8.1. The left half of the plate is first divided into intervals in both *x*- and *y*-directions as shown in Fig. 8.12. The unknowns are at the interior grid points and the grid points along the right model boundary. Thus, there is a total of only 12 unknowns for the half model of the plate.

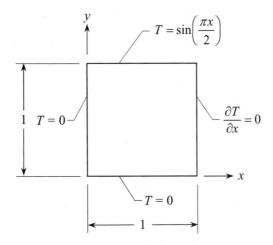

Figure 8.11 Determination of temperature distribution using only the left half of the plate due to the solution symmetry.

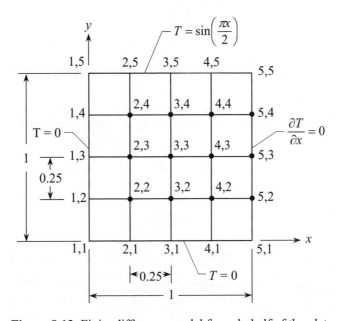

Figure 8.12 Finite difference model for only half of the plate.

The boundary conditions along the four edges of the half model in Fig. 8.12 are:

$$T(x, y = 0) \;\; = \;\; 0 \tag{8.28a}$$

$$T(x = 0, y) \;\; = \;\; 0 \tag{8.28b}$$

$$\frac{\partial T}{\partial x}(x = 1, y) \;\; = \;\; 0 \tag{8.28c}$$

$$T(x, y = 1) \;\; = \;\; \sin\frac{\pi x}{2} \tag{8.28d}$$

The stencil form as shown in Eq. (8.19) can be used to establish the algebraic equations for the interior grid points except those on the right boundary. A new stencil form must be developed for the grid points along the right boundary. Figure 8.13 shows the grid points along the right boundary with a fictitious grid point at $i+1, j$.

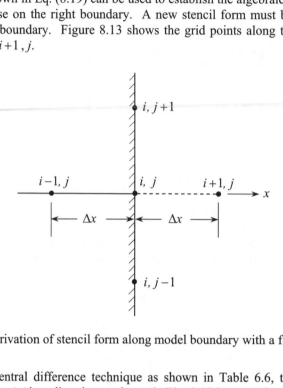

Figure 8.13 Derivation of stencil form along model boundary with a fictitious grid point.

By using the central difference technique as shown in Table 6.6, the first-derivative of the temperature at grid point i, j in x-direction as shown in Fig. 8.13 is approximate by

$$\frac{\partial T}{\partial x} = \frac{T_{i+1, j} - T_{i-1, j}}{2\Delta x} \tag{8.29}$$

Thus, the temperature at grid point $i+1, j$ is

$$T_{i+1, j} = 2\Delta x \frac{\partial T}{\partial x} + T_{i-1, j} \tag{8.30}$$

By substituting Eq. (8.30) into Eq. (8.18), the temperature at the grid point along the right boundary can be determined from

$$2\Delta x \frac{\partial T}{\partial x} + T_{i-1, j} + T_{i-1, j} + T_{i, j+1} + T_{i, j-1} - 4T_{i, j} = 0$$

$$T_{i, j} = \left(2\Delta x \frac{\partial T}{\partial x} + 2T_{i-1, j} + T_{i, j+1} + T_{i, j-1}\right)\Big/4 \tag{8.31}$$

It is noted that if the right boundary of the model is subjected to a specified heating, heat convection or radiation, the term $\partial T / \partial x$ in Eq. (8.31) must be replaced by an appropriate value according to the heat mode. But for this example, $\partial T / \partial x$ is zero along the right boundary, thus the equation for determining the temperature for a grid point along the right boundary reduces to

$$T_{i, j} = \left(2T_{i-1, j} + T_{i, j+1} + T_{i, j-1}\right)\Big/4 \tag{8.32}$$

The computer program as shown in Fig. 8.10 can be modified slightly to solve for the temperatures for only half of the plate. The stencil form in Eq. (8.19) is used for the interior grid points, while Eq. (8.32) is applied to the grid points along the right boundary. The computed temperatures at the grid points are shown in Table 8.2. These temperature solutions are identical to those obtained from using the full plate model as shown in Table 8.1. The later example 8.2 demonstrates that if a problem has solution symmetry, the finite difference model that reflects such the solution symmetry should be used. The symmetry model helps reducing both the number of unknowns and computational time.

Table 8.2 Comparison of temperatures between the finite difference and exact solutions (values in bracket) for half of the plate. Locations of numbers correspond to grid point locations in Fig. 8.12.

0.000	0.383	0.707	0.924	1.000
(0.000)	(0.383)	(0.707)	(0.924)	(1.000)
0.000	0.245	0.453	0.592	0.641
(0.000)	(0.244)	(0.452)	(0.590)	(0.639)
0.000	0.145	0.269	0.351	0.380
(0.000)	(0.144)	(0.267)	(0.349)	(0.377)
0.000	0.068	0.125	0.163	0.177
(0.000)	(0.067)	(0.124)	(0.162)	(0.175)
0.000	0.000	0.000	0.000	0.000
(0.000)	(0.000)	(0.000)	(0.000)	(0.000)

8.3 Parabolic Equation

8.3.1 Differential equation

Parabolic equation is classified as a type of the partial differential equations that occurs in many scientific and engineering applications. One of the simplest examples used for studying the equation is the transient heat conduction in a bar as shown in Fig. 8.14. The bar with length L in x-coordinate direction is made from a material that has the thermal conductivity coefficient k, the mass density ρ and the specific heat c. The temperature varies with x-coordinate along the bar and time t. The partial differential equation in the form of parabolic equation can be derived from the conservation of energy by using a small element of length Δx,

$$\text{Energy in} - \text{Energy out} = \text{Energy stored} \qquad (8.33)$$

i.e.,
$$q_x - q_{x+\Delta x} = \rho c A \Delta x \frac{\partial T}{\partial t} \qquad (8.34)$$

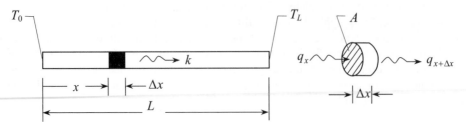

Figure 8.14 Transient heat conduction in a bar.

where q_x and $q_{x+\Delta x}$ are the heat fluxes going into and out from the cross-sectional area A of the small element as shown in the figure, respectively. The amount of heat flux depends on the temperature gradient from the Fourier's law,

$$q_x = -kA\frac{\partial T}{\partial x} \tag{8.35}$$

By substituting Eq. (8.35) into and Eq. (8.34) and applying the Taylor series expansion to the heat flux term $q_{x+\Delta x}$, Eq. (8.34) becomes,

$$-kA\frac{\partial T}{\partial x} - \left[-kA\frac{\partial T}{\partial x} - \frac{\partial}{\partial x}\left(kA\frac{\partial T}{\partial x}\right)\Delta x - \frac{1}{2}\frac{\partial^2}{\partial x^2}\left(kA\frac{\partial T}{\partial x}\right)(\Delta x)^2 - \cdots\right] = \rho cA\Delta x\frac{\partial T}{\partial t}$$

After the first and second terms are cancelled, the equation is divided through by $A\Delta x$. Then, by letting $\Delta x \to 0$, the equation reduces to

$$\frac{\partial}{\partial x}\left(k\frac{\partial T}{\partial x}\right) = \rho c\frac{\partial T}{\partial t} \tag{8.36}$$

If the thermal conductivity coefficient k is constant, Eq. (8.36) leads to a parabolic partial differential equation representing the transient heat conduction in a bar as,

$$\frac{k}{\rho c}\frac{\partial^2 T}{\partial x^2} = \frac{\partial T}{\partial t} \tag{8.37}$$

Different methods for solving the parabolic equation in the form of Eq. (8.37) are presented in the following sections. Advantages and disadvantages of each method are explained. Examples and corresponding computer programs are also presented to demonstrate the efficiency of each method. Understanding details of these methods can help analyzing large practical problems more effectively.

8.3.2 Explicit method

The explicit method is considered as the simplest method for solving the parabolic equation. The method starts from dividing the bar into equal intervals, each has the length of Δx, as shown in Fig. 8.15. These intervals are connected at grid points $i-1$, i, $i+1$ for which the temperatures are unknown and to be determined.

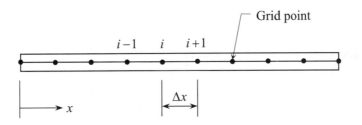

Figure 8.15 A bar is divided into intervals connected at grid points.

The temperatures T that vary with time t at grid points are determined from the parabolic differential Eq. (8.37). The forward difference approximation is applied to the first-order time derivative term as,

$$\frac{\partial T}{\partial t} = \frac{T_i^{n+1} - T_i^n}{\Delta t} \tag{8.38}$$

Such approximation has the error of order $O(\Delta t)$ where Δt is the time step. The superscript n denotes the nth time step in the computation. Similarly, the central difference approximation can be applied to the second-order spatial derivative term,

$$\frac{\partial^2 T}{\partial x^2} = \frac{T_{i+1}^n - 2T_i^n + T_{i-1}^n}{(\Delta x)^2} \tag{8.39}$$

for which the error is of order $O(\Delta x^2)$. It is noted that the grid point temperatures in Eq. (8.39) are the values at the nth time step. By substituting Eqs. (8.38) and (8.39) into the differential Eq. (8.37), an algebraic equation representing the partial differential equation is obtained in the form,

$$\frac{k}{\rho c} \frac{T_{i+1}^n - 2T_i^n + T_{i-1}^n}{(\Delta x)^2} = \frac{T_i^{n+1} - T_i^n}{\Delta t} \tag{8.40}$$

The terms in Eq. (8.40) are rearranged such that the grid point temperatures at time $n+1$ can be determined directly, i.e.,

$$T_i^{n+1} = T_i^n + \alpha\left(T_{i+1}^n - 2T_i^n + T_{i-1}^n\right) \tag{8.41}$$

where

$$\alpha = \frac{k \Delta t}{\rho c (\Delta x)^2} \tag{8.42}$$

Equation (8.41) shows that the unknown temperature at grid point i and at time step $n+1$ is determined from the temperatures at the three grid points $i-1$, i, and $i+1$ which were computed and are known from the time step n. Such computational procedure can be described by the schematic diagram as shown in Fig. 8.16.

Because the unknown temperature at time $n+1$ can be determined directly from the known temperatures at time n by using Eq. (8.41), the method is called the *explicit method*. It should be noted that even though the method is very convenient, it is limited by the value of $\alpha \leq 1/2$ which is governed by the time step Δt. In another word, the requirement of the time step,

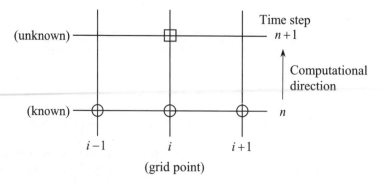

Figure 8.16 Schematic computational diagram of the explicit method.

$$\Delta t \quad \leq \quad \frac{\rho c (\Delta x)^2}{2k} \tag{8.43}$$

must be satisfied. The maximum allowable time step in Eq. (8.43) is called the *critical time step* which depends on the grid spacing Δx and the material properties. If the time step Δt used in the computation is greater than the critical time step, the computed solution will diverge or grow without bound. Because the time step Δt varies with the square of Δx, the time step will be very small for a fine mesh model with small Δx.

Example 8.3 Solve the parabolic Eq. (8.37) for the transient temperature distribution along the bar of length 1 unit. The bar is made from a material such that $k/\rho c = 1$ with the initial temperature of $\sin(\pi x)$ along its length. Determine the grid point temperatures by dividing the bar length into 10 equal intervals ($\Delta x = 0.1$).

The problem statement of this example can be summarized as follows:

$$\frac{k}{\rho c} \frac{\partial^2 T}{\partial x^2} \quad = \quad \frac{\partial T}{\partial t} \qquad\qquad 0 \leq x \leq 1 \tag{8.44}$$

$$T(0, t) \quad = \quad T(1, t) \quad = \quad 0 \tag{8.45}$$

$$T(x, 0) \quad = \quad \sin(\pi x) \tag{8.46}$$

If $k/\rho c = 1$, the exact solution is

$$T(x, t) \quad = \quad e^{-\pi^2 t} \sin(\pi x) \tag{8.47}$$

Figure 8.17 shows the bar dividing into 10 equal intervals with grid point numbers. Each interval has the length of $\Delta x = 0.1$. There are 11 grid points for which the temperatures at grid point numbers 1 and 11 are maintained at zero degree throughout the computation.

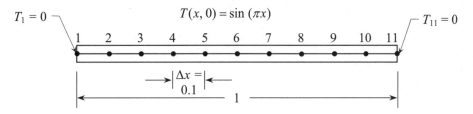

Figure 8.17 A bar with 10 equal intervals and grid point numbers.

The transient temperatures for grid point numbers 2 through 10 can be determined by using the derived Eq. (8.41). For example, the new temperature at grid point 2 is determined from the old temperatures at grid points 1, 2 and 3 as,

$$T_2^1 = T_2^0 + \alpha\left(T_3^0 - 2T_2^0 + T_1^0\right) \tag{8.48}$$

where the subscripts denote the grid point numbers and the superscripts represent the time step. Thus, the terms on the right-hand side of the equation are,

T_1^0 = temperature at grid point 1 at time $t = 0$ which is $= 0$

T_2^0 = temperature at grid point 2 at time $t = 0$ which is $= \sin(0.1\pi) = 0.3090$

T_3^0 = temperature at grid point 3 at time $t = 0$ which is $= \sin(0.2\pi) = 0.5878$

If the time step is specified as $\Delta t = 0.005$, then, from Eq. (8.42)

$$\alpha = \frac{k}{\rho c}\frac{\Delta t}{(\Delta x)^2} = (1)\frac{0.005}{(0.1)^2} = 0.5 \tag{8.49}$$

Thus, the temperature at grid point 2 at time $t = 0.005$ from Eq. (8.48) is,

$$T_2^1 = 0.3090 + (0.5)[0.5878 - 2(0.3090) + 0] = 0.2939$$

The computational procedure for determining transient temperatures at grid points can be used for developing a computer program. Figure 8.18 shows a computer program for determining the transient temperature response as explained in the example. The time step Δt and the total number of steps are specified at the beginning of the program. The temperatures at grid points are determined and are shown as output at each time step.

With the time step of $\Delta t = 0.005$, table 8.3 shows the computed grid point temperatures at different times as compared to the exact solutions of Eq. (8.47). The table shows the computed temperatures for the grid points only on the left-half of the bar due to symmetry of the solution. Figure 8.19 shows the transient temperature response along the bar at different times. The bar temperature drops from the initial temperature of the specified sine function as time increases. As time increases, the bar temperature approaches zero because the temperature at both ends of the bar are maintained at zero.

Fortran

```
       PROGRAz  PARAEXP
C.....A FINITE DIFFERENCE PROGRAM FOR SOLVING
C.....TRANSIENT TEMPERATURE DISTRIBUTION IN A
C.....ROD USING EXPLICIT METHOD
       DIMENSION  TOLD(11), TNEW(11)
C.....ASSIGN TIME STEP AND NO. OF TIME STEPS:
       DTIME = .005
       NSTEPS = 40
C.....SET UP INITIAL AND BOUNDARY CONDITIONS:
       X  = 0.
       DX = .1
       PI = 4.*ATAN(1.)
       DO 10  I=1,11
       TOLD(I) = SIN(PI*X)
       TNEW(I) = TOLD(I)
       X = X + DX
   10  CONTINUE
C.....SOLVE FOR TEMPERATURE RESPONSE:
       ALPHA = DTIME/(DX*DX)
       TIME = DTIME
       DO 100   ISTEP=1,NSTEPS
       DO 200   I=2,10
       TNEW(I) = TOLD(I) +
     1 ALPHA*(TOLD(I+1) - 2.*TOLD(I) + TOLD(I-
1))
  200  CONTINUE
C.....PRINT OUT AT EVERY 4 STEPS:
       IP = ISTEP - (ISTEP/4)*4
       IF(IP.EQ.0)
     *  WRITE(6,300)  TIME, (TNEW(I), I=1,11)
  300  FORMAT(F6.2, 1X, 11F6.4)
       DO 400   I=2,10
       TOLD(I) = TNEW(I)
  400  CONTINUE
       TIME = TIME + DTIME
  100  CONTINUE
       STOP
       END
```

MATLAB

```
%   PROGRAM   PARAEXP
%   A FINITE DIFFERENCE PROGRAM FOR SOLVING
%   TRANSIENT TEMPERATURE DISTRIBUTION IN A
%   ROD USING EXPLICIT METHOD
%   ASSIGN TIME STEP AND NO. OF TIME STEPS:
dtime = 0.005;
nsteps = 40;
%   SET UP INITIAL AND BOUNDARY CONDITIONS:
x  = 0.;
dx = 0.1;
for i = 1:11
    told(i) = sin(pi*x);
    tnew(i) = told(i);
    x = x + dx;
end
%   SOLVE FOR TEMPERATURE RESPONSE:
alpha = dtime/(dx^2);
time = dtime;
for istep = 1:nsteps
    for i = 2:10
        tnew(i) = told(i) + alpha*(told(i+1) - ... 2.*told(i) +
told(i-1));
    end
%   PRINT OUT AT EVERY 4 STEPS:
    ip = istep - floor(istep/4)*4;
    if ip == 0.
        fprintf('\n%6.2f',time);
        for i = 1:11
            fprintf('   %6.4f',tnew(i));
        end
    end
    for i = 2:10
        told(i) = tnew(i);
    end
    time = time + dtime;
end
```

Figure 8.18 A computer program for determining transient temperature response of a bar by using the explicit method.

Table 8.3 Comparison between the computed temperatures at grid points using time step of $\Delta t = 0.005$ from the explicit method and the exact temperatures (values in brackets) at different times.

Time t	1	2	3	4	5	6
0.04	0.0000	0.2068	0.3934	0.5415	0.6366	0.6693
	(0.0000)	(0.2082)	(0.3961)	(0.5451)	(0.6408)	(0.6738)
0.08	0.0000	0.1384	0.2633	0.3625	0.4261	0.4480
	(0.0000)	(0.1403)	(0.2669)	(0.3673)	(0.4318)	(0.4540)
0.12	0.0000	0.0927	0.1763	0.2426	0.2852	0.2999
	(0.0000)	(0.0945)	(0.1798)	(0.2475)	(0.2910)	(0.3059)
0.16	0.0000	0.0620	0.1180	0.1624	0.1909	0.2007
	(0.0000)	(0.0637)	(0.1212)	(0.1668)	(0.1961)	(0.2062)
0.20	0.0000	0.0415	0.0790	0.1087	0.1278	0.1344
	(0.0000)	(0.0429)	(0.0816)	(0.1124)	(0.1321)	(0.1389)

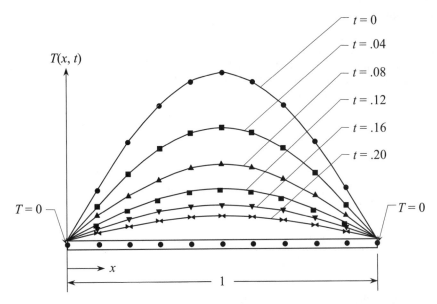

T(x, t)

t = 0
t = .04
t = .08
t = .12
t = .16
t = .20

T = 0

T = 0

x

1

Figure 8.19 Computed transient temperature response at different times
by using the explicit method.

It should be noted that the time step of $\Delta t = 0.005$ used to obtain the solutions is the critical time step for the mesh in the example. Higher solution accuracy can be obtained by reducing the time step Δt used in the computation. Table 8.4 shows the more accurate temperatures at the same grid points computed by using a smaller time step of $\Delta t = 0.00002$. Table 8.4 also shows the diverged solutions of the computed temperatures when the time step of $\Delta t = 0.01$ (which is larger than the critical time step) is used. Such diverged solution from using too large time step Δt is the main disadvantage of the explicit method. Other methods that can alleviate the restriction of too small time step Δt are presented in the following sections.

Table 8.4 Comparison among the computed temperatures at grid points using time step of $\Delta t = 0.00002$, $\Delta t = 0.01$ (values in square brackets) from the explicit method and the exact temperatures (values in round brackets) at different times.

Time t	1	2	3	4	5	6
0.04	0.0000	0.2089	0.3973	0.5469	0.6429	0.6760
	(0.0000)	(0.2082)	(0.3961)	(0.5451)	(0.6408)	(0.6738)
	[0.0000]	[0.2047]	[0.3893]	[0.5358]	[0.6299]	[0.6623]
0.08	0.0000	0.1412	0.2686	0.3697	0.4346	0.4570
	(0.0000)	(0.1403)	(0.2669)	(0.3673)	(0.4318)	(0.4540)
	[0.0000]	[0.1355]	[0.2578]	[0.3549]	[0.4171]	[0.4386]
0.12	0.0000	0.0955	0.1816	0.2499	0.2938	0.3089
	(0.0000)	(0.0945)	(0.1798)	(0.2475)	(0.2910)	(0.3059)
	[0.0000]	[0.0900]	[0.1703]	[0.2356]	[0.2756]	[0.2913]
0.16	0.0000	0.0645	0.1227	0.1689	0.1986	0.2088
	(0.0000)	(0.0637)	(0.1212)	(0.1668)	(0.1961)	(0.2062)
	[0.0000]	[0.0751]	[0.0829]	[0.1986]	[0.1295]	[0.2531]
0.20	0.0000	0.0436	0.0830	0.1142	0.1342	0.1412
	(0.0000)	(0.0429)	(0.0816)	(0.1124)	(0.1321)	(0.1389)
	[0.0000]	[1.2042]	[-2.1852]	[3.3179]	[-3.8300]	[4.5065]

8.3.3 Implicit method

In order to avoid the diverged solution that may occur from using too large time step Δt in the explicit method, an implicit method is presented in this section. The implicit method applies the central difference approximation to the second-order spatial derivative term while the nodal temperatures are determined at time $n+1$ as,

$$\frac{\partial^2 T}{\partial x^2} = \frac{T_{i+1}^{n+1} - 2T_i^{n+1} + T_{i-1}^{n+1}}{(\Delta x)^2} \tag{8.50}$$

By substituting Eqs. (8.50) and (8.38) into the parabolic differential Eq. (8.37), a corresponding algebraic equation is obtained,

$$\frac{k}{\rho c} \frac{T_{i+1}^{n+1} - 2T_i^{n+1} + T_{i-1}^{n+1}}{(\Delta x)^2} = \frac{T_i^{n+1} - T_i^n}{\Delta t} \tag{8.51}$$

which can be written in the form,

$$-\alpha T_{i-1}^{n+1} + (1+2\alpha) T_i^{n+1} - \alpha T_{i+1}^{n+1} = T_i^n \tag{8.52}$$

where the parameter α is defined in Eq. (8.42). The algebraic equation in Eq. (8.52) consists of the unknown temperatures at grid points at $i-1$, i and $i+1$ at the new time step $n+1$ on the left-hand side of the equation. The schematic computational diagram is shown in Fig. 8.20.

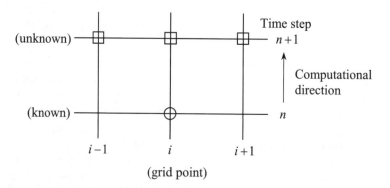

Figure 8.20 Schematic computational diagram of the implicit method.

Application of Eq. (8.52) to the grid points with unknown temperatures leads to a tridiagonal system of algebraic equations. The algebraic equations are coupled and the implicit method thus requires more computational time than that for the explicit method. However, the implicit method does not have restriction of the time step Δt used in the computation, i.e., the method will not produce a diverged solution. But a large time step Δt does produce inaccurate solution. Thus, analysts should understand the solution behaviors obtained from these methods. The example below shows the solution behavior obtained from the example of transient conduction heat transfer by using the implicit method.

Example 8.4 Solve the temperature response of Example 8.3 again but by using the implicit method. Use the time step of $\Delta t = 0.01$, which is larger than the critical time step of the problem which is $\Delta t = 0.005$. Compare the computed transient temperatures with the exact solution for $0 < t < 0.2$.

By using the time step of $\Delta t = 0.01$, Eq. (8.42) gives

$$\alpha = \frac{k}{\rho c} \frac{\Delta t}{(\Delta x)^2} = (1)\frac{0.01}{(0.1)^2} = 1 \tag{8.53}$$

The derived difference Eq. (8.52) with $\alpha = 1$ is

$$-T_{i-1}^{n+1} + 3T_i^{n+1} - T_{i+1}^{n+1} = T_i^n \tag{8.54}$$

Equation (8.54) is applied to all the grid points 2 to 10 for which their temperatures are unknowns. The application leads to a tridiagonal system of equations in the form,

$$\begin{bmatrix} 3 & -1 & & & & \\ -1 & 3 & -1 & & & \\ & -1 & 3 & -1 & & \\ & & \ddots & \ddots & \ddots & \\ & & & -1 & 3 & -1 \\ & & & & -1 & 3 \end{bmatrix} \begin{Bmatrix} T_2 \\ T_3 \\ T_4 \\ \vdots \\ T_9 \\ T_{10} \end{Bmatrix}^{n+1} = \begin{Bmatrix} T_2 \\ T_3 \\ T_4 \\ \vdots \\ T_9 \\ T_{10} \end{Bmatrix}^{n} \tag{8.55}$$

It is noted that the boundary conditions of zero temperatures at grid points 1 and 11 have been imposed into the above triagonal system of Eq. (8.55). Equation (8.55) is then solved for the grid point temperatures at time step $n+1$.

Figure 8.21 shows the computer program for solving the temperature response at grid points by using the implicit method. The program is similar to that of the explicit method except the set of coupled algebraic equations are solved from a subroutine for a tridiagonal system of equations.

The computed transient temperatures at grid points are compared with the exact solution as shown in Table 8.5. The table shows that the implicit method does not produce diverged solutions even though the time step Δt used in the computation is larger than the critical time step of the problem. However, the computed solutions at grid points contain large errors as compared to those obtained from the explicit method. This is mainly due to too large time step Δt is used in the analysis. It should be noted that both methods have the first-order of accuracy in time $O(\Delta t)$ and second-order of accuracy in space $O(\Delta x^2)$. In the next section, another implicit method that has the second-order of accuracy in both time and space is presented.

Fortran

```
      PROGRAM  PARAIMP
C.....A FINITE DIFFERENCE PROGRAM FOR SOLVING
C.....TRANSIENT TEMPERATURE DISTRIBUTION IN A
C.....ROD USING IMPLICIT METHOD
      DIMENSION   TEMP(11)
      DIMENSION  A(9), B(9), C(9), D(9), E(9)
C.....ASSIGN TIME STEP AND NO. OF TIME STEPS:
      DTIME = .01
      NSTEPS = 20
C.....SET UP INITIAL AND BOUNDARY CONDITIONS:
      X = 0.
      DX = .1
      PI = 4.*ATAN(1.)
      DO 10 I=1,11
      TEMP(I) = SIN(PI*X)
      X = X + DX
   10 CONTINUE
      ALPHA = DTIME/(DX*DX)
      COEF = 1. + 2.*ALPHA
      TIME = DTIME
      N = 9
      DO 100   ISTEP=1,NSTEPS
C.....FORM UP TRIDIAGONAL SYSTEM OF N EQUATIONS
C.....FOR INTERIOR GRIDS (HERE N=9):
      B(1) = COEF
      C(1) = -ALPHA
      D(1) = TEMP(2)
      DO 20 I=2,N-1
      A(I) = -ALPHA
      B(I) = COEF
      C(I) = -ALPHA
      D(I) = TEMP(I+1)
   20 CONTINUE
      A(N) = -ALPHA
      B(N) = COEF
      D(N) = TEMP(N+1)
C.....SOLVE SUCH TRIDIAGONAL SYSTEM OF N EQUATIONS
C.....FOR TEMPERATURES AT INTERIOR GRIDS, RETURN
C.....SOLUTION IN E( ):
      DO 30 I=2,N
      A(I) = A(I)/B(I-1)
      B(I) = B(I) - A(I)*C(I-1)
   30 CONTINUE
      DO 35 I=2,N
      D(I) = D(I) - A(I)*D(I-1)
   35 CONTINUE
      E(N) = D(N)/B(N)
      DO 40 I=N-1,1,-1
      E(I) = (D(I) - C(I)*E(I+1))/B(I)
   40 CONTINUE
      DO 50 I=2,10
      TEMP(I) = E(I-1)
   50 CONTINUE
C.....PRINT OUT AT EVERY 2 STEPS:
      IP = ISTEP - (ISTEP/2)*2
      IF(IP.EQ.0)
     *  WRITE(6,60)  TIME, (TEMP(I), I=1,11)
   60 FORMAT(F6.2, 1X, 11F6.4)
      TIME = TIME + DTIME
  100 CONTINUE
      STOP
      END
```

MATLAB

```
%  PROGRAM  PARAIMP
%  A FINITE DIFFERENCE PROGRAM FOR SOLVING
%  TRANSIENT TEMPERATURE DISTRIBUTION IN A
%  ROD USING IMPLICIT METHOD
%  ASSIGN TIME STEP AND NO. OF TIME STEPS:
dtime = 0.01;
nsteps = 20;
% SET UP INITIAL AND BOUNDARY CONDITIONS:
x = 0.;
dx = 0.1;
for i = 1:11
    temp(i) = sin(pi*x);
    x = x + dx;
end
alpha = dtime/(dx^2);
coef = 1 + 2*alpha;
time = dtime;
n = 9;
for istep = 1:nsteps
%  FORM UP TRIDIAGONAL SYSTEM OF N EQUATIONS
%  FOR INTERIOR GRIDS (HERE N=9):
    b(1) = coef;
    c(1) = -alpha;
    d(1) = temp(2);
    for i = 2:n-1
        a(i) = -alpha;
        b(i) =  coef;
        c(i) = -alpha;
        d(i) =  temp(i+1);
    end
    a(n) = -alpha;
    b(n) =  coef;
    d(n) =  temp(n+1);
%  SOLVE SUCH TRIDIAGONAL SYSTEM OF N EQUATIONS
%  FOR TEMPERATURES AT INTERIOR GRIDS, RETURN
%  SOLUTION IN E( ):
    for i = 2:n
        a(i) = a(i)/b(i-1);
        b(i) = b(i) - a(i)*c(i-1);
    end
    for i = 2:n
        d(i) = d(i) - a(i)*d(i-1);
    end
    e(n) = d(n)/b(n);
    for i = n-1:-1:1
        e(i) = (d(i) - c(i)*e(i+1))/b(i);
    end
    for i = 2:10
        temp(i) = e(i-1);
    end
%  PRINT OUT AT EVERY 1 STEPS:
    ip = istep - floor(istep/2)*2;
    if ip == 0.
        fprintf('\n%6.2f',time);
        for i = 1:11
            fprintf('   %6.4f',temp(i));
        end
    end
    time = time + dtime;
end
```

Figure 8.21 A computer program for determining transient temperature response of a bar by using the implicit method.

Table 8.5 Comparison between the computed temperatures at grid points using time step of $\Delta t = 0.01$ from the implicit method and the exact temperatures (values in brackets) at different times.

Time t	1	2	3	4	5	6
0.04	0.0000	0.2127	0.4046	0.5568	0.6546	0.6883
	(0.0000)	(0.2082)	(0.3961)	(0.5451)	(0.6408)	(0.6738)
0.08	0.0000	0.1464	0.2785	0.3833	0.4506	0.4737
	(0.0000)	(0.1403)	(0.2669)	(0.3673)	(0.4318)	(0.4540)
0.12	0.0000	0.1008	0.1917	0.2638	0.3101	0.3261
	(0.0000)	(0.0945)	(0.1798)	(0.2475)	(0.2910)	(0.3059)
0.16	0.0000	0.0694	0.1319	0.1816	0.2134	0.2244
	(0.0000)	(0.0637)	(0.1212)	(0.1668)	(0.1961)	(0.2062)
0.20	0.0000	0.0477	0.0908	0.1250	0.1469	0.1545
	(0.0000)	(0.0429)	(0.0816)	(0.1124)	(0.1321)	(0.1389)

8.3.4 Crank-Nicolson method

The Crank-Nicolson method provides solution with second-order solution accuracy in both space and time. The method starts from approximating the first-order time derivative term in the form of the difference between the solutions at time steps n and $n+1$ as,

$$\frac{\partial T}{\partial t} = \frac{T_i^{n+1} - T_i^n}{\Delta t} \tag{8.56}$$

The method approximates the second-order spatial derivative term using,

$$\frac{\partial^2 T}{\partial x^2} = \frac{1}{2}\left(\frac{T_{i+1}^{n+1} - 2T_i^{n+1} + T_{i-1}^{n+1}}{(\Delta x)^2} + \frac{T_{i+1}^n - 2T_i^n + T_{i-1}^n}{(\Delta x)^2}\right) \tag{8.57}$$

By substituting Eqs. (8.56) and (8.57) into the parabolic differential Eq. (8.37), the algebraic equation representing transient heat conduction in a bar is obtained as,

$$\frac{k}{2\rho c(\Delta x)^2}\left(T_{i+1}^{n+1} - 2T_i^{n+1} + T_{i-1}^{n+1} + T_{i+1}^n - 2T_i^n + T_{i-1}^n\right) = \frac{T_i^{n+1} - T_i^n}{\Delta t} \tag{8.58}$$

With the parameter α defined in Eq. (8.42), the above equations reduces to,

$$-\alpha T_{i-1}^{n+1} + 2(1+\alpha)T_i^{n+1} - \alpha T_{i+1}^{n+1} = \alpha T_{i-1}^n + 2(1-\alpha)T_i^n + \alpha T_{i+1}^n \tag{8.59}$$

The left-hand side of Eq. (8.59) consists of the unknown temperatures at grid points $i-1$. i and $i+1$ at the new time step $n+1$, while the right-hand side contains the known temperatures at time step n. The schematic computational diagrams corresponding to Eq. (8.59) is shown in Fig 8.22.

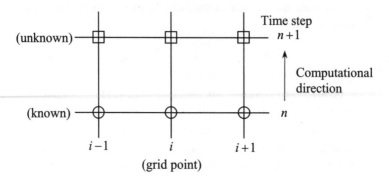

Figure 8.22 Schematic computational diagram of the Crank-Nicolson method.

Equation (8.59) is applied to all grid points for which the temperatures are unknown. The application leads to a tridiagonal set of equations that can be solved for the temperature solutions.

Example 8.5 Use the Crank-Nicolson method to determine the transient temperature response at grid points as described in Example 8.3. Employ the time step of $\Delta t = 0.02$ for $0 < t < 0.2$. Compare the computed grid point temperatures with the exact solution.

By using the time step $\Delta t = 0.02$, the parameter α from Eq. (8.42) is,

$$\alpha = \frac{k}{\rho c}\frac{\Delta t}{(\Delta x)^2} = (1)\frac{0.02}{(0.1)^2} = 2 \tag{8.60}$$

Then, Eq. (8.59) becomes,

$$-2T_{i-1}^{n+1} + 6T_i^{n+1} - 2T_{i+1}^{n+1} = 2T_{i-1}^n - 2T_i^n + 2T_{i+1}^n \tag{8.61}$$

Equation (8.61) is applied to grid points 2 to 10 for which their temperatures are unknown. The application leads to a tridiagonal set of equations similar to that shown in Eq. (8.55). The diagonal set of equations is then solved for the temperature solutions for the grid points at the new time step $n+1$.

The corresponding computer program of the Crank-Nicolson method for solving transient temperature response in the bar is shown in Fig. 8.23. The program is slightly more complicated than that for the implicit method. The computed temperature solutions at grid points, however, are more accurate. Table 8.6 shows the comparison of the computed temperature solutions obtained from the Crank-Nicolson method and the exact solutions for grid points and at 5 different times. Because the Crank-Nicolson method can provide higher solution accuracy as compared to the explicit and implicit methods, the method is widely used for analyzing transient problems in scientific and engineering applications.

Fortran

```
         PROGRAM  PARACN
C.....A FINITE DIFFERENCE PROGRAM FOR SOLVING
C.....TRANSIENT TEMPERATURE DISTRIBUTION IN A
C.....ROD USING CRANK-NICOLSON METHOD
         DIMENSION  TEMP(11)
         DIMENSION  A(9), B(9), C(9), D(9), E(9)
C.....ASSIGN TIME STEP AND NO. OF TIME STEPS:
         DTIME = .02
         NSTEPS = 10
C.....SET UP INITIAL AND BOUNDARY CONDITIONS:
         X = 0.
         DX = .1
         PI = 4.*ATAN(1.)
         DO 10  I=1,11
         TEMP(I) = SIN(PI*X)
         X = X + DX
  10     CONTINUE
         ALPHA = DTIME/(DX*DX)
         COEFP = 2.*(1. + ALPHA)
         COEFM = 2.*(1. - ALPHA)
         TIME = DTIME
         N = 9
         DO 100  ISTEP=1,NSTEPS
C.....FORM UP TRIDIAGONAL SYSTEM OF N EQUATIONS
C.....FOR INTERIOR GRIDS (HERE N=9):
         B(1) = COEFP
         C(1) = -ALPHA
         D(1) = COEFM*TEMP(2) + ALPHA*TEMP(3)
         DO 20  I=2,N-1
         A(I) = -ALPHA
         B(I) = COEFP
         C(I) = -ALPHA
         D(I) = ALPHA*TEMP(I) + COEFM*TEMP(I+1)
     1       + ALPHA*TEMP(I+2)
  20     CONTINUE
         A(N) = -ALPHA
         B(N) = COEFP
         D(N) = ALPHA*TEMP(N) + COEFM*TEMP(N+1)
C.....SOLVE SUCH TRIDIAGONAL SYSTEM OF N EQUATIONS
C.....FOR TEMPERATURES AT INTERIOR GRIDS, RETURN
C.....SOLUTION IN E( ):
         DO 30  I=2,N
         A(I) = A(I)/B(I-1)
         B(I) = B(I) - A(I)*C(I-1)
  30     CONTINUE
         DO 35  I=2,N
         D(I) = D(I) - A(I)*D(I-1)
  35     CONTINUE
         E(N) = D(N)/B(N)
         DO 40  I=N-1,1,-1
         E(I) = (D(I) - C(I)*E(I+1))/B(I)
  40     CONTINUE
         DO 50  I=2,10
         TEMP(I) = E(I-1)
  50     CONTINUE
C.....PRINT OUT AT EVERY 1 STEPS:
         IP = ISTEP - (ISTEP/1)*1
         IF(IP.EQ.0)
     *   WRITE(6,60)  TIME, (TEMP(I), I=1,11)
  60     FORMAT(F6.2, 1X, 11F6.4)
         TIME = TIME + DTIME
 100     CONTINUE
         STOP
         END
```

MATLAB

```
%    PROGRAM  PARACN
%    A FINITE DIFFERENCE PROGRAM FOR SOLVING
%    TRANSIENT TEMPERATURE DISTRIBUTION IN A
%    ROD USING CRANK-NICOLSON METHOD
%    ASSIGN TIME STEP AND NO. OF TIME STEPS:
dtime = 0.02;
nsteps = 10;
%    SET UP INITIAL AND BOUNDARY CONDITIONS:
x = 0;
dx = 0.1;
for i = 1:11
     temp(i) = sin(pi*x);
     x = x + dx;
end
alpha = dtime/(dx^2);
coefp = 2*(1 + alpha);
coefm = 2*(1 - alpha);
time = dtime;
n = 9;
for istep = 1:nsteps
%    FORM UP TRIDIAGONAL SYSTEM OF N EQUATIONS
%    FOR INTERIOR GRIDS (HERE N=9):
     b(1) = coefp;
     c(1) = -alpha;
     d(1) = coefm*temp(2) + alpha*temp(3);
     for i = 2:n-1
          a(i) = -alpha;
          b(i) = coefp;
          c(i) = -alpha;
          d(i) = alpha*temp(i) + coefm*temp(i+1) + ...
alpha*temp(i+2);
     end
     a(n) = -alpha;
     b(n) = coefp;
     d(n) = alpha*temp(n) + coefm*temp(n+1);
%    SOLVE SUCH TRIDIAGONAL SYSTEM OF N EQUATIONS
%    FOR TEMPERATURES AT INTERIOR GRIDS, RETURN
%    SOLUTION IN E( ):
     for i = 2:n
          a(i) = a(i)/b(i-1);
          b(i) = b(i) - a(i)*c(i-1);
     end
     for i = 2:n
          d(i) = d(i) - a(i)*d(i-1);
     end
     e(n) = d(n)/b(n);
     for i = n-1:-1:1
          e(i) = (d(i) - c(i)*e(i+1))/b(i);
     end
     for i = 2:10
          temp(i) = e(i-1);
     end
%    PRINT OUT AT EVERY 1 STEPS:
     ip = istep - (istep/1)*1;
     if ip == 0.
          fprintf('\n%6.2f  ',time);
          for i = 1:11
               fprintf('  %6.4f  ',temp(i));
          end
     end
     time = time + dtime;
end
```

Figure 8.23 A computer program for determining transient temperature response of a bar by using the Crank-Nicolson method.

Table 8.6 Comparison between the computed temperatures at grid points using time step of $\Delta t = 0.02$ from the Crank-Nicolson method and the exact temperatures (values in brackets) at different times.

time t	1	2	3	4	5	6
0.04	0.0000	0.2086	0.3968	0.5462	0.6421	0.6752
	(0.0000)	(0.2082)	(0.3961)	(0.5451)	(0.6408)	(0.6738)
0.08	0.0000	0.1409	0.2679	0.3688	0.4335	0.4558
	(0.0000)	(0.1403)	(0.2669)	(0.3673)	(0.4318)	(0.4540)
0.12	0.0000	0.0951	0.1809	0.2490	0.2927	0.3078
	(0.0000)	(0.0945)	(0.1798)	(0.2475)	(0.2910)	(0.3059)
0.16	0.0000	0.0642	0.1221	0.1681	0.1976	0.2078
	(0.0000)	(0.0637)	(0.1212)	(0.1668)	(0.1961)	(0.2062)
0.20	0.0000	0.0434	0.0825	0.1135	0.1334	0.1403
	(0.0000)	(0.0429)	(0.0816)	(0.1124)	(0.1321)	(0.1389)

In order to summarize the solution accuracy obtained from the explicit, implicit and Crank-Nicolson methods in Examples 8.3, 8.4 and 8.5, respectively, their computed temperature solutions at the bar center obtained from using different time steps Δt are compared. The solutions are also compared with the exact solutions at time $t = 0.2$ as shown in Table 8.7. The table shows that the explicit method yields large solution errors even though the method provides an advantage for solving each equation explicitly. The method produces diverged solutions if the time step Δt used in the computation is larger than the critical time step, e.g., when $\alpha = 1.0$ as shown in the Table. The implicit method does not produce any diverged solution but the computed solutions are not accurate. As shown in the Table, the Crank-Nicolson method provides accurate solutions for different time steps Δt. As the time step Δt decreases, the method yields a solution that converges to the value of 0.1412. It is noted that the exact temperature solution at the middle of the bar is 0.1389. The difference between the Crank-Nicolson and exact solutions is from the number of grid points used for modeling the bar. If the number of grid points is increased by reducing the interval length Δx, then the computed solutions from the Crank-Nicolson method will be close to the exact solution. Such numerical experiments for solution convergence of these methods to the exact solutions are left as exercises.

Table 8.7 Comparison of the computed temperatures at the bar center at time $t = 0.2$ from the three methods by using different time steps Δt. Note that the exact solution is 0.1389.

Time step Δt	α	Method Explicit	Implicit	Crank-Nicolson
0.0100	1.00	4.5065	0.1545	0.1410
0.0050	0.50	0.1344	0.1479	0.1411
0.0010	0.10	0.1398	0.1425	0.1412
0.0005	0.05	0.1405	0.1419	0.1412
0.0001	0.01	0.1410	0.1413	0.1412

8.4 Hyperbolic Equation

8.4.1 Differential equation

Hyperbolic equation is the partial differential equation that normally describes propagation behaviors from one point to another. Examples of such propagation behaviors are the oscillation of a string fixed at both ends, shock wave propagation that occurs in a tube from different air densities, traveling of a wave in a solid from an impact loading, etc. Many of these behaviors consist of sudden changes in the solution. Sudden changes in the solution require a fine mesh with small interval Δx to produce accurate solution. Such accurate solution are thus difficult to obtained, so many realistic problems that are governed by the hyperbolic equation are still under research development. In this section, a hyperbolic equation that arises from a simple phenomenon is considered. Figure 8.24 shows the oscillation behavior of a string with length L which is fixed at both ends.

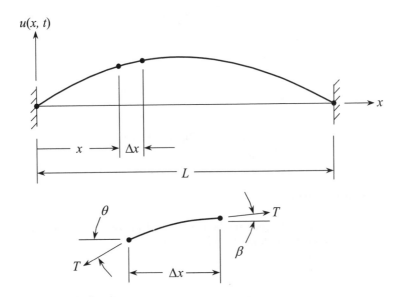

Figure 8.24 Oscillation behavior of a string and its portion under equilibrium.

The governing differential equation for the string deflection $u(x, t)$ that varies with the string x-coordinate and time t can be derived by considering the lower insert of Fig. 8.24. For a small oscillation, the deflecting angles θ and β are small while the string tension T can be assumed constant. By neglecting the weight of the string, the Newton's second law is applied to the small segment of the string to yield

$$T\sin\beta - T\sin\theta \;=\; \frac{w}{g}(\Delta x)\frac{\partial^2 u}{\partial t^2} \tag{8.62}$$

where w is the string weight per unit length and g is the gravitational acceleration constant. With the small angles θ and β, then $\sin\theta = \tan\theta$ and $\sin\beta = \tan\beta$, thus Eq. (8.62) becomes

$$T \tan \beta - T \tan \theta = \frac{w}{g}(\Delta x)\frac{\partial^2 u}{\partial t^2} \tag{8.63}$$

But *tan β* and *tan θ* represent the slopes at the ends of the segment, Eq. (8.63) can be written as

$$T\left(\frac{\partial u}{\partial x}\bigg|_{x+\Delta x} - \frac{\partial u}{\partial x}\bigg|_{x}\right) = \frac{w}{g}(\Delta x)\frac{\partial^2 u}{\partial t^2}$$

Or,

$$\frac{\frac{\partial u}{\partial x}\bigg|_{x+\Delta x} - \frac{\partial u}{\partial x}\bigg|_{x}}{\Delta x} = \frac{w}{Tg}\frac{\partial^2 u}{\partial t^2} \tag{8.64}$$

Then, by letting $\Delta x \to 0$, Eq. (8.64) becomes,

$$\frac{\partial^2 u}{\partial x^2} = \frac{w}{Tg}\frac{\partial^2 u}{\partial t^2} \tag{8.65}$$

Or,

$$\frac{\partial^2 u}{\partial t^2} = k^2 \frac{\partial^2 u}{\partial x^2} \tag{8.66}$$

where

$$k^2 = \frac{Tg}{w} \tag{8.67}$$

It is noted that Eq. (8.66) is in the form of Eq. (8.9), which is the standard form of the hyperbolic equation in one dimension where k^2 is always positive.

8.4.2 Computational procedures

In this section, the finite difference method is applied to solve the hyperbolic equation. The string is first divided into intervals, with the length of Δx each. These intervals are connected by grid points at $i-1$, i, $i+1$ as shown in Fig. 8.25.

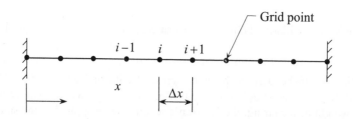

Figure 8.25 A string is divided into intervals connected at grid points.

At these grid points, their deflections u which depend on time t are to be determined. The central differencing is used to approximate the second-order time derivative term,

$$\frac{\partial^2 u}{\partial t^2} = \frac{u_i^{n+1} - 2u_i^n + u_i^{n-1}}{(\Delta t)^2} \tag{8.68}$$

where the subscript i represents the grid point number and the superscript n denotes the time step. The central differencing is also used to approximate the second-order spatial derivative term,

$$\frac{\partial^2 u}{\partial x^2} = \frac{u_{i+1}^n - 2u_i^n + u_{i-1}^n}{(\Delta x)^2} \tag{8.69}$$

By substituting Eqs. (8.68) and (8.69) into Eq. (8.66), the algebraic equation representing the hyperbolic equation is obtained,

$$\frac{u_i^{n+1} - 2u_i^n + u_i^{n-1}}{(\Delta t)^2} = k^2 \frac{u_{i+1}^n - 2u_i^n + u_{i-1}^n}{(\Delta x)^2}$$

After rearranging terms, the deflection for grid point i at time step $n+1$ can be determined from

$$u_i^{n+1} = 2u_i^n - u_i^{n-1} + C\left(u_{i+1}^n - 2u_i^n + u_{i-1}^n\right) \tag{8.70}$$

where

$$C = \frac{k^2(\Delta t)^2}{(\Delta x)^2} \tag{8.71}$$

The parameter C in Eq. (8.71) is called the Courant number. It has been found that accurate solution is obtained when the Courant number C is close to unity. If $C > 1$, the computed solution may diverge from the actual solution. If $C < 1$, the computed solution is less accurate. Study of different values of C that affect the solution accuracy is left as an exercise. With the Courant number as unity, appropriate time step Δt can be determined from the known grid point spacing Δx in the finite difference model.

Figure 8.26 shows the schematic computational diagram for the derived finite difference Eq. (8.70) representing the string oscillation. The unknown deflection for grid point i at time step $n+1$ is determined from the known deflections for grid points $i-1$, i, $i+1$ at time step n and the known deflection for grid point i at time step $n-1$.

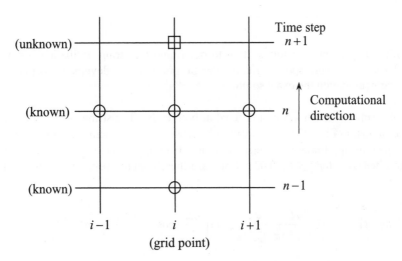

Figure 8.26 Schematic computational diagram for solving the hyperbolic equation.

At the beginning of the computational step 1 ($n = 0$), Eq. (8.70) needs the known solutions at steps 0 and -1. The deflection at step -1 can be determined from the initial condition of the problem. For example, if the initial velocity is given as zero, i.,e.,

$$\frac{\partial u}{\partial t}(x, t = 0) \; = \; 0 \tag{8.72}$$

Then, the central difference approximation for Eq. (8.72) is,

$$\frac{u_i^1 - u_i^{-1}}{2(\Delta t)} \; = \; 0 \tag{8.73}$$

Or,

$$u_i^{-1} \; = \; u_i^1 \tag{8.74}$$

By substituting Eq. (8.74) into Eq. (8.70) at $n = 0$,

$$u_i^1 \; = \; 2u_i^0 \; - \; u_i^1 \; + \; C\left(u_{i+1}^0 - 2u_i^0 + u_{i-1}^0\right)$$

Or,

$$u_i^1 \; = \; u_i^0 \; + \; \frac{C}{2}\left(u_{i+1}^0 - 2u_i^0 + u_{i-1}^0\right) \tag{8.75}$$

In summary, the deflections at grid points along the string are first determined at $n = 0$ by using Eq. (8.75). Equation (8.70) is then used for determining the grid point deflections at the later time steps. Since the Courant number is assigned as unity, the deflections at grid points can be determined by using Eqs. (8.75) and (8.70) as follows,

$$u_i^1 \; = \; \left(u_{i+1}^0 + u_{i-1}^0\right)/2 \qquad\qquad \text{when } n = 0 \tag{8.76a}$$

$$u_i^{n+1} \; = \; -u_i^{n-1} + u_{i+1}^n + u_{i-1}^n \qquad\qquad \text{when } n > 0 \tag{8.76b}$$

8.4.3 Example

An example is presented in this section to determine the string oscillation behavior from the derived Eqs. (8.76a-b). A corresponding computer program is also developed so that the computed solution can be compared with the exact solution.

Example 8.6 A string of length 1.5 unit is fixed at both ends. The string is released from the initial configuration as shown in Fig. 8.27. If the values $k = 100$, the grid spacing $\Delta x = 0.25$ and the Courant number $C = 1$, then the appropriate time step computed from Eq. (8.71) is $\Delta t = 0.0025$. Determine the string oscillation behavior for $0 \le t \le 0.03$. Compare the computed solutions at grid points with the exact solution of

$$u(x, t) \; = \; \frac{2aL^2}{b(L-b)\pi^2} \sum_{n=1}^{\infty} \frac{1}{n^2} \sin\left(\frac{n\pi}{L}\right) \cos\left(\frac{100n\pi t}{L}\right) \sin\left(\frac{n\pi x}{L}\right) \tag{8.77}$$

where a and b are the deflection and distance of the initial string configuration as shown in Fig. 8.27.

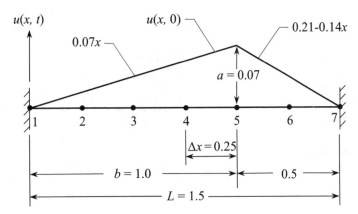

Figure 8.27 A string is divided into 6 intervals and 7 grid points with its initial configuration.

Equations (8.76a-b) are used to determine the string deflection for grid point i at any time step n. For example, the deflection for grid point 5 at the first time step ($t = 0.0025$) is determined from Eq. (8.76a) as,

$$u_5^1 = (u_6^0 + u_4^0)/2$$
$$= (0.0350 + 0.0525)/2 \qquad = 0.04375$$

Similarly, the deflections for grid points 4 and 6 at the first time step are

$$u_4^1 = 0.0525 \quad \text{and} \quad u_6^1 = 0.0350$$

Then, the deflection for grid point 5 at the second time step ($t = 0.0050$) are determined by using Eq. (8.76b) as,

$$u_5^2 = -u_5^0 + u_6^1 + u_4^1$$
$$= -0.07 + 0.0350 + 0.0525 \qquad = 0.0175$$

The computational procedure explained above is used to develop a computer program as shown in Fig. 8.28. The program starts from assigning the time step Δt and the total number of steps for determining the string oscillation behavior. After initial and boundary conditions are specified, the program employs Eq. (8.76a) to determine the grid point deflections at the first time step. Then, Eq. (8.76b) is used for determining the grid point deflections for the rest of the time steps.

Table 8.8 shows the comparison between the computed deflections at grid points and the exact solution. In this example, the computed solutions are exact and the string oscillation behaviors at different times are shown in Fig. 8.29. The figure shows the string deflection shapes for the first half cycle $0 \le t \le 0.015$. The string deflection shapes are reversed for the second half cycle $0.015 \le t \le 0.030$. The deflection shape becomes the initial shape again at time $t = 0.030$ after one cycle of the oscillation is complete.

Fortran

```
      PROGRAM  HYPER
C.....A FINITE DIFFERENCE PROGRAM FOR SOLVING
C.....VIBRATION IN STRING
      DIMENSION  UNP1(7), UN(7), UNM1(7)
C.....ASSIGN TIME STEP AND NO. OF TIME STEPS:
      DTIME = .0025
      NSTEPS = 12
C.....SET UP INITIAL AND BOUNDARY CONDITIONS:
      X = 0.
      DX = .25
      TIME = 0.
      DO 10  I=1,5
      UN(I) = .07*X
      X = X + DX
   10 CONTINUE
      X = 5.*DX
      DO 20  I=6,7
      UN(I) = .21 - .14*X
      X = X + DX
   20 CONTINUE
      WRITE(6,30)  TIME, (UN(I), I=1,7)
   30 FORMAT(F6.4, 1X, 7F8.5)
      TIME = TIME + DTIME
C.....COMPUTE DISPLACEMENTS AT FIRST TIME STEP:
      UNP1(1) = UN(1)
      DO 40  I=2,6
      UNP1(I) = 0.5*(UN(I+1) + UN(I-1))
   40 CONTINUE
      UNP1(7) = UN(7)
      WRITE(6,30)  TIME, (UNP1(I), I=1,7)
C.....COMPUTE DISPLACEMENTS AFTER FIRST TIME STEP:
      TIME = TIME + DTIME
      DO 100 ISTEP=2,NSTEPS
      DO 50  I=1,7
      UNM1(I) = UN(I)
      UN(I)   = UNP1(I)
   50 CONTINUE
      DO 60  I=2,6
      UNP1(I) = -UNM1(I) + UN(I+1) + UN(I-1)
   60 CONTINUE
      WRITE(6,30)  TIME, (UNP1(I), I=1,7)
      TIME = TIME + DTIME
  100 CONTINUE
      STOP
      END
```

MATLAB

```
%   PROGRAM  HYPER
%   A FINITE DIFFERENCE PROGRAM FOR SOLVING
%   VIBRATION IN STRING
%   ASSIGN TIME STEP AND NO. OF TIME STEPS:
dtime  = 0.0025;
nsteps = 12;
%   SET UP INITIAL AND BOUNDARY CONDITIONS:
x  = 0.0;
dx = 0.25;
time = 0.0;
for i = 1:5
    un(i) = 0.07*x;
    x = x + dx;
end
x = 5*dx;
for i = 6:7
    un(i) = 0.21 - 0.14*x;
    x = x + dx;
end
fprintf('\n%6.4f   ',time);
for i = 1:7
    fprintf(' %8.5f   ',un(i));
end
time = time + dtime;
%   COMPUTE DISPLACEMENTS AT FIRST TIME STEP:
unp1(1) = un(1);
for i = 2:6
    unp1(i) = 0.5*(un(i+1) + un(i-1));
end
unp1(7) = un(7);
fprintf('\n%6.4f   ',time);
for i = 1:7
    fprintf(' %8.5f   ',unp1(i));
end
%   COMPUTE DISPLACEMENTS AFTER FIRST TIME STEP:
time = time + dtime;
for istep = 2:nsteps
    for i = 1:7
        unm1(i) = un(i);
        un(i)   = unp1(i);
    end
    for i = 2:6
        unp1(i) = -unm1(i) + un(i+1) + un(i-1);
    end
    fprintf('\n%6.4f   ',time);
    for i = 1:7
        fprintf(' %8.5f   ',unp1(i));
    end
    time = time + dtime;
end
```

Figure 8.28 A computer program for determining deflection of a string fixed at both ends.

Table 8.8 Comparison between the computed deflections at grid points using time step of $\Delta t = 0.0025$ and the exact solutions (values in brackets) at different times. Note that the deflections at grid points 1 and 7 are zero.

	Deflections at grid points				
Time t	**2**	**3**	**4**	**5**	**6**
0.000	0.01750 (0.01750)	0.03500 (0.03500)	0.05250 (0.05250)	0.07000 (0.07000)	0.03500 (0.03500)
0.005	0.01750 (0.01750)	0.03500 (0.03500)	0.02625 (0.02625)	0.01750 (0.01750)	0.00875 (0.00875)
0.010	-0.00875 (-0.00875)	-0.01750 (-0.01750)	-0.02625 (-0.02625)	-0.03500 (-0.03500)	-0.01750 (-0.01750)
0.015	-0.03500 (-0.03500)	-0.07000 (-0.07000)	-0.05250 (-0.05250)	-0.03500 (-0.03500)	-0.01750 (-0.01750)
0.020	-0.00875 (-0.00875)	-0.01750 (-0.01750)	-0.02625 (-0.02625)	-0.03500 (-0.03500)	-0.01750 (-0.01750)
0.025	0.01750 (0.01750)	0.03500 (0.03500)	0.02625 (0.02625)	0.01750 (0.01750)	0.00875 (0.00875)
0.030	0.01750 (0.01750)	0.03500 (0.03500)	0.05250 (0.05250)	0.07000 (0.07000)	0.03500 (0.03500)

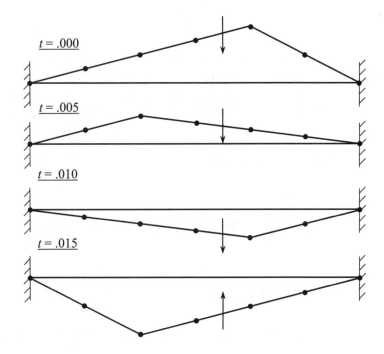

Figure 8.29 String deflection shapes at different times for $0 \le t \le 0.015$. The shapes are reversed for $0.015 \le t \le 0.030$ when the string deflects back to initial configuration at time 0.030.

8.5 Closure

The finite difference method for solving the partial differential equations is presented in this chapter. The partial differential equations are classified into three types of elliptic, parabolic and hyperbolic equations. Appropriate boundary and initial conditions for different types of the partial differential equations were described. Simple examples associated with these types of equations are explained with their physical meanings and solution behaviors.

The elliptic equation is the simplest one among the three types of partial differential equations. The finite difference method for solving the elliptic equation is easy to understand. Steady-state heat conduction in a plate is an example that can be solved conveniently by using the finite difference method. The method divides the plate into a number of intervals where the unknowns are at the grid points connecting these intervals. The elliptic partial differential equation is transformed into an algebraic equation which applies at the grid points. Such application leads to a set of equations that can be solved by the Gauss-Seidel iteration method for the solutions at these grid points.

In the process for solving the parabolic equation, the example of transient conduction heat transfer in a bar was used. Three types of the finite difference methods, which are the explicit, the implicit and the Crank-Nicolson methods, were explained. Examples were presented to highlight the solution accuracy obtained from these methods. Among these methods, the Crank-Nicolson method provides higher solution accuracy than the other two methods. The Crank-Nicolson method is thus widely used in commercial software for solving the parabolic equation.

The hyperbolic equation is the most difficult differential equation for obtaining solutions as compared to the elliptic and parabolic equations. This is mainly because its solutions normally change suddenly within a narrow region. A large number of short intervals is thus needed to produce an accurate solution. Such large number of intervals leads to many unknowns at grid points. Solving the hyperbolic equation thus requires a large computational effort in both the computer time and memory.

All computational techniques presented in this chapter show the simplicity of the finite difference method for solving the partial differential equations. The corresponding computer programs can also be developed straightforwardly. Because the method divides a computational domain into intervals with rectangular shapes, the method is not suitable for problems that have complex geometry. The problems with complex geometry can be handled conveniently by using the finite element method presented in the following chapter.

Exercises

1.   Identify each of the following partial differential equations whether it is the elliptic, parabolic or hyperbolic equation. Then, suggest appropriate computational procedure for solving each of them.

(a)   A differential equation for inviscid compressible flow in two-dimensional x-y coordinates is governed by

$$\left(1 - M^2\right)\frac{\partial^2 \phi}{\partial x^2} + \frac{\partial^2 \phi}{\partial y^2} = 0$$

where M is the Mach number which is less than unity for the subsonic flow and ϕ is the velocity potential.

(b) A one-dimensional unsteady Navier-Stokes equation is given by

$$\frac{\partial u}{\partial t} = v\frac{\partial^2 u}{\partial x^2}$$

where u is the velocity that varies with x-coordinate and time t, and v is the fluid viscosity.

(c) A differential equation which is in the form

$$\left(x^2 - 1\right)\frac{\partial^2 u}{\partial x^2} + 2y\frac{\partial^2 u}{\partial x\partial y} - \frac{\partial^2 u}{\partial y^2} = 0$$

where u is the dependent variable that varies with x and y.

(d) A differential equation that governs a steady-state viscous flow is in the form

$$\frac{\partial^2 \phi}{\partial x^2} + \frac{\partial^2 \phi}{\partial y^2} = -\frac{c}{v}$$

where ϕ is the flow velocity that varies with x- and y-coordinates, c is the pressure difference in the flow direction and v is the fluid viscosity.

(e) A differential equation representing transient heat conduction in a plate with surface convection is

$$-a\frac{\partial^2 T}{\partial x^2} + b\frac{\partial T}{\partial x} + cT + \frac{\partial T}{\partial t} = 0$$

where T is the temperature and a, b, c are positive integers.

2. Use the finite difference method to solve the Laplace's equation

$$\frac{\partial^2 u}{\partial x^2} + \frac{\partial^2 u}{\partial y^2} = 0 \qquad\qquad 0 \le x \le 1, 0 \le y \le 1$$

with the boundary conditions of

$u(x, 0) = 0$,	$u(x, 1) = x$,	$0 \le x \le 1$
$u(0, y) = 0$,	$u(1, y) = y$,	$0 \le y \le 1$

Divide the unit square region into 3×3 intervals in both x- and y-directions. Compare the computed solutions at grid points with the exact solution of $u(x, y) = xy$.

3. Solve Problem 2 again but by dividing the unit square region into 100×100 intervals and developing a computer program. Compare the computed solutions at grid points with the exact solution.

4. Modify the computer program in Fig. 8.10 to solve the plate temperature distribution by dividing the plate into 200 and 100 intervals in x- and y-directions, respectively. Compare the computed solutions at grid points with the exact solution in Eq. (8.25).

5. Because the temperature distribution in Problem 4 has symmetry over the plate, thus only the left half of the plate can be used for modeling. Divide the left half of the plate into 100×100 intervals in both x- and y-directions. Compare the computed solutions at grid points with those obtained from the full model.

6. Consider Fig. 8.13 for constructing an algebraic equation of grid points along the insulated boundary where

$$\frac{\partial T}{\partial x} = 0$$

Then, develop a new algebraic equation when the boundary has surface convection to a surrounding medium temperature T_∞ where

$$\frac{\partial T}{\partial x} = \frac{T - T_\infty}{\Delta x}$$

Explain how to implement the derived algebraic equation to the computer program in Fig. 8.10.

7. Develop the algebraic equation for solving the Poisson's equation

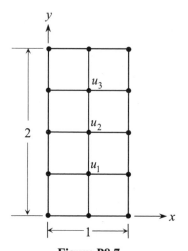

$$\frac{\partial^2 u}{\partial x^2} + \frac{\partial^2 u}{\partial y^2} = 4 \qquad 0 \le x \le 1, 0 \le y \le 2$$

with the boundary conditions of

$$u(x, 0) = x^2, u(x, 2) = (x-2)^2, \qquad 0 \le x \le 1$$

$$u(0, y) = y^2, u(1, y) = (y-1)^2, \qquad 0 \le y \le 2$$

Divide the domain into 2 and 4 intervals in x- and y-directions, respectively, as shown in Fig. P8.7. Compare the computed solutions at grid points with the exact solution of $u(x, y) = (x-y)^2$.

Figure P8.7.

8. Solve Problem 7 again but by developing a computer program. Divide the domain into 50 and 100 intervals in x- and y-directions, respectively. Plot to compare the computed temperature distribution with the exact solution.

9. Steady-state heat conduction in a plate with internal heat generation is governed by the partial differential equation

$$\frac{\partial^2 T}{\partial x^2} + \frac{\partial^2 T}{\partial y^2} = -\frac{Q}{k}$$

where T is the temperature that varies with the plate x-y coordinates, Q is the internal heat generation rate per unit volume and k is the thermal conductivity coefficient of the plate material. Use the central differencing approximation to derive the algebraic equation in a stencil form similar to Eq. (8.19). Then, explain how to apply the stencil form to solve for the plate temperature distribution.

10. Use the central difference method to derive the algebraic equation corresponding to the Poisson's equation

$$\frac{\partial^2 u}{\partial x^2} + \frac{\partial^2 u}{\partial y^2} = \left(x^2 + y^2\right)e^{xy} \qquad\qquad 0 \le x \le 2, \, 0 \le y \le 1$$

with the boundary conditions of

$$u(x, 0) = 1, u(x, 1) = e^x, \qquad 0 \le x \le 2$$
$$u(0, y) = 1, u(2, y) = e^{2y}, \qquad 0 \le y \le 1$$

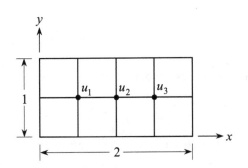

Figure P8.10.

Divide the domain into 4×2 intervals in x- and y-directions, respectively, as shown in Fig. P8.10. Compare the comput-ed solutions with the exact solution of $u(x, y) = e^{xy}$.

11. Solve Problem 10 again but by developing a computer program. Then use the program to determine the unknowns at grid points by dividing the domain into

    (a)   10 and 5 intervals in x- and y-directions, respectively,
    (b)   20 and 10 intervals in x- and y-directions, respectively,
    (c)   40 and 20 intervals in x- and y-directions, respectively,

    Plot to compare the computed solutions obtained from each model with the exact solution. Then, give comments on the convergence of the computed solutions to the exact solution as the model is refined by using more intervals.

12. Use the central difference method to derive the algebraic equation corresponding to the Poisson's equation in Problem 10 but with different interval lengths of Δx and Δy. Then, implement the derived equation on a computer program to solve the same problem by using

    (a) $\Delta x = 0.2$ and $\Delta y = 0.1$
    (b) $\Delta x = 0.1$ and $\Delta y = 0.05$

Plot and compare the computed solutions at grid points with the exact solution. Give comments on the benefit of using different interval lengths Δx and Δy. Provide examples that can provide the benefit of using different interval lengths.

13. Use the central difference method to derive the algebraic equation corresponding to the Laplace's equation in the form

$$\pi^2 \frac{\partial^2 u}{\partial x^2} + \frac{\partial^2 u}{\partial y^2} = 0 \qquad\qquad 0 \le x \le 1, 0 \le y \le 1$$

with the boundary conditions of

$$u(x, 0) = 0, \qquad u(x, 1) = 0, \qquad\qquad 0 \le x \le 1$$

$$u(0, y) = \sin \pi y, \qquad u(1, y) = e \sin \pi y, \qquad\qquad 0 \le y \le 1$$

Develop a computer program for solving the problem by dividing the unit square domain into different intervals as follows

(a) $\Delta x = \Delta y = 0.2$

(b) $\Delta x = \Delta y = 0.1$

(c) $\Delta x = \Delta y = 0.05$

Then, plot to compare the computed solutions with the exact solution of

$$u(x, y) = y^3 \sin x$$

14. Use the central difference method to derive the algebraic equation corresponding to the Poisson's equation in the form

$$\frac{\partial^2 u}{\partial x^2} + \frac{\partial^2 u}{\partial y^2} = 2x\left(x^3 - 6xy + 6xy^2 - 1\right) \qquad\qquad 0 \le x \le 1, 0 \le y \le 1$$

with the boundary conditions of $u(x, y) = 0$ along the four edges. Divide the unit square domain into 3 equal intervals in both x- and y-directions. Compare the computed solutions at grid points with the exact solution of

$$u(x, y) = \left(x - x^4\right)\left(y - y^2\right)$$

15. Use the central difference method to derive the algebraic equation corresponding to the Poisson's equation in the form

$$\frac{\partial^2 u}{\partial x^2} + \frac{\partial^2 u}{\partial y^2} = -2\left(x^2 + y^2\right) \qquad\qquad 0 \le x \le 1, 0 \le y \le 1$$

with the boundary conditions of

$$u(x, 0) = 1 - x^2, \qquad u(x, 1) = 2\left(1 - x^2\right), \qquad\qquad 0 \le x \le 1$$

$$u(0, y) = 1 + y^2, \qquad u(1, y) = 0, \qquad\qquad 0 \le y \le 1$$

Divide the unit square domain into 3 equal intervals in both x- and y-directions. Compare the computed solutions at grid points with the exact solution of

$$u(x, y) = (1-x^2)(1+y^2)$$

16. Apply the central difference method to derive the algebraic equation for solving the one-dimensional boundary value problem governed by the ordinary differential equation

$$\frac{d^2u}{dx^2} + u = 0 \qquad\qquad 0 \le x \le \pi/2$$

with the boundary conditions of $u(0) = u(\pi/2) = 1$. Determine the solutions at grid points by using two models with the interval lengths of: (a) $\Delta x = \pi/4$ and (b) $\Delta x = \pi/10$. Compare the solutions at grid points with the exact solution of

$$u(x) = \sin x + \cos x$$

17. Apply the central difference method to derive the algebraic equation for solving the one-dimensional boundary value problem governed by the ordinary differential equation

$$\frac{d^2u}{dx^2} + u = \sin x \qquad\qquad 0 \le x \le \pi/2$$

with the boundary conditions of $u(0) = u(\pi/2) = 0$. Determine the solutions at grid points by using two models with the interval lengths of: (a) $\Delta x = \pi/4$ and (b) $\Delta x = \pi/10$. Compare the solutions at grid points with the exact solution of

$$u(x) = -\frac{x}{2}\cos x$$

18. The Helmholz equation is in the form

$$\frac{\partial^2 u}{\partial x^2} + \frac{\partial^2 u}{\partial y^2} + a(x, y)u = f(x, y)$$

If $a = -2$ and the function

$$f(x, y) = xy\left[(x^2-7)(1-y^2)+(1-x^2)(y^2-7)\right]$$

in the domain of $0 \le x \le 1$ and $0 \le y \le 1$ with $u = 0$ along the four boundaries, then the exact solution is

$$u(x, y) = (x-x^3)(y-y^3)$$

Use the central difference method to derive the corresponding algebraic equation. Develop a computer program to solve for the solutions at grid points by using the intervals with

(a) $\Delta x = \Delta y = 0.25$

(b) $\Delta x = \Delta y = 0.1$

Plot to compare the computed solutions at grid points with the exact solution above.

19.   A square insulator with the inner and outer surface temperatures of 100 and 0 degrees, respectively, is shown in Fig. P8.19. Due to symmetry of the temperature distri-bution, only the lower-left quarter of the insulator can be used for modeling. The steady-state temperature distribution can be determined by solving the Laplace's Eq. (8.15). Apply the algebraic equation corresponding to the Laplace's equation to the grid points of the model as shown in the figure. Such application leads to a set of equations that can be solved for the temperatures at grid points. Plot the computed temperature distribution by using contour lines.

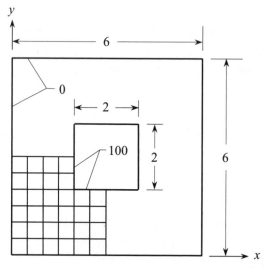

Figure P8.19.

20.   Solve the parabolic equation in Example 8.3 again but by using only the left half of the bar because the temperature distribution is symmetric. Derive appropriate algebraic equation corresponding to the parabolic equation by using the explicit method. Then, develop a computer program to solve for the transient temperature response along the bar and compare the computed temperature at the 6 grid points with Table 8.3.

21.   Employ the computer program developed in Problem 20 to solve the transient temperature response in the bar by using
        (a)   26 grid points ($\Delta x = 0.02$)
        (b)   51 grid points ($\Delta x = 0.01$)
        (c)   101 grid points ($\Delta x = 0.005$)
Compare the solution obtained from each case with the exact solution. Give comments on the solution convergence as the model is refined.

22.   Use the Crank-Nicolson method to solve the problem in Example 8.5 again but by modeling only the left half of the bar because its temperature distribution is symmetric. Develop a corresponding computer program to solve for the transient temperature response at the 6 grid points. Compare the computed solutions with those shown in Table 8.6.

23. The computer programs in Figs. 8.18, 8.21 and 8.23 are for the analysis of transient heat conduction in a bar by using the explicit, implicit and Crank-Nicolson methods, respectively. Employ these computer programs to study the convergence rates of the solutions obtained from the three methods by using the four models with the intervals of:

(a) $\Delta x = 0.02$ (c) $\Delta x = 0.005$

(b) $\Delta x = 0.01$ (d) $\Delta x = 0.001$

Note that the exact temperature at the middle of the bar at time $t = 0.20$ is 0.1389.

24. Use the explicit method to solve the parabolic equation

$$\frac{\partial u}{\partial t} - \frac{1}{\pi^2} \frac{\partial^2 u}{\partial x^2} = 0 \qquad\qquad 0 \le x \le 1, t > 0$$

with the boundary conditions of

$$u(0, t) = u(1, t) = 0 \qquad\qquad t > 0$$

and the initial condition of

$$u(x, 0) = \cos \pi (x - 0.5) \qquad\qquad 0 \le x \le 1$$

Divide the domain into 4 equal intervals with 5 grid points. Use the time step $\Delta t = 0.2$ for $0 \le t \le 1$. Plot to compare the computed solution with the exact solution of $u(x, t) = e^{-t} \cos \pi (x - 0.5)$.

25. Solve Problem 24 again but by using a developed computer program. Divide the domain into: (a) 10 intervals, (b) 20 intervals and (c) 50 intervals. In each case, use the time step Δt about half of its critical time step. Plot to compare the solutions obtained with the exact solution. Give comments on the solution convergence to the exact solution.

26. Use the implicit method as explained in section 8.3.3 to solve the parabolic equation in Problem 24. Divide the domain into 4 equal intervals with 5 grid points. Use the time step of $\Delta t = 0.2$ for $0 \le t \le 1$. Set up a table to compare the computed solutions at grid points with the exact solution.

27. Use the algebraic equation derived in Problem 26 to develop a computer program. Then, employ the program to solve for the transient temperature response by dividing the domain into: (a) 10 intervals, (b) 20 intervals and (c) 50 intervals. Use appropriate time step Δt for the computation in each case. Give comments on the solution behaviors and their convergence to the exact solution.

28. Use the Crank-Nicolson method to establish the algebraic equation corresponding to the parabolic equation in Problem 24. Then, solve the problem by dividing the domain into 4 equal intervals with 5 grid points. Use the time step $\Delta t = 0.2$ for $0 \le t \le 1$. Compare the computed solutions at grid points with the exact solution.

29.   Use the algebraic equation derived in Problem 28 to develop a computer program. Then, employ the program to solve the problem by dividing the domain into: (a) 10 intervals, (b) 20 intervals and (c) 50 intervals. Use appropriate time step Δt for each case and compared the computed solutions with the exact solution.

30.   Derive the algebraic equation for solving the parabolic equation in the form

$$\frac{\partial u}{\partial t} - \frac{\partial^2 u}{\partial x^2} = 2 \qquad\qquad 0 \le x \le 1, t > 0$$

with the boundary conditions of

$$u(0, t) = u(x, t) = 0 \qquad\qquad t > 0$$

and the initial condition of

$$u(x, 0) = \sin \pi x + x(1 - x) \qquad\qquad 0 \le x \le 1$$

by using: (a) the explicit method and (b) the Crank-Nicolson method. Divide the domain into 4 equal intervals with 5 grid points and use the time step $\Delta t = 0.02$ for $0 \le t \le 0.1$. Set up a table to compare the computed solutions with the exact solution of

$$u(x, t) = e^{-\pi^2 t} \sin \pi x + x(1 - x)$$

31.   Use the algebraic equation derived in Problem 30 to develop a computer program. Employ the program to solve the problem by using: (a) the explicit method and (b) the Crank-Nicolson method. Divide the domain into 20 and 50 equal intervals and use the time step Δt about a half of the critical time for each case. Explain advantages and disadvantages of the two methods for solving the problem

32.   Develop the algebraic equation for solving the parabolic equation in the form

$$\frac{\partial u}{\partial t} = \frac{\partial^2 u}{\partial x^2} \qquad\qquad 0 \le x \le 1, t > 0$$

with the boundary conditions of

$$u(0, t) = 1, \qquad u(1, t) = 0, \qquad\qquad t > 0$$

and the initial condition of

$$u(x, 0) = 1 - x - \frac{1}{\pi} \sin(2\pi x), \qquad\qquad 0 \le x \le 1$$

by using: (a) the explicit method and (b) the Crank-Nicolson method. Divide the domain into 4 equal intervals with 5 grid points and use the time step $\Delta t = 0.01$ for $0 \le t \le 0.1$. Plot to compare the computed solutions with the exact solution of

$$u(x, t) = 1 - x - \frac{1}{\pi} e^{-4\pi^2 t} \sin(2\pi x)$$

33. Solve Problem 32 again but by developing a computer program. Divide the domain into 20 equal intervals and use appropriate time step Δt. Plot to compare the computed solutions with the exact solution.

34. Develop the algebraic equation for solving the parabolic equation

$$\frac{\partial u}{\partial t} = \frac{\partial^2 u}{\partial x^2} \qquad\qquad 0 \le x \le \pi, \ t > 0$$

with the boundary conditions of

$$\frac{\partial u}{\partial x}(0, t) = \frac{\partial u}{\partial x}(\pi, t) = 0 \qquad\qquad t > 0$$

and the initial condition of

$$u(x, 0) = \cos x \qquad\qquad 0 \le x \le \pi$$

by using: (a) the explicit method and (b) the Crank-Nicolson method. Divide the domain into 4 equal intervals with 5 grid points and use the time step $\Delta t = 0.05$ for $0 \le t \le 0.2$. Plot to compare the computed solutions with the exact solution of

$$u(x, t) = e^{-t} \cos x$$

35. Solve Problem 34 again but by developing a computer program. Divide the domain into 30 equal intervals and use appropriate time step Δt. Plot to compare the computed solutions with the exact solution.

36. Solve the hyperbolic equation representing the vibration of the string again by using the Courant numbers $C = 0.6$, 0.8 and 1.1, respectively. Then, compare the computed solutions with the exact solution in Table 8.8.

37. Equation (8.75) that was derived for solving the string deflection after releasing it from the initial configuration is based on the condition of zero velocity. Re-derive the equation if the initial velocity is

$$\frac{\partial u}{\partial t}(x, t = 0) = g(x)$$

where $g(x)$ is any given function.

38. Modify the computer program in Fig. 8.28 for the analysis of string vibration by dividing the string into 15 intervals with $\Delta x = 0.1$. Compare the computed solutions with the exact solution in Eq. (8.77). Plot the string deflections similar to those shown in Fig. 8.29 for $0 \le t \le 0.03$.

39. Derive the algebraic equation corresponding to the hyperbolic equation

$$\frac{\partial^2 u}{\partial t^2} = \frac{\partial^2 u}{\partial x^2} \qquad\qquad 0 \le x \le 1, \ t \ge 0$$

with the boundary conditions of

$$u(0, t) \ = \ u(1, t) \ = \ 0 \qquad\qquad t \geq 0$$

and the initial conditions of

$$u(x, 0) \ = \ \sin \pi x \qquad\qquad 0 \leq x \leq 1$$

$$\frac{\partial u}{\partial t}(x, 0) \ = \ 0 \qquad\qquad 0 \leq x \leq 1$$

Divide the domain into 4 equal intervals with 5 grid points and use the time step $\Delta t = 0.25$ for $0 \leq t \leq 1$. Set up a table to compare the computed solutions with the exact solution of

$$u(x, t) \ = \ \sin \pi x \cos \pi t$$

40.   Use the algebraic equation derived in Problem 39 to develop a computer program. Then, employ the program to solve the problem by dividing the domain into: (a) 10 intervals, (b) 20 intervals and (c) 50 intervals. In each case, use appropriate time step Δt for determining the solution. Plot to compare the solution obtained from each model with the exact solution. Study the solutions and give comments on the solution convergence to the exact solution.

41.   Derive the algebraic equation corresponding to the hyperbolic equation

$$\frac{\partial^2 u}{\partial t^2} \ = \ \frac{\partial^2 u}{\partial x^2} + 2e^{-t}\sin x \qquad\qquad 0 \leq x \leq \pi , \, t \geq 0$$

with the boundary conditions of

$$u(0, t) \ = \ u(\pi, t) \ = \ 0 \qquad\qquad t \geq 0$$

and the initial conditions of

$$u(x, 0) \ = \ \sin x \qquad\qquad 0 \leq x \leq \pi$$

$$\frac{\partial u}{\partial t}(x, 0) \ = \ -\sin x \qquad\qquad 0 \leq x \leq \pi$$

Divide the domain into 5 equal intervals with 6 grid points and use the time step $\Delta t = 0.2$ for $0 \leq t \leq 1$. Set up a table to compare the computed solutions with the exact solution of

$$u(x, t) \ = \ e^{-t}\sin x$$

42.   Use the algebraic equation derived in Problem 41 to develop a computer program. Then, employ the program to solve the problem by dividing the domain into: (a) 10 intervals and (b) 20 intervals by using the time step of $\Delta t = 0.1$ and 0.05, respectively. Plot to compare the solution obtained from each model with the exact solution. Compute and tabulate the solution errors at grid points for each case.

43.   Derive the algebraic equation corresponding to the hyperbolic equation

$$\frac{\partial^2 u}{\partial t^2} \ = \ \frac{\partial^2 u}{\partial x^2} \qquad\qquad 0 \leq x \leq 1, \, t \geq 0$$

with the boundary conditions of

$$u(0, t) \ = \ u(1, t) \ = \ 0 \qquad\qquad t \geq 0$$

and the initial conditions of

$$u(x, 0) \ = \ \sin 2\pi x \qquad\qquad 0 \leq x \leq 1$$

$$\frac{\partial u}{\partial t}(x, 0) \ = \ 2\pi \sin 2\pi x \qquad\qquad 0 \leq x \leq 1$$

Divide the domain into 4 equal intervals with 5 grid points and use the time step $\Delta t = 0.25$ for $0 \leq t \leq 1$. Set up a table to compare the computed solutions with the exact solution of

$$u(x, t) \ = \ \sin 2\pi x \left(\sin 2\pi t + \cos 2\pi t \right)$$

44. Use the algebraic equation derived in Problem 43 to develop a computer program. Then, employ the program to solve the problem by dividing the domain into: (a) 10 intervals and (b) 20 intervals. Use appropriate time step Δt for each case. Compare the computed solution obtained from each model with the exact solution. Also plot to compare the solutions that vary with x-coordinate and time t with the exact solution.

Chapter
9

Finite Element Method

9.1 Introduction

The ability to solve engineering and scientific problems plays an important role in shaping and improving our everyday life. Most phenomena that occur around us can be explained through the laws of physics in the form of differential and integral equations. For example, determination of the temperature distribution in a car engine starts with the differential equation that describes the conservation of energy for heat transfer. Stresses in airplane wings could be determined using other differential equations that explain the equilibrium of the structure under varying pressure during the flight. Even the wind direction and speed in a typhoon may be predicted from the differential equations that describe the conservation laws for fluid flow.

Differential equations that explain behaviors for various physical phenomena are normally not difficult to derive. However, derivation of their exact solutions is usually difficult, and sometimes even impossible. Therefore, several methods have been developed for obtaining approximate solutions. One of the most commonly used methods since the last few decades is the finite difference method as explained in the preceding chapter.

The main concept of the finite difference method is to transform the differential equations into a set of algebraic equations that use addition, subtraction, multiplication, and division operations. The advantage of the finite difference method is the ease in learning and understanding. Its solution can be obtained conveniently by developing computer program for performing numerical calculations. However, the drawback of the finite difference method is the difficulty in applying arbitrary boundary conditions, as well as modeling complex geometry.

Figure 9.1 illustrates a typical geometry of an aluminum plate commonly used for supporting equipments. This aluminum plate has both straight and curve edges with three circular holes inside. The finite difference method may be used to analyze the stress distribution in this aluminum plate under an applied load. The aluminum plate is first discretized into small square grids as shown in Fig. 9.2. These squares are connected at the grid points which are located at the square corners. The size of the problem or the number of unknowns depends on the number of these grid points.

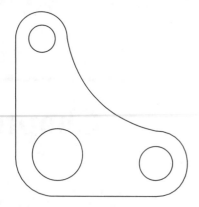

Figure 9.1 An aluminum plate
with arbitrary geometry.

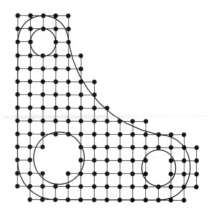

Figure 9.2 A finite difference model
for the aluminum plate.

Figure 9.2 shows that the grids of the finite difference model cannot represent the original configuration accurately. To improve the finite difference model for better representation of the plate geometry, smaller squares are needed. However, the use of smaller squares also increases the number of grid points and consequently increases the number of finite difference equations. Larger number of equations thus requires more computational time and computer memory in the solving process. Such inconvenience and difficulty led to the development of another approximate solution technique, the so called finite element method, or FEM. The method is suitable for solving problems with complex geometry because it can represent the arbitrary geometry more closely. The concept of the finite element method is similar to the finite difference method in that both methods discretize the original geometry into small pieces or elements. For the finite element method, these small pieces can be in the shapes of triangle, rectangle, or combination of shapes as shown in Fig. 9.3.

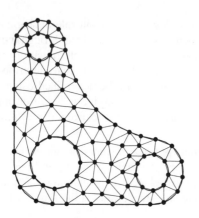

Figure 9.3 A finite element
model for the
aluminum plate.

The advantage of the finite element method is clearly shown in Fig. 9.3 as compared to the finite different method in Fig. 9.2 because the finite element method can closely represent the original geometry of the aluminum piece. Such better representation of the original geometry will lead to the accuracy improvement of the computed solution.

9.2 What is the Finite Element Method?

9.2.1 Basic understanding

Most of engineering problems are normally governed by the differential equations and boundary conditions. The derived exact solution is always valid throughout the domain and provides solutions at an infinite number of locations. As mentioned earlier, exact solutions are very difficult to derive, especially for problems with complex geometry. Approximate or numerical solutions that can approximately satisfy the same set of the differential equations and boundary conditions are alternatively considered. Instead of providing the infinite number of solutions as in the exact solutions, the finite element concept is to determine solutions only at some finite locations. This is done by first discretizing the geometry of the model into a number of finite elements as previously shown in Fig. 9.3. These elements are connected at grid points or nodes at which the unknowns are to be determined.

The key idea of the finite element method is to transform the differential equations into a set of algebraic equations for each element. The finite element equations from all elements are then combined together to form a large set of simultaneous equations. The boundary conditions of the problem are applied prior to solving for the unknowns at all nodes.

From this brief explanation, the accuracy of the approximate solution thus depends on the size and number of elements in the analysis. The accuracy of the solution also depends on the element interpolation functions that will be explained in details later. Element interpolation functions could be in linear, quadratic, or cubic forms, etc., depending on the type of the selected elements. For example, if the temperatures at the three corner nodes of a triangular element with linear interpolation functions are 30, 40 and 50 °C respectively, the temperature inside this element will distribute as a flat plane with values ranging from 30 to 50 °C.

9.2.2 The Differences between Finite Difference Method and Finite Element Method

From the example of the finite difference method in the preceding chapter, it can be seen that the **finite difference method** is a numerical technique that finds an approximate solution of a given problem. The concept of this method is to replace the derivatives that appear in the differential equation by an algebraic approximation. The unknowns of the approximate algebraic equations are the dependent variables at the grid points.

The **Finite element method** is also a numerical technique that is used to find an approximate solution of a given problem. The geometry is discretized into small segments called elements. These elements are connected at the nodes where the unknowns of the problem are to be determined.

To understand the finite element method more clearly, the following section describes the general procedures of the method for analyzing problems. Examples will be presented for solving one- and two-dimensional problems in the later sections.

9.3 General Procedures of the Finite Element Method

Analyzing procedure of a given problem by the finite element method normally consists of six main steps as follows.

Step 1 *Discretization of the computational domain into a number of elements.* The domain could represent the elasticity, heat transfer, or fluid flow problems, etc., as shown in Fig. 9.4.

Step 2 *Selecting element types and their interpolation functions.* For example, a typical triangular element in Fig. 9.4 contains three nodes. This typical element is shown again in Fig. 9.5 with the node numbers 1, 2 and 3. At these three nodes, the nodal unknowns are denoted by ϕ_1, ϕ_2 and ϕ_3 respectively. The nodal unknowns can be the nodal displacements in an elasticity problem, the nodal temperatures in a heat transfer problem, or the nodal flow velocities in a fluid problem. The distribution of unknowns over the element is written in the form of the element interpolation functions and the nodal unknowns as follows

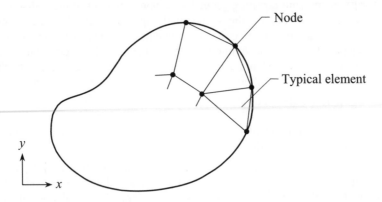

Figure 9.4 Discretization of the computational domain
into a number of elements.

$$\phi(x, y) \ = \ N_1(x, y)\phi_1 + N_2(x, y)\phi_2 + N_3(x, y)\phi_3 \tag{9.1}$$

where $N_i(x, y)$, $i = 1, 2, 3$ are the element interpolation functions

Equation (9.1) can be written in matrix form as

$$\phi(x, y) \ = \ \lfloor N_1 \quad N_2 \quad N_3 \rfloor \begin{Bmatrix} \phi_1 \\ \phi_2 \\ \phi_3 \end{Bmatrix} \ = \ \underset{(1\times3)\ (3\times1)}{\lfloor N \rfloor \{\phi\}} \tag{9.2}$$

where $\lfloor N \rfloor$ is the element interpolation function matrix and $\{\phi\}$ is the vector of the element nodal unknowns. In this textbook, the symbol $\lfloor \ \rfloor$ denotes a row matrix and $\{ \ \}$ denotes a column matrix which is normally called a vector. More information about the matrix, its properties and manipulation can be found in Appendix A.

Step 3 *Deriving the finite element equations.* For example, the finite element equations for the triangular element as shown in Fig. 9.5 are in the form

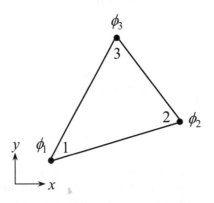

Figure 9.5 Typical triangular element with unknowns at nodes.

$$\begin{bmatrix} k_{11} & k_{12} & k_{13} \\ k_{21} & k_{22} & k_{23} \\ k_{31} & k_{32} & k_{33} \end{bmatrix}_e \begin{Bmatrix} \phi_1 \\ \phi_2 \\ \phi_3 \end{Bmatrix}_e = \begin{Bmatrix} F_1 \\ F_2 \\ F_3 \end{Bmatrix}_e \tag{9.3}$$

Or, in short

$$[K]_e \{\phi\}_e = \{F\}_e \tag{9.4}$$

This third step is the most important step in the finite element analysis. The finite element equations in the form of Eq. (9.4) are derived by using the method of weighted residuals. The method is considered as one of the most general methods for deriving the finite element equations.

Step 4 *Assembling the element equations to form a system of equations.* The system of equations is in the form of simultaneous algebraic equations

$$\sum (element\ equations) \quad \Longrightarrow \quad [K]_{sys} \{\phi\}_{sys} = \{F\}_{sys} \tag{9.5}$$

Step 5 *Applying the boundary conditions and solving for the nodal solutions.* Boundary conditions of the problem are then applied to the system of equations, Eq. (9.5), before solving for the nodal unknowns in $\{\phi\}_{sys}$. The nodal unknowns could be displacements, temperatures, or fluid velocities for solid mechanics, heat transfer, or fluid flow problems, respectively.

Step 6 *Computing other quantities if needed.* After the nodal unknowns are determined, other quantities of interest can be further computed. For examples, the stresses, heat fluxes, and the flow rates can be computed from the now-known displacements, temperatures, and the flow velocities, respectively.

From the 6 steps described above, it is clear that there is a routine procedure in the finite element method. The most important step is the derivation of the finite element equations in step 3. The derivation of the finite element equations for one- and two-dimensional problems will be explained in details in the later sections. Several examples will be presented to ensure clear understanding of the method.

9.4 One-dimensional Problems

9.4.1 Differential equation

In this section, the finite element method is presented for analyzing one-dimensional problems. The problems are governed by the Poisson's equation,

$$\frac{d^2u}{dx^2} = f(x) \tag{9.6}$$

where u is the dependent variable that varies with x-coordinate and $f(x)$ is a given function. Such differential equation may represent the equilibrium condition at any point in a bar due to its own weight, or the conservation of energy at any location along a rod subjected to an internal heat generation. For the rod subjected to an internal heat generation Q as shown in Fig. 9.6, the detailed governing differential equation is

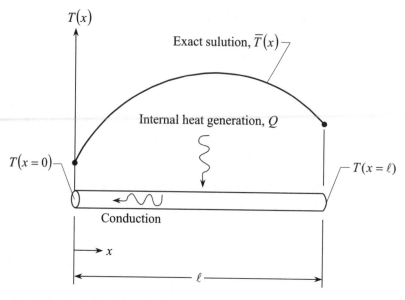

Figure 9.6 Conduction with internal heat generation in a rod.

$$kA\frac{d^2\overline{T}}{dx^2} = -QA \qquad (9.7)$$

where k is the thermal conductivity coefficient, A is the rod cross-sectional area and $\overline{T}$ is the exact temperature distribution that varies along the x-direction of the rod.

The governing differential Eq. (9.7) is may be solved together with the Dirichlet boundary conditions at both ends of the rod as

$$T(x=0) = T_0 \qquad (9.8a)$$

$$T(x=\ell) = T_\ell \qquad (9.8b)$$

Boundary condition at the ends of the rod may be the Neumann type. For example, if a heat flux q is specified at $x = \ell$, then according to the Fourier's law,

$$-kA\frac{dT}{dx}(x=\ell) = q \qquad (9.9)$$

It is noted that if $q = 0$, Eq. (9.9) represents the insulated boundary condition.

9.4.2 Interpolation functions

As explained in section 9.3, the first step of the finite element method is to discretize the computational domain into a number of elements. The rod can be divided into four elements as shown in Fig. 9.7. These elements are connected at nodes where unknowns are the nodal temperatures.

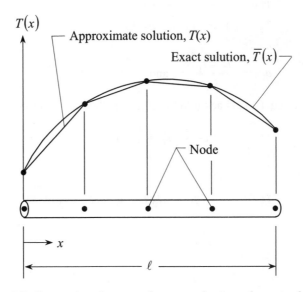

Figure 9.7 Comparison between the approximate and exact solutions.

Within each element, a simple linear function may be used as shown in the figure. An enlarged element in Fig. 9.8 of length L has two nodes 1 and 2 with the nodal temperatures T_1 and T_2, respectively.

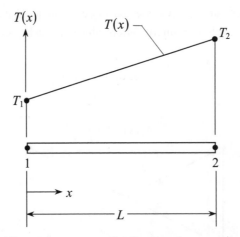

Figure 9.8 Assumed linear temperature distribution along a typical two-node element.

The temperature distribution $T(x)$ is assumed to be linear along the element length as

$$T(x) = ax + b \tag{9.10}$$

where a and b are constants that can be determined from the conditions at nodes, at $x = 0$: $T(x = 0) = T_1 = b$ (9.11a)

at $x = L$: $T(x = L) = T_2 = aL + b$ (9.11b)

which yield

$$a = \frac{T_2 - T_1}{L} \quad \text{and} \quad b = T_1 \tag{9.12}$$

Substituting a and b from Eq. (9.12) back into Eq. (9.10) and rearrange terms,

$$T(x) = \left(1 - \frac{x}{L}\right) T_1 + \left(\frac{x}{L}\right) T_2 \tag{9.13}$$

which can be written as

$$T(x) = N_1 T_1 + N_2 T_2 = \lfloor N_1 \quad N_2 \rfloor \begin{Bmatrix} T_1 \\ T_2 \end{Bmatrix}$$

$$= \underset{(1\times2)\ (2\times1)}{\lfloor N \rfloor \{T\}} \tag{9.14}$$

where

$$N_1 = 1 - \frac{x}{L} \tag{9.15a}$$

and

$$N_2 = \frac{x}{L}$$

$$\tag{9.15b}$$

are called the element interpolation functions.

    If the interpolation functions in Eqs. (9.15a-b) are considered, their distributions are linear as shown in Fig. 9.9. The interpolation function N_1 is equal to one at node 1, and is equal to zero at node 2. Likewise, the interpolation function N_2 is equal to one at node 2, and is equal to zero at node 1. Thus, from the properties of the element interpolations above, it can be concluded that

$$N_i = \begin{cases} 1 & \text{at node } i \\ 0 & \text{at the other nodes} \end{cases} \tag{9.16}$$

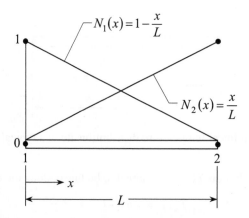

Figure 9.9 Element interpolation functions for the two-node element.

9.4.3 Finite element equations

As explained in section 9.3 that the most important step in studying the finite element method is the derivation of the finite element equations. This section shows the derivation of the finite element equations corresponding to the differential Eq. (9.7),

$$kA\frac{d^2\bar{T}}{dx^2} + QA = 0 \tag{9.17}$$

It is noted that if the exact temperature solution $\bar{T}$ can be derived and substituted back into the differential equation, then the right-hand side of the equation must be zero.

However, the exact solution may not be available for a general problem. Thus, if the approximate solution T in the form of linear distribution is substituted into the differential equation, the right-hand side of the equation is not zero but is equal to a residual,

$$kA\frac{d^2T}{dx^2} + QA = R \tag{9.18}$$

where R denotes the residual. The method of weighted residuals is applied to derive the finite element equations. The residual R is multiplied by a weighting function W. The expression is integrated over the length of the element and set to zero, i.e.,

$$\int_0^L W_i R \, dx = 0 \tag{9.19}$$

Since there are two nodal unknowns per an element, two equations are needed. These two equations can be obtained by using two weighting functions with $i = 1, 2$ in Eq. (9.19). Then, by substituting R from Eq. (9.18) into Eq. (9.19),

$$\int_0^L W_i \left(kA\frac{d^2T}{dx^2} + QA\right) dx = 0$$

Or,
$$\int_0^L W_i \, kA\frac{d^2T}{dx^2} \, dx + \int_0^L W_i \, QA \, dx = 0 \qquad i = 1, 2 \tag{9.20}$$

The first term in Eq. (9.20) can be integrated by parts so that the boundary term is introduced,

$$\int_0^L \underbrace{W_i}_{u} \ \underbrace{kA\frac{d^2T}{dx^2} \, dx}_{dv} = W_i \, kA\frac{dT}{dx}\Big|_0^L - \int_0^L kA\frac{dT}{dx}\frac{dW_i}{dx} \, dx \tag{9.21}$$

It is noted that the equation above is obtained from the integrating by parts formula,

$$\int_0^L u \, dv = uv\Big|_0^L - \int_0^L v \, du \tag{9.22}$$

Herein $\qquad u = W_i \qquad$ then $\qquad du = \dfrac{dW_i}{dx} \, dx$

and
$$dv = kA\frac{d^2T}{dx^2}dx \qquad \text{then} \qquad v = kA\frac{dT}{dx}$$

By substituting Eq. (9.21) into Eq. (9.20) and rearranging terms,

$$\int_0^L kA\frac{dW_i}{dx}\frac{dT}{dx}dx = W_i kA\frac{dT}{dx}\Big|_0^L + \int_0^L W_i QA\,dx$$

Because the element temperature distribution is assumed by Eq. (9.14), then,

$$\frac{dT}{dx} = \left\lfloor \frac{dN_1}{dx}\ \ \frac{dN_2}{dx} \right\rfloor \left\{\begin{matrix} T_1 \\ T_2 \end{matrix}\right\} \tag{9.23}$$

By substituting Eq. (9.23) into the above equation with $i = 1, 2$, then

$$i = 1 \ : \qquad \int_0^L kA\frac{dW_1}{dx}\left\lfloor \frac{dN_1}{dx}\ \ \frac{dN_2}{dx} \right\rfloor \left\{\begin{matrix} T_1 \\ T_2 \end{matrix}\right\} dx = W_1 kA\frac{dT}{dx}\Big|_0^L + \int_0^L W_1 QA\,dx$$

$$i = 2 \ : \qquad \int_0^L kA\frac{dW_2}{dx}\left\lfloor \frac{dN_1}{dx}\ \ \frac{dN_2}{dx} \right\rfloor \left\{\begin{matrix} T_1 \\ T_2 \end{matrix}\right\} dx = W_2 kA\frac{dT}{dx}\Big|_0^L + \int_0^L W_2 QA\,dx$$

If the Bubnov-Galerkin method is used such that $W_i = N_i$, these two equations can be written together as,

$$\underbrace{\int_0^L kA\left\{\begin{matrix} \frac{dN_1}{dx} \\ \frac{dN_2}{dx} \end{matrix}\right\}\left\lfloor \frac{dN_1}{dx}\ \ \frac{dN_2}{dx} \right\rfloor dx}_{[K_c]}\ \underbrace{\left\{\begin{matrix} T_1 \\ T_2 \end{matrix}\right\}}_{\{T\}} = \underbrace{\left(\left\{\begin{matrix} N_1 \\ N_2 \end{matrix}\right\}kA\frac{dT}{dx}\right)\Big|_0^L}_{\{Q_c\}} + \underbrace{\int_0^L \left\{\begin{matrix} N_1 \\ N_2 \end{matrix}\right\}QA\,dx}_{\{Q_Q\}}$$

Or
$$[K_c]\{T\} = \{Q_c\}+\{Q_Q\} \tag{9.24}$$

Equation (9.24) is the finite element equations of this problem where $[K_c]$ is the conduction matrix, $\{T\}$ is the vector of element nodal temperatures, $\{Q_c\}$ is the conduction load vector, and $\{Q_Q\}$ is the load vector due to internal heat generation. From Eq. (9.15)

$$N_1 = 1-\frac{x}{L} \qquad \text{and} \qquad N_2 = \frac{x}{L}$$

then
$$\frac{dN_1}{dx} = -\frac{1}{L} \qquad \text{and} \qquad \frac{dN_2}{dx} = \frac{1}{L}$$

The finite element matrices in Eq. (9.24) can be derived in closed-form as follows.

Conduction matrix

$$[K_c] = \int_0^L k\,A \begin{Bmatrix} \dfrac{dN_1}{dx} \\ \dfrac{dN_2}{dx} \end{Bmatrix} \begin{bmatrix} \dfrac{dN_1}{dx} & \dfrac{dN_2}{dx} \end{bmatrix} dx$$

$$= \int_0^L k\,A \begin{Bmatrix} -\dfrac{1}{L} \\ \dfrac{1}{L} \end{Bmatrix} \begin{bmatrix} -\dfrac{1}{L} & \dfrac{1}{L} \end{bmatrix} dx = \int_0^L \dfrac{k\,A}{L^2} \begin{bmatrix} 1 & -1 \\ -1 & 1 \end{bmatrix} dx$$

If the thermal conductivity coefficient k and the conduction area A are constant,

$$[K_c] = \dfrac{k\,A}{L} \begin{bmatrix} 1 & -1 \\ -1 & 1 \end{bmatrix} \tag{9.25}$$

Load vector due to internal heat generation

$$\{Q_Q\} = \int_0^L \begin{Bmatrix} N_1 \\ N_2 \end{Bmatrix} Q\,A\,dx = \int_0^L \begin{Bmatrix} 1 - \dfrac{x}{L} \\ \dfrac{x}{L} \end{Bmatrix} Q\,A\,dx$$

$$= Q\,A\,L \begin{Bmatrix} \dfrac{1}{2} \\ \dfrac{1}{2} \end{Bmatrix} \tag{9.26}$$

Conduction load vector

$$\{Q_c\} = \left(\begin{Bmatrix} N_1 \\ N_2 \end{Bmatrix} k\,A\,\dfrac{dT}{dx} \right)\Bigg|_0^L = \begin{Bmatrix} \left(N_1\,k\,A\,\dfrac{dT}{dx} \right)\Big|_0^L \\ \left(N_2\,k\,A\,\dfrac{dT}{dx} \right)\Big|_0^L \end{Bmatrix}$$

$$= \begin{Bmatrix} N_1(L)\,k\,A\,\dfrac{dT}{dx}(L) - N_1(0)\,k\,A\,\dfrac{dT}{dx}(0) \\ N_2(L)\,k\,A\,\dfrac{dT}{dx}(L) - N_2(0)\,k\,A\,\dfrac{dT}{dx}(0) \end{Bmatrix}$$

but $N_1(0) = 1$, $N_1(L) = 0$, $N_2(0) = 0$, $N_2(L) = 1$, then

$$\{Q_c\} = \begin{Bmatrix} -k\,A\,\dfrac{dT}{dx}(0) \\ k\,A\,\dfrac{dT}{dx}(L) \end{Bmatrix} \tag{9.27}$$

The conduction load vector consists of two quantities representing the heat flux at nodes 1 and 2, respectively, as shown in Fig. 9.10.

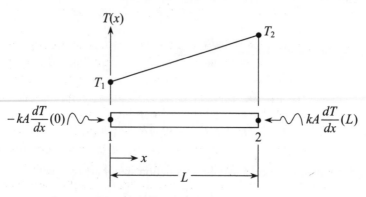

Figure 9.10 Conduction heat fluxes at the two nodes.

Thus, the finite element equations for this example are

$$\frac{k\,A}{L}\begin{bmatrix} 1 & -1 \\ -1 & 1 \end{bmatrix}\begin{Bmatrix} T_1 \\ T_2 \end{Bmatrix} = \begin{Bmatrix} -k\,A\dfrac{dT}{dx}(0) \\ k\,A\dfrac{dT}{dx}(L) \end{Bmatrix} + QAL\begin{Bmatrix} \dfrac{1}{2} \\ \dfrac{1}{2} \end{Bmatrix}$$ (9.28)

Equation (9.28) represents the typical finite element equations for conduction heat transfer in a rod subjected to an internal heat generation. The finite element equations are derived for all elements in the model before assembling them together to become a set of simultaneous equations. Boundary conditions of the problem are then applied to the set of simultaneous equations prior to solving them for the solutions of all nodal temperatures.

9.4.4 Example

Example 9.1 A steel bar of length ℓ subjected to an internal heat generation Q, as shown in Fig. 9.6, has the specified temperatures at both ends as

$$T(x = 0) \;=\; 0$$ (9.29a)

$$T(x = \ell) \;=\; T_\ell$$ (9.29b)

The bar is under a steady-state conduction heat transfer.

(a) Use two linear elements to compute the temperature at the center of the bar $(x = \ell/2)$.

(b) Recompute the temperature at the center of the bar if $T_\ell = 0$.

(c) If $T_\ell = 0$, the temperature distribution along the bar is symmetry. Use only one element to compute the temperature at the center of the bar again, and compare it with the solution obtained in (b).

Solution *(a) Use two linear elements to compute the temperature at the center of the bar* $(x = \ell/2)$, as shown in Fig. 9.11.

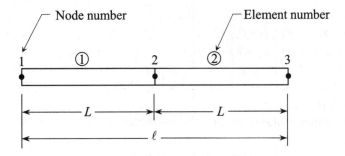

Figure 9.11(a) The steel bar discretized into two elements.

Each element has the length L which is equal to $\ell/2$. From the general form of the derived finite element Eq. (9.28), the finite element equations for element ① are

$$\frac{kA}{L}\begin{bmatrix} 1 & -1 \\ -1 & 1 \end{bmatrix}\begin{Bmatrix} T_1 \\ T_2 \end{Bmatrix} = \begin{Bmatrix} \left(-kA\dfrac{dT}{dx}(0)\right)_① \\ \left(kA\dfrac{dT}{dx}(L)\right)_① \end{Bmatrix} + \frac{QAL}{2}\begin{Bmatrix} 1 \\ 1 \end{Bmatrix} \tag{9.30}$$

Similarly, the finite element equations for element ② are

$$\frac{kA}{L}\begin{bmatrix} 1 & -1 \\ -1 & 1 \end{bmatrix}\begin{Bmatrix} T_2 \\ T_3 \end{Bmatrix} = \begin{Bmatrix} \left(-kA\dfrac{dT}{dx}(0)\right)_② \\ \left(kA\dfrac{dT}{dx}(L)\right)_② \end{Bmatrix} + \frac{QAL}{2}\begin{Bmatrix} 1 \\ 1 \end{Bmatrix} \tag{9.31}$$

The finite element Eqs. (9.30) and (9.31) are assembled together to become a system of equations. The physical meaning for assembling the element equations is equivalent to joining the two elements together as shown in Fig. 9.11(b).

$$\frac{kA}{L}\begin{bmatrix} 1 & -1 & 0 \\ -1 & 1+1 & -1 \\ 0 & -1 & 1 \end{bmatrix}\begin{Bmatrix} T_1 \\ T_2 \\ T_3 \end{Bmatrix} = \begin{Bmatrix} \left(-kA\dfrac{dT}{dx}(0)\right)_① \\ \left(kA\dfrac{dT}{dx}(L)\right)_① + \left(-kA\dfrac{dT}{dx}(0)\right)_② \\ \left(kA\dfrac{dT}{dx}(L)\right)_② \end{Bmatrix} = \frac{QAL}{2}\begin{Bmatrix} 1 \\ 2 \\ 1 \end{Bmatrix} \tag{9.32}$$

As the heat fluxes conducted between elements ① and ② at node 2 must be continuous, then

$$\left(kA\frac{dT}{dx}(L)\right)_① + \left(-kA\frac{dT}{dx}(0)\right)_② = 0$$

Since the temperatures at nodes 1 and 3 are respectively zero and T_ℓ, the system of equations (9.32) becomes

$$\frac{kA}{L}\begin{bmatrix} 1 & -1 & 0 \\ -1 & 2 & -1 \\ 0 & -1 & 1 \end{bmatrix}\begin{Bmatrix} T_1 = 0 \\ T_2 = ? \\ T_3 = T_\ell \end{Bmatrix} = \begin{Bmatrix} \left(-kA\dfrac{dT}{dx}(0)\right)_{\textcircled{1}} \\ 0 \\ \left(kA\dfrac{dT}{dx}(L)\right)_{\textcircled{2}} \end{Bmatrix} + QAL\begin{Bmatrix} \dfrac{1}{2} \\ 1 \\ \dfrac{1}{2} \end{Bmatrix} \tag{9.33}$$

The second equation of Eq. (9.33) is used to solve for the temperature T_2 at node 2 such that

$$\frac{kA}{L}(0 + 2T_2 - T_\ell) = 0 + QAL$$

$$T_2 = \frac{QL^2}{2k} + \frac{T_\ell}{2} \tag{9.34}$$

Then, the first and the third equation of Eq. (9.33) are used to determine the heat fluxes at nodes 1 and 3, respectively.

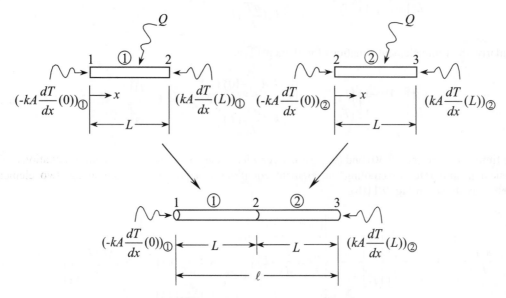

Figure 9.11(b) Joining two elements together to represent the problem.

(b) Recompute the temperature at the center of the bar if $T_\ell = 0$. If $T_\ell = 0$, Eq. (9.34) becomes

$$T_2 = \frac{QL^2}{2k} \tag{4.37}$$

In this case, the finite element temperature distribution along the length of the bar is symmetry (Fig. 9.12) with the exact temperature distribution,

$$\bar{T}(x) = \frac{QL^2}{2k}\left(2\frac{x}{L} - \frac{x^2}{L^2}\right) \tag{9.36}$$

Such exact solution can be easily derived by integrating the differential equation Eq. (9.7) twice and applying the boundary conditions of zero temperature at both ends of the bar.

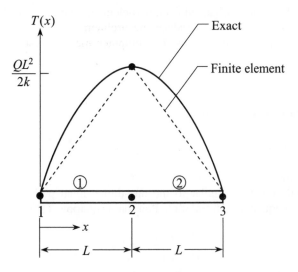

Figure 9.12 Comparative exact and finite element temperature distributions.

(c) If $T_\ell = 0$, the temperature distribution along the bar is symmetry. Use only one element to compute the temperature at the center of the bar again, and compare it with the solution obtained in (b). If only element ① is used, the finite element equations are

$$\frac{kA}{L}\begin{bmatrix} 1 & -1 \\ -1 & 1 \end{bmatrix} \begin{Bmatrix} T_1 \\ T_2 \end{Bmatrix} = \begin{Bmatrix} \left(-kA\dfrac{dT}{dx}(0)\right)_① \\ \left(kA\dfrac{dT}{dx}(L)\right)_① \end{Bmatrix} + QAL \begin{Bmatrix} \frac{1}{2} \\ \frac{1}{2} \end{Bmatrix} \tag{9.37}$$

Due to the symmetry of the temperature solution as shown in Fig. 9.12, the heat flux conducted through node 2 must be zero. Thus

$$\left(kA\frac{dT}{dx}(L)\right)_① = 0 \tag{9.38}$$

By substituting Eq. (9.38) into Eq. (9.37) and applying the boundary condition of zero temperature at node 1,

$$\frac{kA}{L}\begin{bmatrix} 1 & -1 \\ -1 & 1 \end{bmatrix} \begin{Bmatrix} T_1 = 0 \\ T_2 = ? \end{Bmatrix} = \begin{Bmatrix} \left(-kA\dfrac{dT}{dx}(0)\right)_① \\ 0 \end{Bmatrix} + QAL \begin{Bmatrix} \frac{1}{2} \\ \frac{1}{2} \end{Bmatrix} \tag{9.39}$$

The temperature at node 2 can be solved from the second equation of Eq. (9.39) as

$$\frac{kA}{L}(0+T_2) = 0 + \frac{QAL}{2}$$

$$T_2 = \frac{QL^2}{2k} \tag{9.40}$$

which is identical to the solution obtained from two-element model in (b). This example demonstrates that the use of solution symmetry can reduce the problem size. Such utilization can significantly reduce the computational time as well as the computer memory, especially for three-dimensional problems.

9.5 Two-dimensional Problems

9.5.1 Differential equation

The finite element equation for two-dimensional problems can be derived directly from the governing differential equations similar to the one-dimensional problems in the preceding section. Herein, the differential equation in form of the Poisson's equation is studied,

$$\frac{\partial^2 u}{\partial x^2} + \frac{\partial^2 u}{\partial y^2} = f(x, y) \tag{9.41}$$

where u is the dependent variable that varies with x- and y-coordinates, and $f(x, y)$ is a given function. Such differential equation in the form of Poisson's equation occurs in many engineering problems. For example, steady-state heat conduction in a plate subjected to an internal heat generation, as shown in Fig. 9.13, is governed by the differential equation,

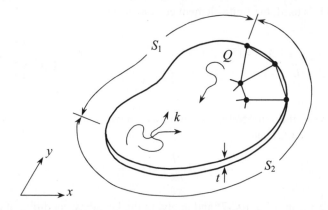

Figure 9.13 Conduction heat transfer in a plate with internal heat generation.

$$\frac{\partial}{\partial x}\left(k\frac{\partial \overline{T}}{\partial x}\right) + \frac{\partial}{\partial y}\left(k\frac{\partial \overline{T}}{\partial y}\right) = -Q \tag{9.42}$$

where k is the thermal conductivity coefficient, $\overline{T}$ is the exact temperature distribution, and Q is the internal heat generation per unit volume.

The temperature distribution in the plate also depends on the boundary conditions along the outer edges. Typical boundary conditions are:

(a) Specified temperature along the edge S_1

$$T(x, y) = T_1(x, y) \tag{9.43}$$

(b) Specified heat flux along the edge S_2

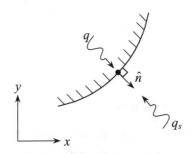

Figure 9.14 Specified heat flux along the outer edge.

From Fourier's law,

$$q = -k\frac{\partial T}{\partial x}n_x - k\frac{\partial T}{\partial y}n_y \tag{9.44}$$

where n_x and n_y are the direction cosines of the unit vector $\hat{n}$ normal to the edge

$$\hat{n} = n_x\hat{i} + n_y\hat{j} \tag{9.45}$$

while $\hat{i}$ and $\hat{j}$ are the unit vectors in x- and y-direction, respectively. If q_s is the specified heat flux into the edge as shown in Fig. 9.14,

then

$$q_s = -q = k\frac{\partial T}{\partial x}n_x + k\frac{\partial T}{\partial y}n_y \tag{9.46}$$

It is noted that, for an insulated edge, Eq. (9.46) becomes

$$k\frac{\partial T}{\partial x}n_x + k\frac{\partial T}{\partial y}n_y = 0 \tag{9.47}$$

9.5.2 Element interpolation functions

For two-dimensional problems, the triangular finite element is widely used because any two-dimensional domain can be discretized conveniently. Their element equations and matrices, in addition, are easy to derive and understand. The basic triangular element consists of 3 nodes which are numbered in the counter-clockwise direction as shown in Fig. 9.15. These nodes are at coordinates (x_i, y_i), $i = 1, 2, 3$ with the temperature unknowns of T_i at the nodal locations.

The element interpolation functions for a triangular element can be derived by the same procedure as for a two-node linear element in section 9.4.2. The derivation starts by assuming a flat plane distribution of the approximate solution over the element as

$$T(x, y) = \alpha_1 + \alpha_2 x + \alpha_3 y \tag{9.48}$$

where α_i, $i = 1, 2, 3$ are constants that can be determined from the conditions at nodes

node 1: $\quad T(x_1, y_1) = T_1 = \alpha_1 + \alpha_2 x_1 + \alpha_3 y_1$

node 2: $\quad T(x_2, y_2) = T_2 = \alpha_1 + \alpha_2 x_2 + \alpha_3 y_2$

node 3: $\quad T(x_3, y_3) = T_3 = \alpha_1 + \alpha_2 x_3 + \alpha_3 y_3$

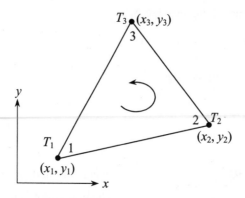

Figure 9.15 Three-node triangular element.

The constants $\alpha_i, i = 1, 2, 3$ can be determined from the three equations above in forms of the nodal temperatures T_i and nodal coordinates (x_i, y_i). By substituting these constants back into Eq. (9.48) and rearranging terms, the temperature distribution over the element is obtained in the form

$$T(x, y) = \lfloor N_1(x, y) \quad N_2(x, y) \quad N_3(x, y) \rfloor \begin{Bmatrix} T_1 \\ T_2 \\ T_3 \end{Bmatrix} = \underset{(1\times3)}{\lfloor N \rfloor} \underset{(3\times1)}{\{T\}} \qquad (9.49)$$

where $\lfloor N \rfloor$ is the element interpolation matrix and $\{T\}$ is the vector of nodal temperature. Herein

$$N_i(x, y) = \frac{1}{2A}(a_i + b_i x + c_i y) \qquad i = 1, 2, 3 \qquad (9.50)$$

where

$$A = \text{area of the triangle}$$

$$= \frac{1}{2}[x_2(y_3 - y_1) + x_1(y_2 - y_3) + x_3(y_1 - y_2)]$$

The coefficients a_i, b_i and c_i, $i = 1,2,3$ in Eq. (9.50) are

$$\begin{array}{lll} a_1 = x_2 y_3 - x_3 y_2 & b_1 = y_2 - y_3 & c_1 = x_3 - x_2 \\ a_2 = x_3 y_1 - x_1 y_3 & b_2 = y_3 - y_1 & c_2 = x_1 - x_3 \\ a_3 = x_1 y_2 - x_2 y_1 & b_3 = y_1 - y_2 & c_3 = x_2 - x_1 \end{array} \qquad (9.52)$$

To understand the element interpolation functions derived above, a typical element that has the coordinates of nodes 1, 2 and 3 as shown in Fig. 9.16 is considered,

$$\begin{array}{ll} x_1 = 0 & y_1 = 0 \\ x_2 = 1 & y_2 = 0 \\ x_3 = 0 & y_3 = 2 \end{array}$$

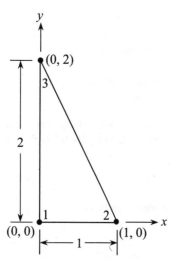

Figure 9.16 Example of a triangular element.

From Eq. (9.51), the element area is

$$A = \frac{1}{2}[1(2-0) + 0(0-2) + 0(0-0)] = 1$$

From Eq. (9.52), the coefficients a_i, b_i, c_i, $i = 1, 2, 3$ are

$$a_1 = (1)(2)-(0)(0) = 2 \qquad b_1 = 0-2 = -2 \qquad c_1 = 0-1 = -1$$

$$a_2 = (0)(0)-(0)(2) = 0 \qquad b_2 = 2-0 = 2 \qquad c_2 = 0-0 = 0$$

$$a_3 = (0)(0)-(1)(0) = 0 \qquad b_3 = 0-0 = 0 \qquad c_3 = 1-0 = 0$$

Therefore, the element interpolation functions for this element are

$$N_1 = \frac{1}{2(1)}(2+(-2)x+(-1)y) = 1-x-\frac{y}{2}$$

$$N_2 = \frac{1}{2(1)}(0+(2)x+(0)y) = x$$

$$N_3 = \frac{1}{2(1)}(0+(0)x+(1)y) = \frac{y}{2}$$

which are represented by flat plane distributions as shown in Fig. 9.17. According to Eq. (9.49), the summation of the products of the above element interpolation functions N_i and the nodal values T_i is the distribution of the approximate solution over the element as shown in Fig. 9.18.

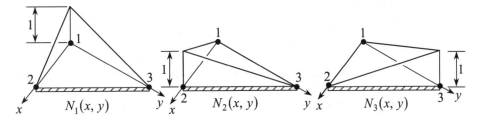

Figure 9.17 Distributions of element interpolation functions.

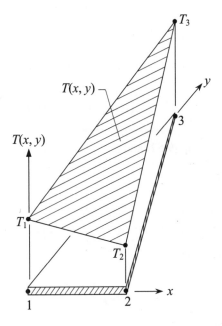

Figure 9.18 Temperature distribution over the element.

9.5.3 Finite element equations

In this section, derivation of finite element equations from a two-dimensional Poisson's equation is presented. The Poisson's equation that represents conduction heat transfer in a plate with an internal heat generation as shown in Eq. (9.42) is

$$\frac{\partial}{\partial x}\left(k\frac{\partial \overline{T}}{\partial x}\right) + \frac{\partial}{\partial y}\left(k\frac{\partial \overline{T}}{\partial y}\right) + Q = 0 \tag{9.53}$$

where $\overline{T}$ is the exact temperature distribution.

By substituting an approximate temperature T into the differential Eq. (9.53), the right-hand side of the equation is not zero but equal to a residual R, i.e.,

$$\frac{\partial}{\partial x}\left(k\frac{\partial T}{\partial x}\right) + \frac{\partial}{\partial y}\left(k\frac{\partial T}{\partial y}\right) + Q = R \tag{9.54}$$

The finite element equations can be derived by applying the method of weighted residuals. The method is to multiply the residual R by a weighting function W, perform integration over the element domain $\Omega^{(e)}$, and set the result to zero,

$$\int_{\Omega^{(e)}} W_i\, R\, d\Omega = 0 \tag{9.55}$$

Since there are three unknowns for the three-node triangular element, three equations per an element are needed, i.e., $i = 1, 2, 3$ in Eq. (9.55). By substituting Eq. (9.54) into Eq. (9.55),

$$\int_{\Omega^{(e)}} W_i\left(\frac{\partial}{\partial x}\left(k\frac{\partial T}{\partial x}\right) + \frac{\partial}{\partial y}\left(k\frac{\partial T}{\partial y}\right) + Q\right) d\Omega = 0$$

$$\int_{\Omega^{(e)}} W_i\left(\frac{\partial}{\partial x}\left(k\frac{\partial T}{\partial x}\right) + \frac{\partial}{\partial y}\left(k\frac{\partial T}{\partial y}\right)\right) d\Omega + \int_{\Omega^{(e)}} W_i\, Q\, d\Omega = 0 \tag{9.56}$$

To introduce the boundary term, the first term in Eq. (9.56) is integrated by parts by employing the Gauss's theorem which states that

$$\int_{\Omega^{(e)}} u\left(\nabla \cdot \vec{V}\right) d\Omega = \int_{\Gamma^{(e)}} u\left(\vec{V} \cdot \hat{n}\right) d\Gamma - \int_{\Omega^{(e)}} \left(\nabla u \cdot \vec{V}\right) d\Omega \tag{9.57}$$

If the term on the left-hand side of Eq. (9.57) and the first term in Eq. (9.56) are compared

$$u = W_i$$

$$\left.\begin{array}{l} \nabla = \dfrac{\partial}{\partial x}\hat{i} + \dfrac{\partial}{\partial y}\hat{j} \\[2mm] \vec{V} = k\dfrac{\partial T}{\partial x}\hat{i} + k\dfrac{\partial T}{\partial y}\hat{j} \end{array}\right\} \quad \left(\nabla \cdot \vec{V}\right) = \left(\dfrac{\partial}{\partial x}\left(k\dfrac{\partial T}{\partial x}\right) + \dfrac{\partial}{\partial y}\left(k\dfrac{\partial T}{\partial y}\right)\right)$$

Because $\hat{n} = n_x\hat{i} + n_y\hat{j}$, then

$$\vec{V} \cdot \hat{n} \quad = \quad k\frac{\partial T}{\partial x}n_x + k\frac{\partial T}{\partial y}n_y$$

$$u\left(\vec{V} \cdot \hat{n}\right) \quad = \quad W_i\left(k\frac{\partial T}{\partial x}n_x + k\frac{\partial T}{\partial y}n_y\right)$$

$$\nabla u \quad = \quad \frac{\partial W_i}{\partial x}\hat{i} + \frac{\partial W_i}{\partial y}\hat{j}$$

$$\nabla u \cdot \vec{V} \quad = \quad \frac{\partial W_i}{\partial x}k\frac{\partial T}{\partial x} + \frac{\partial W_i}{\partial y}k\frac{\partial T}{\partial y}$$

and if $W_i = N_i$, Eq. (9.56) becomes

$$\int_{\Gamma^{(e)}} N_i\left(k\frac{\partial T}{\partial x}n_x + k\frac{\partial T}{\partial y}n_y\right)d\Gamma - \int_{\Omega^{(e)}}\left(\frac{\partial N_i}{\partial x}k\frac{\partial T}{\partial x} + \frac{\partial N_i}{\partial y}k\frac{\partial T}{\partial y}\right)d\Omega$$

$$+ \int_{\Omega^{(e)}} N_i Q\, d\Omega \quad = \quad 0 \qquad\qquad i = 1, 2, 3 \qquad (9.58)$$

Next, the boundary integral term associated with the element edge $\Gamma^{(e)}$ is replaced by the boundary conditions of that problem. The first term in Eq. (9.58) represents the heat flux that transfers in and out of the element edges. If the element is located inside the plate, this conduction heat flux term will be cancelled with the similar terms from the adjacent elements. If the element is located along the outer boundary of the plate, this term is replaced by the appropriate conditions in Eq. (9.46) or (9.47). Thus, Eq. (9.58) can be written in a more general form for an element at any location on the plate as

$$\int_{S_2^{(e)}} N_i q_s\, dS + \int_{\Gamma^{(e)}} N_i\left(k\frac{\partial T}{\partial x}n_x + k\frac{\partial T}{\partial y}n_y\right)d\Gamma$$

$$- \int_{\Omega^{(e)}}\left(\frac{\partial N_i}{\partial x}k\frac{\partial T}{\partial x} + \frac{\partial N_i}{\partial y}k\frac{\partial T}{\partial y}\right)d\Omega + \int_{\Omega^{(e)}} N_i Q\, d\Omega \quad = \quad 0 \qquad i = 1, 2, 3$$

These element equations can be written in matrix form as

$$\int_{\Omega^{(e)}}\left(\left\{\frac{\partial N}{\partial x}\right\}k\frac{\partial T}{\partial x} + \left\{\frac{\partial N}{\partial y}\right\}k\frac{\partial T}{\partial y}\right)d\Omega \quad = \quad \int_{\Gamma^{(e)}} \{N\}\left(k\frac{\partial T}{\partial x}n_x + k\frac{\partial T}{\partial y}n_y\right)d\Gamma$$

$$+ \int_{\Omega^{(e)}} \{N\}Q\, d\Omega + \int_{S_2^{(e)}} \{N\}q_s\, dS$$

Since, for each element, the temperature distribution is assumed in the form of Eq. (9.49) as

$$T \quad = \quad T(x, y) \quad = \quad \underset{(1\times3)\ (3\times1)}{\lfloor N \rfloor\{T\}}$$

Therefore, $\qquad \dfrac{\partial T}{\partial x} = \underset{(1\times3)\ (3\times1)}{\left\lfloor\dfrac{\partial N}{\partial x}\right\rfloor\{T\}} \quad$ and $\quad \dfrac{\partial T}{\partial y} = \underset{(1\times3)\ (3\times1)}{\left\lfloor\dfrac{\partial N}{\partial y}\right\rfloor\{T\}}$

Thus, the finite element equations become

$$\underbrace{\int_{\Omega^{(e)}} \left(\underset{(3\times1)}{\left\{\frac{\partial N}{\partial x}\right\}} k \underset{(1\times3)}{\left\lfloor\frac{\partial N}{\partial x}\right\rfloor} + \underset{(3\times1)}{\left\{\frac{\partial N}{\partial y}\right\}} k \underset{(1\times3)}{\left\lfloor\frac{\partial N}{\partial y}\right\rfloor} \right) d\Omega}_{\substack{[K_c] \\ (3\times3)}} \underset{(3\times1)}{\{T\}}$$

$$= \underbrace{\int_{\Gamma^{(e)}} \underset{(3\times1)}{\{N\}} \left(k\frac{\partial T}{\partial x} n_x + k\frac{\partial T}{\partial y} n_y \right) d\Gamma}_{\substack{\{Q_c\} \\ (3\times1)}} + \underbrace{\int_{\Omega^{(e)}} \underset{(3\times1)}{\{N\}} Q\, d\Omega}_{\substack{\{Q_Q\} \\ (3\times1)}} + \underbrace{\int_{S_2^{(e)}} \underset{(3\times1)}{\{N\}} q_s\, dS}_{\substack{\{Q_q\} \\ (3\times1)}} \qquad (9.59a)$$

Or, in short,

$$[K_c]\{T\} = \{Q_c\} + \{Q_Q\} + \{Q_q\} \qquad (9.59b)$$

where $[K_c]$ is the conduction matrix, $\{T\}$ is the vector of nodal temperatures, $\{Q_c\}$ is the conduction load vector, $\{Q_Q\}$ is the load vector due to internal heat generation and $\{Q_q\}$ is the load vector from the specified heating.

These finite element matrices can be derived in closed-form expressions so that they are ready for computer programming. Since the interpolation functions for the three-node triangular element are

$$N_i(x, y) = \frac{1}{2A}(a_i + b_i x + c_i y) \qquad i = 1, 2, 3$$

Then $\qquad \dfrac{\partial N_i}{\partial x} = \dfrac{b_i}{2A} \qquad$ and $\qquad \dfrac{\partial N_i}{\partial y} = \dfrac{c_i}{2A} \qquad\qquad (9.60)$

Conduction matrix

The coefficients in the conduction matrix in Eq. (9.59) can be written in the form

$$K_{ij} = \int_A k\left(\frac{\partial N_i}{\partial x}\frac{\partial N_j}{\partial x} + \frac{\partial N_i}{\partial y}\frac{\partial N_j}{\partial y}\right) t\, dx\, dy \qquad i, j = 1, 2, 3 \qquad (9.61)$$

If the thermal conductivity coefficient k and the element thickness t are constant, then Eq. (9.61) becomes

$$K_{ij} = kt \int_A \left(\frac{\partial N_i}{\partial x}\frac{\partial N_j}{\partial x} + \frac{\partial N_i}{\partial y}\frac{\partial N_j}{\partial y}\right) dx\, dy \qquad (9.62)$$

By substituting Eq. (9.60) into Eq. (9.62)

$$K_{ij} = kt \int_A \left(\frac{b_i}{2A}\frac{b_j}{2A} + \frac{c_i}{2A}\frac{c_j}{2A}\right) dx\, dy$$

$$= \frac{kt}{4A^2}(b_i b_j + c_i c_j) \int_A dx\, dy$$

$$K_{ij} = \frac{kt}{4A}(b_i b_j + c_i c_j) \qquad i, j = 1, 2, 3 \qquad (9.63)$$

It is noted that the conduction matrix above can be written as

$$\underset{(3\times3)}{[K_c]} = kAt \underset{(3\times2)}{[B]^T} \underset{(2\times3)}{[B]} \tag{9.64}$$

where

$$[B] = \frac{1}{2A}\begin{bmatrix} b_1 & b_2 & b_3 \\ c_1 & c_2 & c_3 \end{bmatrix} \tag{9.65}$$

Load vector due to internal heat generation

The coefficients in the load vector due to an internal heat generation as shown in Eq. (9.59) are

$$Q_i = \int_A N_i\, Q\, t\, dx\, dy \qquad\qquad i, j = 1, 2, 3 \tag{9.66}$$

If the internal heat generation Q and the element thickness are constant, then Eq. (9.66) becomes

$$Q_i = Qt \int_A N_i\, dx\, dy$$

$$= Qt \int_A \frac{1}{2A}\left(a_i + b_i\, x + c_i\, y\right) dx\, dy$$

$$Q_i = \frac{QAt}{3} \tag{9.67a}$$

Integrations in Eq. (9.67a) can be performed easily by using the formula

$$\int_A N_1^\alpha\, N_2^\beta\, N_3^\gamma\, dx\, dy = \frac{\alpha!\, \beta!\, \gamma!}{(\alpha+\beta+\gamma+2)!}\, 2A \tag{9.68}$$

For example, if $\alpha = 1, \beta = \gamma = 0$, then

$$\int_A N_1\, dx\, dy = \frac{1!\, 0!\, 0!}{(1+0+0+2)!}\, 2A$$

$$= \frac{2A}{3!} = \frac{A}{3}$$

Thus, the load vector due to an internal heat generation is

$$\{Q_Q\} = \frac{QAt}{3}\begin{Bmatrix} 1 \\ 1 \\ 1 \end{Bmatrix} \tag{9.67b}$$

Load vector from specified edge heating

The load vector from a specified heating along an element edge as shown in Eq. (9.59) which is

$$\{Q_q\} = \int_{S_2} \{N\} q_s\, dS \tag{9.69}$$

can also be derived in closed-form expressions. Detailed derivation is omitted herein so that it can be used as an exercise. As an example, the load vector from a specified heating along the element edge that connects between nodes 1 and 2 is

$$\{Q_q\} \;=\; \frac{q_s t \, \ell_{12}}{2} \begin{Bmatrix} 1 \\ 1 \\ 0 \end{Bmatrix} \tag{9.70}$$

where q_s is the specified heating, t is the element thickness and ℓ_{12} is the edge length between nodes 1 and 2.

Conduction load vector

The conduction load vector as shown in Eq. (9.59) is

$$\{Q_c\} \;=\; \int_{\Gamma^{(e)}} \{N\} \left(k \frac{\partial T}{\partial x} n_x + k \frac{\partial T}{\partial y} n_y \right) d\Gamma \tag{9.71}$$

The conduction load vector consists of three terms representing heat fluxes at the three nodes of an element. If a node is located inside the computational domain, the net conduction flux at that node is zero after all elements are assembled. If the node is located at the domain boundary, the conduction flux must be replaced by the specified heating. Thus, the load vector in the form of Eq. (9.71) is never been implemented in the finite element computer program. The explanation above can be understood clearly by studying example in the section below.

9.5.4 Example

The finite element equations and their matrices derived in the preceding section can be used to analyze two-dimensional problem as shown in the following example.

Example 9.2 An equilateral triangular plate under steady-state heat conduction has a thermal conductivity coefficient k and internal heat generation Q. Zero temperature is specified along all outer edges as shown in Fig. 9.19. Use three triangular elements to determine the temperature at the plate centroid.

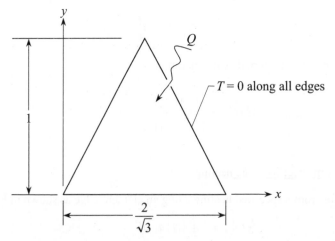

Figure 9.19 Equilateral triangular plate with internal heat generation.

Solution The plate is discretized by using 3 elements with 4 nodes as shown in Fig. 9.20.

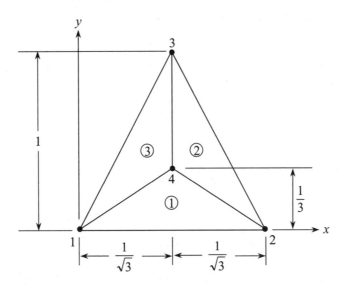

Figure 9.20 Finite element model of the plate using 3 elements and 4 nodes.

Since the problem is heat conduction with internal heat generation, then the finite element Eq. (9.59b) reduces to

$$[K_c]\{T\} = \{Q_c\} + \{Q_Q\} \tag{9.72}$$

The finite element Eq. (9.72) can be written in full for a typical element, such as for element ①, as

$$\begin{bmatrix} k_{11} & k_{12} & k_{13} \\ k_{21} & k_{22} & k_{23} \\ k_{31} & k_{32} & k_{33} \end{bmatrix} \begin{Bmatrix} T_1 \\ T_2 \\ T_3 \end{Bmatrix} = \begin{Bmatrix} Q_{c_1} \\ Q_{c_2} \\ Q_{c_3} \end{Bmatrix} + \frac{QAt}{3} \begin{Bmatrix} 1 \\ 1 \\ 1 \end{Bmatrix} \tag{9.73}$$

where $K_{ij}, i, j = 1, 2, 3$ can be determined from the element properties and nodal coordinates. Starting from element ① which is connected by node numbers 1, 2 and 4, the actual node numbers are matched to the standard node numbers 1, 2 and 3 as depicted in Fig. 9.21.

From the figure and Eqs. (9.51)-(9.52),

$$x_1 = 0 \qquad y_1 = 0 \qquad b_1 = -\frac{1}{3} \qquad c_1 = -\frac{1}{\sqrt{3}}$$

$$x_2 = \frac{2}{\sqrt{3}} \qquad y_2 = 0 \qquad b_2 = \frac{1}{3} \qquad c_2 = -\frac{1}{\sqrt{3}}$$

$$x_3 = \frac{1}{\sqrt{3}} \qquad y_3 = \frac{1}{3} \qquad b_3 = 0 \qquad c_3 = \frac{2}{\sqrt{3}}$$

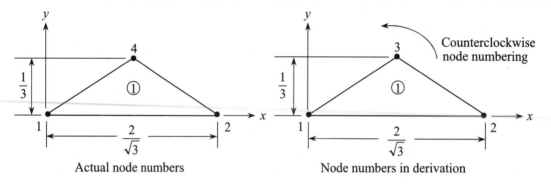

Figure 9.21 Element number ① and its node numbers.

and the element area is $A = 1/3\sqrt{3}$. Then Eq. (9.63) is used to determine the coefficients in the element matrices. For example, when $i = 2, j = 3$, the coefficient in the element conduction matrix is

$$K_{23} = \frac{kt}{4A}(b_2 b_3 + c_2 c_3)$$

$$= \frac{kt}{4\left(\dfrac{1}{3\sqrt{3}}\right)}\left(\left(\frac{1}{3}\right)(0) + \left(-\frac{1}{\sqrt{3}}\right)\left(\frac{2}{\sqrt{3}}\right)\right)$$

$$= kt\left(-\frac{3}{2\sqrt{3}}\right)$$

Therefore, the finite element equations for element ① are

$$\frac{kt}{2\sqrt{3}}\begin{bmatrix} 2 & 1 & -3 \\ 1 & 2 & -3 \\ -3 & -3 & 6 \end{bmatrix}\begin{Bmatrix} T_1 \\ T_2 \\ T_4 \end{Bmatrix} = \begin{Bmatrix} (Q_{c1})_1 \\ (Q_{c2})_1 \\ (Q_{c4})_1 \end{Bmatrix} + \frac{Qt}{9\sqrt{3}}\begin{Bmatrix} 1 \\ 1 \\ 1 \end{Bmatrix} \qquad (9.74)$$

Similarly, the finite element equations for element ② are

$$\frac{kt}{2\sqrt{3}}\begin{bmatrix} 2 & 1 & -3 \\ 1 & 2 & -3 \\ -3 & -3 & 6 \end{bmatrix}\begin{Bmatrix} T_2 \\ T_3 \\ T_4 \end{Bmatrix} = \begin{Bmatrix} (Q_{c2})_2 \\ (Q_{c3})_2 \\ (Q_{c4})_2 \end{Bmatrix} + \frac{Qt}{9\sqrt{3}}\begin{Bmatrix} 1 \\ 1 \\ 1 \end{Bmatrix} \qquad (9.75)$$

and the finite element equations for element ③ are

$$\frac{kt}{2\sqrt{3}}\begin{bmatrix} 2 & 1 & -3 \\ 1 & 2 & -3 \\ -3 & -3 & 6 \end{bmatrix}\begin{Bmatrix} T_3 \\ T_1 \\ T_4 \end{Bmatrix} = \begin{Bmatrix} (Q_{c3})_3 \\ (Q_{c1})_3 \\ (Q_{c4})_3 \end{Bmatrix} + \frac{Qt}{9\sqrt{3}}\begin{Bmatrix} 1 \\ 1 \\ 1 \end{Bmatrix} \qquad (9.76)$$

By assembling Eqs. (9.74) - (9.76) together, the system of equations is

$$\frac{kt}{2\sqrt{3}}\begin{bmatrix} 2+2 & 1 & 1 & -3-3 \\ 1 & 2+2 & 1 & -3-3 \\ 1 & 1 & 2+2 & -3-3 \\ -3-3 & -3-3 & -3-3 & 6+6+6 \end{bmatrix}\begin{Bmatrix} T_1 \\ T_2 \\ T_3 \\ T_4 \end{Bmatrix}$$

$$= \begin{Bmatrix} (Q_{c1})_1 + (Q_{c1})_3 \\ (Q_{c2})_1 + (Q_{c2})_2 \\ (Q_{c3})_2 + (Q_{c3})_3 \\ (Q_{c4})_1 + (Q_{c4})_2 + (Q_{c4})_3 \end{Bmatrix} + \frac{Qt}{9\sqrt{3}}\begin{Bmatrix} 1+1 \\ 1+1 \\ 1+1 \\ 1+1+1 \end{Bmatrix} \tag{9.77}$$

Then, the boundary conditions of the specified nodal temperatures $T_1 = T_2 = T_3 = 0$ are applied. The system of equations

$$\frac{kt}{2\sqrt{3}}\begin{bmatrix} 4 & 1 & 1 & -6 \\ 1 & 4 & 1 & -6 \\ 1 & 1 & 4 & -6 \\ -6 & -6 & -6 & 18 \end{bmatrix}\begin{Bmatrix} 0 \\ 0 \\ 0 \\ T_4 \end{Bmatrix} = \begin{Bmatrix} Q_1 \\ Q_2 \\ Q_3 \\ 0 \end{Bmatrix} + \frac{Qt}{9\sqrt{3}}\begin{Bmatrix} 2 \\ 2 \\ 2 \\ 3 \end{Bmatrix} \tag{9.78}$$

It is noted that the first term on the right-hand side of the last equation is zero from the continuity of heat flux at node 4. Thus, the temperature at node 4 can be determined by using the fourth equation as

$$\frac{kt}{2\sqrt{3}}(0+0+0+18T_4) = 0 + \frac{Qt}{9\sqrt{3}}(3)$$

$$T_4 = \frac{1}{27}\frac{Q}{k}$$

After the temperature at node 4 is known, the conduction heat fluxes at nodes 1, 2 and 3 can be determined by using the first, second and third equation, respectively as

$$Q_1 = Q_2 = Q_3 = -\frac{6kt}{2\sqrt{3}}T_4 - \frac{2Qt}{9\sqrt{3}}$$

$$= -\frac{Qt}{3\sqrt{3}}$$

A finite element computer program is developed in the next section. The program can be used to re-analyze the example but by using many elements. Solution accuracy is increased as the plate is modeled by using more elements with smaller sizes. The accuracy of the finite element solution can be examined by comparing with the exact solution of

$$\overline{T}(x, y) = \frac{Q}{4k}(y-2+\sqrt{3}x)(y-\sqrt{3}x)y \tag{9.79}$$

9.6 Computer Program

One of the important aspects in studying the finite element method is to understand development and use of finite element computer programs. The finite element equations derived for different types of elements as presented in the preceding sections can be used for developing finite element computer programs directly. The programs can then be employed to analyze problems that have complex geometry. Solution accuracy increases if more elements are used in the finite element models. Efficiency of a finite element computer program depends on programming experience of developer.

Finite element computer program for heat transfer analysis is simple to develop and easy to understand as compared to the others for analyzing solid and fluid problems. This is mainly because the heat transfer problem is governed by a single energy equation that has only a dependent variable of the temperature. Thus each node has only one unknown of the nodal temperature. The total number of equations for the system of equations is equal to the number of nodes in the model. The heat transfer analysis program presented herein is for two-dimensional heat conduction in a plate that may have arbitrary geometry. The plate may be subjected to an internal heat generation with the boundary conditions of specified nodal temperatures. Examples of the finite element computer program for analyzing two-dimensional heat transfer problems are presented.

9.6.1 Problem statement

Steady-state temperature distribution T on a plate that lies in x-y coordinates as shown in Fig. 9.13 is determined from the governing differential equation

$$\frac{\partial}{\partial x}\left(k\frac{\partial T}{\partial x} \right) + \frac{\partial}{\partial y}\left(k\frac{\partial T}{\partial y} \right) = -Q \tag{9.80}$$

where k is the thermal conductivity coefficient, T is the temperature and Q is the internal heat generation rate per unit volume. The boundary conditions may consist of specifying temperature along the edge S_1 as shown in Eq. (9.43) and zero heat flux (insulated) along the edge S_2 as given by Eq. (9.47)

The three-node triangular element is used in the development of a computer program. The corresponding finite element equations necessary for developing the computer program for analyzing the problem statement above are

$$\underset{(3\times3)\ (3\times1)}{[K_c]\{T\}} = \underset{(3\times1)}{\{Q_Q\}} \tag{9.81}$$

where $[K_c]$ is the conduction matrix and $\{Q_Q\}$ is the load vector due to internal heat generation. Details of these two matrices are presented in Eqs. (9.64) and (9.67), respectively.

9.6.2 Details of computer program FINITE

A heat transfer analysis computer program according to the finite element formulation above is developed and named as FINITE. Listing of the program is provided in Appendix D. The program consists of a main program and many subroutines. Computational procedure of the program is as follows:

(a) Start from reading input data such as the numbers of nodes, elements, material properties, nodal coordinates, element nodal connections, etc., in the main program.

(b) Compute element matrices by calling subroutine TRI and assemble them together by calling subroutine ASSMBLE.

(c) Apply boundary conditions of the problem onto the system of equations by calling subroutine APPLYBC.

(d) Solve the set of equations for all nodal temperatures by calling subroutine GAUSS.

(e) Print the computed nodal temperatures into an output file.

The computational procedure above can be summarized by the flowchart in Fig. 9.22.

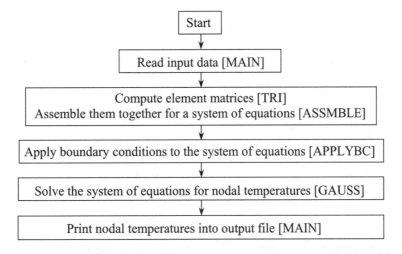

Figure 9.22 Computational procedure of the heat transfer analysis program FINITE. The word in [ ] denotes the subroutine names.

Main Program

Maximum numbers of the nodes (MXPOI) and elements (MXELE) are assigned at the beginning of the main program. Users can adjust these values for problems that may have different sizes. At the beginning of computation, the program asks for an input data file containing all information of the problem. During the computation, the program shows the status of performing different tasks such as determining element matrices, assembling element equations, applying boundary conditions, solving for solution of nodal temperatures, etc. After the analysis is completed, the program asks user to provide the output file names for storing the model information with its computed solutions.

Subroutine TRI

Subroutine TRI determines the element conduction matrix $[K_c]$ and the load vector due to internal heat generation $\{Q_Q\}$ as shown in Eqs. (9.64) and (9.67), respectively. The subroutine calls subroutine ASSMBLE to assemble these element matrices from all elements in the model together. The procedure used in assembling the element matrices is identical to that explained in example 9.2.

Subroutine ASSMBLE

Subroutine ASSMBLE combines all element matrices together to yield a system of equations by using the same procedure as shown in Eq. (9.77) of Example 9.2. By assigning the equation numbers as the node numbers, the subroutine can assemble element equations for a model that has a large number of elements.

Subroutine APPLYBC

Subroutine APPLYBC applies the boundary conditions to the system of equations. For example, if a system of equations consists of 4 equations as follow

$$\begin{bmatrix} 4 & 3 & 2 & 1 \\ 3 & 5 & 4 & 3 \\ 2 & 4 & 6 & 5 \\ 1 & 3 & 5 & 7 \end{bmatrix} \begin{Bmatrix} T_1 \\ T_2 \\ T_3 \\ T_4 \end{Bmatrix} = \begin{Bmatrix} 10 \\ 20 \\ 30 \\ 40 \end{Bmatrix} \tag{9.82}$$

and if the temperature at node 3 is specified as $T_3 = 100$, then the subroutine APPLYBC modifies the system of Eq. (9.82) to

$$\begin{bmatrix} 4 & 3 & 0 & 1 \\ 3 & 5 & 0 & 3 \\ 0 & 0 & 1 & 0 \\ 1 & 3 & 0 & 7 \end{bmatrix} \begin{Bmatrix} T_1 \\ T_2 \\ T_3 \\ T_4 \end{Bmatrix} = \begin{Bmatrix} 10-2(100) \\ 20-4(100) \\ 100 \\ 40-5(100) \end{Bmatrix} \tag{9.83}$$

so that the computed temperature at node 3 is $T_3 = 100$ after calling the subroutine GAUSS below.

Subroutine GAUSS

Subroutine GAUSS solves for the nodal temperatures from the system of equations. The subroutine uses the Gauss elimination technique for solving the set of algebraic equations. The subroutine calls the two subroutines SCALE and PIVOT that perform scaling and pivoting as explained in Chapter 3 during the solving process.

From the explanation of the subroutines above, it can be seen that the computational procedure used in the finite element computer program is the same as presented in the examples. The main advantage for using the program is the ability in analyzing problems with a large number of elements and unknowns. In addition, the program can be modified to solve different types of problems in the other fields. Users can only modify the subroutine for determining the finite element matrices that correspond to the new problems.

9.6.3 Input data for computer program

Input data required by FINITE program is divided into five parts as explained below.

Part 1   Problem description:
First line   Number of lines for describing problem
Next lines   Words contained within the number of lines prescribed above

Example: 2
CONDUCTION IN PLATE WITH SPECIFIED TOP
EDGE TEMPERATURE AS SINE FUNCTION

Part 2 Size of problem:
First line   Words for number of nodes and elements
Second line   Number of nodes and elements
Example: NPOIN   NELEM
              25   32

Part 3 Material properties and plate thickness:
First line   Words for material properties and plate thickness
Second line   Values for the thermal conductivity coefficient and plate thickness
Example: TK   THICK
            1.0   0.1

Part 4 Nodal information:
First line   Words for nodal information
Next lines   Node number, boundary condition, *x*- and *y*-coordinates, and temperature
Example: NODE   IBC   X   Y   T
              1   1   0.00   0.00   0.0
              2   0   0.25   0.00   0.0
              ⋮   ⋮   ⋮   ⋮   ⋮
              25   1   1.00   1.00   0.8

Note on the boundary condition codes:
          IBC = 1   Temperature is prescribed
          IBC = 0   Temperature is to be determined

Part 5 Element information:
First line   Words for element information
Next lines   Element number, three node numbers, internal heat generation
Example: ELEMENT   I   J   K   Q
              1   7   3   4   0.0
              2   15   8   11   0.2
              ⋮   ⋮   ⋮   ⋮   ⋮
              32   13   21   24   0.0

Note that the element nodal numbers (I, J, K) are given in the counterclockwise direction.

The Explanation of the input data information above is apparent by considering the example below.

9.6.4 Example

Example 9.3 The finite element program FINITE is used to solve the equilateral triangular plate subjected to an internal heat generation as shown in Example 9.2. The plate is modeled by using 3 triangular elements with 4 nodes. Nodes 1, 2 and 3 are at the three corners of the plate for which the temperature is specified as zero. The only unknown is the temperature at node 4 which is at the plate centroid. By assuming the thermal conduction coefficient $k = 1$ and the internal heat generation $Q = 1$, then the temperature at node 4 obtained from using the three-element model is 1/27 or 0.037037.

The input data corresponds to the three element model of Fig. 9.20 is presented in the file "EX1.DAT" as shown in Fig. 9.23.

```
2
TRIANGULAR PLATE WITH INTERNAL HEAT GENERATION.
CRUDE MESH WITH 4 NODES AND 3 ELEMENTS.
   NPOIN    NELEM
      4        3
      TK    THICK
      1.       .1
NODAL BOUNDARY CONDITIONS AND COORDINATES [4]:
      1      1     0.00000     0.00000     0.
      2      1     1.15470     0.00000     0.
      3      1     0.57735     1.00000     0.
      4      0     0.57735     0.33333     0.
ELEMENT NODAL CONNECTIONS AND HEAT GEN. [3]:
      1      1     2      4     1.
      2      2     3      4     1.
      3      3     1      4     1.
```

Figure 9.23 Input data file "EX1.DAT".

The program FINITE starts the execution by asking for an input data file name. With the information in the input data file, the program performs calculation and the computational status is presented on the monitor screen as shown in Fig. 9.24. After the calculation is completed, the program asks user to enter the solution file name. User may enter the file name as "SOL.OUT" as shown in the figure.

```
>FINITE      <Enter>

PLEASE ENTER THE INPUT FILE NAME:
EX1.DAT

THE F.E. MODEL INCLUDES THE FOLLOWING HEAT TRANSFER MODE(S):
   --   HEAT CONDUCTION
   --   INTERNAL HEAT GENERATION

*** THE FINITE ELEMENT MODEL CONSISTS OF   4 NODES AND  3
    ELEMENTS ***

*** ESTABLISHING ELEMENT MATRICES AND ASSEMBLING ELEMENT
    EQUATIONS ***

*** APPLYING BOUNDARY CONDITIONS OF NODAL TEMPERATURES ***

*** SOLVING A SET OF SIMULTANEOUS EQUATIONS FOR TEMPERATURE
    SOLUTIONS ***
    ( TOTAL OF   4 EQUATIONS TO BE SOLVED )

PLEASE ENTER FILE NAME FOR TEMPERATURE SOLUTIONS:
SOL.OUT

Stop - Program terminated
```

Figure 9.24 Computational status that appears on screen monitor
during executing the program FINITE.

The solution output file "SOL.OUT" containing all nodal temperatures is shown in Fig. 9.25.

```
NODAL TEMPERATURE SOLUTIONS [    4]:

   NODE    TEMPERATURE

      1    .000000E+00
      2    .000000E+00
      3    .000000E+00
      4    .370370E-01
```

Figure 9.25 Solution output file "SOL.OUT".

It can be seen from the output file "SOL.OUT" in Fig. 9.25 that the computed temperature at node 4 is 0.037037. The computed nodal temperature is identical to that obtained in Example 9.2.

To improve the solution accuracy, the triangular plate has to be modeled by using more elements. The input data file "EX1.DAT" must be modified to include information of the refined finite element model. The refined model is left for an exercise.

Example 9.4 The finite difference technique was used in Example 8.2 to solve for the temperature distribution in a square plate as shown in Fig. 9.26. The same problem is re-analyzed herein by using the finite element computer program FINITE. The computed finite element solution is compared with the exact solution in Eq. (8.25).

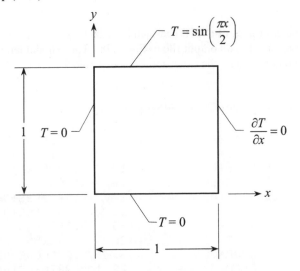

Figure 9.26 A square plate with specified temperatures and zero heat flux along the four edges.

Figure 9.27 shows a finite element model that consists of 32 triangular elements and 25 nodes. An input data file similar to "EX1.DAT" can be created.

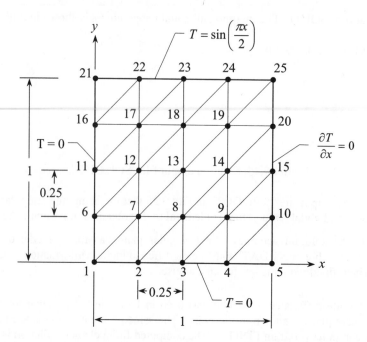

Figure 9.27 A finite element model with 32 triangular elements and 25 nodes.

The finite element computer program FINITE can then be used to analyze the problem. The computed nodal temperatures are presented in an output file in Fig. 9.28. These nodal temperatures are plotted as shown in Fig. 9.29 and compared with the exact solutions in Table 9.1.

```
NODAL   TEMPERATURE SOLUTIONS [  25]:

NODE      TEMPERATURE          NODE      TEMPERATURE
  1       .000000E+00           14       .351000E+00
  2       .000000E+00           15       .379919E+00
  3       .000000E+00           16       .000000E+00
  4       .000000E+00           17       .245359E+00
  5       .000000E+00           18       .453365E+00
  6       .000000E+00           19       .592350E+00
  7       .675522E-01           20       .641155E+00
  8       .124820E+00           21       .000000E+00
  9       .163086E+00           22       .382680E+00
 10       .176523E+00           23       .707110E+00
 11       .000000E+00           24       .923880E+00
 12       .145389E+00           25       .100000E+01
 13       .268643E+00
```

Figure 9.28 Nodal temperatures of the square plate in Fig. 9.27 obtained from using the finite element computer program FINITE.

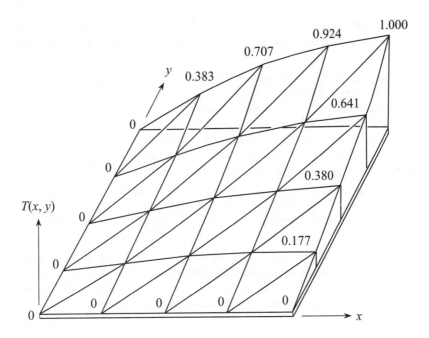

Figure 9.29 Carpet plot of the temperature distribution over the square plate obtained from using the finite element computer program FINITE.

Table 9.1 Comparison of the finite element and exact solutions for the square plate in Fig. 9.27. Numbers in brackets are the exact solutions.

0.000	0.383	0.707	0.924	1.000
(0.000)	(0.383)	(0.707)	(0.924)	(1.000)
0.000	0.245	0.453	0.592	0.641
(0.000)	(0.244)	(0.452)	(0.590)	(0.639)
0.000	0.145	0.269	0.351	0.380
(0.000)	(0.144)	(0.267)	(0.349)	(0.377)
0.000	0.068	0.125	0.163	0.177
(0.000)	(0.067)	(0.124)	(0.162)	(0.175)
0.000	0.000	0.000	0.000	0.000
(0.000)	(0.000)	(0.000)	(0.000)	(0.000)

9.7 Closure

In this chapter, overview of the finite element method and its applications were presented. The chapter started with the explanation of the finite difference method because of it basic properties are similar to the finite element method. The finite difference method transforms differential equations into approximate algebraic equations. The first step of the finite difference method is to divide the domain into small rectangles. From the explanation, the method is simple and easy to understand.

However, rectangles cannot accurately represent the arbitrary geometry with curves and circles. The method can not provide accurate solutions for practical problems of which geometries are usually complex.

General procedure of the finite element method that consists of six steps was then explained. The first step is to discretize the computational domain into a number of elements. Element shapes could be triangle or quadrilateral which can model arbitrary geometry more precisely. These elements are connected by nodes at which the unknowns are to be determined. The most important step of the finite element method is the derivation of the finite element equations that correspond to the given problem. The element equations from each element are then assembled, leading to a large set of algebraic equations. Boundary conditions are applied prior to solving for solutions at all nodes.

The finite element equations and their matrices were derived for both one- and two-dimensional elements. Examples of the finite element method for analyzing one- and two-dimensional problems were presented. These examples help readers to understand the finite element method more clearly.

A finite element computer program was developed and presented in the last section of the chapter. The program solves two-dimensional Poisson's equation subjected to arbitrary boundary conditions. Listing of the program, that follows the formulation presented in the text, was provided in an appendix. The program can be modified easily to solve other problems that have complex geometry by adding new element types with proper element equations and matrices.

Exercises

1.   Given the ordinary differential equation

$$\frac{d^2\phi}{dx^2} + \alpha\phi + \beta = 0 \qquad\qquad a < x < b$$

where α and β are constants, derive the finite element equations from the given differential equation for a linear two-node element of length L in Fig. 9.8.

2.   Use the finite element equations derived in Problem 1 to develop a system of equations for the finite element model with m elements and the boundary conditions in Fig. P9.2.

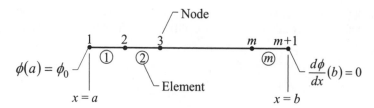

Figure P9.2

3.   Describe the alternation of the system of equations developed in Problem 2 if the boundary condition at the right end where $x = b$ is changed to

$$\frac{d\phi}{dx}(b) + C\,\phi(b) = D$$

where C and D are constants.

4.   Analyze the problem in Example 9.1 again but by using 4 elements. Give comments on the accuracy of the computed nodal temperatures as compared to the exact solution.

5.   Use the weighted residuals method to show that the finite element equations for the two-node linear element with the length L corresponding to the differential equation

$$\frac{d}{dx}\left(a\frac{du}{dx}\right) + cu = f$$

are $[K]\{u\} = \{F\}$

where $[K] = \dfrac{a}{L}\begin{bmatrix} 1 & -1 \\ -1 & 1 \end{bmatrix} + \dfrac{cL}{6}\begin{bmatrix} 2 & 1 \\ 1 & 2 \end{bmatrix}$

and $\{F\} = \dfrac{fL}{2}\begin{Bmatrix} 1 \\ 1 \end{Bmatrix}$

6.   Apply the weighted residuals method to derive the finite element equations from the differential equation

$$-\frac{d^2u}{dx^2} + \pi^2 u = 2\pi^2 \sin(\pi x)$$

by using the two-node element of length L. Then, use the derived finite element equations to solve for the solutions of u_2, u_3 and u_4 in Fig. P9.6. Plot to compare the computed solution with the exact solution of $u(x) = \sin(\pi x)$.

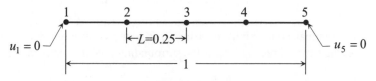

Figure P9.6

7.   Apply the weighted residuals method to derive the finite element equations from the differential equation

$$\frac{d^2u}{dx^2} = u + 4e^x$$

by using the two-node element of length L. Then, use the derived finite element equations to solve for the solutions of u_2, u_3, u_4 and u_5 in Fig. P9.7. Plot to compare the computed solution with the exact solution of $u(x) = (2x+1)e^x$.

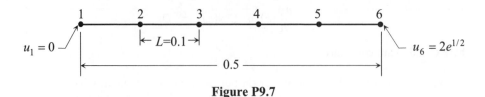

$u_1 = 0$ $\leftarrow L{=}0.1 \rightarrow$ $u_6 = 2e^{1/2}$

0.5

Figure P9.7

8.   Conduction heat transfer in a tapered rod with internal heat generation Q as shown in Fig. P9.8 is governed by the differential equation

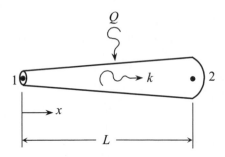

Figure P9.8

$$\frac{d}{dx}\left(k\,A(x)\frac{dT}{dx} \right) = -Q\,A(x)$$

$$A(x) = N_1 A_1 + N_2 A_2$$

$$= \left(1 - \frac{x}{L} \right) A_1 + \left(\frac{x}{L} \right) A_2$$

where A_1 and A_2 are the areas at nodes 1 and 2, respectively. Derive the finite element equations and their element matrices in closed-form so that they can be used for computer programming directly.

9.   Develop the system of the finite element equations for heat conduction in a fin with surface convection along it length and end surface at node 3 with the area of A as shown in Fig. P9.9. The governing differential equation of this problem is

$$kA\,\frac{d^2T}{dx^2} - h\,p\,(T - T_\infty) = 0$$

where k is the thermal conductivity coefficient, h is the convection coefficient, p is the fin perimeter, T is the temperature, and T_∞ is the surrounding medium temperature.

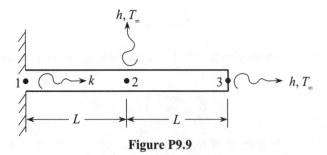

Figure P9.9

10.   A tapered bar hanging from ceiling, shown in Fig. P9.10, deforms due to its own weight. The governing differential equation for the equilibrium along the bar is

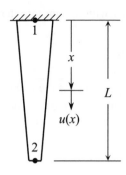

$$\frac{d}{dx}\left(E\,A(x)\frac{du}{dx}\right) = -\gamma\,A(x)$$

where E is the modulus of elasticity, γ is the specific weight, and $A(x)$ is the cross-sectional area

$$A(x) = N_1 A_1 + N_2 A_2 = \left(1-\frac{x}{L}\right) A_1 + \left(\frac{x}{L}\right) A_2$$

where A_1 and A_2 are the cross-sectional areas at nodes 1 and 2, respectively. Derive the finite element equations and apply the appropriate boundary conditions to solve for the displacement at node 2 and the element stress.

Figure P9.10

11. Heat conduction in a ring with inner radius a and outer radius b, as shown in Fig. P9.11, is governed by the differential equation

$$k\frac{d}{dr}\left(r\frac{dT}{dr}\right) = 0 \qquad\qquad a < r < b$$

where r is the radius, k is the thermal conductivity coefficient, and T is the temperature. Derive the finite element equations and their element matrices by assuming the element temperature distribution in the form

(a) $\quad T(r) = C_1 + C_2 r$

(b) $\quad T(r) = C_1 + C_2 \ln r$

where C_1 and C_2 are constants. It should be noted that the assumed element temperature distribution in (b) provides exact solution to the problem.

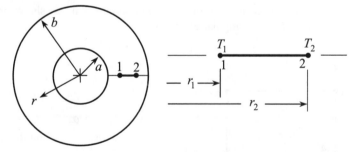

Figure P9.11

12. If the outer edge of the ring in Fig. P9.11 has surface heat convection to the surrounding medium, the heat flux along this edge is

$$-k\frac{dT}{dr}(b) = h(T(b)-T_\infty)$$

where h is the convection coefficient, and T_∞ is the surrounding medium temperature. Derive the finite element equations to show how the surface heat convection alters the solutions as compared to Problem 11.

13.   The differential equation that describes the displacement in a bar with non-uniform temperature distribution is

$$\frac{d}{dx}\left[EA\left(\frac{du}{dx}-\alpha T\right)\right] = 0$$

where E is the modulus of elasticity, A is the cross-sectional area, $u(x)$ is the displacement, α is the coefficient of thermal expansion, and T is the temperature. Derive the corresponding finite element equations and the element matrices by using a two-node linear element as shown in Fig. 9.8 for the case of:

    (a)   constant temperature T

    (b)   linear temperature $T = \left(1-\dfrac{x}{L}\right)T_1 + \left(\dfrac{x}{L}\right)T_2$

where T_1 and T_2 are the temperatures at nodes 1 and 2, respectively.

14.   Show that the finite element equations for a three-node triangular element subjected to an internal heat generation and specified heating on its surface are in the same form as those shown in Eq. (9.59b), i.e.

$$[K_c]\{T\} = \{Q_c\}+\{Q_Q\}+\{Q_q\}+\{Q_S\}$$

where $\{Q_S\}$ is the heat load vector due to surface heating. Such vector can be derived in closed-form as

$$\{Q_S\} = \frac{qA}{3}\begin{Bmatrix}1\\1\\1\end{Bmatrix}$$

where q is the surface heating rate per unit area and A is the area of the triangle.

15.   Determine the temperature at the centroid of a unit square plate as shown in Fig. 9.15. The plate is discretized by using 4 triangular elements.

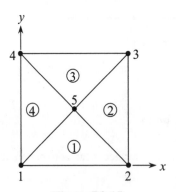

$$\frac{\partial^2 u}{\partial x^2} + \frac{\partial^2 u}{\partial y^2} = -1$$

with $u = 0$ along the four edges.

Figure P9.15

Compare the computed solution with the exact solution of

$$u(x, y) = \frac{16}{\pi^4}\sum_{i,j=1,3,5,\dots}^{\infty}\frac{\sin(i\pi x)\sin(j\pi y)}{i^3 j^2 + i^2 j^3}$$

16. Show that the finite element equations for a three-node triangular element subjected to an internal heat generation and surface convection are in the same form as those shown in Eq. (9.59b), i.e.

$$[[K_c]+[K_h]]\{T\} = \{Q_c\}+\{Q_Q\}+\{Q_q\}+\{Q_h\}$$

The convection matrix and load vector can be derived in closed form as

$$[K_h] = \frac{hA}{12}\begin{bmatrix} 2 & 1 & 1 \\ 1 & 2 & 1 \\ 1 & 1 & 2 \end{bmatrix} \quad \text{and} \quad \{Q_h\} = \frac{hAT_\infty}{3}\begin{Bmatrix} 1 \\ 1 \\ 1 \end{Bmatrix}$$

where h is the convection coefficient, A is the element area and T_∞ is the surrounding medium temperature.

17. The load vector associated with specified heating along an element edge can be written in the form of Eq. (9.69). Derive such load vector in a closed-form as shown in Eq. (9.70) so that it can be used in computer programming directly.

18. Show the derivation of Eq. (9.59) from Eq. (9.58) in details. Draw figures of possible boundaries to help the explanation.

19. Derive the element interpolation functions for a four-node rectangular element as shown in Fig. P9.19. The element has four nodal unknowns of temperature as T_1, T_2, T_3 and T_4. Start the derivation by assuming the element temperature distribution in the bilinear form

$$T(x, y) = \alpha_1 + \alpha_2 x + \alpha_3 y + \alpha_4 xy$$

in order to obtain the element interpolation matrix $\lfloor N \rfloor$ so that

$$T(x, y) = \underset{(1\times4)\ (4\times1)}{\lfloor N \rfloor \{T\}}$$

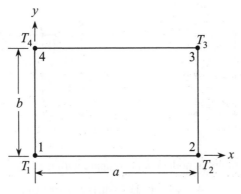

Figure P9.19

20.   Use the element interpolation matrix obtained from Problem 19 to derive the finite element matrices in Eq. (9.59) for the 4-node rectangular element. Show the derivation in details and explain physical meaning of each matrix.

21.   Apply the finite element matrices derived in Problem 20 to solve the problem in Example 9.4 again by discretizing the plate into four equal elements. Compare the computed solution with the exact solution in Eq. (8.25).

22.   The differential equation for a bar under torsion is

$$\frac{\partial^2 \phi}{\partial x^2} + \frac{\partial^2 \phi}{\partial y^2} = -2G\theta$$

where ϕ is the stress function, G is the shearing modulus, and θ is the torsional angle. Use the method of weighted residuals to derive the finite element equations. Then, derive their element matrices in closed-form for a three-node finite element as shown in Fig. 9.15 so that the matrices can be used in the development of a finite element program directly.

23.   The problem in Fig. P9.23 has a symmetrical solution due to its geometry and boundary conditions.

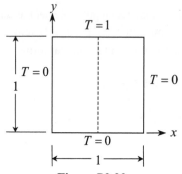

Figure P9.23

$$\frac{\partial^2 T}{\partial x^2} + \frac{\partial^2 T}{\partial y^2} = 0$$

$T = 1$ on upper edge

$T = 0$ on other three edges

Use the three finite element models below

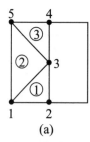

(a)

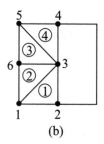

(b)

to determine the temperature at node 3, and compare with the exact solution of

$$T = \frac{4}{\pi} \sum_{n=1,3,5,...}^{\infty} \frac{(-1)^{n+1} \sin(n\pi x)\, \sinh(n\pi y)}{n \sinh(n\pi)}$$

24. Use the finite element computer program FINITE to determine the temperature distribution in the equilateral triangular plate subjected to internal heat generation in Example 9.3 again but by using a finite element model with at least 50 elements. Comment on the improved solution accuracy by using more elements. Compare the computed solution with the exact solution as shown in Eq. (9.79).

25. Solve for the temperature distribution in the rectangular plate as shown in Example 9.4 by using the finite element computer program FINITE. Divide the plate into 8 equal intervals in both x- and y-directions. Compare the computed solution with the exact solution of Eq. (8.25) and the finite element solution in Table 9.1.

26. A 2×2 unit square plate as shown in Fig. P9.26 has a specified zero temperature along the four edges. If both the thermal conductivity coefficient k and the internal heat generation rate Q of the plate are specified as one unit, then the exact temperature distribution over the plate is

$$T(x,y) = \frac{(1-x^2)}{2} - \frac{16}{\pi^3} \sum_{n=0}^{\infty} \frac{(-1)^n \cos((2n+1)\pi x/2) \cosh((2n+1)\pi y/2)}{(2n+1)^3 \cosh((2n+1)\pi/2)}$$

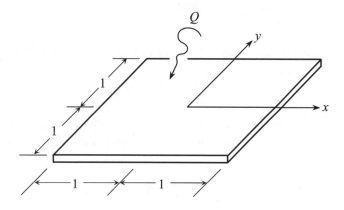

Figure P9.26

Use the finite element program FINITE with an appropriate mesh to solve for the temperature distribution. Compare the computed temperature distribution with the exact solution given above.

27. An insulator with the dimensions as shown in Fig. P9.27 has its thermal conductivity coefficient of 0.12 W/m-°C and the specified temperature along the inner and outer edges as shown in the figure. Use the program FINITE with an appropriate mesh to determine the temperature distribution. Plot the computed temperature distribution by using contour lines.

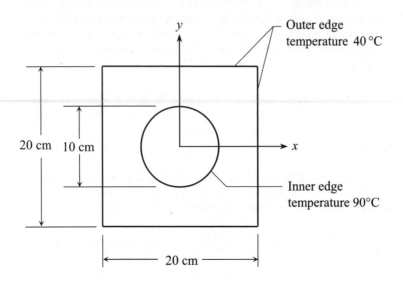

Figure P9.27

28.   Fluid flow through a porous media is governed by Laplace equation in the form

$$\frac{\partial^2 h}{\partial x^2} + \frac{\partial^2 h}{\partial y^2} = 0$$

where h is the head of the fluid. Use the program FINITE with an appropriate mesh to determine distribution of the fluid head for the geometry and boundary conditions as shown in Fig. 9.28.

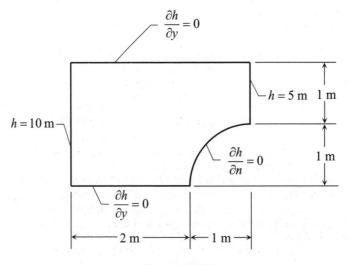

Figure P9.27

29. Study the finite element computer program FINITE in details. Then, explain how to add other element types, such as the one-dimensional two-node linear element and the two-dimensional four-node bilinear element, into the program. Draw the flow chart of the modified program and explain the computational tasks of new subroutines associated with the new elements.

30. Improve the finite element computer program FINITE to have additional capability for analysis of surface convection heat transfer. Use the program to determine the temperature that occurs in a rectangular plate with surface convection. The plate with the length a of 0.5 m, the width b of 0.3 m and the thickness t of 0.0025 m, has a specified edge temperature T_i of 80°C. The plate has surface convection to a surrounding medium that has the temperature T_∞ of 0°C. If the thermal conductivity coefficient of the plate is $k = 38.0$ W/m-°C and its surface convection coefficient is $h = 11.4$ W/m-°C, use the modified program with an appropriate mesh to determine the plate temperature distribution. Then compare the computed plate temperature distribution with the exact solution of

$$T(x,y) = \frac{2}{\pi} \sum_{n=1}^{\infty} A_n \left[\frac{\sin \lambda_n \, x \, (\sinh \alpha_n y + \sinh \alpha_n (b-y))}{n \sinh \alpha_n b} + \frac{\sin \gamma_n y \, (\sinh \beta_n x + \sinh \beta_n (a-x))}{n \sinh \beta_n a} \right]$$

where

$$A_n = T_i - (-1)^n T_i \quad ; \quad \lambda_n = \frac{n\pi}{a} \quad ; \quad \gamma_n = \frac{n\pi}{b}$$

$$\alpha_n = \left(\kappa^2 + \frac{n^2 \pi^2}{a^2} \right)^{1/2} \quad ; \quad \beta_n = \left(\kappa^2 + \frac{n^2 \pi^2}{b^2} \right)^{1/2} \quad ; \quad \kappa^2 = \frac{h}{kt}$$

31. Use the finite element program FINITE to solve the partial differential equation

$$\frac{\partial^2 u}{\partial x^2} + \frac{\partial^2 u}{\partial x^2} = 0$$

on a unit square domain. Divide the domain into 10 equal intervals in both x- and y-directions. The boundary conditions of the square domain are

$$u(x, 0) = 0 \quad ; \quad u(0, y) = \frac{y}{1 + y^2}$$

$$u(x, 1) = \frac{1}{(1+x)^2 + 1} \quad ; \quad u(1, y) = \frac{y}{4 + y^2}$$

Plot to compare the computed solution with the exact solution of

$$u(x, y) = \frac{y}{(1+x)^2 + y^2}$$

32. Use the finite element computer program FINITE to solve the partial differential equation

$$\frac{\partial^2 u}{\partial x^2} + \frac{\partial^2 u}{\partial y^2} = 0$$

on a rectangular domain of 2×1 units. Divide the domain into 20 and 10 equal intervals in x- and y-direction, respectively. The boundary conditions along the edges of the domain are

$$u(x, 0) = \ln\left((x+1)^2 + 1\right) ; \qquad u(0, y) = \ln\left(1 + (y+1)^2\right)$$

$$u(x, 1) = \ln\left((x+1)^2 + 4\right) ; \qquad u(2, y) = \ln\left(9 + (y+1)^2\right)$$

Plot to compare the computed solution with the exact solution of

$$u(x, y) = \ln\left((x+1)^2 + (y+1)^2\right)$$

33.   Prepare input data for the finite element computer program FINITE for solving the partial differential equation

$$\frac{\partial^2 u}{\partial x^2} + \frac{\partial^2 u}{\partial y^2} = -2\pi^2 \sin(\pi x)\sin(\pi y)$$

on a unit square domain. The boundary condition along the four edges of the domain is $u = 0$. Use the three finite element meshes with (a) 5×5, (b) 10×10 and (c) 20×20 equal intervals to study the solution convergence. Compare the finite element solutions with the exact solution of $u(x, y) = \sin(\pi x)\sin(\pi y)$.

34.   Prepare input data for the finite element computer program FINITE for solving the partial differential equation

$$\frac{\partial^2 u}{\partial x^2} + \frac{\partial^2 u}{\partial y^2} = 2x\left(x^3 - 6xy + 6xy^2 - 1\right)$$

on a unit square domain. The boundary condition along the four edges of the domain is $u = 0$. Use the three finite element meshes with (a) 5×5, (b) 10×10 and (c) 20×20 equal intervals to study the solution convergence. Compare the finite element solutions with the exact solution of $u(x, y) = \left(x - x^4\right)\left(y - y^2\right)$. ·

35.   Set up input data for the finite element computer program FINITE for solving the partial differential equation

$$\frac{\partial^2 u}{\partial x^2} + \frac{\partial^2 u}{\partial y^2} = -2\left(x^2 + y^2\right)$$

on a unit square domain. The boundary conditions along the four edges of the domain are

$$u(x, 0) = 1 - x^2 \qquad\quad ; \; u(0, y) \quad = 1 + y^2$$

$$u(x, 1) = 2\left(1 - x^2\right) \qquad ; \; u(1, y) \quad = 0$$

Use the three finite element meshes with (a) 5×5, (b) 10×10 and (c) 20×20 equal intervals to study the solution convergence. Compare the finite element solutions with the exact solution of $u(x, y) = \left(1 - x^2\right)\left(1 + y^2\right)$.

36. Modify the finite element computer program FINITE for solving the partial differential equation

$$\pi^2 \frac{\partial^2 u}{\partial x^2} + \frac{\partial^2 u}{\partial y^2} = 0$$

on a unit square domain. The boundary conditions along the four edges are

$$u(x, 0) = 0 \quad ; \quad u(0, y) = \sin(\pi y)$$

$$u(x, 1) = 0 \quad ; \quad u(1, y) = e \sin(\pi y)$$

Use the two finite element meshes with (a) 5×5 and (b) 10×10 equal intervals to obtain the finite element solutions. Compare the finite element solutions with the exact solution of $u(x, y) = e^x \sin(\pi y)$.

37. Use the weighted residuals method to derive the finite element equations and their matrices corresponding to the partial differential equation

$$\frac{\partial^2 u}{\partial x^2} + \frac{\partial^2 u}{\partial y^2} + 2u = g(x, y)$$

where

$$g(x, y) = (xy+1)(xy-x-y) + x^2 + y^2$$

Show the derivation in details for the three-node triangular element.

38. Modify the finite element computer program FINITE by incorporating the finite element matrices derived in Problem 37. Then, employ the modify program to solve for the solution on a unit square domain with the boundary condition of $u = 0$ along the four edges. Use the two finite element meshes with (a) 5×5 and (b) 10×10 equal intervals to obtain the finite element solutions. Compare the finite element solutions with the exact solution of $u(x, y) = xy(x-1)(y-1)/2$.

Bibliography

Atkinson, K. and Han, W., *Elementary Numerical Analysis*, Second Edition, John Wiley & Sons, New York, 2004.

Borse, G. J., *FORTRAN77 and Numerical Methods for Engineers*, Second Edition, PWS-Kent Publishing, Boston, 1991.

Bradie, B., *A Friendly Introduction to Numerical Analysis*, Pearson Education International, 2004.

Buchanan, J. L. and Turner, P. R., *Numerical Methods and Analysis*, McGraw-Hill, New York, 1992.

Burden, R. L. and Faires, J. D., *Numerical Analysis*, Fifth Edition, PWS Publishing, Boston, 1993.

Carnahan, B., Luther, H. A. and Wilkes, J. O., *Applied Numerical Methods*, John Wiley & Sons, New York, 1969.

Carslaw, H. S. and Jaeger, J. C., *Conduction of Heat in Solids*, Second Edition, Oxford University Press, London, 1959.

Chapman, S. J., *MATLAB Programming for Engineers*, Third Edition, Thomson International Edition, 2004.

Chapra, S. C. and Canale, R. P., *Numerical Methods for Engineers*, Fifth Edition, McGraw-Hill International, 2006.

Cheney, W. and Kincaid, D., *Numerical Mathematics and Computing*, Sixth Edition, Thomson International Edition, 2008.

Davies, A. J., *The Finite Element Method: A First Approach*, Clarendon Press, Oxford, 1980.

Davis, G. B. and Hoffmann, T. R., *FORTRAN77: A Structured, Disciplined Style*, Third Edition, McGraw-Hill, New York, 1988.

Dechaumphai, P., *Finite Element Method: Fundamentals and Applications*, Alpha Science International, Oxford, 2010.

Dechaumphai, P. and Phongthanapanich, S., *Easy Finite Element Method with Software*, Alpha Science International, Oxford, 2009.

Dorn, W. S. and McCracken, D. D., *Numerical Methods with FORTRAN IV Case Studies*, John Wiley & Sons, New York, 1972.

Fausett, L. V., *Applied Numerical Analysis Using MATLAB*, Prentice Hall, New Jersey, 1999.

Ferziger, J. H., *Numerical Methods for Engineering Application*, Second Edition, John Wiley & Sons, New York, 1998.

Fox, R. W., McDonald, A. T. and Pritchard, P. J., *Introduction to Fluid Mechanics*, Sixth Edition, John Wiley & Sons, New York, 2005.

Gerald, C. F. and Wheatley, P. O., *Applied Numerical Analysis*, Seventh Edition, Pearson Education International, 2004.

Gilat, A., *MATLAB: An Introduction with Applications*, Second Edition, John Wiley & Sons, New York, 2005.

Gilat, A. and Subramaniam, V., *Numerical Methods for Engineers and Scientist: An Introduction with Applications Using MATLAB*, John Wiley & Sons, New York, 2007.

Hoffman, J. D., *Numerical Methods for Engineers and Scientists*, McGraw-Hill, New York, 1993.

Holman, J. P., *Heat Transfer*, Ninth Edition, McGraw-Hill, New York, 2006.

Hombeck, R. W., *Numerical Methods*, Prentice-Hall, New Jersey, 1975.

Kahaner, D., Moler, C. and Nash, S., *Numerical Methods and Software*, Prentice Hall, New Jersey, 1989.

Kaplan, W., *Advanced Calculus*, Fifth Edition, Addison-Wesley, Massachusetts, 2003.

Lam, C. Y., *Applied Numerical Methods for Partial Differential Equations*, Prentice Hall, New York, 1994.

Lapidus, L. and Pinder, G. F., *Numerical Solutions of Partial Differential Equations in Science and Engineering*, John Wiley & Sons, New York, 1981.

Mathews, J. H. and Fink, K. D., *Numerical Methods Using MATLAB*, Fourth Edition, Pearson Education International, 2004.

MATLAB Reference Guide, The MathWorks, Inc, Massachusetts, 1992.

Moore, H., *MATLAB for Engineers*, Second Edition, Pearson Education International, 2009.

Nakamura, S., *Applied Numerical Methods with Software*, Prentice Hall International, 1991.

Palm, W. J., *A Concise Introduction to MATLAB*, McGraw-Hill International, 2008.

Press, W. H., Flannery, B. P., Teukolsky, S. A. and Vetterling, W. T., *Numerical Recipes - The Art of Scientific Computing*, Cambridge University Press, Cambridge, 1989.

Rao, S. S., *Applied Numerical Methods for Engineers and Scientists*, Pearson Education International, 2002.

Rice, R. J., *Numerical Methods, Software, and Analysis*, Second Edition, Academic Press, San Diego, 1993.

Shewchuk, J. R., *An Introduction to the Conjugate Gradient Method Without the Agonizing Pain*, School of Computer Science, Carnegie Mellon University, 1994.

Thomas, G. B. and Finney, R. L., *Calculus and Analytical Geometry*, Fifth Edition, Addison-Wesley, Reading, 1979.

Trainor, T. N. and Krasnewich, D., *Computers*, Third Edition, McGraw-Hill, New York, 1992.

Ugural, A. C. and Fenster, S. K., *Advanced Strength and Applied Elasticity*, Fourth Edition, Prentice Hall, New York, 2003.

Zienkiewicz, O. C., Taylor, R. L. and Zhu, J. Z., *The Finite Element Method: Its Basis and Fundamentals*, Sixth Edition. Elsevier, Oxford, 2005.

Appendix A

Matrices

Understanding concepts of matrices and their uses are important in the study of the numerical methods. Matrices are used in the topics of solving a set of simultaneous equations, interpolation functions, least-square regressions, finite difference and finite element methods. Concepts of matrices are employed in the development of the corresponding computer programs. In this appendix, definitions, properties and basic operations of matrices are presented.

A.1 Definitions

Matrices provide a shorthand scheme for dealing with systems of linear algebraic equations. For example, a set of three linear algebraic equations can be expressed in the form

$$
\begin{aligned}
a_{11}x_1 + a_{12}x_2 + a_{13}x_3 &= b_1 \\
a_{21}x_1 + a_{22}x_2 + a_{23}x_3 &= b_2 \\
a_{31}x_1 + a_{32}x_2 + a_{33}x_3 &= b_3
\end{aligned}
\tag{A.1}
$$

These equations can be written in matrix form as

$$
\begin{bmatrix} a_{11} & a_{12} & a_{13} \\ a_{21} & a_{22} & a_{23} \\ a_{31} & a_{32} & a_{33} \end{bmatrix}
\begin{Bmatrix} x_1 \\ x_2 \\ x_3 \end{Bmatrix}
=
\begin{Bmatrix} b_1 \\ b_2 \\ b_3 \end{Bmatrix}
\tag{A.2}
$$

Or, in short

$$
[A]\{X\} = \{B\}
\tag{A.3}
$$

where
$$[A] = \begin{bmatrix} a_{11} & a_{12} & a_{13} \\ a_{21} & a_{22} & a_{23} \\ a_{31} & a_{32} & a_{33} \end{bmatrix} \tag{A.4}$$

is a 3×3 matrix, i.e., having three rows and three columns. The coefficients a_{ij}, $i, j = 1, 2, 3$, in $[A]$ matrix are known. Similarly, the vector $\{B\}$ is a 3×1 matrix that contains the coefficients b_i, $i = 1, 2, 3$ as

$$\{B\} = \begin{Bmatrix} b_1 \\ b_2 \\ b_3 \end{Bmatrix} \tag{A.5}$$

In Eq. (A.3), the matrix $\{X\}$ is

$$\{X\} = \begin{Bmatrix} x_1 \\ x_2 \\ x_3 \end{Bmatrix} \tag{A.6}$$

which has the same size as the matrix $\{B\}$ but consists of unknowns x_i, where $i = 1, 2, 3$.

For a practical problem, a set of algebraic equations in the form of Eq. (A.3) may be assembled from a large number of equations. If the set of algebraic equations consists of 1,000 equations, the matrix $[A]$ has 1,000 rows and 1,000 columns. The total number of coefficients in the matrix $[A]$ is 1,000,000 while the vectors $\{B\}$ and $\{X\}$ contain 1,000 coefficients and unknowns, respectively.

Often, the matrix $[A]$ is symmetric. In this case, it may be written shortly as

$$\begin{bmatrix} a_{11} & a_{12} & a_{13} \\ & a_{22} & a_{23} \\ \text{Symmetric} & & a_{33} \end{bmatrix} \tag{A.7}$$

Furthermore, if the coefficients in the symmetric matrix $[A]$ in Eq. (a.7) are

$$a_{ij} = \begin{cases} 0 & \text{when } i \neq j \\ 1 & \text{when } i = j \end{cases}$$

the matrix $[A]$ becomes

$$\begin{bmatrix} 1 & 0 & 0 \\ 0 & 1 & 0 \\ 0 & 0 & 1 \end{bmatrix} \tag{A.8}$$

which is called an identity matrix and normally denoted as $[I]$.

If a matrix has only one row, it is called a row matrix. For example

$$\lfloor D \rfloor = \begin{bmatrix} x & 2x^2 & \dfrac{x^3}{3} \end{bmatrix} \tag{A.9}$$

Similarly, if a matrix has only one column, it is called a column matrix. The matrices $\{B\}$ and $\{X\}$ in Eqs. (A.5) and (A.6) are column matrices.

A.2 Matrix Addition and Subtraction

Matrices can be added or subtracted if they have the same numbers of rows and columns. For example, as matrices $[Q]$ and $[R]$ both have the size of 2×3,

$$\underset{(2\times3)}{[P]} \;=\; \underset{(2\times3)}{[Q]} + \underset{(2\times3)}{[R]} \tag{A.10}$$

Matrix addition yields the matrix $[P]$ with the same size of 2×3. Coefficients of the matrices in Eq. (A.10) can be written in tensor form as

$$P_{ij} \;=\; Q_{ij} + R_{ij}$$

where i = 1, 2 and j = 1, 2, 3. Such matrix addition can be written for computer programming in Fortran language as

```
        DO 10   I=1,2
        DO 10   J=1,3
    10    P(I,J)  =  Q(I,J)  +  R(I,J)
```

A.3 Matrix Multiplication

A scalar and a matrix can be multiplied directly. As an example,

$$\alpha \begin{bmatrix} c_{11} & c_{12} \\ c_{21} & c_{22} \end{bmatrix} = \begin{bmatrix} \alpha c_{11} & \alpha c_{12} \\ \alpha c_{21} & \alpha c_{22} \end{bmatrix} \tag{A.11}$$

However, for multiplication of two matrices such as $[Q]$ and $[R]$ with sizes of $i{\times}j$ and $k{\times}l$, the number of columns (j) in matrix $[Q]$ must be equal to the number of rows (k) in matrix $[R]$, i.e.,

$$\underset{(i\times\ell)}{[P]} \;=\; \underset{(i\times j)}{[Q]} \underset{(j\times\ell)}{[R]} \tag{A.12}$$

This leads to the resulting matrix $[P]$ with the size of $i{\times}\ell$ as shown in Eq. (A.12). Coefficients in the matrices can be written in tensor form as

$$P_{i\ell} \;=\; \sum_{m=1}^{j} Q_{im}\, R_{m\ell}$$

As an example, if the matrix $[Q]$ has the size of (2×3) while the matrix $[R]$ has the size of (3×4), the matrix $[P]$ will then have the size of (2×4). Such matrix multiplication can be written for computer programming in Fortran language as

```
        DO 10   I = 1,2
        DO 10   L = 1,4
        DO 10   J = 1,3
   10   P(I,L) = P(I,L) + Q(I,J)*R(J,L)
```

It is noted that the positions of matrices during their multiplication lead to different results, i.e.,

$$[Q][R] \neq [R][Q]$$

Thus, pre- or post-multiplication of a matrix by another matrix must be performed carefully.

A.4 Matrix Transpose

If we have a matrix $[C]$ defined by

$$[C] = \begin{bmatrix} c_{11} & c_{12} \\ c_{21} & c_{22} \end{bmatrix} \tag{A.13}$$

Then its transpose is

$$[C]^T = \begin{bmatrix} c_{11} & c_{21} \\ c_{12} & c_{22} \end{bmatrix} \tag{A.14}$$

i.e., coefficients in the rows and columns are interchanged so that row i of the matrix becomes column i in its transpose.

The following properties of matrix transpose are frequently used in the derivation of the finite element equations

$$([Q]+[R])^T = [Q]^T + [R]^T \tag{A.15}$$

$$([Q][R])^T = [R]^T [Q]^T \tag{A.16}$$

A.5 Matrix Inverse

If $[A]$ is a square and non-singular matrix, then

$$[A]^{-1}[A] = [I] \tag{A.17}$$

where $[A]^{-1}$ is the inverse matrix of matrix $[A]$. For a small set of algebraic equations (A.3), the solution $\{X\}$ can be easily obtained with the inverse matrix of matrix $[A]$ as shown below

$$[A]^{-1}[A]\{X\} = [A]^{-1}\{B\}$$

$$[I]\{X\} = [A]^{-1}\{B\} \tag{A.18}$$

$$\{X\} = [A]^{-1}\{B\}$$

However, this procedure is not used to solve the unknown $\{X\}$ in practical problems with a large set of algebraic equations.   This is mainly because the determination of $[A]^{-1}$ consumes a large computational time and computer memory as compared to other solution methods.

A.6 Matrix Partitioning

Matrix partitioning can simplify the application of boundary conditions on the set of algebraic equations. For example, a set of four algebraic equations is considered,

$$
\begin{bmatrix}
a_{11} & a_{12} & a_{13} & a_{14} \\
a_{21} & a_{22} & a_{23} & a_{24} \\
a_{31} & a_{32} & a_{33} & a_{34} \\
a_{41} & a_{42} & a_{43} & a_{44}
\end{bmatrix}
\begin{Bmatrix} x_1 \\ x_2 \\ x_3 \\ x_4 \end{Bmatrix}
=
\begin{Bmatrix} b_1 \\ b_2 \\ b_3 \\ b_4 \end{Bmatrix}
\tag{A.19}
$$

If x_1, x_2, b_3, b_4 are known quantities, Eq. (A.19) can be written in a form of sub-matrices as follow

$$
\begin{bmatrix}
\underset{(2\times2)}{[A_{11}]} & \underset{(2\times2)}{[A_{12}]} \\
\underset{(2\times2)}{[A_{21}]} & \underset{(2\times2)}{[A_{22}]}
\end{bmatrix}
\begin{Bmatrix} \underset{(2\times1)}{\{X_1\}} \\ \underset{(2\times1)}{\{X_2\}} \end{Bmatrix}
=
\begin{Bmatrix} \underset{(2\times1)}{\{B_1\}} \\ \underset{(2\times1)}{\{B_2\}} \end{Bmatrix}
\tag{A.20}
$$

where the matrices $\{X_1\}$ and $\{B_2\}$ are known while the matrices $\{X_2\}$ and $\{B_1\}$ are unknown.   The unknown matrix $\{X_2\}$ can be determined by using the lower set of equations in the system of Eq. (A.20) as follows

$$
[A_{21}]\{X_1\} + [A_{22}]\{X_2\} = \{B_2\}
$$

$$
[A_{22}]\{X_2\} = \{B_2\} - [A_{21}]\{X_1\}
$$

$$
\underset{(2\times1)}{\{X_2\}} = \underset{(2\times2)}{[A_{22}]^{-1}} \left(\underset{(2\times1)}{\{B_2\}} - \underset{(2\times2)}{[A_{21}]}\underset{(2\times1)}{\{X_1\}} \right)
\tag{A.21}
$$

After obtaining the matrix $\{X_2\}$, the upper set of equations in the system of Eq. (A.20) is used to solve for the matrix $\{B_1\}$ as

$$
\underset{(2\times1)}{\{B_1\}} = \underset{(2\times2)}{[A_{11}]}\underset{(2\times1)}{\{X_1\}} + \underset{(2\times2)}{[A_{12}]}\underset{(2\times1)}{\{X_2\}}
\tag{A.22}
$$

A.7 Calculus of Matrices

Coefficients in a matrix can be differentiated or integrated in a usual manner. As an example, if a matrix $[A]$ is

$$[A] = \begin{bmatrix} x & x^2 \\ 2x^2 & 3x \end{bmatrix} \qquad (A.23)$$

Then the derivative of the matrix $[A]$ with respect to x is

$$\frac{d}{dx}[A] = \begin{bmatrix} 1 & 2x \\ 4x & 3 \end{bmatrix} \qquad (A.24)$$

Similarly, the integration of the matrix $[A]$ from a lower limit 0 to an upper limit L is

$$\int_0^L [A]\, dx = \begin{bmatrix} \dfrac{x^2}{2} & \dfrac{x^3}{3} \\ \dfrac{2x^3}{3} & \dfrac{3x^2}{2} \end{bmatrix}_0^L$$

$$= \begin{bmatrix} \dfrac{L^2}{2} & \dfrac{L^3}{3} \\ \dfrac{2L^3}{3} & \dfrac{3L^2}{2} \end{bmatrix} \qquad (A.25)$$

Appendix B

MATLAB Fundamentals

B.1 Introduction

MATLAB is a powerful software for solving scientific and engineering problems. The software contains a large number of mathematical functions and commands that can be used easily and conveniently. MATLAB provides many advantages as compared to many conventional high-level computer languages, such as Fortran, Pascal and C, commonly used for developing computer software. MATLAB has built-in graphic and visualization capability that can be used in a friendly, non-intimidating fashion.

MATLAB stands for *Mat*rix *Lab*oratory because the basic data that used in computation are from the elements in matrices. The software is widely used in colleges and universities especially in learning science and engineering courses. The software is also employed in design and development of new products in industry.

B.2 MATLAB Environment

When the MATLAB software is open, a window called "MATLAB desktop" appears as shown in Fig. B.1. The window consists of many sub-windows which are:

a) Command Window
b) Command History Window
c) Edit/Debug Window
d) Figure Window
e) Workspace Window

B.2.1 Command Window

The Command Window is on the right-hand side of the MATLAB desktop as shown in Fig. B.1. In this window, users can directly type commands at the command prompt (>>). For example, the volume of a sphere with a radius of 3.5 can be determined by typing the following command,

Workspace Window   Command Window

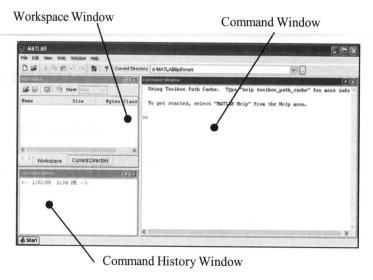

Command History Window

Figure B.1 MATLAB desktop and sub-windows.

```
>> volume = 4/3*pi*3.5^3
volume =
   179.5944
```

After pressing the Enter key, MATLAB calculates the answer, saves it in a variable `volume` (array size 1x1) and displays the value of the variable on the screen as shown in Fig. B.2. If users do not want to show the result from the calculation on the screen, the semi-colon (;) must be added at the end of the command line before pressing the Enter key as follow,

```
>> area = 4*pi*3.5^2;
```

Figure B.2 Example of a command entered with its result displayed.

In the above case, MATLAB stores the result from calculation in a variable called `area` and does not display the result on the screen as shown in Fig. B.3. Ii is noted that the value of π that was used in the above example is predefined by MATLAB with the variable named `pi`.

Figure B.3 Use of semi-colon at the end of command line to hide the displayed result.

For a set of commands, users can gather them in a single file. When that set of commands is needed, users can simply type the name of that file and press the Enter key. MATLAB will then execute all the commands in that file, line by line. The file that contains a set of commands is called the "script file". The file is commonly known as the "m-file" because its extension is symbolized by ".m".

B.2.2 Command History Window

The command history window shows a list of commands that have been executed. If any command in the list is needed to re-execute, users may simply left-clicks on that command in the Command History Window and press the Enter key or double-click directly on that command. To delete the command from the list, users can right-click on that command or select the "Delete Selection" from the popup menu as shown in Fig. B.4.

Figure B.4 Deletion of the used command in the Command History Window.

B.2.3 Edit/Debug Window

The Edit/Debug Window allows users to create or edit the existing m-file. An m-file can be created by selecting File>New>M-file from the desktop menu or clicking the icon ◻ on tool bar menu. The existing m-file can be edited by selecting File>Open or clicking the icon ☞ on tool bar menu.

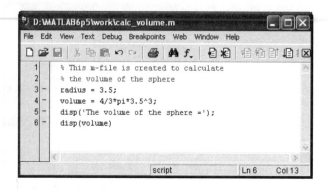

Figure B.5 Example of an m-file.

Figure B.5 shows example of an m-file that was created in the Edit/Debug Window. The m-file calculates the volume of a sphere that has a radius of 3.5 and displays result on the screen. If this m-file is saved as "`calc_volume.m`", the file can be executed by typing its name in the Command Window as follow,

```
>> calc_volume
The volume of the sphere =
   179.5944
```

B.2.4 Figure Window

The Figure Window displays different styles of graphics in two and three dimensions. As an example, a cosine function can be displayed by creating a file "`cos_x.m`" that contains commands as follows,

```
% cos_x.m: This file is created to calculate
%          and plot function cos(x)
x = 0:0.1:8;
y = cos(x);
plot(x,y);
```

When the above m-file is executed, the Figure Window displays the plot of the function cos(x) as shown in Fig. B.6.

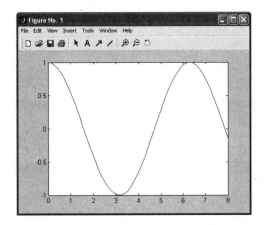

Figure B.6 Plotting the function cos(x) by using MATLAB.

B.2.5 Workspace Window

Variables are stored in of MATLAB memory called "workspace". The workspace is displayed in Workspace Window on the top left corner of the desktop. An example of the workspace, after executing `calc_volume.m` and `cos_x.m`, is shown in Fig. B.7. The figure shows that the sizes of the variables `radius` and `volume` are 1×1 while the sizes for the variables `x` and `y` are 1×81.

Name	Size	Bytes	Class
▦ radius	1x1	8	double array
▦ volume	1x1	8	double array
▦ x	1x81	648	double array
▦ y	1x81	648	double array

Workspace   Current Directory

Figure B.7 Workspace Window.

Users can delete variables from the MATLAB memory by left-clicking on that variable and pressing the keyboard Delete button, or right-clicking on the variable and selecting the Delete command from the popup menu.

B.2.6 Help Window

Useful information is available in the Help Window. To open the window as shown in Fig. B.8, users may left-click on the icon **?** in the toolbar menu or type "helpdesk" or "helpwin" in the Command Window and press Enter.

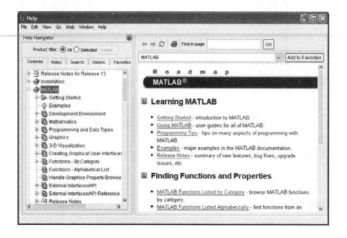

Figure B.8 Help Window.

B.3 Variables in MATLAB

Scalar and vector variables can be used in MATLAB explained below.

B.3.1 Scalar variable

The variable named 'a' for storing the value of 4 can be declared as,

```
>> a = 4
a =
     4
```

After pressing the Enter key, MATLAB displays value of the variable 'a'. To assign values to many variables by using only one command line, comma (,) or semi-colon (;) is used between commands as,

```
>> a = 4; A = 6, x = 5;
A =
     6
```

From the above example, if the semi-colon (;) is used at the end of the command, MATLAB will not display the value of that variable. It is noted that the variable names are case-sensitive in MATLAB. As shown in the example above, MATLAB assigns the value of 4 to the lower-case letter variable 'a' and 6 to the upper-case letter variable 'A'.

B.3.2 Vector variable

Values of variables are stored in the matrix form or array. For example, the scalar variable as shown in B.3.1 is stored in an array with size 1×1. Array is a set of data normally stored in row and column form. If only one row or one column of an array is used to store a set of data, the array is called a vector. If both row and column of an array are used to store a set of data, the array is called a matrix. In this section, the arrays representing the vector and matrix are explained. To create a vector variable, users can simply enter the set of data in the square parentheses. For example,

```
>> a = [1 2 3 4 5]
a =
    1   2   3   4   5
```

The vector variable 'a' is then created to store the numbers 1 to 5 in a row. If users want to store these numbers in a column style, the semi-colon is used to separate the data as follow,

```
>> b = [1; 2; 3; 4; 5]
b =
    1
    2
    3
    4
    5
```

A row vector can be transposed to become a column vector by using the apostrophe (') symbol as,

```
>> b = [1 2 3 4 5]'
b =
    1
    2
    3
    4
    5
```

A matrix, with size 3×3 for example, can be created by using the command,

```
>> c = [1 2 3; 4 5 6; 7 8 9]
c =
    1   2   3
    4   5   6
    7   8   9
```

A data can be taken from the created vector or matrix. For example, the third data in the vector 'a', can be retrieved by typing,

```
>> a(3)
ans =
    3
```

Similarly, the data from the third row and second column of the matrix 'c' can be retrieved by

```
>> c(3,2)
ans =
      8
```

MATLAB contains built-in functions that can help users to create special matrices. For example, to create the null matrix with the size of 3×4, the built-in function zeros may be used as,

```
>> d = zeros(3,4)
d =
      0    0    0    0
      0    0    0    0
      0    0    0    0
```

The built-in function ones is used to create the unity matrix. For example,

```
>> e = ones(2,4)
e =
      1    1    1    1
      1    1    1    1
```

B.3.3 Use of colon symbol

The colon (:) symbol between two numbers, while assigning values to a variable, leads to a set of data in arithmetic series from the first to the second number with the step size of one. For example,

```
>> s = 2:6
s =
      2    3    4    5    6
```

To create a set of data with a specific step size, the colon symbols are used between the three numbers. A set of data from the first to the third number with the step size equal to the second number is created as shown in the examples of the variables t, u and v below.

```
>> t = 2:0.5:4
t =
      2.0000    2.5000    3.0000    3.5000    4.0000

>> u = 4:-0.5:2
u =
      4.0000    3.5000    3.0000    2.5000    2.0000

>> v = 1:-0.75:-1.5
v =
      1.000     0.2500    -0.5000    -1.2500
```

B.3.4 Displaying data

MATLAB normally displays the computed values with four decimal digits. For example,

```
>>  2.5 + 3.37
ans =
      5.8700
```

More decimal digits can be displayed by using the command,

```
>>  format long
```

So that the computed value is displayed with fourteen decimal digits,

```
>>  2.5 + 3.37
ans =
      5.87000000000000
```

To display values with the original format again, the short format command is used,

```
>>  format short
```

Another format that is normally used in calculation is the scientific format. To display a computed value in such format, the commands `format short e` and `format long e` may be used. For examples,

```
>>  format short e
>>  2.5 + 3.37
ans =
      5.8700e+000

>>  format long e
>>  2.5 + 3.37
ans =
      5.870000000000000e+000
```

B.3.5 Use of long commands

If a long command is not fitted within a single command line, users can use the ellipsis, three periods (. . .), at the end of the line before pressing the Enter key to pause execution. Users can then continue typing the rest of the command on the new line as shown in the example below.

```
>>  fx = exp(-1.5/4)*...
(2-1.5)-1

fx =
     -6.563553606045139e-001
```

B.4 Mathematical Operation

Mathematical operators in MATLAB are represented by the symbols + − * / and ^. Mathematical expressions are executed from left to right by following the order of precedence. Meanings of the symbols and order of precedence are explained in Tables B.1 and B.2.

Table B.1 Arithmetic Operations.

Symbol	Operation
^	Exponentiation
*	Multiplication
/	Division
+	Addition
−	Subtraction

The use of mathematical operators is explained by the following examples.

```
>> pi^2
ans =
      9.8696
>> x = 2*pi;
>> x/2.5
ans =
      2.5133

>> x = -3^2
x =
      -9

>> x = (-3)^2
x =
      9
```

Table B.2 Order of precedence.

Precedence	Operations
First	Parentheses, evaluated from innermost pair.
Second	Exponentiation, evaluated from left to right.
Third	Multiplication and division (same precedence) evaluated from left to right.
Fourth	Addition and subtraction (same precedence) evaluated from left to right.

These mathematical operators are used for matrix operations such as the scalar-matrix and matrix-matrix multiplications. For example,

```
>> a = [1 2 3];
>> b = [2; 4; 6];
>> a*b
ans =
      28

>> b*a
ans =
      2    4    6
      4    8   12
      6   12   18
```

The matrix c obtained earlier can be multiplied by itself using the command,

```
>> c*c
ans =
      30    34    42
      66    81    96
     102   126   150
```

or by using the exponentiation operator to obtain the same result as,

```
>> c^2
ans =
      30    34    42
      66    81    96
     102   126   150
```

To multiply matrix A by matrix B, the number of columns in matrix A must be equal to the number of rows in matrix B. If users try to multiply two matrices that have inappropriate sizes, such as multiplying matrix s (1×5) by matrix c (3×3), MATLAB will display the error massage as follow,

```
>> s*c
??? Error using ==> *
Inner matrix dimensions must agree.
```

The mathematical operators can also be used for scalar-matrix operation. For example,

```
>> c*2
ans =
      2    4    6
      8   10   12
     14   16   18
```

To multiply element-by-element between two matrices, the period (.) must be used before the operator symbol. For example,

```
>> c.^2
ans =
       1     4     9
      16    25    36
      49    64    81
```

B.5 Built-in Functions

There are numerous built-in functions in MATLAB that can be used for scientific and engineering calculation as shown in Table B.3.

Table B.3 Some commonly used mathematical functions.

Function	Syntax
Natural logarithmic, $ln(x)$	`log(x)`
Exponential, e^x	`exp(x)`
Square root, $\sqrt{x}$	`sqrt(x)`
Absolute value, $\lvert x \rvert$	`abs(x)`
Sine function, $sin(x)$	`sin(x)`
Cosine function, $cos(x)$	`cos(x)`
Hyperbolic tangent, $tanh(x)$	`tanh(x)`
Etc.	

MATLAB also contains built-in functions for rounding numbers, such as `round`, `ceil` and `floor`. These functions can be used by studying the following examples.

```
>> h = [-1.3   -1.6   1.4   1.7];
```

Function `round` is used to round numbers to the nearest integers as follow

```
>> round(h)
ans =
      -1    -2    1    2
```

Function `ceil` is used to round numbers to the nearest integers toward ∞ as follow

```
>> ceil(h)
ans =
      -1    -1    2    2
```

Function `floor` is used to round numbers to the nearest integers toward -∞ as follow
```
>> floor(h)
ans =
      -2    -2    1    1
```

Some statistic built-in functions can be used conveniently as follows.

```
>> f = [3 5 4 7 6];
>> sum(f)
ans =
      25
>> min(f), max(f), mean(f), sort(f)
ans =
      3
ans =
      7
ans =
      5
ans =
      3    4    5    6    7
```

B.6 Plotting Graphs

MATLAB contains a number of functions for plotting graphs. For example, to plot a graph from the two sets of data,

```
>> x = [0:0.1:20];
>> y = 0.25.*x.*cos(x);
```

the function `plot` can be employed as,

```
>> plot(x,y)
```

MATLAB generates a plot with x values on the horizontal axis and the corresponding y values on the vertical axis as shown in Fig. B.9. Users can copy and use the plot for reports and presentation conveniently. Moreover, users can include additional information into the plot as shown in Fig. B.10, such as the name of the graph or the labels for x and y axes by typing the following commands,

```
>> title('Plot of y versus x')
>> xlabel('Values of x')
>> ylabel('Values of y')
>> grid on
```

Normally, the graph created by using the function `plot` is displayed by continuous lines. Users can change the type or color of lines by adding symbols in function `plot`. For example, to plot a graph of vector x versus vector y with **red dashed line**, users may use the command,

```
>> plot(x,y,'--r')
```

where the symbol (--) in the function `plot` refers to a dashed line and the letter (r) indicates the red color.

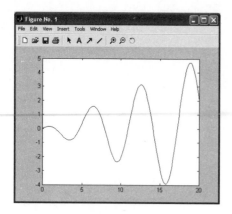

Figure B.9 Graph generated by
the `plot` function.

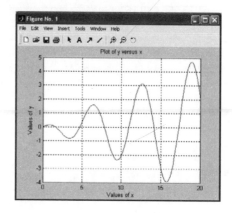

Figure B.10 Graph with additional
information.

Users can plot a graph and mark each data point with a data marker. For example, to plot a graph of vector x versus vector y with a continuous line and mark each data point with a small circle, the following command is used,

```
>> plot(x,y,x,y,'o')
```

MATLAB will display the graph as shown in Fig. B.11. List of symbols for controlling the line types, colors and data makers are presented in Table B.4.

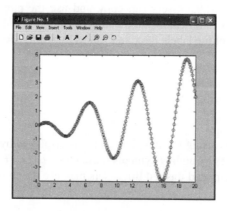

Figure B.11 Graph with data markers.

Table B.4 Symbols for line types, colors and data markers.

Line types		Data markers		Colors	
Continuous line	—	Dot	.	Black	k
Dotted line	:	Circle	o	Magenta	m
Dash-dotted line	—.	Cross	×	Blue	b
Dashed line	— —	Plus sign	+	Cyan	c
		Triangle	∧	Green	g
		Triangle (down)	∨	Yellow	y
		Triangle (left)	<	Red	r
		Triangle (right)	>		
		Square	s		
		Diamond	d		

By using the function `plot`, the Figure Window is clear before executing the new command. To create new figures from a given array, users can use the function `subplot` with its syntax of,

$$\text{subplot (u, v, w)}$$

The command divides the Figure Window into an array with u rows and v columns. The variable w locates the position for displaying the output from the function `plot` when the function subplot is executed. For example, `subplot(2,2,3)` creates an array of four panes (two rows and two columns) and directs the next plot to the third pane (the lower-left corner). An example for using this function is as follows,

```
>> subplot(1,2,1); plot(x,y)
>> axis square
>> subplot(1,2,2); plot(x,y, 'o')
>> axis square
```

The corresponding plots are shown in Fig. B.12.

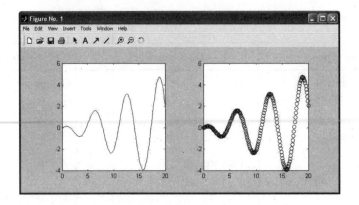

Figure B.12 Plots created by `subplot` command.

To plot a three-dimensional graph, users may use the function `plot3` which has the syntax of,

$$\text{plot3}(x, y, z)$$

For example, the following equations of x, y and z generate two three-dimensional curves that vary with t,

curve 1:	$x = (1 - t^2) \sin(t);$	$y = 1 + \cos(t);$	$z = t$
curve 2:	$x = \sin(t);$	$y = \cos(t);$	$z = t$

By using the following commands, the plots are shown in Fig. B.13.

```
>> t = 0:pi/50:10*pi;
>> subplot(1,2,1); plot3((1-t.^2).*sin(t), 1+cos(t), t)
>> axis square
>> grid on
>> subplot(1,2,2); plot3(sin(t), cos(t), t)
>> axis square
>> grid on
```

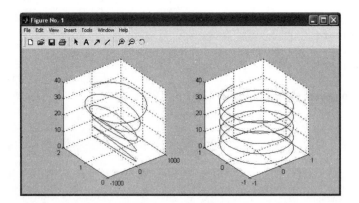

Figure B.13 Three-dimensional curves plotted by using the `plot3` function and `subplot` command.

Sometimes, users may need to plot a function with two variables, such as $z = f(x, y)$. The function represents a surface when plotted on *x-y-z* coordinates, such as,

$$f(x, y) \;=\; x(1-x)\,y(1-y)\tan^{-1}\!\left(100\!\left(\frac{x+y}{\sqrt{2}} - 0.8\right)\right)$$

where $0 \le x \le 1$ and $0 \le y \le 1$. To plot such surface, the boundaries in *x*- and *y*-directions are firstly defined,

```
>> x = 0:0.05:1;
>> y = 0:0.05:1;
```

Then, the grid points on *x-y* plane are generated by using the function `meshgrid`,

```
>> [X,Y] = meshgrid(x,y);
```

Next, the function $f(x,y)$ is evaluated at each grid point,

```
>> Z = X.*(1-X).*Y.*(1-Y).*atan(100*((X+Y)/sqrt(2)-0.8));
```

Finally, the function `mesh` is used to create the surface plot,

```
>> mesh(X,Y,Z)
>> view(248,8.5)  % control direction of viewpoint
```

The plot of the surface is shown in Fig.B.14. Another method to visualize the shape of a function is by using contour plot. A contour line in the contour plot represents a level or an elevation of the function. MATLAB uses the function `contour` to create the contour plots with the syntax of

```
contour(x,y,z,n)
```

where x, y and z are the matrix that can be prepared in the same way as those required by the `mesh` function and n is the number of contour lines.

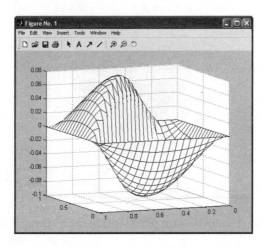

Figure B.14 A surface plot by using the `mesh` command.

Users can create the contour plot of the previous function with 20 contour lines by using the commands below. The resulting plot is shown in Fig. B.15.

```
>> contour(X,Y,Z,20)
>> axis square
```

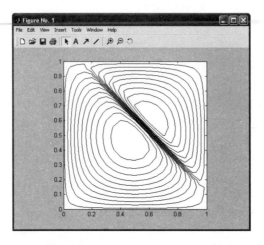

Figure B.15 A contour plot by using the `contour` command.

To create a contour plot with fill-in color, the `contourf` command may be used,

```
>> contourf(X,Y,Z,15)
>> axis square
```

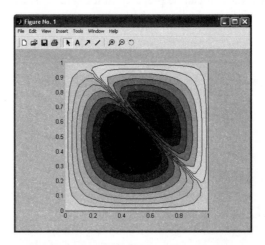

Figure B.16 A color contour plot by using the `contourf` command.

B.7 Programming

A set of command lines can be included in a file so that they can be executed later. The file that contains a set of command lines is called an 'm-file'. Two types of the m-file presented herein are the Script and Function files.

B.7.1 Script file

Script file is a file that contains a set of command lines. By executing this file, it is equivalent to typing each line, one at a time, for execution. For example, a script file `calvel.m` that contains the two commands below can be created by selecting File>New>M-file on the window menu,

```
g = 9.81; m = 60.2; t = 12; Cd = 0.25;
v = sqrt(g*m/Cd)*tanh(sqrt(g*Cd/m)*t)
```

After saving it, the file can be executed by typing its name at the Command Window as,

```
>> calvel
v =
      47.8435
```

MATLAB executes the commands in the file line-by-line. First, the program assigns values to the variable g, m, t and Cd. The program then calculates the values of variable v and displays them on the screen. All values used in script file are stored in the memory. Users can verify any value, such as that value of Cd (created by script file), by typing its name at the command prompt and press the Enter key as,

```
>> Cd
Cd =
      0.25
```

B.7.2 Function file

Another type of the m-file is the function file. Unlike the script file, MATLAB deletes values of all variables created by the function file after execution. The function file is useful when the set of commands is executed repeatedly.

The first line of a function file must begin with the `function` command and a list of input and output variables. The structure of function file is as follows:

```
Function [output variable] = function_name(input variables)
% comment line
% other comment lines
...
Executable commands
...
(End of executable command)
```

As an example of a function file named as `fcalvel.m` is listed below,

```
function v = vel(m, t, Cd)
% Compute velocity from given data
% Input: mass(m), time(t), and drag coefficient(Cd)
% Output: velocity(v)
g = 9.81;
v = sqrt(g*m/Cd)*tanh(sqrt(g*Cd/m)*t);
```

To execute the function file above, the file's name and its input variables are typed. For example, to compute the variable v (m/sec) by using the variable m = 60.2 kg, t = 12 sec and Cd = 0.25 kg/sec, the following command is used,

```
>> fcalvel(60.2, 12, 0.25)
ans =
     47.8435
```

It is noted that, if users try to access the variable g created by the function file, the error massage below is displayed.

```
>> g
??? Undefined function or variable 'g'.
```

B.7.3 Input and output commands

From the example of the script file above, if the variable v is to be determined by using the new value of m, users must open the script file and edit the value of variable before execute it. It will be better if the script file can wait for the users to input new data while the program is running. The input function can help users by displaying the massage on the screen and waiting for users to enter new data from the keyboard. The input data will be stored in the specified variable. The syntax for the input function is,

```
variable = input ('massage to interact with users')
```

For example, the command

```
t = input ('Time(t):   ')
```

is used for users to input the value of Time(t). When the statement is executed, the massage

```
Time(t):
```

appears on the screen waiting for the users to type in the value. Users may prefer to store the input data as a character string. In this latter case, the second argument in the input function must be added as shown in the following syntax,

```
variable = input ('massage to interact with users'),'s')
```

Another command that can help users to interact with the program is by using the disp function. The function displays text messages or values of variables on the screen. The syntax of this function is

```
disp (A)
```

where A is the variable name or the text message in a single quote. For example, the m-file below shows the use of different commands explained above.

```
function velocity
% Example of interactive function
% The function is created for
% calculating the velocity
g = 9.81;
m  = input('Mass(kg)   : ');
t  = input('Time(sec) : ');
Cd = input('Drag coefficient (kg/m) : ');
disp(' ')
disp('Velocity (m/s):')
disp(sqrt(g*m/Cd)*tanh(sqrt(g*Cd/m)*t))
```

If the file is saved as velocity.m, users can execute the file by typing its name in the Command Window. After pressing the Enter key, the program will ask the users to enter the value of each variable before determining and displaying the value of velocity solution as follows,

```
>> velocity
Mass(kg)   : 60.2
Time(sec) : 12
Drag coefficient (kg/m) : 0.25

Velocity (m/s):
     47.8435
```

To display the output data on the screen with a specific format, the fprintf function should be used instead of disp function. The syntax of this function is

```
fprintf ('format', variable)
```

where format contains the text to be printed on the screen and the special characters that describe format of the variable. To display the variable v which is equal 47.8435 on the screen with the floating point format in a field of eight characters wide, including three digits after the decimal point, the following command is used,

```
>> fprintf('The velocity is %8.3f m/s \n', v)
The velocity is   47.843 m/s
```

From the example of fprintf function, MATLAB displays the text inside the single quote until the program execution reaches the special character % or \. MATLAB then checks the meaning of that character and displays the variable according to the specified format. The special characters starting with the symbol % are called the conversion characters while the characters starting with \ are called the escape characters. Examples of these special characters are described in Tables B.5 and B.6.

Table B.5. Some conversion characters for fprintf function.

Character	Description
%d	Display value as an integer
%e	Display value in exponential format
%f	Display value in floating point format
etc.	

Table B.6. Some escape characters for fprintf function.

Character	Description
\n	Skip to a new line
\t	Horizontal tab
etc.	

It is noted that when the program is executed to the escape character (\n) in the above example, MATLAB will move the command prompt to the new line and wait for the next command. To understand how the conversion characters affect the displaying format, the following statements are considered.

```
>> z = 7;
>> fprintf('The velocity is %18.4e m/s \n', z)
The velocity is          7.0000e+000 m/s
>> fprintf('The velocity is %8d m/s \n , z)
The velocity is          7 m/s
>> fprintf('The velocity is %8.4f m/s \n', z)
The velocity is    7.0000 m/s
```

Users can use the fprintf function to display multiple variables as,

```
>> a = 100; b = pi; c = 3*pi;
>> fprintf(' %5d %10.3f %8.5e \n ', a, b, c)
    100      3.142 9.42478e+000
```

B.7.4 Read and write data file

To read or write data on a file, the file must be firstly identified. The file is open by using the fopen function. The syntax of this function is,

```
fid = fopen ('filename', 'permission')
```

where `fid` (short for file identification) is a number assigned to the file when it is opened. The `filename` is the name of the file including its extension and `permission` is the character for specifying the mode for opening file. For example, if the permission character is `r`, MATLAB will open the existing file for reading only. Example of the permission characters are shown in Table B.7.

Table B.7. Permissions used in `fopen` function.

File permission	Description
r	Open existing file for reading only
r+	Open existing file for reading and writing
w	Delete all data in existing file (or create a new file) and allow for writing only
w+	Delete all data in existing file (or create a new file) and allow for reading and writing.
etc.	

In addition, users can close the opened file by using the `fclose` function with the syntax of

```
fclose (fid)
```

After the file is opened, the data in that file can be read. MATLAB uses the function `fscanf` to read formatted data from the open file with the syntax of

```
variable = fscanf (fid, 'format', size)
```

where `fid` must be the same file identification of the file open, `format` is the pattern of the data, and `size` specifies the total number of data to be read from the file. There are three types of this argument:

- `n`   Read data for n values and store them in a column vector.
- `[n m]`   Read data for n×m values and store them in a matrix size n×m.
- `Inf`   Read data until reaching the end of the file.

In the proceeding section, the method for writing the formatted data on the screen by using `fprintf` function was explained. Users can also use the function to write formatted data into a file by first opening the file and using `fprintf` function with the syntax,

```
fprintf (fid, 'format', variable)
```

To understand how to use `fopen`, `fclose`, `fscanf` and `fprintf` function clearly, the following script file is considered.

```
% EXAMPLE PROGRAM
% OPEN INPUT FILE AND READ DATA
fid = fopen('input.dat', 'r');
neq = fscanf(fid,'%f', 1);
a = fscanf(fid,'%f',[neq neq]);
a = a.';
fclose(fid);
% OPEN OUTPUT FILE AND WRITE DATA
fid = fopen('output.dat', 'w');
fprintf(fid,'\n   THIS TEXT WILL BE WRITTEN TO OUTPUT.DAT \n');
fprintf(    '\n   THIS TEXT WILL BE WRITTEN ON SCREEN      \n');
fprintf(fid,' %14.6e   %14.6e   %14.6e \n', a(2,1), a(2,2), a(2,3));
fprintf(    ' %14.6e   %14.6e   %14.6e \n', a(2,1), a(2,2), a(2,3));
fclose(fid);
```

The `input.dat` file contains the data,

```
3.
4.    -4.     0.
-1.    4.    -2.
0.    -2.     4.
```

When the script file above is executed, MATLAB opens the file `input.dat`, reads the first and only one data which is the number 3 and stores it in the variable `neq` which has its size defined as one. The program then reads the rest of the data from the input file and stores them into the variable `a` which is a matrix with the size of $(neq \times neq)$. To read the data in matrix form, MATLAB reads the first row from left to right and store them into the first column of variable `a`. After reading the data in such fashion, users must transpose the matrix so that it is in the same form as the input file. The program closes the `input.dat` file and opens another file for writing output data. The output file name is `output.dat`. The program then prints the text inside a single quote into the output file and on the screen by using the `fprintf` function. Next, the program prints values of the matrix `a` into the output file and on the screen. Finally, the program closes the output file and stop executing the script file. The `output.dat` file generated from the program contains the information as shown below,

```
    THIS TEXT WILL BE WRITTEN TO OUTPUT.DAT
-1.000000e+000    4.000000e+000   -2.000000e+000
```

while the messages on the screen are,

```
    THIS TEXT WILL BE WRITTEN ON SCREEN
-1.000000e+000    4.000000e+000   -2.000000e+000
```

B.7.5 Programming commands

Like other computer programs written in high-level languages, commands inside a MATLAB program are executed line-by-line from top to bottom. Executing series of commands, line by line, is not effective in solving scientific and engineering problems. Additional commands for making decision and performing iteration should be included to increase the efficiency of the computation.

B.7.5.1 Decision commands

To make a decision, the `if` command is used with its structure,

```
if logical expression
    statements
end
```

The command asks a question by using a logical expression for making decision. If the answer is true, the statements inside the `if` command is executed. Examples of the logical operators used in logical expressions are shown in Table. B.8.

Table B.8. Logical operators.

Logical operators	Example	Operation
==	x == 4	Equal to
~=	x ~= 4	Not equal to
<	y < x	Less than
>	y > 5	Greater than
<=	10 <= a/3	Less than or equal to
>=	t >= a	Greater than or equal to
~	~(t >= a)	Logical NOT
&	(x==4)&(y>x)	Logical AND
|	(a<x)|(x<y)	Logical OR

To understand operations of the `if` command, the following function file is considered.

```
function grade(score)
% Determines whether score is good, fair or poor
% Input: numerical value of score (1-100)
% Output: display message (GOOD, FAIR or POOR)
if score >= 80
    disp('GOOD')
```

```
end
if score >= 50 & score < 80
    disp('FAIR')
end
if score < 50
    disp('POOR')
end
```

If the program is executed three times with the scores of 87, 64 and 30, the results are,

```
>> grade(87)
GOOD
>> grade(64)
FAIR
>> grade(30)
POOR
```

Another command for making decision is the `if...else` command with its structure of

```
if logical expression
  statements #1
else
  statements #2
end
```

From the `if...else` command above, the statements in #1 are executed when the logical expression is true while the statements in #2 are executed when the logical expression is false. For example, the example below shows the use of the `if...else` command.

```
function grade2(score)
% Determines whether score is pass or fail
% Input: numerical value of score (1-100)
% Output: display message (PASS or FAIL)
if score >= 50
    disp('PASS')
else
    disp('FAIL')
end
```

The results from executing the program are shown as below.

```
>> grade2(75)
PASS
>> grade2(49)
FAIL
```

Another form of the commands that can make a decision is the `if...elseif` command with its structure of,

```
If   logical expression #1
        statements #1
elseif logical expression #2
        statements #2
elseif logical expression #3
        statements #3
                .
                .
                .

else
        statements
end
```

The command allows users to use multiple logical expressions for making complex decision. The example for using the if...elseif command is shown below.

```
function grade3(score)
% Determines whether score is good, fair or poor
% Input: numerical value of score (1-100)
% Output: display message (GOOD, FAIR or POOR)
if score >= 80
   disp('GOOD')
elseif score >= 50 & score < 80
   disp('FAIR')
else
   disp('POOR')
end
```

The results after executing the program are,

```
>> grade3(25)
POOR
>> grade3(89)
GOOD
>> grade3(57)
FAIR
```

B.7.5.2 Iteration commands

The two commands commonly used for making iteration are the for and while commands. The while command terminates the iteration by using a logical expression. The for command stops the iteration when the number of iterations is reached

- **for command**

The structure of the for command is

```
for   index = first:increment:last
          statement 1
              .
              .
          statement n
    end
```

Operations of the `for` command (if `increment` value is positive) are as follows.

1. At the beginning of execution, a set of number which starts from the `first` to the `last` number with the step of `increment` is created. The increment value is set to one if it is not assigned.

2. The program then assigns the first number to the variable `index` and executes the statements inside the `for` command.

3. After all statements in the `for` command are executed, the program then checks the value of the `index` variable. If the value of the `index` variable is greater than or equal to the `last` number, the operation terminated. If it is not, the program assigns the next number created from step 1 to the variable `index`.

4. The program executes the statements inside the `for` command again by repeating step 3 until the `index` variable is equal to the `last` number.

To understand the operations of the `for` command more clearly, the following example is studied.

```
for i = 1:5
    t = i*2;
    disp(t)
end
```

In this case, the program creates a set of number which starts from 1 to 5. The loop index `i` is assigned as one and the statements inside the `for` command are executed. At this step, the value of the index `i` is not yet equal to 5, so the program assigns the new value of 2 to the index `i` and the operation is repeated. The iteration is performed until the loop index `i` is 5 which is equal the `last` value. The `for` command is terminated and the program continues to execute the statement after the `end` statement. If the `increment` value is negative as shown in the example below,

```
for j = 10:-2:1
    disp(j)
end
```

The `for` command is terminated when the loop index `j` is less than or equal the `last` number. The example displays the numbers of 10, 8, 6, 4 and 2 respectively on the screen.

while command

The structure of the `while` command is,

```
while   logical expression
        statement 1
                .
                .
        statement n
end
```

The `while` command keeps executing statements inside its loop as long as the `logical expression` is true. Thus, the number of iteration is not known in advance. An example for using the `while` loop is shown below.

```
t = 10;
while t > 0
    t = t - 3;
    disp(t)
end
```

The displayed results from the example are 7, 4, 1 and -2 respectively.

- **pause command**

The `pause` command causes the program to stop and wait for user to press any key before continuing. The syntax for this command is,

```
pause  or  pause(n)
```

The `pause` command halts execution temporarily while the `pause(n)` suspends the execution for `n` seconds before continuing. Examples for the two commands above are shown below.

```
for i = 1:3
    pause
    disp(i)
end
```

The program uses the `for` command to display the numbers 1 to 3 on the screen but these numbers are displayed after a key on the keyboard is pressed.

```
for i = 1:3
    pause(5)
    disp(i)
end
```

The program displays the number 1 and waits for 5 second before showing the number 2. The program then waits for another 5 seconds before displaying the number 3.

MATLAB commands and examples presented in this Appendix highlight the capability of the software for helping scientists and engineers to solve their problems more effectively. The MATLAB commands presented herein are fundamental and essential. Further details for using MATLAB can be found in textbooks listed in bibliography.

Appendix C

Derivation of Fourth-Order Runge-Kutta Formula

The Runge-Kutta method is widely used to solve the ordinary differential equation. The most popular form is the fourth-order Runge-Kutta formula as shown in Eq. (7.55), which is

$$y_{i+1} = y_i + \left[\frac{1}{6}(k_1 + 2k_2 + 2k_3 + k_4) \right]h \tag{C.1}$$

where

$$k_1 = f(x_i, y_i) \tag{C.2a}$$

$$k_2 = f\left(x_i + \frac{1}{2}h, \, y_i + \frac{1}{2}hk_1 \right) \tag{C.2b}$$

$$k_3 = f\left(x_i + \frac{1}{2}h, \, y_i + \frac{1}{2}hk_2 \right) \tag{C.2c}$$

$$k_4 = f(x_i + h, \, y_i + hk_3) \tag{C.2d}$$

The Runge-Kutta formulas provide accurate solutions to the ordinary differential equations because their coefficients are determined by matching the formulas with the Taylor series. Chapter 7 shows the derivation for the coefficients of the second-order Runge-Kutta formula which are derived by matching with the Taylor series containing the terms up to h^2. The coefficients of the fourth-order Runge-Kutta formula are determined in the same way by using the Taylor series that contains the terms up to h^4. The orders of the Runge-Kutta formulas are thus identified by the highest order of h for the terms in the Taylor series used for matching.

The coefficients for the terms in the fourth-order Runge-Kutta formula can be derived by first writing Eqs. (C.1) - (C.2) in the general form of Eq. (7.38) as,

$$y_{i+1} = y_i + (\alpha_1 k_1 + \alpha_2 k_2 + \alpha_3 k_3 + \alpha_3 k_4)h \tag{C.3}$$

where
$$k_1 = f(x_i, y_i) \tag{C.4a}$$
$$k_2 = f(x_i + \beta_1 h, y_i + \beta_2 h k_1) \tag{C.4b}$$
$$k_3 = f(x_i + \beta_3 h, y_i + \beta_4 h k_1 + \beta_5 h k_2) \tag{C.4c}$$
$$k_4 = f(x_i + \beta_6 h, y_i + \beta_7 h k_1 + \beta_8 h k_2 + \beta_9 h k_3) \tag{C.4d}$$

Equations (C.3) - (C.4) contains 13 unknowns of $\alpha_i, i = 1$ to 4 and $\beta_j, j = 1$ to 9. These coefficients are determined by matching the fourth-order Runge-Kutta formula in (C.3)-(C.4) with the Taylor series that contains 5 terms (up to h^4). The matching leads to only 11 equations. Thus, 2 unknown coefficients are assigned so that the remaining 11 coefficients can be determined from 11 equations.

By assigning the 2 unknown coefficients, the fourth-order Runge-Kutta formula can be written in many forms. The most popular form is shown by Eqs. (C.1) - (C.2). Derivation of the fourth-order Runge-Kutta formula in this form starts by writing Eqs. (C.1) - (C.2) as

$$y_{i+1} = y_i + (a k_1 + b k_2 + c k_3 + d k_4)h \tag{C.5}$$

where
$$k_1 = f(x_i, y_i) \tag{C.6a}$$
$$k_2 = f(x_i + mh, y_i + mh k_1) \tag{C.6b}$$
$$k_3 = f(x_i + nh, y_i + nh k_2) \tag{C.6c}$$
$$k_4 = f(x_i + ph, y_i + ph k_3) \tag{C.6d}$$

There are 7 unknown coefficients of a, b, c, d, m, n, p. In order to simplify the derivation, the following variables are defined,

$$f = f(x_i, y_i) \quad ; \quad f_x = \frac{\partial f}{\partial x} \quad ; \quad f_y = \frac{\partial f}{\partial y} \tag{C.7}$$

and
$$F_1 = f_x + f f_y \tag{C.8a}$$
$$F_2 = f_{xx} + 2f f_{xy} + f^2 f_{yy} \tag{C.8b}$$
$$F_3 = f_{xxx} + 3f f_{xxy} + 3f^2 f_{xyy} + f^3 f_{yyy} \tag{C.8c}$$

Since the Taylor series with the first 5 terms is,

$$y_{i+1} = y_i + f h + f'\frac{h^2}{2!} + f''\frac{h^3}{3!} + f'''\frac{h^4}{4!} + \dots \tag{C.9}$$

By starting with $y' = f$, then,

$$f' = f_x + f_y y' = f_x + f_y f = F_1$$

$$f'' = f_{xx} + 2f f_{xy} + f^2 f_{yy} + f_y(f_x + f f_y) = F_2 + f_y F_1$$

$$f''' = f_{xxx} + 3f f_{xxy} + 3f^2 f_{xyy} + f^3 f_{yyy} + f_y(f_{xx} + 2f f_{xy} + f^2 f_{yy})$$
$$+ 3(f_x + f f_y)(f_{xy} + f f_{yy}) + f_y^2(f_x + f f_y)$$
$$= F_3 + f_y F_2 + 3F_1(f_{xy} + f f_{yy}) + f_y^2 F_1$$

and the Taylor series in Eq. (C.9) becomes,

$$y_{i+1} = y_i + f h + \frac{1}{2}F_1 h^2 + \frac{1}{6}(F_2 + f_y F_1) h^3$$
$$+ \frac{1}{24}\left[F_3 + f_y F_2 + 3(f_{xy} + f f_{yy})F_1 + f_y^2 F_1\right] h^4 + \dots \qquad (C.10)$$

By applying the two-variable Taylor series expansion according to Eq. (7.45) to k_2, k_3, k_4 in Eq. (C.6),

$$k_1 = f$$

$$k_2 = f + mh F_1 + \frac{1}{2}m^2 h^2 F_2 + \frac{1}{6}m^3 h^3 F_3 + \dots$$

$$k_3 = f + nh F_1 + \frac{1}{2}h^2\left(n^2 F_2 + 2mn f_y F_1\right)$$
$$+ \frac{1}{6}h^3\left(n^3 F_3 + 3m^2 n f_y F_2 + 6mn^2(f_{xy} + f f_{yy})F_1\right) + \dots$$

$$k_4 = f + ph F_1 + \frac{1}{2}h^2\left(p^2 F_2 + 2np f_y F_1\right)$$
$$+ \frac{1}{6}h^3\left(p^3 F_3 + 3n^2 p f_y F_2 + 6np^2(f_{xy} + f f_{yy})F_1 + 6mnp f_y^2 F_1\right) + \dots$$

Then, substituting k_1, k_2, k_3, k_4 into Eq. (C.5) to yield,

$$y_{i+1} = y_i + (a + b + c + d)h f + (bm + cn + dp)h^2 F_1$$
$$+ \frac{1}{2}(bm^2 + cn^2 + dp^2)h^3 F_2 + \frac{1}{6}(bm^3 + cn^3 + dp^3)h^4 F_3$$
$$+ (cmn + dnp)h^3 f_y F_1 + \frac{1}{2}(cm^2 n + dn^2 p)h^4 f_y F_2$$
$$+ (cmn^2 + dnp^2)h^4(f_{xy} + f f_{yy})F_1 + dmnph^4 f_y^2 F_1 + \dots \qquad (C.11)$$

Finally, by matching the coefficients in Eqs. (C.10) and (C.11), a total of 8 nonlinear algebraic equations is obtained as,

$$a + b + c + d = 1$$
$$bm + cn + dp = 1/2$$
$$bm^2 + cn^2 + dp^2 = 1/3$$

$$bm^3 + cn^3 + dp^3 \;=\; 1/4$$

$$cmn + dnp \;=\; 1/6$$

$$cmn^2 + dnp^2 \;=\; 1/8$$

$$cm^2 n + dn^2 p \;=\; 1/12$$

$$dmnp \;=\; 1/24$$

leading to the coefficient values of,

$$a \;=\; d \;=\; 1/6 \qquad ; \qquad b \;=\; c \;=\; 1/3$$

$$m \;=\; n \;=\; 1/2 \qquad ; \qquad p \;=\; 1$$

Thus, the fourth-order Runge-Kutta formula in Eqs. (C.5) - (C.6) becomes,

$$y_{i+1} \;=\; y_i + \left[\frac{1}{6} k_1 + \frac{1}{3} k_2 + \frac{1}{3} k_3 + \frac{1}{6} k_4 \right] h$$

$$k_1 \;=\; f(x_i, y_i)$$

$$k_2 \;=\; f\left(x_i + \frac{1}{2} h, \; y_i + \frac{1}{2} h k_1 \right)$$

$$k_3 \;=\; f\left(x_i + \frac{1}{2} h, \; y_i + \frac{1}{2} h k_2 \right)$$

$$k_4 \;=\; f(x_i + h, \; y_i + h k_3)$$

which is the most popular form of the Runge-Kutta formula as shown in Eqs. (C.1) - (C.2).

Appendix D

Finite Element Computer Program

The finite element method for solving two-dimensional heat transfer problems that are governed by the Poisson's equation is presented in Chapter 9. The chapter shows the derivation of the finite element equations and their matrices for the three-node triangular element. Examples are presented by using the finite element computer program listed in this Appendix. The computer program can be employed to solve problems that have complex geometry. The computer program can also be modified to include other element types, such as the four-node quadrilateral element. The programs written in Fortran and Matlab are presented in this appendix while those written in C and Pascal are shown in Appendix F. These finite element computer programs can be extended to solve three-dimensional problems by adding subroutines for the three-dimensional elements, such as the four-node tetrahedral element or the eight-node hexahedral element into them.

Listing of the Finite Element Computer Program in Fortran

```
C      PROGRAM  FINITE
C
C      A FINITE ELEMENT COMPUTER PROGRAM FOR SOLVING PARTIAL
C      DIFFERENTIAL EQUATION IN THE FORM OF POISSON'S EQUATION
C      FOR TWO-DIMENSIONAL STEADY-STATE HEAT CONDUCTION WITH
C      INTERNAL HEAT GENERATION.
C                                     PROF. DR. PRAMOTE DECHAUMPHAI
C                                     FACULTY OF ENGINEERING
C                                     CHULALONGKORN UNIVERSITY
C
C      THE VALUES DECLARED IN THE PARAMETER STATEMENT BELOW SHOULD
C      BE ADJUSTED ACCORDING TO THE SIZE OF THE PROBLEMS AND TYPES
C      OF COMPUTERS:
C          MXPOI = MAXIMUM NUMBER OF NODES IN THE MODEL
C          MXELE = MAXIMUM NUMBER OF ELEMENTS IN THE MODEL
C
       PARAMETER (MXPOI=150, MXELE=300)
C
       IMPLICIT REAL*8 (A-H,O-Z)
       DIMENSION  COORD(MXPOI,2), TEMP(MXPOI), TEXT(20)
       DIMENSION  SYSK(MXPOI,MXPOI), SYSQ(MXPOI), QELE(MXELE)
       CHARACTER*20  NAME1, NAME2
C
       INTEGER  INTMAT(MXELE,3), IBC(MXPOI)
C
   10  WRITE(6,20)
   20  FORMAT(/, ' PLEASE ENTER THE INPUT FILE NAME:')
       READ(5, '(A)', ERR=10)  NAME1
       OPEN(UNIT=7, FILE=NAME1, STATUS='OLD', ERR=10)
C
C      READ TITLE OF COMPUTATION:
C
       READ(7,*)  NLINES
       DO 100  ILINE=1,NLINES
       READ(7,1)  TEXT
    1  FORMAT(20A4)
  100  CONTINUE
C
C      READ INPUT DATA:
C
       READ(7,1)  TEXT

       READ(7,*)  NPOIN, NELEM
       IF(NPOIN.GT.MXPOI)  WRITE(6,110)  NPOIN
  110  FORMAT(/,' PLEASE INCREASE THE PARAMETER MXPOI TO ', I5)
       IF(NPOIN.GT.MXPOI)  STOP
       IF(NELEM.GT.MXELE)  WRITE(6,120)  NELEM
  120  FORMAT(/,' PLEASE INCREASE THE PARAMETER MXELE TO ', I5)
       IF(NELEM.GT.MXELE)  STOP
       READ(7,1)  TEXT
       READ(7,*)  TK, THICK
       READ(7,1)  TEXT
       DO 130  IP=1,NPOIN
       READ(7,*)  I, IBC(I), (COORD(I,K), K=1,2), TEMP(I)
       IF(I.NE.IP)  WRITE(6,135)  IP

  135  FORMAT(/, ' NODE NO.', I5, ' IN DATA FILE IS MISSING')
       IF(I.NE.IP)  STOP
  130  CONTINUE
       IQ = 0
       READ(7,1)  TEXT
       DO 140  IE=1,NELEM
       READ(7,*)  I, (INTMAT(I,J), J=1,3), QELE(I)
       IF(I.NE.IE)  WRITE(6,150)  IE
```

```
  150 FORMAT(/, ' ELEMENT NO.', I5, ' IN DATA FILE IS MISSING')
      IF(I.NE.IE)  STOP
      IF(QELE(I).NE.0.)  IQ = 1
  140 CONTINUE
      WRITE(6,160)
  160 FORMAT(/,' THE F.E. MODEL INCLUDES THE FOLLOWING',
     *          ' HEAT TRANSFER MODE(S):',
     *        /,'    -- HEAT CONDUCTION                ')
      IF(IQ.EQ.1)  WRITE(6,170)
  170 FORMAT(  '    -- INTERNAL HEAT GENERATION       ')
C
      NEQ  = NPOIN
      DO 180  I=1,NEQ
      SYSQ(I) = 0.
  180 CONTINUE
      DO 190  I=1,NEQ
      DO 190  J=1,NEQ
      SYSK(I,J) = 0.
  190 CONTINUE
C
      WRITE(6,200)  NPOIN, NELEM
  200 FORMAT(/,' *** THE FINITE ELEMENT MODEL CONSISTS OF', I5,
     *          ' NODES AND', I5,' ELEMENTS ***')
C
C     ESTABLISH ALL ELEMENT MATRICES AND ASSEMBLE THEM TO FORM
C     FORM UP SYSTEM EQUATIONS
C
      WRITE(6,210)
  210 FORMAT(/,' *** ESTABLISHING ELEMENT MATRICES AND',
     *          ' ASSEMBLING ELEMENT EQUATIONS ***'    )
      CALL TRI(NELEM, INTMAT, COORD,    TK, QELE, THICK,
     *         SYSK,   SYSQ, MXPOI, MXELE            )
C
      WRITE(6,220)
  220 FORMAT(/,' *** APPLYING BOUNDARY CONDITIONS OF NODAL',
     *          ' TEMERATURES ***'                      )
      CALL APPLYBC(NPOIN, IBC, TEMP, SYSK, SYSQ, MXPOI)
C
      WRITE(6,230)
  230 FORMAT(/,' *** SOLVING A SET OF SIMULTANEOUS EQUATIONS',
     *          ' FOR TEMPERATURE SOLUTIONS ***'            )
      WRITE(6,240)  NEQ
  240 FORMAT(5X,'( TOTAL OF', I5,' EQUATIONS TO BE SOLVED )')
      CALL GAUSS(NEQ, SYSK, SYSQ, TEMP, MXPOI)
C
C     PRINT OUT NODAL TEMPERATURE SOLUTIONS:
C
  250 WRITE(6,260)
  260 FORMAT(/, ' PLEASE ENTER FILE NAME FOR TEMPERATURE'
     *          ' SOLUTIONS:'                          )
      READ(5, '(A)', ERR=250)  NAME2
      OPEN(UNIT=8, FILE=NAME2, STATUS='NEW', ERR=250)
      WRITE(8,270)  NPOIN
  270 FORMAT(' NODAL TEMPERATURE SOLUTIONS [', I5,']:',
     *        //, 2X, 'NODE', 3X, 'TEMPERATURE', /      )
      DO 280  IP=1,NPOIN
      WRITE(8,290)  IP, TEMP(IP)
  290 FORMAT(I6, E14.6)
  280 CONTINUE
C
      STOP
      END
C
C--------------------------------------------------------------------
C
      SUBROUTINE APPLYBC(NPOIN, IBC, TEMP, SYSK, SYSQ, MXPOI)
C
```

```
C      APPLY TEMPERATURE BOUNDARY CONDITIONS WITH CONDITION CODES OF:
C             0 = FREE TO CHANGE  (TO BE COMPUTED)
C             1 = FIXED AS SPECIFIED
C
       IMPLICIT REAL*8 (A-H,O-Z)
       DIMENSION  SYSK(MXPOI,MXPOI), SYSQ(MXPOI), TEMP(MXPOI)
C
       INTEGER  IBC(MXPOI)
C
       DO 100  IEQ=1,NPOIN
       IF(IBC(IEQ).EQ.0)  GO TO 100
C
       DO 200  IR=1,NPOIN
       IF(IR.EQ.IEQ)  GO TO 200
       SYSQ(IR) = SYSQ(IR)  - SYSK(IR,IEQ)*TEMP(IEQ)
       SYSK(IR,IEQ) = 0.
  200 CONTINUE
C
       DO 300  IC=1,NPOIN
       SYSK(IEQ,IC) = 0.
  300 CONTINUE
       SYSK(IEQ,IEQ) = 1.
       SYSQ(IEQ) = TEMP(IEQ)
C
  100 CONTINUE
C
       RETURN
       END
C
C-------------------------------------------------------------------
C
       SUBROUTINE ASSMBLE(   IE, INTMAT, AKC, QQ, SYSK, SYSQ,
      *                      MXPOI,  MXELE                    )
C
C      ASSEMBLE ELEMENT EQUATIONS INTO SYSTEM EQUATIONS
C
       IMPLICIT REAL*8 (A-H,O-Z)
       DIMENSION  AKC(3,3), QQ(3)
       DIMENSION  SYSK(MXPOI,MXPOI), SYSQ(MXPOI)
C
       INTEGER  INTMAT(MXELE,3)
C
       NNODE = 3
C
       DO 100  IR=1,NNODE
       DO 200  IC=1,NNODE
       IROW = INTMAT(IE,IR)
       ICOL = INTMAT(IE,IC)
       SYSK(IROW,ICOL) = SYSK(IROW,ICOL) + AKC(IR,IC)
  200 CONTINUE
       SYSQ(IROW) = SYSQ(IROW) + QQ(IR)
  100 CONTINUE
C
       RETURN
       END
C
C-------------------------------------------------------------------
C
       SUBROUTINE GAUSS(N, A, B, X, MXPOI)
       IMPLICIT REAL*8 (A-H,O-Z)
       DIMENSION  A(MXPOI,MXPOI), B(MXPOI), X(MXPOI)
C
C      PERFORM SCALING:
C
       CALL SCALE(N, A, B, MXPOI)
C
C      FORWARD ELIMINATION:
```

```
C
C     PERFORM ACCORDING TO ORDER OF 'PRIME' FROM 1 TO N-1:
C
      DO 100  IP=1,N-1
C
C     PERFORM PARTIAL PIVOTING:
C
      CALL PIVOT(N, A, B, MXPOI, IP)
C
C     LOOP OVER EACH EQUATION STARTING FROM THE ONE THAT CORRESPONDS
C     WITH THE ORDER OF 'PRIME' PLUS ONE:
C
      DO 200  IE=IP+1,N
      RATIO = A(IE,IP)/A(IP,IP)
C
C     COMPUTE NEW COEFFICIENTS OF THE EQUATION CONSIDERED:
C
      DO 300  IC=IP+1,N
      A(IE,IC) = A(IE,IC) - RATIO*A(IP,IC)
  300 CONTINUE
      B(IE) = B(IE) - RATIO*B(IP)
  200 CONTINUE
C
C     SET COEFFICIENTS ON LOWER LEFT PORTION TO ZERO:
C
      DO 400  IE=IP+1,N
      A(IE,IP) = 0.
  400 CONTINUE
  100 CONTINUE
C
C     BACK SUBSTITUTION:
C
C     COMPUTE SOLUTION OF THE LAST EQUATION:
C
      X(N) = B(N)/A(N,N)
C
C     THEN COMPUTE SOLUTIONS FROM EQUATION N-1 TO 1:
C
      DO 500  IE=N-1,1,-1
      SUM = 0.
      DO 600  IC=IE+1,N
      SUM = SUM + A(IE,IC)*X(IC)
  600 CONTINUE
      X(IE) = (B(IE) - SUM)/A(IE,IE)
  500 CONTINUE
      RETURN
      END
C
C-------------------------------------------------------------------
C
      SUBROUTINE PIVOT(N, A, B, MXPOI, IP)
      IMPLICIT REAL*8 (A-H,O-Z)
      DIMENSION A(MXPOI,MXPOI), B(MXPOI)
C
C     PERFORM PARTIAL PIVOTING:
C
      JP = IP
      BIG = ABS(A(IP,IP))
      DO 10  I=IP+1,N
      AMAX = ABS(A(I,IP))
      IF(AMAX.GT.BIG)  THEN
         BIG = AMAX
         JP  = I
      ENDIF
   10 CONTINUE
      IF(JP.NE.IP)  THEN
      DO 20  J=IP,N
```

```
      DUMY    = A(JP,J)
      A(JP,J) = A(IP,J)
      A(IP,J) = DUMY
   20 CONTINUE
      DUMY  = B(JP)
      B(JP) = B(IP)
      B(IP) = DUMY
         ENDIF
      RETURN
      END
C
C------------------------------------------------------------------
C
      SUBROUTINE SCALE(N, A, B, MXPOI)
      IMPLICIT REAL*8 (A-H,O-Z)
      DIMENSION  A(MXPOI,MXPOI), B(MXPOI)
C
C     PERFORM SCALING:
C
      DO 10  IE=1,N
      BIG = ABS(A(IE,1))
      DO 20  IC=2,N
      AMAX = ABS(A(IE,IC))
      IF(AMAX.GT.BIG)  BIG = AMAX
   20 CONTINUE
      DO 30  IC=1,N
      A(IE,IC) = A(IE,IC)/BIG
   30 CONTINUE
      B(IE) = B(IE)/BIG
   10 CONTINUE
      RETURN
      END
C
C------------------------------------------------------------------
C
      SUBROUTINE TRI(NELEM, INTMAT, COORD,    TK, QELE, THICK,
     *                   SYSK,   SYSQ, MXPOI, MXELE              )
C
C     ESTABLISH ALL ELEMENT MATRICES AND ASSEMBLE THEM TO FORM
C     UP SYSTEM EQUATIONS
C
      IMPLICIT REAL*8 (A-H,O-Z)
      DIMENSION  COORD(MXPOI,2), SYSK(MXPOI,MXPOI), SYSQ(MXPOI)
      DIMENSION  QELE(MXELE)
      DIMENSION  AKC(3,3), QQ(3), B(2,3), BT(3,2)
C
      INTEGER  INTMAT(MXELE,3)
C
C     LOOP OVER THE NUMBER OF ELEMENTS:
C
      DO 500  IE=1,NELEM
C
C     FIND ELEMENT LOCAL COORDINATES:
C
      II = INTMAT(IE,1)
      JJ = INTMAT(IE,2)
      KK = INTMAT(IE,3)
C
      XG1 = COORD(II,1)
      XG2 = COORD(JJ,1)
      XG3 = COORD(KK,1)
      YG1 = COORD(II,2)
      YG2 = COORD(JJ,2)
      YG3 = COORD(KK,2)
      AREA= 0.5*(XG2*(YG3-YG1) + XG1*(YG2-YG3) + XG3*(YG1-YG2))
      IF(AREA.LE.0.)  WRITE(6,5)  IE
    5 FORMAT(/,'  !!! ERROR !!!  ELEMENT NO.', I5,
```

```
      *             ' HAS NEGATIVE OR ZERO AREA ', /,
      *             ' --- CHECK F.E. MODEL FOR NODAL COORDINATES',
      *             ' AND ELEMENT NODAL CONNECTIONS ---'            )
      IF(AREA.LE.0.)   STOP
C
      B1 = YG2 - YG3
      B2 = YG3 - YG1
      B3 = YG1 - YG2
      C1 = XG3 - XG2
      C2 = XG1 - XG3
      C3 = XG2 - XG1
C
      DO 10  I=1,2
      DO 10  J=1,3
      B(I,J) = 0.
   10 CONTINUE
C
      B(1,1) = B1
      B(1,2) = B2
      B(1,3) = B3
      B(2,1) = C1
      B(2,2) = C2
      B(2,3) = C3
C
      DO 20  I=1,2
      DO 30  J=1,3
      B(I,J)  = B(I,J)/(2.*AREA)
      BT(J,I) = B(I,J)
   30 CONTINUE
   20 CONTINUE
C
C     ELEMENT CONDUCTION MATRIX:
C
      DO 100   I=1,3
      DO 100   J=1,3
      AKC(I,J) = 0.
      DO 110   K=1,2
      AKC(I,J) = AKC(I,J) + BT(I,K)*B(K,J)
  110 CONTINUE
      AKC(I,J) = TK*AREA*THICK*AKC(I,J)
  100 CONTINUE
C
C     ELEMENT LOAD VECTOR DUE TO INTERNAL HEAT GENERATION:
C
      FAC = QELE(IE)*AREA*THICK/3.
      DO 200   I=1,3
      QQ(I) = FAC
  200 CONTINUE
C
C     ASSEMBLE THESE ELEMENT MATRICES TO FORM SYSTEM EQUATIONS:
C
      CALL ASSMBLE(   IE, INTMAT,     AKC, QQ, SYSK, SYSQ,
      *              MXPOI,   MXELE                         )
C
  500 CONTINUE
C
      RETURN
      END
```

Listing of the Finite Element Computer Program in Matlab

```
%   PROGRAM FINITE
%   A FINITE ELEMENT COMPUTER PROGRAM FOR SOLVING PARTIAL
%   DIFFERENTIAL EQUATION IN THE FORM OF POISSON'S EQUATION
%   FOR TWO-DIMENTIONAL STEADY-STATE HEAT CONDUCTION WITH
%   INTERNAL HEAT GENERATION.
%                                   PROF. DR. PRAMOTE DECHAUMPHAI
%                                   SUTHEE TRAIVIVATANA
%                                   FACULTY OF ENGINEERING
%                                   CHULALONGKORN UNIVERSITY
%
filename = input('PLEASE ENTER THE INPUT FILE NAME: ', 's');
fid = fopen(filename, 'r');
%
%   READ TITLE OF COMPUTATION:
%
ntitle = fscanf(fid,'%i',1);
for i = 1:ntitle+1
    text = fgetl(fid);
end
%
%   READ INPUT DATA:
%
text  = fgetl(fid);
npoin = fscanf(fid,'%i',1);
nelem = fscanf(fid,'%i',1);
for i=1:2
    text = fgetl(fid);
end
tk    = fscanf(fid,'%f',1);
thick = fscanf(fid,'%f',1);
for i=1:2
    text = fgetl(fid);
end
nodmat = fscanf(fid,'%i %i %f %f %f',[5 npoin]);
nodmat = nodmat';

nodid = squeeze(nodmat(:,1));
ibc   = squeeze(nodmat(:,2));
coord(nodid,1) = squeeze(nodmat(:,3));
coord(nodid,2) = squeeze(nodmat(:,4));
temp  = squeeze(nodmat(:,5));

for ip = 1:npoin
    if nodmat(ip,1) ~= ip
        fprintf(' node no. %5d in data file is missing', ip);
        pause;
        break
    end
end

iq = 0;
for i=1:2
    text = fgetl(fid);
end
elemat = fscanf(fid,'%i %i %i %i %f',[5 nelem]);
elemat = elemat';

eleid = squeeze(elemat(:,1));
intmat(eleid,1) = squeeze(elemat(:,2));
intmat(eleid,2) = squeeze(elemat(:,3));
intmat(eleid,3) = squeeze(elemat(:,4));
qele  = squeeze(elemat(:,5));
```

```
for ie = 1:nelem
    if elemat(ie,1) ~= ie
        fprintf(' element no. %5d in data file is missing', ie);
        pause;
        break
    end
    if qele(ie) ~= 0., iq = 1;, end
end
fprintf('\n THE F.E. MODEL INCLUDES THE FOLLOWING HEAT TRANSFER MODE(S):');
fprintf('\n    -- HEAT CONDUCTION');
if iq==1
    fprintf('\n    -- INTERNAL HEAT GENERATION');
end

fclose(fid);

neq  = npoin;
sysq = zeros(neq, 1);
sysk = zeros(neq);
fprintf('\n *** THE FINITE ELEMENT MODEL CONSISTS OF %5d NODES AND %5d ELEMENTS
***', npoin, nelem);
%
%  ESTABLISH ALL ELEMENT MATRICES AND ASSEMBLE THEM
%  TO FORM UP SYSTEM EQUATIONS
%
fprintf('\n *** ESTABLISHING ELEMENT MATRICES AND ASSEMBLING ELEMENT EQUATIONS
***');
[sysk, sysq] = tri(nelem, intmat, coord, tk, qele, thick, sysk, sysq);
fprintf('\n *** APPLY BOUNDARY CONDITIONS OF NODAL TEMPERATURES ***');
[sysk, sysq] = applybc(npoin, ibc, temp, sysk, sysq);
fprintf('\n *** SOLVING A SET OF SIMULTANEOUS EQUATIONS FOR TEMPERATURE SOLUTIONS
***');
fprintf('\n     ( TOTAL OF %5d EQUATIONS TO BE SOLVED )\n'   , neq);
temp = gauss(neq, sysk, sysq);
%
%  PRINT OUT NODAL TEMPERATURES:
%
filename = input('PLEASE ENTER FILE NAME FOR TEMPERATURE SOLUTIONS: ', 's');
fid = fopen(filename, 'w');
fprintf(fid,' NODAL TEMPERATURE SOLUTION [%5d]:\n', npoin);
fprintf(    ' NODAL TEMPERATURE SOLUTION [%5d]:\n', npoin);
fprintf(fid,'\n   NODE    TEMPERATURE\n');
fprintf(    '\n   NODE    TEMPERATURE\n');
for ip = 1:npoin
    fprintf(fid,' %6d  %14.6e\n', ip, temp(ip));
    fprintf(    ' %6d  %14.6e\n', ip, temp(ip));
end

fclose(fid);
```

```
function [sysk, sysq] = tri(nelem, intmat, coord, tk, qele, thick, sysk, sysq)
%
%  ESTABLISH ALL ELEMENT MATRICES AND ASSEMBLE THEM
%  TO FORM UP SYSTEM EQUATIONS
%
%  LOOP OVER THE NUMBER OF ELEMENTS:
%
for ie = 1:nelem
%
%  FIND ELEMENT LOCAL COORDINATES:
%
    ii = intmat(ie,1);
    jj = intmat(ie,2);
    kk = intmat(ie,3);
```

```
    xg1 = coord(ii,1);
    xg2 = coord(jj,1);
    xg3 = coord(kk,1);
    yg1 = coord(ii,2);
    yg2 = coord(jj,2);
    yg3 = coord(kk,2);
    area = 0.5*(xg2*(yg3 - yg1) + xg1*(yg2 - yg3) + xg3*(yg1 - yg2));
    if area <= 0.
        fprintf('\n  !!! ERROR !!! ELEMENT NO. %5d HAS NEGATIVE OR ZERO...
                 AREA', ie);
        fprintf('\n  --- CHECK F.E. MODEL FOR NODAL COORDINATES AND ELEMENT...
                 NODAL CONNECTIONS ---');
        pause
    end

    b1 = yg2 - yg3;
    b2 = yg3 - yg1;
    b3 = yg1 - yg2;
    c1 = xg3 - xg2;
    c2 = xg1 - xg3;
    c3 = xg2 - xg1;

    b = zeros(2,3);

    b(1,1) = b1;
    b(1,2) = b2;
    b(1,3) = b3;
    b(2,1) = c1;
    b(2,2) = c2;
    b(2,3) = c3;

    b  = b/(2.*area);
    bt = b.';
%
%   ELEMENT CONDUCTION MATRIX:
%
    akc = zeros(3);
    akc = bt*b;
    akc = tk*area*thick*akc;
%
%   ELEMENT LOAD VECTOR DUE TO INTERNAL HEAT GENERATION:
%
    fac = qele(ie)*area*thick/3.;
    qq(1:3) = fac;
%
%   ASSEMBLE THESE ELEMENT MATRICES TO FORM SYSTEM EQUATIONS:
%
    [sysk, sysq] = assemble(ie, intmat, akc, qq, sysk, sysq);
end
% ----------------------------------------------
%
%   FUNCTION ASSEMBLE
%
function [sysk, sysq] = assemble(ie, intmat, akc, qq, sysk, sysq)
%
%   ASSEMBLE ELEMENT EQUATIONS INTO SYSTEM EQUATIONS
%
nnode = 3;
for ir = 1:nnode
    for ic = 1:nnode
        irow = intmat(ie,ir);
        icol = intmat(ie,ic);
        sysk(irow,icol) = sysk(irow,icol) + akc(ir,ic);
    end
    sysq(irow) = sysq(irow) + qq(ir);
end
```

```
function [sysk, sysq] = applybc(npoin, ibc, temp, sysk, sysq)
%
%  APPLY TEMPERATURE BOUNDARY CONDITIONS WITH CONDITION CODES OF:
%        0 = FREE TO CHANGE (TO BE COMPUTED)
%        1 = FIXED AS SPECIFIED
%
for ieq = 1:npoin
    if ibc(ieq) == 0, continue;, end

    for ir = 1:npoin
        if ir == ieq, continue;, end
        sysq(ir) = sysq(ir) - sysk(ir,ieq)*temp(ieq);
        sysk(ir,ieq) = 0.;
    end

    for ic = 1:npoin
        sysk(ieq,ic) = 0.;
    end
    sysk(ieq,ieq) = 1.;
    sysq(ieq) = temp(ieq);
end
```

```
function x = gauss(n, a, b)
%  PERFORM SCALING
[a, b] = scale(n, a, b);
%  FORWARD ELIMINATION: PERFORM ACCORDING TO
%  THE ORDER OF 'PRIME' FROM 1 TO N-1:
for ip = 1:n-1
%  PERFORM PARTIAL PIVOTING
[a, b] = pivot(n, a, b, ip);
%  LOOP OVER EACH EQUATION STARTING FROM THE
%  ONE THAT COORESPONDS WITH THE ORDER OF
%  'PRIME' PLUS ONE:
    for ie = ip+1:n
        ratio = a(ie,ip)/a(ip,ip);
%  COMPUTE NEW COEFF. OF THE EQ. CONSIDERED:
        for ic = ip+1:n
            a(ie,ic) = a(ie,ic) - ratio*a(ip,ic);
        end
        b(ie) = b(ie) - ratio*b(ip);
    end
%  SET COEFF. ON LOWER LEFT PORTION TO ZERO:
    for ie = ip+1:n
        a(ie,ip) = 0.;
    end
end
%  BACK SUBSTITUTION:
%  COMPUTE SOLUTION OF THE LAST EQUATION:
x(n) = b(n)/a(n,n);
%  COMPUTE SOLUTIONS FROM EQUATION N-1 TO 1:
for ie = n-1:-1:1
    sum = 0.;
    for ic = ie+1:n
        sum = sum + a(ie,ic)*x(ic);
    end
    x(ie) = (b(ie) - sum)/a(ie,ie);
end
% ----------------------------------------------
%  FUNCTION SCALING
function [a, b] = scale(n, a, b)
for ie = 1:n
    big = abs(a(ie,1));
    for ic = 2:n
        amax = abs(a(ie,ic));
```

```
        if amax > big
            big = amax;
        end
    end
    for ic = 1:n
        a(ie,ic) = a(ie,ic)/big;
    end
    b(ie) = b(ie)/big;
end
% ----------------------------------------------
%  FUNCTION PIVOTING
function [a, b] = pivot(n, a, b, ip)
jp = ip;
big = abs(a(ip,ip));
for i = ip+1:n
    amax = abs(a(i,ip));
    if amax > big
        big = amax;
        jp  = i;
    end
end
if jp ~= ip
    for j = ip:n
        dumy     = a(jp,j);
        a(jp,j) = a(ip,j);
        a(ip,j) = dumy;
    end
    dumy   = b(jp);
    b(jp) = b(ip);
    b(ip) = dumy;
end
```

Appendix E

Details of Computer Programs

Details of computer programs presented in all chapters in the book and their usage are explained in this appendix. Many of these computer programs are for solving specific examples shown in the chapters, while some of them are general and can be used to solve other problems. Some of these computer programs can be extended and employed for research work. Details of these computer programs in different chapters of the book are as follows.

Computer programs in
Chapter 1: First Step toward Numerical Methods

SHUTTLE   *Page 10*

The program determines the shuttle velocity that varies with time in an example for introducing numerical methods. The program demonstrates the simplicity for using a numerical method to analyze a typical problem as compared to an analytical technique that requires finding an exact solution.

SINCOS   *Page 11*

The program determines the sine and cosine functions by using the infinite series. The program shows how to compute the infinite series that often occur in science and engineering problems.

Computer programs in
Chapter 2: Roots of Equations

BISECT   *Page 24*

The program determines the root of the transcendental Eq. (2.12) by using the bisection method.

FALPOS   *Page 28*

The program determines the root of the transcendental Eq. (2.12) by using the false-position method.

ONEPT   *Page 31*

The program determines the root of the transcendental Eq. (2.12) by using the one-point iteration method.

NEWRAP   *Page 39*

The program determines the root of the transcendental Eq. (2.12) 34 by using the Newton-Raphson iteration method.

SECANT   *Page 41*

The program determines the root of the transcendental Eq. (2.12) by using the secant method.

Computer programs in
Chapter 3: Solving System of Equations

NGELIM   *Page 66*

The program solves the system of equations as shown in Eq. (3.10) by using the Gauss elimination method. Table 3.1 shows the program input data and the output solution.

GELIM   *Page 71*

The program solves the system of equations as shown in Eq. (3.10) by using the Gauss elimination method. The scaling and pivoting techniques are added into the **NGELIM** program to improve the efficiency of the Gauss elimination method. Table 3.2 shows the program input data and the output solution.

TRIDG   *Page 73*

The program solves a system of equations by using the Gauss-elimination method when $[A]$ is a tri-diagonal matrix.

LUDCOM   *Page 82*

The program solves the system of equations as shown in Eq. (3.10) by using the LU decomposition method. Table 3.3 shows the program input data and the output solution.

CHOLESKY   *Page 87*

The program solves the system of equations $[A]\{X\} = \{B\}$ when $[A]$ is a symmetric matrix. The program uses Eq. (3.65) as an example to solve for its solution. The input data of the example and the output solution from the program are shown in Table 3.4.

JACOBI   *Page 91*

The program solves the system of equations as shown in Eq. (3.10) by using the Jacobi iteration method. Table 3.5 shows the program input data and the output solution.

GSEIDEL   *Page 93*

The program solves the system of equations as shown in Eq. (3.10) by using the Gauss-Seidel iteration method. Table 3.6 shows the program input data and the output solution.

SOR *Page 95*

The program solves the system of equations as shown in Eq. (3.10) by using the successive over-relaxation method. Table 3.7 shows the program input data and the output solution.

CG *Page 102*

The program solves the system of equations by using the conjugate gradient method. The system of equations as shown in Eq. (3.100) is used as an example to solve for its solution.

CGNEW *Page 105*

The program solves the system of equations by using the improved conjugate gradient method. The system of equations, Eq. (3.10), is used as an example to solve for its solution.

Computer programs in
Chapter 4: Interpolation Functions

NEWDIV *Page 125*

The program interpolates a value at a desired location from a set of data by using the Newton's divided difference method. Figure 4.5 shows a set of input data and the interpolated value at the desired location obtained from the program.

LAGPOL *Page 132*

The program interpolates a value at a desired location from a set of data by using the Lagrange interpolating method. Figure 4.13 shows a set of input data and the interpolated value at the desired location obtained from the program.

CUBSPLN *Page 141*

The program interpolates a value at a desired location from a set of data by using the cubic spline interpolating method. Example 4.7 shows the use of the program to fit a set of data by the method.

Computer programs in
Chapter 5: Least-Squares Curve Fitting

LREGRES *Page 156*

The program determines the two coefficients of a straight line fitting for a given set of data by using the linear regression method. Table 5.3 shows a typical set of data and the two coefficients obtained from the program.

PREGRES *Page 164*

The program determines coefficients of a polynomial function for fitting a given set of data by using the polynomial regression method. Table 5.5 shows a typical set of data and the computed coefficients are shown in Eq. (5.28).

MREGRES *Page 170*

The program determines coefficients of a linear function that varies with many variables for fitting for a given set of data by using the multiple linear regression method. Table 5.6 shows a typical set of data and the coefficients obtained from the program.

Computer programs in
Chapter 6: Numerical Integration and Differentiation

TRAPEZ *Page 196*

The program performs numerical integration of a given function by using the multiple-segment trapezoidal rule. An example of the function is given by Eq. (6.12) and the computed solutions using different numbers of segments are shown in Table 6.1.

SIMPSON *Page 200*

The program performs numerical integration of a given function by using the multiple-segment Simpson's rule. An example of the function is given by Eq. (6.48) and the computed solutions using different numbers of segments are shown in Table 6.2.

ROMBERG *Page 210*

The program performs numerical integration of a given function by using the Romberg method. The method is presented in details as explained in Example 6.10. The computation procedures inside the program follow those presented in the example.

GAUSINT   *Page 218*

The program performs numerical integration of a given function by using the Gauss-Legendre integration method. An example of the function is given by Eq. (6.12) and the computed solutions from using different numbers of Gauss points are shown in Fig. 6.16.

NUMDIF   *Page 230*

The program determines the first-order derivatives of a given function at different points. An example of the function is given by Eq. (6.148) and the computed solutions obtained from the program are presented in Table 6.10.

Computer programs in
Chapter 7: Ordinary Differential Equations

EULER   *Page 247*

The program solves the first-order ordinary differential equations by using the Euler's method. An example of the differential equations is given by Eq. (7.13) and the solutions obtained using different step sizes are shown in Table 7.1.

HEUN   *Page 251*

The program solves the first-order ordinary differential equations by using the Heun's method. An example of the differential equations is given by Eq. (7.13) and the computed solutions are shown in Table 7.2.

MEULER   *Page 254*

The program solves the first-order ordinary differential equations by using the modified Euler's method. An example of the differential equations is given by Eq. (7.13) and the computed solutions are shown in Table 7.2.

RK3   *Page 259*

The program solves the first-order ordinary differential equations by using the third-order Runge-Kutta method. An example of the differential equations is given by Eq. (7.13) and the computed solutions are shown in Table 7.3.

RK4   *Page 261*

The program solves the first-order ordinary differential equations by using the fourth-order Runge-Kutta method. An example of the differential equations is given by Eq. (7.13) and the computed solutions are shown in Table 7.3.

SYSEUL *Page 263*

The program solves a system of first-order differential equations by using the Euler's method. An example of the system of differential equations is given by Eq. (7.60) and the computed solutions are shown in Table 7.4.

SYSRK4 *Page 265*

The program solves a system of first-order differential equations by using the fourth-order Runge-Kutta method. An example of the system of differential equation is given by Eq. (7.60) and the computed solutions are shown in Table 7.4.

ADAMBAS *Page 274*

The program employs the multistep method with the fourth-order Adams-Bashforth technique to solve a system of first-order differential equations. An example of the system of differential equations is given by Eq. (7.13) and the computed solutions are shown in Table 7.9.

Computer programs in
Chapter 8: Partial Differential Equations

ELLIP *Page 296*

The program employs the finite difference method to solve the elliptic partial differential equation representing steady-state heat conduction in a plate. The differential equation is given by Eq. (8.15) and the finite difference model is shown in Fig. 8.7. The temperature solutions obtained from the program are shown in Table 8.1.

PARAEXP *Page 304*

The program employs the explicit finite difference method to solve the parabolic partial differential equation representing transient heat conduction in a bar. The differential equation is given by Eq. (8.37) and the finite difference model is shown in Fig. 8.17. The transient temperature solutions at different times obtained from the program are shown in Table 8.3 and plotted in Fig. 8.19.

PARAIMP *Page 308*

The program employs the implicit finite difference method to solve the parabolic partial differential equation representing transient heat conduction in a bar. The differential equation is given by Eq. (8.37) and the finite difference model is shown in Fig. 8.17. The transient temperature solutions at different times obtained from the program are shown in Table 8.5.

PARACN   *Page 311*

The program employs the Crank-Nicolson finite difference method to solve the parabolic partial differential equation representing transient heat conduction in a bar. The differential equation is given by Eq. (8.37) and the finite difference model is shown in Fig. 8.17. The transient temperature solutions at different times obtained from the program are shown in Table 8.6.

HYPER   *Page 318*

The program employs the finite difference method to solve the hyperbolic partial differential equation representing an oscillation in a string. The differential equation is given by Eq. (8.66) and the finite difference model is shown in Fig. 8.27. The string oscillating solutions at different times obtained from the program are shown in Table 8.8 and plotted in Fig. 8.29.

Computer program in
Chapter 9: Finite Element Method

FINITE   *Page 421*

The program solves the partial differential equation in the form of the Poisson's equation representing steady-state heat conduction in plate subjected to an internal heat generation. The Poisson's equation is given by Eq. (9.80). The computational procedures of the program are shown by the flow chart in Fig. 9.22. Examples 9.3 and 9.4 show the typical input data files required by the program and the output solutions obtained from the program.

Appendix F

Computer Programs in C and Pascal

Computer programs presented in the chapters of the book are in Fortran and Matlab languages. These computer programs are translated into C and Pascal languages as shown in this appendix. Details of these computer programs and their usage are explained in Appendix E.

Computer programs in C

```c
/* PROGRAM SHUTTLE   */
#include <stdio.h>
#define DT 30.0
main()
{
    int I;
    float T = 0.0 ;
    float V = 0.0 ;
    for ( I = 1; I <= 50; I = I + 1 ) {
        V = V + DT*(9.8 - 0.005*v);
        T = T + DT;
        printf(" %12.0f %12.0f \n", T, V);
    }
    return 0;
}
```

```c
/*    PROGRAM SINCOS  */
/*    PROGRAM FOR COMPUTING SIN AND COSINE */
/*    FUNCTIONS FOR ANGLES FROM 0 TO 180 DEGRESS  */
/*    WITH THE INCREMENT AT EVERY 10 DEGREES  */
#include <stdio.h>
#include <math.h>
#define PI  4.0*atan(1.0)
#define DEL 10.0
main()
{
    int IDEG, N, MS, MC;
    float X, SUMS, SUMC, TERMS, TERMC, SIGN;
    float DEG = 0.0;
    printf("\n    DEGREE       SIN       COS\n\n");
    for (IDEG = 1; IDEG <= 19; IDEG = IDEG + 1) {
        X = PI*DEG/180.;
        SUMS = X;
        SUMC = 1.;
        TERMS = X;
        TERMC = 1.;
        SIGN = -1.;
        for (N = 1; N <= 100; N = N + 1) {
            MS = 2*N + 1;
            MC = 2*N;
            TERMS = TERMS*X*X/(MS*(MS-1));
            TERMC = TERMC*X*X/(MC*(MC-1));
            SUMS = SUMS + SIGN*TERMS;
            SUMC = SUMC + SIGN*TERMC;
            SIGN = -SIGN;
        }
        printf("%10.0f %16.6f %16.6f\n", DEG, SUMS, SUMC);
        DEG = DEG + DEL;
    }
    return 0;
}
```

```c
/*    PROGRAM BISECT    */
//....PROGRAM FOR COMPUTING ROOT OF NONLINEAR    */
//....EQUATION USING THE BISECTION METHOD    */
//....DEFINE THE FOLLOWING GIVEN VALUES:
/*....    XL = LEFT VALUE OF X    */
/*....    XR = RIGHT VALUE OF X    */
/*....    ES = STOPPING CRITERION TOLERANCE (%)    */
#include <stdio.h>
#include <math.h>
float FUNC( float X );
main()
{
    int ITER;
    float XL,XR,XM,XN,ES,FXL,FXR,FXM,AA,TOL;
    XL = 0.0;
    XR = 2.0;
    ES = 0.001;
/*....CHECK WHETHER THE ROOT IS IN GIVEN RANGE   */
    FXL = FUNC(XL);
    FXR = FUNC(XR);
    AA = FXL*FXR;
    if ( AA >= 0.0 ) {
        printf("\n ROOT IS NOT IN THE GIVEN RANGE\n");
        return 0;
    }
    printf("\n    ITERATION NO.         XN");
    for (ITER = 1; ITER <= 500; ITER = ITER + 1) {
        XM = (XL+XR)/2.;
        FXM = FUNC(XM);
        FXR = FUNC(XR);
        AA = FXM*FXR;
        if ( AA > 0.0 )
/*....CASE  A:   XL < ROOT < XM   */
            XR = XM;
        else
/*....CASE  B:   XM < ROOT < XR   */
            XL = XM;
/*....CHECK FOR TOLERANCE:   */
        XN = (XL+XR)/2.;
        printf(" %8d         %14.6f\n", ITER,XN);
        TOL = fabs((XN-XM)*100./XN);
        if (TOL < ES)
            goto jump1;
    }
    printf("\n ROOT CAN NOT BE REACHED FOR\n");
    printf(" THE GIVEN CONDITIONS\n");
    goto jump2;
jump1:  printf("\n THE ROOT IS %14.6f\n", XN);
jump2:  return 0;
}
float FUNC( float X )
{
    float p;
    p = exp(-X/4.)*(2.-X) - 1.;
    return p;
}
```

```c
/*    PROGRAM FALPOS    */
//....PROGRAM FOR COMPUTING ROOT OF NONLINEAR    */
//....EQUATION USING THE FALSE-POSITION METHOD    */
//....DEFINE THE FOLLOWING GIVEN VALUES:
/*....    XL = LEFT VALUE OF X    */
/*....    XR = RIGHT VALUE OF X    */
/*....    ES = STOPPING CRITERION TOLERANCE (%)    */
#include <stdio.h>
```

```c
#include <math.h>
float FUNC( float X );
main()
{
    int ITER;
    float XL,XR,X1,X1OLD,ES,FXL,FXR,FX1,AA,TOL;
    XL = 0.0;
    XR = 2.0;
    ES = 0.001;
/*.....CHECK WHETHER THE ROOT IS IN GIVEN RANGE    */
    FXL = FUNC(XL);
    FXR = FUNC(XR);
    AA = FXL*FXR;
    if ( AA >= 0. ) {
        printf("\n ROOT IS NOT IN THE GIVEN RANGE\n");
        return 0;
    }
    printf("\n  ITERATION NO.          X\n");
    for (ITER = 1; ITER <= 500; ITER = ITER + 1) {
        FXL = FUNC(XL);
        FXR = FUNC(XR);
        X1 = (XL*FXR - XR*FXL)/(FXR - FXL);
        FX1 = FUNC(X1);
        AA = FX1*FXR;
        if ( AA < 0. )
/*.....CASE A:    X1 < ROOT < XR    */
            XL = X1;
        else
/*.....CASE B:    XL < ROOT < X1    */
            XR = X1;
/*.....CHECK FOR TOLERANCE:    */
        printf(" %8d        %14.6f\n", ITER, X1);
        TOL = fabs((X1-X1OLD)*100./X1);
        if ( TOL < ES )
            goto jump1;
        X1OLD = X1;
    }
    printf("\n ROOT CAN NOT BE REACHED FOR");
    printf("\n THE GIVEN CONDITION \n");
    goto jump2;
jump1:  printf("\n    THE ROOT IS %14.6f", X1);
jump2:  return 0;
}
float FUNC( float X )
{
    float p;
    p = exp(-X/4.)*(2.-X) - 1.;
    return p;
}
```

```c
/*   PROGRAM ONEPT     */
#include <stdio.h>
#include <math.h>
main()
{
    int I;
    float XNEW,XOLD,ES,TOL;
    XOLD = 0.;
    ES = 0.001;
    printf("\n  ITERATION NO.        X\n");
    for ( I = 1; I <= 100; I = I + 1 ) {
        XNEW = 2. - exp(XOLD/4.);
        printf(" %8d        %14.6f\n", I, XNEW);
        TOL = fabs((XNEW-XOLD)*100./XNEW);
        if ( TOL < ES )
            goto jump;
        XOLD = XNEW;
    }
jump:
    printf("\n    THE ROOT IS %14.6f", XNEW);
    return 0;
}
```

```c
/*     PROGRAM NEWRAP     */
/*....PROGRAM FOR COMPUTING ROOT OF NONLINEAR      */
/*....EQUATION USING THE NEWTON-RAPHSON METHOD     */
/*....DEFINE THE FOLLOWING GIVEN VALUES:           */
/*....    X0 = INITIAL GUESS VALUE OF X            */
/*....    ES = STOPPING CRITERION TOLERANCE  (%)   */
#include <stdio.h>
#include <math.h>
float FUNC( float X );
float DERIV( float X );
main()
{
    int I,ITER;
    float X,X0,ES,F,DX,DF,TOL;
    X0 = 3.;
    ES = 0.001;
    printf("\n  ITERATION NO.          X\n");
    X = X0;
    for (ITER = 1; ITER <= 500; ITER = ITER + 1) {
        F = FUNC(X);
        DF = DERIV(X);
        DX = -F/DF;
        X = X + DX;
        printf(" %8d        %14.6f\n", ITER, X);
        TOL = fabs(DX*100./X);
        if ( TOL < ES )
            goto jump1;
    }
    printf("\n ROOT CAN NOT BE REACHED FOR");
    printf(" THE GIVEN CONDITION\n");
    goto jump2;
jump1:  printf("\n    THE ROOT IS %14.6f\n", X);
jump2:  return 0;
}
float FUNC( float X )
{
    float p;
    p = exp(-X/4.)*(2.-X) - 1.;
    return p;
}
float DERIV( float X )
{
    float d;
    d = -exp(-X/4.)*(1.5-X/4.);
    return d;
}
```

```c
/*      PROGRAM SECANT     */
/*....PROGRAM FOR COMPUTING ROOT OF NONLINEAR       */
/*....EQUATION USING THE SECANT METHOD              */
/*....DEFINE THE FOLLOWING GIVEN VALUES:            */
/*......  X0 = FIRST VALUE OF INITIAL GUESS OF X    */
/*......  X1 = SECOND VALUE OF INITIAL GUESS OF X   */
/*......  ES = STOPPING CRITERION TOLERANCE (%)     */
#include <stdio.h>
#include <math.h>
float FUNC( float X );
main()
{
    int I,ITER;
    float X0,X1,ES,F0,F1,DF,DX,TOL;
    X0 = 3.;
    X1 = 2.;
    ES = 0.001;
    printf("\n    ITERATION NO.        X\n");
    for (ITER = 1; ITER <= 500; ITER++) {
        F0 = FUNC(X0);
        F1 = FUNC(X1);
        DF = (F0-F1)/(X0-X1);
        DX = -F1/DF;
        X0 = X1;
        X1 = X1 + DX;
        printf(" %8d        %14.6f\n", ITER, X1);
        TOL = fabs(DX*100./X1);
        if ( TOL < ES )
            goto jump1;
    }
    printf("\n ROOT CAN NOT BE REACHED FOR");
    printf(" THE GIVEN CONDITION\n");
    goto jump2;
jump1:  printf("\n    THE ROOT IS %14.6f\n", X1);
jump2:  return 0;
}
float FUNC( float X )
{
    float p;
    p = exp(-X/4.)*(2.-X) - 1.;
    return p;
}
```

```c
/*      PROGRAM NGELIM     */
/*....PROGRAM FOR SOLVING A SET OF SIMULTANEOUS     */
/*....LINEAR EQUATIONS [A]{X} = {B} USING           */
/*....NAIVE GAUSS ELIMINATION                       */
#include <stdio.h>
void GAUSS(int N,float A[50][50], float B[50], float X[50]);
void main()
{
    FILE *inpf, *outf;
    float A[50][50],B[50],X[50];
    int I,N,ICOL,IROW;
    inpf = fopen("INPUT.DAT","r");
    outf = fopen("SOL.OUT","w");
/*....READ TOTAL NUMBER OF EQUATIONS TO BE SOLVED: */
    fscanf(inpf,"%d",&N);
/*....READ MATRIX [A] AND VECTOR {B}               */
    for (IROW = 1; IROW <= N; IROW++) {
        for (ICOL = 1; ICOL <= N; ICOL++)
            fscanf(inpf,"%f",&A[IROW][ICOL]);
        fscanf(inpf,"%f",&B[IROW]);
    }
    GAUSS(N, A, B, X);
    fprintf(outf,"\n    EQUATION NO.        SOLUTION X\n");
    for (I = 1; I <= N; I++)
        fprintf(outf,"%12d        %16.6E\n",I,X[I]);
    fclose(inpf);
    fclose(outf);
    return;
}
void GAUSS(int N,float A[50][50], float B[50], float X[50])
{
    int IC,IE,IP;
    float RATIO,SUM;
/*....FORWARD ELIMINATION: PERFORM ACCORDING TO     */
/*....THE ORDER OF 'PRIME' FROM 1 TO N-1:           */
    for (IP = 1; IP <= N-1; IP++) {
/*....LOOP OVER EACH EQUATION STARTING FROM THE     */
/*....ONE THAT CORRESPONDS WITH THE ORDER OF        */
/*....'PRIME' PLUS ONE:                             */
        for (IE = IP+1; IE <= N; IE++) {
            RATIO = A[IE][IP]/A[IP][IP];
/*....COMPUTE NEW COEFF. OF THE EQ. CONSIDERED:     */
            for (IC = IP+1; IC <= N; IC++)
                A[IE][IC] = A[IE][IC] - RATIO*A[IP][IC];
            B[IE] = B[IE] - RATIO*B[IP];
/*....SET COEFF. ON LOWER LEFT PORTION TO ZERO:     */
            for (IE = IP+1; IE <= N; IE++)
                A[IE][IC] = 0.;
        }
    }
/*....BACK SUBSTITUTION                             */
/*....COMPUTE SOLUTION OF THE LAST EQUATION:        */
    X[N] = B[N]/A[N][N];
/*....COMPUTE SOLUTION FROM EQUATION N-1 TO 1       */
    for (IE = N-1; IE >= 1; IE--) {
        SUM = 0.;
        for (IC = IE+1; IC <= N; IC++)
            SUM = SUM + A[IE][IC]*X[IC];
        X[IE] = (B[IE] - SUM)/A[IE][IE];
    }
    return;
}
```

```c
void GAUSS(int N,float A[50][50], float B[50], float X[50])
{
    int IC,IE,IP;
    void SCALE(int N,float A[50][50],float B[50]);
    void PIVOT(int N,float A[50][50], float B[50],int IP);
    float RATIO,SUM;
/*....PERFORM SCALING:                              */
    SCALE(N, A, B);
/*....FORWARD ELIMINATION: PERFORM ACCORDING TO     */
/*....THE ORDER OF 'PRIME' FROM 1 TO N-1:           */
    for (IP = 1; IP <= N-1; IP++) {
/*....PERFORM PARTIAL PIVOTING                      */
        PIVOT(N, A, B, IP);
/*....LOOP OVER EACH EQUATION STARTING FROM THE     */
/*....ONE THAT CORRESPONDS WITH THE ORDER OF        */
```

```c
/*....'PRIME' PLUS ONE:                               */
    for (IE = IP+1;  IE <= N; IE++) {
        :
        :
}
/*----------------------------------------------------*/
void SCALE(int N,float A[50][50],float B[50])
{
    int IC,IE;
    float AMAX,BIG;
/*....PERFORM SCALING                                 */
    for (IE = 1; IE <= N; IE++) {
        BIG = fabs(A[IE][1]);
        for (IC = 2; IC <= N; IC++) {
            AMAX = fabs(A[IE][IC]);
            if (AMAX > BIG) BIG = AMAX;
        }
        for (IC = 1; IC <= N; IC++)
            A[IE][IC] = A[IE][IC]/BIG;
        B[IE] = B[IE]/BIG;
    }
    return;
}
void PIVOT(int N,float A[50][50],float B[50],int IP)
{
    int I,J,JP;
    float AMAX,BIG,DUMY;
/*....PERFORM PARTIAL PIVOTING:                       */
    JP = IP;
    BIG = fabs(A[IP][IP]);
    for (I = IP+1; I <= N; I++) {
        AMAX = fabs(A[I][IP]);
        if (AMAX > BIG) {
            BIG = AMAX;
            JP = I;
        }
    }
    if (JP != IP) {
        for (J = IP; J <= N; J++) {
            DUMY = A[JP][J];
            A[JP][J] = A[IP][J];
            A[IP][J] = DUMY;
        }
        DUMY = B[JP];
        B[JP] = B[IP];
        B[IP] = DUMY;
    }
    return;
}
/*                                                    */
/*....PROGRAM TRISYS                                  */
/*....PROGRAM FOR SOLVING A SET OF SIMULTANEOUS       */
/*....LINEAR EQUATIONS [A]{X} = {B}  USING            */
/*....GAUSS ELIMINATION FOR TRIDIAGONAL SYSTEM        */
#include <stdio.h>
void TRIDG(int N,float A[50], float B[50], float C[50],
float D[50], float X[50]);
void main()
{
    FILE *inpf, *outf;
    float A[50],B[50],C[50],D[50],X[50];
    int I,IROW, N;
    inpf = fopen("INPUT.DAT","r");
    outf = fopen("SOL.OUT","w");
/*....READ TOTAL NUMBER OF EQUATIONS TO BE SOLVED:    */
    fscanf(inpf,"%d",&N);
/*....READ VECTOR {A}, {B}, {C} AND {D}  :            */
    for (IROW = 1; IROW <= N; IROW++) {
        fscanf(inpf,"%f",&A[IROW]);
        fscanf(inpf,"%f",&B[IROW]);
        fscanf(inpf,"%f",&C[IROW]);
        fscanf(inpf,"%f",&D[IROW]);
    }
    TRIDG(N, A, B, C, D, X);
    fprintf(outf,"\n    EQUATION NO.        SOLUTION X\n");
    for (I = 1; I <= N; I++)
        fprintf(outf,"%12d        %16.6E\n",I,X[I]);
    fclose(inpf);
    fclose(outf);
    return;
}
void TRIDG(int N,float A[50], float B[50], float C[50],
float D[50], float X[50])
/*....SOLVE TRIDIAGONAL SYSTEM FOR N EQUATIONS:       */
{
    int I;
/*....PERFORM FORWARD ELIMINATION:                    */
    for (I = 2; I <= N; I++) {
        A[I] = A[I]/B[I-1];
        B[I] = B[I] - A[I]*C[I-1];
        D[I] = D[I] - A[I]*D[I-1];
    }
/*....PERFORM BACKWARD SUBSTITUTION:                  */
    X[N] = D[N]/B[N];
    for (I = N-1; I >= 1; I--)
        X[I] = (D[I] - C[I]*X[I+1])/B[I];
    return;
}
/*....PROGRAM LUDCOM                                  */
/*....PROGRAM FOR SOLVING A SET OF SIMULTANEOUS       */
/*....LINEAR EQUATIONS [A]{X} = {B}                   */
/*....USING NAIVE LU DECOMPOSITION                    */
#include <stdio.h>
void LU(int N,float A[51][51],float B[51],float X[51],float
AL[51][51],float AU[51][51],float Y[51]);
void main()
{
    FILE *inpf, *outf;
    int I,ICOL,IROW,N;
    float A[51][51],B[51],X[51];
    float AL[51][51],AU[51][51],X[51];
    inpf = fopen("INPUT.DAT","r");
    outf = fopen("SOL.OUT","w");
/*....READ TOTAL NUMBER OF EQUATIONS TO BE SOLVED     */
    fscanf(inpf,"%d",&N);
/*....READ MATRIX [A] AND VECTOR {B}                  */
    for (IROW = 1; IROW <= N; IROW++) {
        for (ICOL = 1; ICOL <= N; ICOL++)
            fscanf(inpf,"%f",&A[IROW][ICOL]);
        fscanf(inpf,"%f",&B[IROW]);
    }
    LU(N, A, B, X, AL, AU, Y);
    fprintf(outf,"\n    EQUATION NO.        SOLUTION X\n");
    for (I = 1; I <= N; I++)
```

```c
/*      PROGRAM CHOLESKY                               */
/*....PROGRAM FOR SOLVING A SET OF SIMULTANEOUS        */
/*....LINEAR EQUATIONS [A]{X} = {B} USING              */
/*....CHOLESKY DECOMPOSITION IF [A] IS SYMMETRIC       */
#include <stdio.h>
#include <math.h>

void CHOLES(int N,float A[51][51], float B[51],float X[51],float
AL[51][51],float Y[51]);
void main()
{
    FILE *inpf, *outf;
    float A[51][51],B[51],X[51];
    float AL[51][51],Y[51];
    int I,N,ICOL,IROW;
    inpf = fopen("INPUT2.DAT","r");
    outf = fopen("SOL2.OUT","w");
/*....READ TOTAL NUMBER OF EQUATIONS TO BE SOLVED: */
    fscanf(inpf,"%d",&N);
/*....READ MATRIX [A] AND VECTOR {B} */
    for (IROW = 1; IROW <= N; IROW++) {
        for (ICOL=1; ICOL <= N; ICOL++)
            fscanf(inpf,"%f",&A[IROW][ICOL]);
        fscanf(inpf,"%f",&B[IROW]);
    }
    CHOLES(N, A, B, X, AL, Y);
    fprintf(outf,"\n    EQUATION NO.        SOLUTION X\n");
    for (I = 1; I <= N; I++)
        fprintf(outf,"%12d        %16.6E\n",I,X[I]);
    fclose(inpf);
    fclose(outf);
    return;
}
void CHOLES(int N,float A[51][51],  float B[51],float X[51],float
AL[51][51],float Y[51])
{
    int I,J,K;
    float SUM;
/*....PERFORM DECOMPOSITION  [A] = [L][LT] */
    for (I = 1; I <= 50; I++)
        for (J = 1; J <= 50; J++)
            AL[I][J] = 0.;
    AL[1][1] = sqrt(A[1][1]);
    for (K = 2; K <= N; K++) {
        for (I = 1; I <= K-1; I++) {
            SUM = 0.;
            if (I == 1) goto jump;
            for (J = 1; J <= I-1; J++)
                SUM = SUM + AL[I][J]*AL[K][J];
jump:       AL[K][I] = (A[K][I] - SUM)/AL[I][I];
        }
        SUM = 0.;
        for (J = 1; J <= K-1; J++)
            SUM = SUM + AL[K][J]*AL[K][J];
        AL[K][K] = sqrt(A[K][K] - SUM);
    }
/*....PERFORM FORWARD PASS */
    Y[1] = B[1]/AL[1][1];
    for (I = 2; I <= N; I++) {
        SUM = 0.;
        for (J = 1; J <= I-1; J++)
            SUM = SUM + AL[I][J]*Y[J];
        Y[I] = (B[I] - SUM)/AL[I][I];
    }
/*....PERFORM BACKWARD PASS */
    X[N] = Y[N]/AL[N][N];
    for (I = N-1; I >= 1; I--) {
        SUM = 0.;
        for (J = I+1; J <= N; J++)
            SUM = SUM + AL[J][I]*X[J];
        X[I] = (Y[I] - SUM)/AL[I][I];
    }
    return;
}
```

```c
        fprintf(outf,"%12d        %16.6E\n",I,X[I]);
    return;
}
void LU(int N,float A[51][51],float B[51],float X[51],float
AL[51][51],float AU[51][51],float Y[51])
{
    int I,J,K;
    float SUM;
/*....PERFORM DECOMPOSITION  [A] = [L][U] */
    for (I = 1; I <= 50; I++)
        for (J = 1; J <= 50; J++) {
            AL[I][J] = 0.;
            AU[I][J] = 0.;
        }
    for (I = 1; I <= N; I++)
        AL[I][1] = A[I][1];
    for (J = 2; J <= N; J++)
        AU[1][J] = A[1][J]/AL[1][1];
    for (J = 2; J <= N-1; J++) {
        for (I = J; I <= N; I++) {
            SUM = 0.;
            for (K = 1; K <= J-1; K++)
                SUM = SUM + AL[I][K]*AU[K][J];
            AL[I][J] = A[I][J] - SUM;
        }
        for (K = J+1; K <= N; K++) {
            SUM = 0.;
            for (I = 1; I <= J-1; I++)
                SUM = SUM + AL[J][I]*AU[I][K];
            AU[J][K] = (A[J][K] - SUM)/AL[J][J];
        }
    }
    SUM = 0.;
    for (K = 1; K <= N-1; K++)
        SUM = SUM + AL[N][K]*AU[K][N];
    AL[N][N] = A[N][N] - SUM;
/*....PERFORM FORWARD PASS TO SOLVE [L]{Y} = {B}: */
    Y[1] = B[1]/AL[1][1];
    for (I = 2; I <= N; I++) {
        SUM = 0.;
        for (J = 1; J <= I-1; J++)
            SUM = SUM + AL[I][J]*Y[J];
        Y[I] = (B[I] - SUM)/AL[I][I];
    }
/*....PERFORM BACKWARD PASS TO SOLVE [U]{X} = {Y}: */
    X[N] = Y[N];
    for (I = N-1; I >= 1; I--) {
        SUM = 0.;
        for (J = I+1; J <= N; J++)
            SUM = SUM + AU[I][J]*X[J];
        X[I] = Y[I] - SUM;
    }
    return;
}
```

```c
/*   PROGRAM JACOBI                                        */
/*.... JACOBI ITERATION METHOD FOR EXAMPLE 3.13           */
#include <stdio.h>
#include <math.h>
#define TOL 0.05
void main()
{
    int I,IFLAG,ITER;
    float XOLD[4], XNEW[4];
    float EPS;
    for (I = 1; I <= 3; I++)
        XOLD[I] = 100.;
    printf("  ITERATION NO.        X1        X2        X3\n\n");
    for (ITER = 1; ITER <= 500; ITER++) {
        XNEW[1] = 100. + XOLD[2];
        XNEW[2] = 100. + 0.25*XOLD[1] + 0.5*XOLD[3];
        XNEW[3] = 100. + 0.50*XOLD[2];
        printf("%8d ",ITER);
        for (I = 1; I <= 3; I++)
            printf("%10.0f",XOLD[I]);
        printf("\n");
        IFLAG = 0;
        for (I = 1; I <= 3; I++) {
            EPS = fabs((XNEW[I] - XOLD[I])*100./XNEW[I]);
            if ( EPS >= TOL) IFLAG = 1;
        }
        if (IFLAG == 0) return;
        for (I = 1; I <= 3; I++)
            XOLD[I] = XNEW[I];
    }
    return;
}
```

```c
/*   PROGRAM GSEIDEL                                       */
/*.... GAUSS-SEIDEL ITERATION METHOD FOR EXAMPLE 3.14      */
#include <stdio.h>
#include <math.h>
main()
{
    int I,IFLAG,ITER;
    float XOLD[4], XNEW[4];
    float EPS,TOL;
    TOL = 0.050;
    for (I = 1; I <= 3; I = I + 1)
        XOLD[I] = 100.;
    printf("  ITERATION NO.        X1        X2        X3\n\n");
    for (ITER = 1; ITER <= 500; ITER = ITER + 1) {
        XNEW[1] = 100. + XOLD[2];
        XNEW[2] = 100. + 0.25*XNEW[1] + 0.5*XOLD[3];
        XNEW[3] = 100. + 0.50*XNEW[2];
        printf("%8d ",ITER);
        for (I = 1; I <= 3; I = I + 1)
            printf("%10.0f",XOLD[I]);
        printf("\n");
        IFLAG = 0;
        for (I = 1; I <= 3; I = I + 1) {
            EPS = fabs((XNEW[I] - XOLD[I])*100./XNEW[I]);
            if ( EPS >= TOL) IFLAG = 1;
        }
        if (IFLAG == 0) goto jump;
        for (I = 1; I <= 3; I = I + 1)
            XOLD[I] = XNEW[I];
    }
jump:
    return 0;
}
```

```c
/*   PROGRAM SOR                                           */
/*.... SUCCESSIVE OVER RELAXATION FOR EXAMPLE 3.15         */
#include <stdio.h>
#include <math.h>
void main()
{
    int I,IFLAG,ITER;
    float XOLD[4],XNEW[4];
    float EPS,W,TOL;
    W = 1.2;
    TOL = 0.05;
    for (I = 1; I <= 3; I++)
        XOLD[I] = 100.;
    printf("  ITERATION NO.        X1        X2        X3\n\n");
    for (ITER = 1; ITER <= 20; ITER++) {
        XNEW[1] = 100. + XOLD[2];
        XNEW[1] = W*XNEW[1] + (1.-W)*XOLD[1];
        XNEW[2] = 100. + 0.25*XNEW[1] + 0.50*XOLD[3];
        XNEW[2] = W*XNEW[2] + (1.-W)*XOLD[2];
        XNEW[3] = 100. + 0.50*XNEW[2];
        XNEW[3] = W*XNEW[3] + (1.-W)*XOLD[3];
        printf("%8d ",ITER);
        for (I = 1; I <= 3; I++)
            printf("%10.0f",XOLD[I]);
        printf("\n");
        IFLAG = 0;
        for (I = 1; I <= 3; I++) {
            EPS = fabs((XNEW[I] - XOLD[I])*100./XNEW[I]);
            if (EPS >= TOL)  IFLAG = 1;
        }
        if (IFLAG == 0 ) return;
        for (I = 1; I <= 3; I++)
            XOLD[I] = XNEW[I];
    }
    return;
}
```

```c
/*   PROGRAM ORCG                                          */
/*....PROGRAM FOR SOLVING A SET OF SIMULTANEOUS            */
/*....LINEAR EQUATIONS [A]{X} = {B}  USING                 */
/*....CONJUGATE GRADIENT                                   */
#include <stdio.h>
#include <math.h>
void CG(int N, float A[50][50], float B[50], float X[50]);
void main()
{
    FILE *inpf, *outf;
    float A[50][50],B[50],X[50];
    int I,ICOL,IROW, N;
    inpf = fopen("INPUT.DAT","r");
```

```c
    outf = fopen("SOL.OUT", "w");
/*....READ TOTAL NUMBER OF EQUATIONS TO BE SOLVED:  */
    fscanf(inpf,"%d",&N);
/*....READ MATRIX [A] AND VECTOR {B}                */
    for (IROW = 1; IROW <= N; IROW++) {
        for (ICOL = 1; ICOL <= N; ICOL++)
            fscanf(inpf,"%f",&A[IROW][ICOL]);
        fscanf(inpf,"%f",&B[IROW]);
    }
    CG(N, A, B, X);
    fprintf(outf,"\n    EQUATION NO.      SOLUTION X\n");
    for (I = 1; I <= N; I++)
        fprintf(outf,"%12d       %16.6E\n",I,X[I]);
    fclose(inpf);
    fclose(outf);
    return;
}

void CG(int N, float A[50][50], float B[50], float X[50])
{
    float D[50],R[50];
    float ALAM,ALPHA,DOWN,RES,SUM,TOL,UP,XZERO;
    int I,J,K;
/*....ASSIGN INITIAL VALUES IN VECTOR X:            */
    XZERO = 0.;
    for (I = 1; I <= N; I++)
        X[I] = XZERO;
/*....ASSIGN TOLERANCE FOR STOPPING CRITERION:      */
    TOL = 0.0001;
/*....COMPUTE INITIAL RESIDUAL & SEARCH DIRECTION:  */
    for (I = 1; I <= N; I++) {
        SUM = 0.;
        for (J = 1; J <= N; J++)
            SUM = SUM + A[I][J]*X[J];
        R[I] = SUM - B[I];
        D[I] = -R[I];
    }
/*....ENTER THE ITERATION LOOP:                     */
    for (K = 1; K <= N+1; K++) {
        UP = 0.;
        for (I = 1; I <= N; I++)
            UP = UP + D[I]*R[I];
        DOWN = 0.;
        for (I = 1; I <= N; I++) {
            SUM = 0.;
            for (J = 1; J <= N; J++)
                SUM = SUM + A[I][J]*D[J];
            DOWN = DOWN + D[I]*SUM;

        ALAM = -UP/DOWN;
        for (I = 1; I <= N; I++)
            X[I] = X[I] + ALAM*D[I];
        for (I = 1; I <= N; I++) {
            SUM = 0.;
            for (J = 1; J <= N; J++)
                SUM = SUM + A[I][J]*X[J];
            R[I] = SUM - B[I];
        }
        RES = 0.;
        for (I = 1; I <= N; I++)
            RES = RES + R[I]*R[I];
        RES = sqrt(RES);
        if (RES < TOL) goto jump1;
        UP = 0.;
        DOWN = 0.;
        for (I = 1; I <= N; I++) {
            SUM = 0.;
            for (J = 1; J <= N; J++)
                SUM = SUM + A[I][J]*D[J];
            UP = UP + R[I]*SUM;
            DOWN = DOWN +D[I]*SUM;

    ALPHA = UP/DOWN;
    for (I = 1; I <= N; I++)
        D[I] = -R[I] + ALPHA*D[I];
    }
    jump1:
    return;
}

/*  PROGRAM IMPCG                                    */
/*....PROGRAM FOR SOLVING A SET OF SIMULTANEOUS      */
/*....LINEAR EQUATIONS [A]{X} = {B}  USING           */
/*....IMPROVED CONJUGATE GRADIENT                    */
#include <stdio.h>
void CGNEW(int N,float A[50][50], float B[50], float X[50]);
void main()
{
    FILE *inpf, *outf;
    float A[50][50],B[50],X[50];
    int I,ICOL,IROW, N;
    inpf = fopen("INPUT.DAT", "r");
    outf = fopen("SOL.OUT", "w");
/*....READ TOTAL NUMBER OF EQUATIONS TO BE SOLVED:  */
    fscanf(inpf,"%d",&N);
/*....READ MATRIX [A] AND VECTOR {B}                */
    for (IROW = 1; IROW <= N; IROW++) {
        for (ICOL = 1; ICOL <= N; ICOL++)
            fscanf(inpf,"%f",&A[IROW][ICOL]);
        fscanf(inpf,"%f",&B[IROW]);
    }
    CGNEW(N, A, B, X);
    fprintf(outf,"\n    EQUATION NO.      SOLUTION X\n");
    for (I = 1; I <= N; I++)
        fprintf(outf,"%12d       %16.6E\n",I,X[I]);
    fclose(inpf);
    fclose(outf);
    return;
}

void CGNEW(int N,float A[50][50], float B[50], float X[50])
{
    float D[50],R[50],U[50];
    float ALAM,ALPHA,DEL,DEL1,DOWN,SUM,TOL,XZERO;
    int I,J,K;
/*....ASSIGN INITIAL VALUES IN VECTOR X:            */
    XZERO = 0.;
    for (I = 1; I <= N; I++)
        X[I] = XZERO;
/*....ASSIGN TOLERANCE FOR STOPPING CRITERION:      */
    TOL = 0.0001;
/*....COMPUTE INITIAL RESIDUAL & SEARCH DIRECTION:  */
    for (I = 1; I <= N; I++) {
        SUM = 0.;
        for (J = 1; J <= N; J++)
            SUM = SUM + A[I][J]*X[J];
        R[I] = SUM - B[I];
        D[I] = -R[I];

        DEL = 0.;
```

```c
                                                       */
    for (I = 1; I <= N; I++)
        DEL = DEL + R[I]*R[I];
/*.....ENTER THE ITERATION LOOP:                          */
    for (K = 1; K <= N+1; K++) {
        for (I = 1; I <= N; I++) {
            U[I] = 0.;
            for (J = 1; J <= N; J++)
                U[I] = U[I] + A[I][J]*D[J];
        }
        DOWN = 0.;
        for (I = 1; I <= N; I++)
            DOWN = DOWN + D[I]*U[I];
        ALAM = DEL/DOWN;
        for (I = 1; I <= N; I++) {
            X[I] = X[I] + ALAM*D[I];
            R[I] = R[I] + ALAM*U[I];
        }
        DEL1 = 0.;
        for (I = 1; I <= N; I++)
            DEL1 = DEL1 + R[I]*R[I];
        if (DEL1 < TOL) goto jump1;
        ALPHA = DEL1/DEL;
        for (I = 1; I <= N; I++)
            D[I] = -R[I] + ALPHA*D[I];
        DEL = DEL1;
    }
    jump1:
    return;
}
```

```c
/*.....  PROGRAM LAGPOL                                   */
/*.....  PROGRAM FOR COMPUTING F(X) AT A GIVEN X          */
/*.....  USING LAGRANGE INTERPOLATION                     */
#include <stdio.h>
void main()
{
    int I,J,N;
    float X[11],FX[11];
    float AL,XX,YY;
/*.....  READ NUMBER OF DATA SETS, DATA OF X AND FX       */
    scanf("%d",&N);
    for (I = 1; I <= N; I++)
        scanf("%f %f",&X[I],&FX[I]);
/*.....  COMPUTE DESIRED F(X) AT THE GIVEN X VALUE        */
    scanf("%f",&XX);
    YY = 0.;
    for (I = 1; I <= N; I++) {
        AL = 1.;
        for (J = 1; J <= N; J++)
            if (J != I) AL = AL*(XX-X[J])/(X[I]-X[J]);
        YY = YY + AL*FX[I];
    }
    printf(" VALUE OF F(X) AT X = %10.4f IS %16.7f",XX,YY);
    return;
}
```

```c
/*.....                                                   */
/*.....  PROGRAM NEWDIV                                   */
/*.....  PROGRAM FOR COMPUTING F(X) AT A GIVEN X          */
/*.....  USING NEWTON'S DIVIDED-DIFFERENCE                */
/*.....  INTERPOLATING POLYNOMIALS                        */
#include <stdio.h>
void main()
{
    int I,ICOL,IROW,M,N;
    float X[11],FX[11][11];
    float FAC,FF,XX;
/*.....  READ NUMBER OF DATA SETS, DATA OF X AND FX       */
    scanf("%d",&N);
    for (IROW = 1; IROW <= N; IROW++)
        scanf("%f %f",&X[IROW],&FX[IROW][1]);
/*.....  COMPUTE DIVIDED-DIFFERENCE COEFFICIENTS:         */
    M = N;
    for (ICOL = 2; ICOL <= N; ICOL++) {
        M = M - 1;
        for (IROW = 1; IROW <= M; IROW++) {
            FX[IROW][ICOL] = (FX[IROW+1][ICOL-1] - FX[IROW][ICOL-1])/(X[IROW+ICOL-1]-X[IROW]);
        }
    }
/*.....  COMPUTE DESIRED F(X) AT THE GIVEN X VALUE:       */
    scanf("%f",&XX);
    FF = FX[1][1];
    FAC = 1.;
    for (I = 2 ; I <= N; I++) {
        FAC = FAC*(XX - X[I-1]);
        FF = FF + FX[1][I]*FAC;
    }
    printf(" VALUE OF F(X) AT X = %10.4f IS %16.7f",XX,FF);
    return;
}
```

```c
/*.....  PROGRAM CUBSPLN                                  */
/*.....  A CUBIC SPLINE INTERPOLATING PROGRAM            */
#include <stdio.h>
void main()
{
    int I,N;
    float X[11],FX[11];
    float A[11],B[11],C[11],D[11],E[11];
    float D1,D2,DD,FF,T1,T2,T3,T4,XX;
/*.....  READ NUMBER OF DATA SETS, DATA OF X AND FX,      */
/*.....  THEN THE VALUE OF X THAT F(X) IS NEEDED:         */
    scanf("%d",&N);
    for (I = 1; I <= N; I++)
        scanf("%f %f",&X[I],&FX[I]);
    scanf("%f",&XX);
/*.....  FROM UP TRIDIAGONAL SYSTEM OF N EQUATIONS:       */
    for (I = 2; I <= N-1; I++) {
        A[I] = X[I] - X[I-1];
        B[I] = 2.*(X[I+1] - X[I-1]);
        C[I] = X[I+1] - X[I];
        D[I] = 6.*(FX[I+1]-FX[I])/(X[I+1]-X[I]);
        D[I] = D[I] + 6.*(FX[I-1]-FX[I])/(X[I]-X[I-1]);
    }
    B[1] = 1.;
    C[1] = 0.;
    D[1] = 0.;
    A[N] = 0.;
    D[N] = 0.;
/*.....  SOLVE TRIDIAGONAL SYSTEM OF N EQUATIONS FOR      */
/*.....  2ND DERIVATIVES, RETURN SOLUTION IN E():         */
/*.....  (STANDARD TRIDIAGONAL SYSTEM SOLVER - N EQS)     */
/*.....  COMPUTE F(X) AT THE GIVEN X:                     */
```

```c
/*.....PROGRAM LREGRES                              */
/*.....A LINEAR REGRESSION PROGRAM                  */
#include <stdio.h>
void main()
{
    int I,N;
    float X[101],Y[101];
    float DET,SUMX,SUMY,SUMX2,SUMXY,A0,A1;
/*.....READ NUMBER OF DATA SETS, DATA OF X AND Y:   */
    scanf("%d",&N);
    for (I = 1; I <= N; I++)
        scanf("%f %f", &X[I],&Y[I]);
/*.....COMPUTE SUMMATION TERMS:                     */
    SUMX = 0.;
    SUMY = 0.;
    SUMX2 = 0.;
    SUMXY = 0.;
    for (I = 1; I <= N; I++) {
        SUMX = SUMX + X[I];
        SUMY = SUMY + Y[I];
        SUMX2 = SUMX2 + X[I]*X[I];
        SUMXY = SUMXY + X[I]*Y[I];
    }
/*.....SOLVE FOR COEFFICIENTS:                      */
    DET = N*SUMX2 - SUMX*SUMX;
    A0 = (SUMY*SUMX2 - SUMXY*SUMX)/DET;
    A1 = (N*SUMXY - SUMX*SUMY)/DET;
    printf(" COEFFICIENT A0 = %14.6f\n",A0);
    printf(" COEFFICIENT A1 = %14.6f",A1);
    return;
}
```

```c
/*.....PROGRAM PREGRES                              */
/*.....A POLYNOMIAL REGRESSION PROGRAM              */
#include <stdio.h>
#include <math.h>
float A[11][11];
float power(float X, int N);
void GAUSS(int MP1,float B[],float XX[]);
void main()
{
    FILE *fp;
    int I,IC,IM1,IR,K,M,MP1,N;
    float X[100],Y[100];
    float B[11],XX[11];
/*.....READ NUMBER OF DATA SETS, DATA OF X AND Y:   */
    fp = fopen("INPUT1.DAT","r");
    if ( fp == NULL ) printf(" Can't find file name");
    fscanf(fp,"%d",&N);
    for (I = 1; I <= N; I++)
        fscanf(fp,"%e %e", &X[I],&Y[I]);
/*.....READ ORDER OF POLYNOMIAL NEEDED:             */
    fscanf(fp,"%d",&M);
    for (IR = 1; IR <= 10; IR++) {
        B[IR] = 0.;
    for (IC = 1; IC <= 10; IC++)
        A[IR][IC] = 0.;
/*.....COMPUTE SQUARE MATRIX ON LHS AND             */
/*.....VECTOR ON RHS OF SYSTEM EQUATIONS:           */
    for (IR = 1; IR <= M+1; IR++) {
        for (IC = 1; IC <= M+1; IC++){
            K = IR + IC - 2;
            A[IR][IC] = A[IR][IC] + power(X[I],K);
        }
        B[IR] = B[IR] + Y[I]*power(X[I],IR-1);
    }
/*.....CALL SUBROUTINE FOR SOLVING SYSTEM EQS:      */
    MP1 = M + 1;
    GAUSS(MP1,B,XX);
/*.....PRINT OUT POLYNOMIAL COEFFICIENTS:           */
    printf("\n COEFFICIENTS OF FITTED POLYNOMIAL ARE:\n");
    for (I = 1; I <= M+1; I++) {
        IM1 = I - 1;
        printf("            A(%1d) = %13.7E   ",IM1,XX[I]);
    }
    printf("\n");
    return;
}
```

```c
/*.....PROGRAM MREGRES                              */
/*.....A MULTIPLE LINEAR REGRESSION PROGRAM         */
#include <stdio.h>
void GAUSS(int N,float A[11][11], float B[11], float XX[11]);
void main()
{
    FILE *inpf;
    float X[100][11],Y[100];
    float A[11][11],B[11],XX[11];
    float FC,FR;
    int I,IC,IM1,IR,J,K,KP1,N;
/*.....READ NUMBER OF DATA SETS N,                  */
/*.....NUMBER OF INDEPENDENT VARIABLES K,           */
/*.....AND DATA OF X(I,K) AND Y(I):                 */
    inpf = fopen("INPUT3.DAT","r");
    fscanf(inpf,"%d %d",&N,&K);
    for (I = 1; I <= N; I++) {
        for (J = 1; J <= K; J++)
            fscanf(inpf,"%f",&X[I][J]);
```

```c
      fscanf(inpf,"%f",&Y[I]);
   }
   for (IR = 1; IR <= 10; IR++) {
      B[IR] = 0.;
      for (IC = 1; IC <= 10; IC++)
         A[IR][IC] = 0.;

/*.....COMPUTE SQUARE MATRIX ON LHS AND                      */
/*.....VECTOR ON RHS OF SYSTEM EQUATIONS:                    */
/*.....CALL SUBROUTINE FOR SOLVING SYSTEM EQS:               */
   for (I = 1; I <= N; I++) {
      for (IR = 1; IR <= K+1; IR++) {
         if (IR == 1)  FR = 1.;
         if (IR > 1)   FR = X[I][IR-1];
         for (IC = 1; IC <= K+1; IC++) {
            if (IC == 1)  FC = 1.;
            if (IC > 1)   FC = X[I][IC-1];
            A[IR][IC] = A[IR][IC] + FR*FC;
         }
         B[IR] = B[IR] + FR*Y[I];
      }
   }
   KP1 = K + 1;
   GAUSS(KP1, A, B, XX);
/*.....PRINT OUT COEFFICIENTS:                               */
   printf("\n COEFFICIENTS OF FITTED FUNCTION ARE:\n");
   for (I = 1; I <= K+1; I++) {
      IM1 = I-1;
      printf("  A(%1d) = %13.7E\n",IM1,XX[I]);
   }
   fclose(inpf);
   return;
}

void GAUSS(int N,float A[11][11], float B[11], float XX[11])
{
   int IC,IE,IR,IP;
   float RATIO,SUM;

/*.....FORWARD ELIMINATION:  PERFORM ACCORDING TO            */
/*.....THE ORDER OF 'PRIME' FROM 1 TO N-1:                   */
   for (IP = 1; IP <= N-1; IP++) {
/*.....LOOP OVER EACH EQUATION STARTING FROM THE             */
/*.....ONE THAT CORRESPONDS WITH THE ORDER OF                */
/*.....'PRIME' PLUS ONE:                                     */
      for (IE = IP+1; IE <= N; IE++) {
         RATIO = A[IE][IP]/A[IP][IP];
/*.....COMPUTE NEW COEFF. OF THE EQ. CONSIDERED:             */
         for (IC = IP+1; IC <= N; IC++)
            A[IE][IC] = A[IE][IC] - RATIO*A[IP][IC];
         B[IE] = B[IE] - RATIO*B[IP];
      }
/*.....SET COEFF. ON LOWER LEFT PORTION TO ZERO:             */
      for (IE = IP+1; IE <= N; IE++)
         A[IE][IC] = 0.;
/*.....BACK SUBSTITUTION                                      */
/*.....COMPUTE SOLUTION OF THE LAST EQUATION:                */
   XX[N] = B[N]/A[N][N];
/*.....COMPUTE SOLUTION FROM EQUATION N-1 TO 1               */
   for (IE = N-1; IE >= 1; IE--) {
      SUM = 0.;
      for (IC = IE+1; IC <= N; IC++)
         SUM = SUM + A[IE][IC]*XX[IC];
      XX[IE] = (B[IE] - SUM)/A[IE][IE];
   }
   return;
}
```

```c
/*.....    PROGRAM TRAPEZ                                    */
/*.....    A MULTIPLE-SEGMENT TRAPEZOIDAL PROGRAM            */
/*.....    FOR ESTIMATING INTERGRAL OF F(X)                  */
#include <stdio.h>
float FUNC( float X );
main()
{
   int I,N;
   float A,B,FX,FX0,FXN,H,SOL,SUM,X;
   A = 0.;
   B = 2.;
/*.....    READ NUMBER OF SEGMENTS REQUIRED:                 */
   scanf("%d",&N);
   H = (B - A)/N;
   SUM = 0.;
   X = A + H;
   for (I = 1; I <= N-1 ; I = I + 1) {
      FX = FUNC(X);
      SUM = SUM + FX;
      X = X + H;
   }
   FX0 = FUNC(A);
   FXN = FUNC(B);
   SOL = (FX0 + FXN + 2.*SUM)*H/2.;
   printf(" INTEGRAL OF F(X) USING %3d",N);
   printf(" SEGMENTS IS %10.6f\n",SOL);
   return 0;
}

float FUNC( float X )
{
   float p;
   p = 2.*X*X*X - 5.*X*X + 3.*X + 1.;
   return p;
}
```

```c
/*.....    PROGRAM SIMPSON                                   */
/*.....    A MULTIPLE-SEGMENT SIMPSON'S 1/3 PROGRAM          */
/*.....    FOR ESTIMATING INTERGRAL OF F(X)                  */
#include <stdio.h>
float FUNC( float X );
main()
{
   int I,M,N;
   float A,B,FX,FX0,FXN,H,SOL,SUM,SUM1,SUM2,X;
   A = 0.;
   B = 2.;
/*.....    READ NUMBER OF SEGMENTS REQUIRED:                 */
   scanf("%d",&N);
   M = N - (N/2)*2;
   if (M == 0) goto jump0;
   printf(" NUMBER OF SEGMENTS MUST BE EVEN\n");
   return 0;
jump0:
   H = (B - A)/N;
   SUM1 = 0.;
   X = A + H;
   for (I = 1; I <= N-1; I = I + 2) {
      FX = FUNC(X);
      SUM1 = SUM1 + FX;
      X = X + 2.*H;
```

```c
        int I;
        float FXO,FXN,FX,H,SUM,X;
/*....MULTIPLE-SEGMENT TRAPEZOIDAL RULE            */
        H = (B - A)/N;
        SUM = 0.;
        X = A + H;
        for (I = 1; I <= N-1; I++) {
            FX = FUNC(X);
            SUM = SUM + FX;
            X = X + H;
        }
        FXO = FUNC(A);
        FXN = FUNC(B);
        AREA = (FXO + FXN + 2.*SUM)*H/2.;
        return;
}

float FUNC(float X)
{
        float p;
        p = sin(X);
        return p;
}
```

```c
/*          PROGRAM GAUSSINT                          */
/*....A GAUSS-LEGENDRE INTEGRATION PROGRAM FOR        */
/*....ESTIMATING INTEGRATION OF FUNCTION F(X)         */
/*....USING 1 THROUGH 6 GAUSS POINTS                  */
#include <stdio.h>
float FUNC(float X);
void main ()
{
    int IC,ITERMS,NG;
    float A,AO,AI,AI,B,SUM,X;
    float XI[22] = { 0.0000000,
     0.0000000,-.5773503,+.5773503,-.7745967,
     0.0000000,+.7745967,-.8611363,-.3399810,
    +.3399810,+.8611363,-.9061798,-.5384693,
     0.0000000,+.5384693,+.9061798,-.9324695,
    -.6612094,-.2386192,+.2386192,+.6612094,
    +.9324695 };
    float W[22] = { 0.0000000,
     2.0000000,1.0000000,1.0000000,
     0.5555556,0.8888889,0.5555556,
     0.3478549,0.6521452,0.6521452,
     0.3478549,0.2369269,0.4786287,
     0.5688889,0.4786287,0.2369269,
     0.1713245,0.3607616,0.4679139,
     0.4679139,0.3607616,0.1713245 };
/*....READ LIMITS OF INTEGRATION:
    scanf("%f %f",&A,&B);
    AO = (A + B)/2.;
    AI = (B - A)/2.;
/*....PERFORM 1 THRU 6 GAUSS POINTS COMPUTATION:     */
    IC = 1;
    for (NG = 1; NG <= 6; NG++) {
        SUM = 0.;
        for (ITERMS = 1; ITERMS <= NG; ITERMS++) {
            AI = AO + AI*XI[IC];
            AI = FUNC(X);
            SUM = SUM + W[IC]*AI;
            IC = IC + 1;
        }
        SUM = AI*SUM;
```

```c
        }
        SUM2 = 0.;
        X = A + 2.*H;
        for (I = 2; I <= N-2; I = I + 2) {
            FX = FUNC(X);
            SUM2 = SUM2 + FX;
            X = X + 2.*H;
        }
        FXO = FUNC(A);
        FXN = FUNC(B);
        SOL = (FXO + FXN + 4.*SUM1 + 2.*SUM2)*H/3.;
        printf(" INTEGRAL OF F(X) USING %3d", N);
        printf(" SEGMENTS IS %10.6f\n",SOL);
        return 0;
}

float FUNC( float X )
{
        float p;
        p = X*X*X*X + 2.*X*X*X - 5.*X*X + 3.*X + 1.;
        return p;
}
```

```c
/*          PROGRAM ROMBERG                           */
/*....A ROMBERG INTERGRATING PROGRAM FOR ESTIMATING*/
/*....INTEGRAL OF F(X) WITHIN SPECIFIED % ERROR     */
#include <stdio.h>
#include <math.h>
float AREA;
void TRAP(float A, float B, int N);
float FUNC(float X);
void main ()
{
    int I,IC,IR,K,N;
    float R[11][11];
    float A,B,ERR,EPS,FXO,FXN,PI;
    PI = 4.*atan(1.);
    A = 0.;
    B = PI/2.;
    EPS = .0001;
/*....COMPUTE R(1,1):                               */
    FXO = FUNC(A);
    FXN = FUNC(B);
    R[1][1] = (FXO+FXN)*(B-A)/2.;
/*....LOOP OVER NUMBER OF ROMBERG APPLICATIONS       */
    for (I = 1; I <= 9; I++) {
        N = pow(2,I);
        TRAP(A, B, N);
        R[I+1][1] = AREA;
        for (IC = 2; IC <= I+1; IC++) {
            K = IC - 1;
            IR = 2 + I - IC;
            R[IR][IC] = (pow(4,K)*R[IR+1][K] - R[IR][K])/(pow(4,K)-1);
        }
        ERR = 100.*(R[1][K+1] - R[2][K])/R[1][K+1];
        ERR = fabs(ERR);
        if (ERR <= EPS) break;
    }
    printf(" FINAL INTEGRAL VALUE = %16.10E\n",R[1][K+1]);
    printf(" WITH RELATIVE ERROR = %12.6E \n",ERR);
    return;
}

void TRAP(float A, float B, int N)
{
```

```
    float H,SLOPE,X,Y;
    scanf("%f %f %d %f",&X,&Y,&N,&H);
    printf(" SOLUTION WITH STEP SIZE = %10.4E IS\n",H);
    printf("          X                    Y\n");
    printf("%16.6E %16.6E\n",X,Y);
    for (I = 1; I<= N; I++) {
       SLOPE = FUNC(X,Y);
       Y = Y + SLOPE*H;
       X = X + H;
       printf("%16.6E %16.6E\n",X,Y);
    }
    return;
}
float FUNC(float X,float Y)
{
    float p;
    p = Y*cos(X);
    return p;
}
```

```
/*          PROGRAM HEUN                                        */
/*.....A PROGRAM FOR SOLVING ORDINARY DIFFERENTIAL             */
/*.....EQUATION USING THE HEUN'S METHOD                        */
/*.....READ INITIAL CONDITIONS, NUMBER OF STEPS,              */
/*.....AND STEP SIZE.                                          */
#include <stdio.h>
#include <math.h>
float FUNC(float X, float Y);
void main ()
{
    int I,N;
    float H,S0,S1,SA,X,Y,Y1;
    scanf("%f %f %d %f",&X,&Y,&N,&H);
    printf(" SOLUTION WITH STEP SIZE = %10.4E IS\n",H);
    printf("          X                    Y\n");
    printf("%16.6E %16.6E\n",X,Y);
    for (I = 1; I<= N; I++) {
       S0 = FUNC(X,Y);
       Y1 = Y + S0*H;
       X = X + H;
       S1 = FUNC(X,Y1);
       SA = (S0 + S1)/2.;
       Y = Y + SA*H;
       printf("%16.6E %16.6E\n",X,Y);
    }
    return;
}
float FUNC(float X,float Y)
{
    float p;
    p = Y*cos(X);
    return p;
}
```

```
/*          PROGRAM MEULER                                      */
/*.....A PROGRAM FOR SOLVING ORDINARY DIFFERENTIAL             */
/*.....EQUATION USING THE MODIFIED EULER'S METHOD             */
/*.....READ INITIAL CONDITIONS, NUMBER OF STEPS,              */
/*.....AND STEP SIZE:                                          */
#include <stdio.h>
#include <math.h>
```

```
    printf(" RESULT OF INTEGRATION WITH",
    "%2d GAUSS POINT (S) IS %12.6E\n", NG, SUM);
    return;
}
float FUNC(float X)
{
    float p;
    p = 2.*X*X*X - 5.*X*X + 3.*X + 1.;
    return p;
}
```

```
/*          PROGRAM NUMDIF                                       */
/*.....A NUMERICAL DIFFERENTIATION PROGRAM                      */
#include <stdio.h>
float FUNC( float X );
main()
{
    int I,N;
    float FX[100], DIFF[100];
    float A,B,H,X;
/*.....READ END LOCATIONS AND NO. OF POINTS: */
    scanf("%f %f %d",&A,&B,&N);
    H = (B - A)/(N - 1);
    X = A;
/*.....COMPUTE FUNCTION VALUES AT POINTS     */
    for (I = 1; I <= N; I = I + 1) {
        FX[I] = FUNC(X);
        X = X + H;
    }
/*.....COMPUTE DERIVATIVES AT POINTS         */
    DIFF[1] = (FX[2] - FX[1])/H;
    for (I = 2; I <= N - 1; I = I + 1)
        DIFF[I] = (FX[I+1] - FX[I-1])/(2.*H);
    DIFF[N] = (FX[N] - FX[N-1])/H;
    printf("        X             FX           DERIVATIVE\n");
    X = A;
    for (I = 1; I <= N; I = I + 1) {
        printf(" %5.2f    %10.3f    %10.3f\n",X,FX[I],DIFF[II]);
        X = X + H;
    }
    return 0;
}
float FUNC( float X )
{
    float p;
    p = 2.*X*X*X - 5.*X*X + 3.*X + 1.;
    return p;
}
```

```
/*          PROGRAM EULER                                        */
/*.....A PROGRAM FOR SOLVING ORDINARY DIFFERENTIAL             */
/*.....EQUATION USING THE EULER'S METHOD                       */
/*.....READ INITIAL CONDITIONS, NUMBER OF STEPS,              */
/*.....AND STEP SIZE:                                          */
#include <stdio.h>
#include <math.h>
float FUNC(float X, float Y);
void main ()
{
    int I,N;
```

```c
float FUNC(float X, float Y);
void main ()
{
    int I,N;
    float H,S0,SA,X,X1,Y,Y1;
    scanf("%f %f %d %f",&X,&Y,&N,&H);
    printf("   SOLUTION WITH STEP SIZE = %10.4E IS\n",H);
    printf("        X                    Y\n");
    printf("%16.6E %16.6E\n",X,Y);
    for (I = 1; I<= N; I++) {
        S0 = FUNC(X,Y);
        Y1 = Y + S0*H/2.;
        X1 = X + H/2.;
        SA = FUNC(X1,Y1);
        Y = Y + SA*H;
        X = X + H;
        printf("%16.6E %16.6E\n",X,Y);
    }
    return;
}
float FUNC(float X, float Y)
{
    float p;
    p = Y*cos(X);
    return p;
}
```

```c
/*     PROGRAM RK3                                */
/*....A PROGRAM FOR SOLVING ORDINARY DIFFERENTIAL */
/*....EQUATION OF THIRD-ORDER RUNGE-KUTTA METHOD  */
/*....READ INITIAL CONDITIONS, NUMBER OF STEPS,   */
/*....AND STEP SIZE:                              */
#include <stdio.h>
#include <math.h>
float FUNC(float X, float Y);
void main ()
{
    int I,N;
    float AK1,AK2,AK3,H,X,XX,Y,YY;
    scanf("%f %f %d %f",&X,&Y,&N,&H);
    printf("   SOLUTION WITH STEP SIZE = %10.4E IS\n",H);
    printf("        X                    Y\n");
    printf("%16.6E %16.6E\n",X,Y);
    for (I = 1; I<= N; I++) {
        AK1 = FUNC(X,Y);
        XX = X + H/2.;
        YY = Y + H*AK1/2.;
        AK2 = FUNC(XX,YY);
        XX = X + H;
        YY = Y - H*AK1 + 2.*H*AK2;
        AK3 = FUNC(XX,YY);
        Y = Y + (AK1 + 4.*AK2 + AK3)*H/6.;
        X = X + H;
        printf("%16.6E %16.6E\n",X,Y);
    }
    return;
}
float FUNC(float X,float Y)
{
    float p;
    p = Y*cos(X);
    return p;
}
```

```c
/*     PROGRAM RK4                                 */
/*....A PROGRAM FOR SOLVING ORDINARY DIFFERENTIAL  */
/*....EQUATION OF FOURTH-ORDER RUNGE-KUTTA METHOD  */
/*....READ INITIAL CONDITIONS, NUMBER OF STEPS,    */
/*....AND STEP SIZE:                               */
#include <stdio.h>
#include <math.h>
float FUNC(float X, float Y);
void main ()
{
    int I,N;
    float AK1,AK2,AK3,AK4,H,X,XX,Y,YY;
    scanf("%f %f %d %f",&X,&Y,&N,&H);
    printf("   SOLUTION WITH STEP SIZE = %10.4E IS\n",H);
    printf("        X                    Y\n");
    printf("%16.6E %16.6E\n",X,Y);
    for (I = 1; I<= N; I++) {
        AK1 = FUNC(X,Y);
        XX = X + H/2.;
        YY = Y + H*AK1/2.;
        AK2 = FUNC(XX,YY);
        YY = Y + H*AK2/2.;
        AK3 = FUNC(XX,YY);
        XX = X + H;
        YY = Y + H*AK3;
        AK4 = FUNC(XX,YY);
        Y = Y + (AK1 + 2.*AK2 + 2.*AK3 + AK4)*H/6.;
        X = X + H;
        printf("%16.6E %16.6E\n",X,Y);
    }
    return;
}
float FUNC(float X, float Y)
{
    float p;
    p = Y*cos(X);
    return p;
}
```

```c
/*     PROGRAM SYSEUL                                */
/*....A PROGRAM FOR SOLVING A SET OF TWO ORDINARY    */
/*....FIRST-ORDER DIFFERENTIAL EQUATIONS USING       */
/*....THE EULER'S METHOD                             */
/*....READ INITIAL CONDITIONS, NUMBER OF STEPS,      */
/*....AND STEP SIZE:                                 */
#include <stdio.h>
#include <math.h>
float FUNC1(float Z);
float FUNC2(float Y,float Z);
void main ()
{
    int I,N;
    float F1,F2,H,X,Y,Z;
    scanf("%f %f %f %d %f",&X,&Y,&Z,&N,&H);
    printf("   SOLUTION WITH STEP SIZE = %10.4E IS\n",H);
    printf("        X              Y              Z\n");
    printf("%16.6E %16.6E %16.6E\n",X,Y,Z);
    for (I = 1; I<= N; I++) {
        F1 = FUNC1(Z);
        F2 = FUNC2(Y,Z);
```

```c
    Y   = Y + F1*H;
    Z   = Z + F2*H;
    X   = X + H;
    printf("%16.6E %16.6E %16.6E\n",X,Y,Z);
    return;
}
float FUNC1(float Z)
{
    float p;
    p = Z;
    return p;
}
float FUNC2(float Y,float Z)
{
    float q;
    q = -2.*Z - 4.*Y;
    return q;
}
```

```c
/*      PROGRAM SYSRK4                                   */
/*....A PROGRAM FOR SOLVING A SET OF TWO ORDINARY        */
/*....FIRST-ORDER DIFFERENTIAL EQUATIONS USING           */
/*....THE FOURTH-ORDER RUNGE-KUTTA METHOD                */
/*....READ INITIAL CONDITIONS, NUMBER OF STEPS,          */
/*....AND STEP SIZE:                                     */
#include <stdio.h>
#include <math.h>
float FUNC1(float X, float Y, float Z);
float FUNC2(float X, float Y, float Z);
void main ()
{
    int I,N;
    float AK1Y,AK1Z,AK2Y,AK2Z,AK3Y,AK3Z,AK4Y,AK4Z;
    float H,X,XX,Y,YY,Z,ZZ;
    scanf("%f %f %f %d %f",&X,&Y,&Z,&N,&H);
    printf(" SOLUTION WITH STEP SIZE = %10.4E IS\n",H);
    printf("         X             Y             Z\n");
    printf("%16.6E %16.6E %16.6E\n",X,Y,Z);
    for (I = 1; I<= N; I++) {
        AK1Y = FUNC1(X,Y,Z);
        AK1Z = FUNC2(X,Y,Z);
        XX   = X + H/2.;
        YY   = Y + H*AK1Y/2.;
        ZZ   = Z + H*AK1Z/2.;
        AK2Y = FUNC1(XX,YY,ZZ);
        AK2Z = FUNC2(XX,YY,ZZ);
        YY   = Y + H*AK2Y/2.;
        ZZ   = Z + H*AK2Z/2.;
        AK3Y = FUNC1(XX,YY,ZZ);
        AK3Z = FUNC2(XX,YY,ZZ);
        XX   = X + H;
        YY   = Y + H*AK3Y;
        ZZ   = Z + H*AK3Z;
        AK4Y = FUNC1(XX,YY,ZZ);
        AK4Z = FUNC2(XX,YY,ZZ);
        Y = Y + (AK1Y + 2.*AK2Y + 2.*AK3Y + AK4Y)*H/6.;
        Z = Z + (AK1Z + 2.*AK2Z + 2.*AK3Z + AK4Z)*H/6.;
        X = X + H;
        printf("%16.6E %16.6E %16.6E\n",X,Y,Z);
    }
    return;
}
```

```c
float FUNC1(float X,float Y,float Z)
{
    float p;
    p = Z;
    return p;
}
float FUNC2(float X,float Y,float Z)
{
    float q;
    q = -2.*Z -4.*Y;
    return q;
}
```

```c
/*      PROGRAM ADAMBAS                                  */
/*....A PROGRAM FOR SOLVING ORDINARY DIFFERENTIAL        */
/*....EQ. BY FOURTH-ORDER ADAMS-BASHFORTH METHOD         */
/*....READ INITIAL CONDITIONS OF X0, Y0, Y-1             */
/*....Y-2, Y-3, NUMBER OF STEPS AND STEP SIZE:           */
#include <stdio.h>
#include <math.h>
float FUNC(float X, float Y);
void main ()
{
    int I,N;
    float F0,F1,F2,F3;
    float X,XM1,XM2,XM3,Y,YM1,YM2,YM3,H;
    scanf("%f %f %f %f %f %f %d %f",&X,&Y,&YM1,&YM2,&YM3,&N,&H);
    printf(" SOLUTION WITH STEP SIZE = %10.4E IS\n",H);
    printf("          X                    Y\n");
    printf("%16.6E\n",X,Y);
    for (I = 1; I <= N; I++) {
        F0  = FUNC(X,Y);
        XM1 = X - H;
        F1  = FUNC(XM1,YM1);
        XM2 = X - 2.*H;
        F2  = FUNC(XM2,YM2);
        XM3 = X - 3.*H;
        F3  = FUNC(XM3,YM3);
        YM3 = YM2;
        YM2 = YM1;
        YM1 = Y;
        Y = Y + (55*F0 - 59*F1 + 37*F2 - 9*F3)*H/24.;
        X = X + H;
        printf("%16.6E\n",X,Y);
    }
    return;
}
float FUNC(float X, float Y)
{
    float p;
    p = Y*cos(X);
    return p;
}
```

```c
/*      PROGRAM ELLIP                                    */
/*....A FINITE DIFFERENCE PROGRAM FOR SOLVING            */
/*....TEMPERATURE DISTRIBUTION IN A PLATE                */
#include <stdio.h>
#include <math.h>
void main ()
{
```

```c
    int I,IFLAG,ITER,J,MXITER,N;
    float T[10][6];
    float DIFF,DX,PI,TEMP,TOL,X;
    TOL = .00001;
    MXITER = 100;
/*....ASSIGN TOLERANCE AND MAX. NO. OF ITERATIONS:     */
/*....SET UP BOUNDARY CONDITIONS:                      */
    for (I = 1; I <= 9; I++)
        T[I][1] = 0.;
    for (J = 1; J <= 5; J++) {
        T[1][J] = 0.;
        T[9][J] = 0.;
    }
    X = .25;
    DX = X;
    PI = 4.*atan(1.);
    for (I = 2; I <= 8; I++) {
        T[I][5] = sin(PI*X/2.);
        X = X + DX;
    }
/*....SET UP INITIAL TEMPERATURE VALUES:               */
    for (I = 2; I <= 8; I++)
        for (J = 2; J <= 4; J++)
            T[I][J] = J*T[I][5]/5.;
/*....SOLVE UNKNOWN TEMPERATURES AT GRID POINTS        */
/*....USING GAUSS-SEIDEL ITERATION TECHNIQUE:          */
    for (ITER = 1; ITER <= MXITER; ITER++) {
        IFLAG = 0;
        for (I = 2; I <= 8; I++)
            for (J = 2; J <= 4; J++) {
                TEMP = (T[I-1][J]+T[I+1][J]+T[I][J+1]+T[I][J-1])/4.;
                DIFF = T[I][J] - TEMP;
                if (fabs(DIFF) > TOL) IFLAG = 1;
                T[I][J] = TEMP;
            }

        if (IFLAG == 0) goto jump;
    }
    printf(" SOLUTION NOT CONVERGED WITHIN THE");
    printf(" SPECIFIED NO. OF ITERATIONS & TOLERANCE");
    return;
jump:
/*....PRINT OUT TEMPERATURE AT GRID POINTS IN THE      */
/*....FORMAT CORRESPONDING TO THE PROBLEM FIGURE:      */
    for (J = 5; J >= 1; J--) {
        for (I = 1; I <= 9; I++)
            printf("%6.3f",T[I][J]);
        printf("\n");
    }
    return;
}
```

```c
/*       PROGRAM PARAEXP                               */
/*....A FINITE DIFFERENCE PROGRAM FOR SOLVING          */
/*....TRANSIENT TEMPERATURE DISTRIBUTION IN A          */
/*....ROD USING EXPLICIT METHOD                        */
#include <stdio.h>
#include <math.h>
void main ()
{
    int I,IP,ISTEP,J,NSTEPS;
    float TOLD[12],TNEW[12];
    float ALPHA,DTIME,DX,PI,TIME,X;
    DTIME = .005;
    NSTEPS = 40;
```

```c
/*....SET UP INITIAL AND BOUNDARY CONDITIONS:          */
    X = 0.;
    DX = .1;
    PI = 4.*atan(1.);
    for (I = 1; I <= 11; I++) {
        TOLD[I] = sin(PI*X);
        TNEW[I] = TOLD[I];
        X = X + DX;
    }
/*....SOLVE FOR TEMPERATURE RESPONSE:                  */
    ALPHA = DTIME/(DX*DX);
    TIME = DTIME;
    for (ISTEP = 1; ISTEP <= NSTEPS; ISTEP++) {
        for (I = 2; I <= 10; I++)
            TNEW[I] = TOLD[I] + ALPHA*(TOLD[I+1]-2.*TOLD[I]+TOLD[I-1]);
/*....PRINT OUT AT EVERY 4 STEPS:                      */
        IP = ISTEP - (ISTEP/4)*4;
        if (IP == 0) {
            printf("%6.2f ",TIME);
            for (I = 1; I <= 11; I++)
                printf("%6.4f ",TNEW[I]);
        }
        printf("\n");
        for (I = 2; I <= 10; I++)
            TOLD[I] = TNEW[I];
        TIME = TIME + DTIME;
    }
    return;
}
```

```c
/*       PROGRAM PARAIMP                               */
/*....A FINITE DIFFERENCE PROGRAM FOR SOLVING          */
/*....TRANSIENT TEMPERATURE DISTRIBUTION IN A          */
/*....ROD USING IMPLICIT METHOD                        */
#include <stdio.h>
#include <math.h>
void main ()
{
    int I,IP,ISTEP,J,N,NSTEPS;
    float TEMP[12];
    float A[10],B[10],C[10],D[10],E[10];
    float ALPHA,COEF,DTIME,DX,PI,TIME,X;
    DTIME = .01;
    NSTEPS = 20;
    N = 9;
/*....SET UP INITIAL AND BOUNDARY CONDITIONS:          */
    X = 0.;
    DX = .1;
    PI = 4.*atan(1.);
    for (I = 1; I <= 11; I++) {
        TEMP[I] = sin(PI*X);
        X = X + DX;
    }
/*....SOLVE FOR TEMPERATURE RESPONSE:                  */
    ALPHA = DTIME/(DX*DX);
    COEF = 1. + 2.*ALPHA;
    TIME = DTIME;
    for (ISTEP = 1; ISTEP <= NSTEPS; ISTEP++) {
/*....FORM UP TRIDIAGONAL SYSTEM OF N EQUATIONS        */
/*....FOR INTERIOR GRIDS (HERE N=9)                    */
        B[1] = COEF;
        C[1] = -ALPHA;
        D[1] = TEMP[2];
```

```c
        for (I = 2; I <= N-1; I++) {
            A[I] = -ALPHA;
            B[I] = COEF;
            C[I] = -ALPHA;
            D[I] = TEMP[I+1];
        }
        A[N] = -ALPHA;
        B[N] = COEF;
        D[N] = TEMP[N+1];
/*.....SOLVE SUCH TRIDIAGONAL SYSTEM OF N EQUATIONS  */
/*.....FOR TEMPERATURES AT INTERIOR GRIDS, RETURN    */
/*.....SOLUTION IN E().                              */
        for (I = 2; I <= N; I++) {
            A[I] = A[I]/B[I-1];
            B[I] = B[I] - A[I]*C[I-1];
        }
        for (I = 2; I <= N; I++) {
            D[I] = D[I] - A[I]*D[I-1];
        E[N] = D[N]/B[N];
        for (I = N-1; I >= 1; I--)
            E[I] = (D[I] - C[I]*E[I+1])/B[I];
        for (I = 2; I <= 10; I++)
            TEMP[I] = E[I-1];
/*.....PRINT OUT AT EVERY 2 STEPS:                   */
        IP = ISTEP - (ISTEP/2)*2;
        if (IP == 0) {
            printf("%6.2f ",TIME);
            for (I = 1; I <= 11; I++)
                printf("%6.4f ",TEMP[I]);
        }
        printf("\n");
        TIME = TIME + DTIME;
    }
    return;
}

/*       PROGRAM PARACN                              */
/*.....A FINITE DIFFERENCE PROGRAM FOR SOLVING       */
/*.....TRANSIENT TEMPERATURE DISTRIBUTION IN A       */
/*.....ROD USING CRANK-NICOLSON METHOD               */
#include <stdio.h>
#include <math.h>
void main ()
{
    int I,IP,ISTEP,J,N,NSTEPS;
    float TEMP[12];
    float A[10],B[10],C[10],D[10],E[10];
    float ALPHA,COEFM,COEFP,DTIME,DX,PI,TIME,X;
    DTIME = .02;
    NSTEPS = 10;
/*.....SET UP INITIAL AND BOUNDARY CONDITIONS:       */
    X = 0.;
    DX = .1;
    PI = 4.*atan(1.);
    for (I = 1; I <= 11; I++) {
        TEMP[I] = sin(PI*X);
        X = X + DX;
    }
/*.....SOLVE FOR TEMPERATURE RESPONSE:               */
    ALPHA = DTIME/(DX*DX);
    COEFP = 2.*(1. + ALPHA);
    COEFM = 2.*(1. - ALPHA);
    TIME = DTIME;

    N = 9;
    for (ISTEP = 1; ISTEP <= NSTEPS; ISTEP++) {
/*.....FORM UP TRIDIAGONAL SYSTEM OF N EQUATIONS     */
/*.....FOR INTERIOR GRIDS (HERE N=9)                 */
        B[1] = COEFP;
        C[1] = -ALPHA;
        D[1] = COEFM*TEMP[2] + ALPHA*TEMP[3];
        for (I = 2; I <= N-1; I++) {
            A[I] = -ALPHA;
            B[I] = COEFP;
            C[I] = -ALPHA;
            D[I] = ALPHA*TEMP[I] + COEFM*TEMP[I+1] +
                   ALPHA*TEMP[I+2];
        }
        A[N] = -ALPHA;
        B[N] = COEFP;
        D[N] = ALPHA*TEMP[N] + COEFM*TEMP[N+1];
/*.....SOLVE SUCH TRIDIAGONAL SYSTEM OF N EQUATIONS  */
/*.....FOR TEMPERATURES AT INTERIOR GRIDS, RETURN    */
/*.....SOLUTION IN E().                              */
        for (I = 2; I <= N; I++) {
            A[I] = A[I]/B[I-1];
            B[I] = B[I] - A[I]*C[I-1];
        }
        for (I = 2; I <= N; I++) {
            D[I] = D[I] - A[I]*D[I-1];
        E[N] = D[N]/B[N];
        for (I = N-1; I >= 1; I--)
            E[I] = (D[I] - C[I]*E[I+1])/B[I];
        for (I = 2; I <= 10; I++)
            TEMP[I] = E[I-1];
/*.....PRINT OUT AT EVERY 1 STEPS.                   */
        IP = ISTEP - (ISTEP/1)*1;
        if (IP == 0) {
            printf("%6.2f ",TIME);
            for (I = 1; I <= 11; I++)
                printf("%6.4f ",TEMP[I]);
        }
        printf("\n");
        TIME = TIME + DTIME;
    }
    return;
}

/*       PROGRAM HYPER                               */
/*.....A FINITE DIFFERENCE PROGRAM FOR SOLVING       */
/*.....VIBRATION IN STRING                           */
#include <stdio.h>
void main ()
{
    int I,ISTEP,J,NSTEPS;
    float UN[8],UNM1[8],UNP1[8];
    float DTIME,DX,TIME,X;
/*.....ASSIGN TIME STEP AND NO. OF TIME STEPS:       */
    DTIME = .0025;
    NSTEPS = 12;
/*.....SET UP INITIAL AND BOUNDARY CONDITIONS:       */
    X = 0.;
    DX = .25;
    TIME = 0.;
    for (I = 1; I <= 5; I++) {
        UN[I] = .07*X;
        X = X + DX;
```

```c
    }
    X = 5.*DX;
    for (I = 6; I <= 7; I++) {
        UN[I] = .21 - .14*X;
        X = X + DX;
    }
    printf("%6.4f ",TIME);
    for (I = 1; I <= 7; I++)
        printf("%8.5f",UN[I]);
    printf("\n");
    TIME = TIME + DTIME;
/*.....COMPUTE DISPLACEMENTS AT FIRST TIME STEP          */
    UNP1[1] = UN[1];
    for (I = 2; I <= 6; I++)
        UNP1[I] = 0.5*(UN[I+1] + UN[I-1]);
    UNP1[7] = UN[7];
    printf("%6.4f ",TIME);
    for (I = 1; I <= 7; I++)
        printf("%8.5f",UNP1[I]);
    printf("\n");
/*.....COMPUTE DISPLACEMENTS AFTER FIRST TIME STEP       */
    TIME = TIME + DTIME;
    for (ISTEP = 2; ISTEP <= NSTEPS; ISTEP++) {
        for (I = 1; I <= 7; I++) {
            UNM1[I] = UN[I];
            UN[I] = UNP1[I];
        }
        for (I = 2; I <= 6; I++)
            UNP1[I] = -UNM1[I] + UN[I+1] + UN[I-1];
        printf("%6.4f ",TIME);
        for (I = 1; I <= 7; I++)
            printf("%8.5f",UNP1[I]);
        printf("\n");
        TIME = TIME + DTIME;
    }
    return;
}

/*  ************************************************************  */
/*  PROGRAM FINITE                                              *  */
/*                                                              *  */
/*  A FINITE ELEMENT COMPUTER PROGRAM FOR SOLVING PARTIAL       *  */
/*  DIFFERENTIAL EQUATION IN THE FORM OF POISSON'S EQUATION     *  */
/*  FOR TWO-DIMENSIONAL STEADY-STATE HEAT CONDUCTION WITH       *  */
/*  INTERNAL HEAT GENERATION.                                   *  */
/*                                 PROF. DR. PRAMOTE DECHAUMPHAI *  */
/*                                 FACULTY OF ENGINEERING        *  */
/*                                 CHULALONGKORN UNIVERSITY      *  */
/*                                                              *  */
/*  THE VALUES DECLARED IN THE PARAMETER STATEMENT BELOW SHOULD *  */
/*  BE ADJUSTED ACCORDING TO THE SIZE OF THE PROBLEMS AND TYPES *  */
/*  OF COMPUTERS:                                               *  */
/*  MXPOI = MAXIMUM NUMBER OF NODES IN THE MODEL                *  */
/*  MXELE = MAXIMUM NUMBER OF ELEMENTS IN THE MODEL             *  */
/*  ************************************************************  */
#include <stdio.h>
#include <math.h>
#define MXPOI 100
#define MXELE 300
int ISTOP;
FILE *inpf, *outf;
void GAUSS(int N ,float A[MXPOI][MXPOI],float B[MXPOI],float X[MXPOI]);
void APPLYBC(int NPOIN, int IBC[], float TEMP[], float SYSK[][], float
SYSQ[]);
void TRI(int NELEM,int INTMAT[MXELE][3], float COORD[MXPOI][2], float TK,
float QELE[MXELE],float THICK,float SYSK[MXPOI][MXPOI],float
SYSQ[MXPOI]);
void main()
{
    int I,IE,ILINE,IP,IQ,J,NEQ,NLINES,NPOIN,NELEM,M,MM;
    int INTMAT[MXELE][3], IBC[MXPOI];
    float COORD[MXPOI][2], TEMP[MXPOI];
    float SYSK[MXPOI][MXPOI], SYSQ[MXPOI], QELE[MXELE];
    float TK,THICK;
    char NAME1[20], NAME2[20], TEXT[20];

    printf("\n PLEASE ENTER THE INPUT FILE NAME:");
    scanf("%s",&NAME1);
    inpf = fopen(NAME1,"r");
    if (inpf == NULL) goto jump0;
jump0:
/*                                                            */
/*.....READ TITLE OF COMPUTATION:                            */
/*                                                            */
    fscanf(inpf,"%d",&NLINES);
    for (ILINE = 1; ILINE <= NLINES; ILINE++)
        fscanf(inpf,"%s",&TEXT);

/*                                                            */
/*.....READ INPUT DATA :                                      */
/*                                                            */
    fscanf(inpf,"%s ",&TEXT);
    fscanf(inpf,"%d %d",&NPOIN,&NELEM);
    if (NPOIN > MXPOI)
        printf("\n PLEASE INCREASE THE PARAMETER MXPOI TO%5d",NPOIN);
    if (NPOIN > MXPOI) return;
    if (NELEM > MXELE)
        printf("\n PLEASE INCREASE THE PARAMETER MXELE TO%5d",NELEM);
    if (NELEM > MXELE) return;
    fscanf(inpf,"%s",&TEXT);
    fscanf(inpf,"%f %f",&TK,&THICK);
    fscanf(inpf,"%s",&TEXT);
    for (IP = 1; IP <= NPOIN; IP++) {
        fscanf(inpf,"%d",&I);
        fscanf(inpf,"%d %f %f",&IBC[I],&COORD[I][1],
               &COORD[I][2],&TEMP[I]);

        if (I != IP)
            printf("\n NODE NO. %5d IN DATA FILE MISSING",IP);
        if (I != IP) return;
    }
    IQ = 0;
    fscanf(inpf,"%s",&TEXT);
    for (IE = 1; IE <= NELEM; IE++) {
        fscanf(inpf,"%d",&I);
        for (J = 1; J <= 3; J++)
            fscanf(inpf,"%d",&INTMAT[I][J]);
        fscanf(inpf,"%f",&QELE[I]);
        if (I != IE)
            printf("\n ELEMENT NO. %5d IN DATA FILE IS MISSING",IE);
        if (I != IE) return;
        if (QELE[I] != 0.) IQ = 1;
    }
    printf("\n THE F.E. MODEL INCLUDES THE FOLLOWING");
    printf(" HEAT TRANSFER MODE(S):");
    printf("\n    -- HEAT CONDUCTION");
    if (IQ == 1) printf("\n    -- INTERNAL HEAT GENERATION");

/*                                                            */
    NEQ = NPOIN;
    for (I = 1; I <= NEQ; I++)
        SYSQ[I] = 0.;
    for (I = 1; I <= NEQ; I++)
        for (J = 1; J <= NEQ; J++)
```

```c
        SYSK[I][J] = 0.;

    printf("\n *** THE FINITE ELEMENT MODEL CONSISTS OF");       /*                */
    printf("%5d NODES AND %5d ELEMENTS ***",NPOIN,NELEM);        /*                */

/*... ESTABLISH ALL ELEMENT MATRICS AND ASSEMBLE THEM TO  */
/*... FORM UP SYSTEM EQUATIONS                            */

    printf("\n *** ESTABLISHING ELEMENT MATRICES AND");          /*                */
    printf(" ASSEMBLING ELEMENT EQUATIONS ***");
    ISTOP = 0;
    TRI (NELEM,INTMAT,COORD,TK,QELE,THICK,SYSK,SYSQ);            /*                */

    if (ISTOP == 1) return;
    printf("\n *** APPLYING BOUNDARY CONDITIONS OF NODAL");      /*                */
    printf(" TEMPERATURE ***");
    APPLYBC (NPOIN,IBC,TEMP,SYSK,SYSQ);                          /*                */

    printf("\n *** SOLVING A SET OF SIMULTANEOUS EQUATIONS");    /*                */
    printf(" FOR TEMPERATURE SOLUTIONS ***\n");                  /*                */
    GAUSS (NEQ,SYSK,SYSQ,TEMP);                                  /*                */

/*... PRINT OUT NODAL TEMPERATURE SOLUTIONS:    */

jump1:
    printf("\n PLEASE ENTER FILE NAME FOR TEMPERATURE ");        /*                */
    printf(" SOLUTIONS: ");
    scanf("%s",&NAME2);
    outf = fopen(NAME2,"w");
    if (outf != NULL) goto jump1;
    fprintf(outf," NODAL TEMPERATURE SOLUTIONS [%5d]:",NPOIN);
    fprintf(outf,"\n\n NODE   TEMPERATURE \n");
    for (IP = 1; IP <= NPOIN; IP++)
        fprintf(outf,"%6d %14.6E\n",IP,TEMP[IP]);
    return;
}
/*-----------------------------------------------------------------*/
void APPLYBC(int NPOIN, int IBC[], float TEMP[], float SYSK[][MXPOI],  /*          */
             float SYSQ[])
/*  APPLY TEMPERATURE BOUNDARY CONDITIONS WITH CONDITION CODES OF:*/
/*  0 = FREE TO CHANGE (TO BE COMPUTED)    */
/*  1 = FIXED AS SPECIFIED                 */
{
    int IC,IEQ,IR;

    for (IEQ = 1; IEQ <= NPOIN; IEQ++) {                         /*                */
        if (IBC[IEQ] == 0) goto jump3;

    for (IR = 1; IR <= NPOIN; IR++) {                            /*                */
        if (IR == IEQ) goto jump2;
        SYSQ[IR] = SYSQ[IR] - SYSK[IR][IEQ]*TEMP[IEQ];
        SYSK[IR][IEQ] = 0.;
jump2:      continue;
    }

    for (IC = 1; IC <= NPOIN; IC++)                              /*                */
        SYSK[IEQ][IC] = 0.;
    SYSK[IEQ][IEQ] = 1.;
    SYSQ[IEQ] = TEMP[IEQ];
jump3:
```

```c
    continue;

    return;                                                     /*                */
}
/*-----------------------------------------------------------------*/
void ASSEMBLE (int IE,int INTMAT[][3],float AKC[][3],           /*                */
               float QQ[],float SYSK[][MXPOI],float SYSQ[])
/*  ASSEMBLE ELEMENT EQUATIONS INTO SYSTEM EQUATIONS  */
{
    int IC,IR,ICOL,IROW,NNODE;                                  /*                */

    NNODE = 3;                                                  /*                */

    for (IR = 1; IR <= NNODE; IR++) {                           /*                */
        for (IC = 1; IC <= NNODE; IC++) {
            IROW = INTMAT[IE][IR];
            ICOL = INTMAT[IE][IC];
            SYSK[IROW][ICOL] = SYSK[IROW][ICOL] + AKC[IR][IC];
        }
        SYSQ[IROW] = SYSQ[IROW] + QQ[IR];

    return;                                                     /*                */
}
/*-----------------------------------------------------------------*/
void GAUSS(int N ,float A[][MXPOI],float B[],float X[])         /*                */
{
    int IC,IE,IP;
    void SCALE(int N, float A[MXPOI][MXPOI], float B[MXPOI]);   /*                */
    void PIVOT(int N, float A[MXPOI][MXPOI], float B[MXPOI],int IP);   /*         */
    float RATIO,SUM;

/*... PERFORM SCALING:      */                                  /*                */

    SCALE(N,A,B);

/*... FORWARD ELIMINATION: PERFORM ACCORDING TO  */
/*... THE ORDER OF 'PRIME' FROM 1 TO N-1:        */

    for (IP = 1; IP <= N-1; IP++) {                             /*                */

/*... PERFORM PARTIAL PIVOTING     */

    PIVOT(N,A,B,IP);                                            /*                */

/*... LOOP OVER EACH EQUATION STARTING FROM THE ONE THAT  */
/*... CORRESPONDS WITH THE ORDER OF 'PRIME' PLUS ONE:     */

    for (IE = IP+1; IE <= N; IE++) {                            /*                */
        RATIO = A[IE][IP]/A[IP][IP];                            /*                */

/*... COMPUTE NEW COEFF. OF THE EQ. CONSIDERED:   */

        for (IC = IP+1; IC <= N; IC++)                          /*                */
            A[IE][IC] = A[IE][IC] - RATIO*A[IP][IC];
        B[IE] = B[IE] - RATIO*B[IP];
```

```c
/*....SET COEFF. ON LOWER LEFT PORTION TO ZERO:               */
/*                                                            */
          for (IE = IP+1; IE <= N; IE++)
             A[IE][IC] = 0.;
          }
/*....BACK SUBSTITUTION                                       */
/*....COMPUTE SOLUTION OF THE LAST EQUATION:                  */
          X[N] = B[N]/A[N][N];
/*                                                            */
/*....THEN COMPUTE SOLUTION FROM EQUATION N-1 TO 1            */
          for (IE = N-1; IE >= 1; IE--) {
             SUM = 0.;
             for (IC = IE+1; IC <= N; IC++)
                SUM = SUM + A[IE][IC]*x[IC];
             X[IE] = (B[IE] - SUM)/A[IE][IE];
          }
          return;
/*                                                            */
/*------------------------------------------------------------*/
void PIVOT(int N, float A[][MXPOI], float B[], int IP)
{
          int I,IC,IE,J,JP;
          float AMAX,BIG,DUMY;
/*                                                            */
/*....PERFORM PARTIAL PIVOTING:                       */
          JP = IP;
          BIG = fabs(A[IP][IP]);
          for (I = IP+1; I <= N; I++) {
             AMAX = fabs(A[I][IP]);
             if (AMAX > BIG) {
                BIG = AMAX;
                JP = I;
             }
          }
          if (JP != IP) {
             for (J = IP; J <= N; J++) {
                DUMY = A[JP][J];
                A[JP][J] = A[IP][J];
                A[IP][J] = DUMY;
             }
             DUMY = B[JP];
             B[JP] = B[IP];
             B[IP] = DUMY;
          }
          return;
/*                                                            */
/*------------------------------------------------------------*/
void SCALE(int N, float A[][MXPOI], float B[])
{
          int IC,IE;
          float AMAX,BIG;
/*                                                            */
/*....PERFORM SCALING                                         */
          for (IE = 1; IE <= N; IE++) {
             BIG = fabs(A[IE][1]);
             for (IC = 2; IC <= N; IC++) {
                AMAX = fabs(A[IE][IC]);
                if (AMAX > BIG) BIG = AMAX;
             }
             for (IC = 1; IC <= N; IC++)
                A[IE][IC] = A[IE][IC]/BIG;
             B[IE] = B[IE]/BIG;
          }
          return;
}
/*                                                            */
/*------------------------------------------------------------*/
void TRI(int NELEM,int INTMAT[][3],float COORD[][2], float TK,
         float QELE[],float THICK,float SYSK[][MXPOI],float SYSQ[])
/*....ESTABLISH ALL ELEMENT MATRICES AND ASSEMBLE THEM TO FORM */
/*....UP THE SYSTEM EQUATIONS                                 */
{
   void ASSEMBLE (int IE,int INTMAT[MXELE][3],float AKC[3][3],
   float QQ[3],float SYSK[MXPOI][MXPOI], float SYSQ[MXPOI]);
   int I,IE,II,J,JJ,K,KK;
   float AKC[3][3],B[2][3],BT[3][2],QQ[3];
   float AREA,B1,B2,B3,C1,C2,C3,FAC,XG1,XG2,XG3,YG1,YG2,YG3;
/*                                                            */
/*....LOOP OVER THE NUMBER OF ELEMENTS                        */
/*                                                            */
   for (IE = 1; IE <= NELEM; IE++) {
/*                                                            */
/*....FIND ELEMENT LOCAL COORDINATES:                         */
/*                                                            */
      II = INTMAT[IE][1];
      JJ = INTMAT[IE][2];
      KK = INTMAT[IE][3];

      XG1 = COORD[II][1];
      XG2 = COORD[JJ][1];
      XG3 = COORD[KK][1];
      YG1 = COORD[II][2];
      YG2 = COORD[JJ][2];
      YG3 = COORD[KK][2];
      AREA = 0.5*(XG2*(YG3-YG1) + XG1*(YG2-YG3) + XG3*(YG1-YG2));
      if (AREA <= 0. ) {
         printf("\n !!! ERROR !!! ELEMENT NO. %5d HAS ",IE);
         printf("NEGATIVE OR ZERO AREA \n --- CHECK F.E. MODEL");
         printf(" FOR NODAL COORDINATES AND ELEMENT NODAL ");
         printf("CONNECTIONS ---");
         ISTOP = 1;
         return;
      }
      if (AREA <= 0.) return;
/*                                                            */
      B1 = YG2 - YG3;
      B2 = YG3 - YG1;
      B3 = YG1 - YG2;
      C1 = XG3 - XG2;
      C2 = XG1 - XG3;
      C3 = XG2 - XG1;
/*                                                            */
      for (I = 1; I <= 2; I++)
         for (J = 1; J <= 3; J++)
            B[I][J] = 0.;
      B[1][1] = B1;
      B[1][2] = B2;
```

```c
        B[1][3] = B3;
        B[2][1] = C1;
        B[2][2] = C2;
        B[2][3] = C3;
/*                                                            */
        for (I = 1; I <= 2; I++) {
            for (J = 1; J <= 3; J++) {
                B[I][J] = B[I][J]/(2.*AREA);
                BT[J][I] = B[I][J];
            }

/*                                                            */
/*...ELEMENT CONDUCTION MATRIX:                               */
/*                                                            */

        for (I = 1; I <= 3; I++) {
            for (J = 1; J <= 3; J++) {
                AKC[I][J] = 0.;
                for (K = 1; K <= 2; K++)
                    AKC[I][J] = AKC[I][J] + BT[I][K]*B[K][J];
                AKC[I][J] = TK*AREA*THICK*AKC[I][J];
            }
        }

/*                                                            */
/*...ELEMENT LOAD VECTOR DUE TO INTERNAL HEAT GENERATION:     */
/*                                                            */

        FAC = QELE[IE]*AREA*THICK/3.;
        for (I = 1; I <= 3; I++)
            QQ[I] = FAC;

/*                                                            */
/*...ASSEMBLE THESE ELEMENT MATRICES TO FORM SYSTEM EQUATIONS: */
/*                                                            */

        ASSEMBLE(IE,INTMAT,AKC,QQ,SYSK,SYSQ);

/*                                                            */
    }
    return;
}
```

Computer programs in Pascal

```pascal
PROGRAM SHUTTLE;
VAR T,DT,V : real;
    I : integer;
BEGIN
    T  := 0.;
    DT := 30.;
    V  := 0.;
    FOR I := 1 TO 50 DO
        BEGIN
            V := V + DT*(9.8 - 0.005*V);
            T := T + DT;
            WRITELN(T : 12 : 0, V : 12 : 0)
        END
END.
```

```pascal
PROGRAM SINCOS;
(*....PROGRAM FOR COMPUTING SIN AND COSINE   ....*)
(*....FUNCTION FOR ANGLES FROM 0 TO 180 DEGREES *)
(*....WITH INCREMENT AT EVERY 10 DEGREES      ....*)
VAR PI,DEG,DEL,SUMS,SUMC,TERMS,TERMC,SIGN,X : real;
    IDEG,N,MS,MC : integer;
BEGIN
    PI  := 4.*ARCTAN(1.0);
    DEG := 0.;
    DEL := 10.;
    WRITELN;
    WRITELN('           SIN        DEGREES            COS');
    WRITELN;
    FOR IDEG := 1 TO 19 DO
        BEGIN
            X     := PI*DEG/180.;
            SUMS  := X;
            SUMC  := 1.;
            TERMS := X;
            TERMC := 1.;
            SIGN  := -1.;
            FOR N :=1 TO 100 DO
                BEGIN
                    MS    := 2*N + 1;
                    MC    := 2*N;
                    TERMS := TERMS*X*X/(MS*(MS-1));
                    TERMC := TERMC*X*X/(MC*(MC-1));
                    SUMS  := SUMS + SIGN*TERMS;
                    SUMC  := SUMC + SIGN*TERMC;
                    SIGN  := -SIGN
                END;
            WRITELN(DEG :10 : 0, SUMS : 16 : 6,
                    SUMC : 16 : 6);
            DEG := DEG + DEL
        END
END.
```

```pascal
PROGRAM BISECT;
(*...PROGRAM FOR COMPUTING ROOT OF NONLINEAR...*)
(*...EQUATION USING THE BISECTION METHOD     ...*)
(*...DEFINE THE FOLLOWING GIVEN VALUES:       ...*)
(*...    XL = LEFT VALUE OF X                  ...*)
(*...    XR = RIGHT VALUE OF X                 ...*)
(*...    ES = STOPPING CRITERION TOLERANCE (%)*)
LABEL 200,300.;
VAR XL,XR,XM,XN,ES,FXL,FXR,AA,FXM,TOL : real;
    ITER : integer;
(*---------------------------------------------*)
FUNCTION FUNC(X : real) : real;
BEGIN
    FUNC := EXP(-X/4.0)*(2.-X) - 1.
END;
(*---------------------------------------------*)
BEGIN
    XL := 0.;
    XR := 2.;
    ES := 0.001;
(*...CHECK WHETHER THE ROOT IS IN GIVEN RANGE:*)
    FXL := FUNC(XL);
    FXR := FUNC(XR);
    AA  := FXL*FXR;
    IF AA >= 0. THEN
        BEGIN
            WRITELN;
```

```
    WRITELN(' ROOT IS NOT IN THE GIVEN RANGE');
    GOTO 300
END;
WRITELN;
WRITELN('    ITERATION NO.        X ');
WRITELN;
FOR ITER := 1 TO 500 DO
BEGIN
    XM  := (XL+XR)/2.;
    FXM := FUNC(XM);
    FXR := FUNC(XR);
    AA  := FXM*FXR;
    IF AA > 0. THEN
(*....CASE A:  XL < ROOT < XM        ....*)
        XR := XM
    ELSE
(*....CASE B:  XM < ROOT < XR        ....*)
        XL := XM;
(*....CHECK FOR TOLERANCE:
    XN := (XL+XR)/2.;
    WRITELN(' ',ITER : 8,'          ',XN  : 13);
    TOL := ABS((XN-XM)*100./XN);
    IF TOL < ES THEN GOTO 200
END;
WRITELN;
WRITELN('   ROOT CAN NOT BE REACHED FOR',
        ' THE GIVEN CONDITION');
GOTO 300;
200 : WRITELN;
    WRITELN('        THE ROOT IS ',XN : 13);
300 : WRITELN
END.

PROGRAM FALPOS;
(*....PROGRAM FOR COMPUTING ROOT OF NONLINEAR ..*)
(*....DEFINE THE FOLLOWING GIVEN VALUES :      ..*)
(*....   XL = LEFT VALUE OF X                   ..*)
(*....   XR = RIGHT VALUE OF X                  ..*)
(*....   ES = STOPPING CRITERION TOLERANCE (%) *)
LABEL 200,300;
VAR XL,XR,X1,X1OLD,ES,FXL,FXR,AA,FX1,TOL : real;
    ITER : integer;
(*-----------------------------------------------*)
FUNCTION FUNC(X : real) : real;
BEGIN
    FUNC := EXP(-X/4.0)*(2.-X) - 1.
END;
(*-----------------------------------------------*)
BEGIN
    XL := 0.;
    XR := 2.;
    ES := 0.001;
(*....CHECK WHETHER THE ROOT IS IN GIVEN RANGE: *)
    FXL := FUNC(XL);
    FXR := FUNC(XR);
    AA  := FXL*FXR;
    IF AA >= 0. THEN
    BEGIN
        WRITELN;
        WRITELN(' ROOT IS NOT IN THE GIVEN RANGE');
        GOTO 300
    END;
    X1OLD := XL;
    WRITELN;
    WRITELN('    ITERATION NO.        X ');
    WRITELN;
    FOR ITER := 1 TO 500 DO
    BEGIN
        FXL := FUNC(XL);
        FXR := FUNC(XR);
        X1  := (XL*FXR - XR*FXL)/(FXR - FXL);
        FX1 := FUNC(X1);
        AA  := FX1*FXR;
        IF AA < 0. THEN
(*....CASE A:  X1 < ROOT < XR        ....*)
            XL := X1
        ELSE
(*....CASE B:  XL < ROOT < X1        ....*)
            XR := X1;
(*....CHECK FOR TOLERANCE:
        WRITELN(' ',ITER : 8,'          ',X1 :13);
        TOL := ABS((X1-X1OLD)*100./X1);
        IF TOL < ES THEN GOTO 200;
        X1OLD := X1
    END;
    WRITELN;
    WRITELN('   ROOT CAN NOT BE REACHED FOR THE',
            ' GIVEN CONDITION');
GOTO 300;
200 : WRITELN;
    WRITELN('        THE ROOT IS ',X1 : 13);
300 : WRITELN
END.

PROGRAM ONEPT;
LABEL 30;
VAR XOLD,ES,XNEW,TOL : real;
    I : integer;
BEGIN
    XOLD := 0.;
    ES   := 0.001;
    WRITELN;
    WRITELN('    ITERATION NO.        X ');
    WRITELN;
    FOR I := 1 TO 100 DO
    BEGIN
        XNEW := 2. - EXP(XOLD/4.0);
        WRITELN(' ',I:8,'          ',XNEW:13);
        TOL := ABS((XNEW-XOLD)*100./XNEW);
        IF TOL < ES THEN GOTO 30;
        XOLD := XNEW
    END;
30 : WRITELN;
    WRITELN('        THE ROOT IS ',XNEW : 13)
END.

PROGRAM NEWRAP;
(*....PROGRAM FOR COMPUTING ROOT OF NONLINEAR    *)
```

```
(*....EQUATION USING THE NEWTON-RAPHSON METHOD  *)
(*....DEFINE THE FOLLOWING GIVEN VALUES:        *)
(*....    X0 = INITIAL GUESS VALUE OF X      ...*)
(*....    ES = STOPPING CRITERION TOLERANCE (%)*)
LABEL 200,300;
VAR X0,ES,X,F,DF,DX,TOL : real;
    ITER : integer;
(*----------------------------------------------*)
FUNCTION FUNC(X : real) : real;
BEGIN
   FUNC := EXP(-X/4.0)*(2.-X) - 1.
END;
(*----------------------------------------------*)
FUNCTION DERIV(X : real) : real;
BEGIN
   DERIV := -EXP(-X/4.0)*(1.5-X/4.0)
END;
(*----------------------------------------------*)
BEGIN
   X0 := 3.;
   ES := 0.001;
   WRITELN;
   WRITELN('          ITERATION NO.           X');
   WRITELN;
   X := X0;
   FOR ITER := 1 TO 500 DO
      BEGIN
         F  := FUNC(X);
         DF := DERIV(X);
         DX := -F/DF;
         X  := X + DX;
         WRITELN(' ',ITER : 8,'              ',X : 14);
         TOL := ABS(DX*100./X);
         IF TOL <ES THEN GOTO 200
      END;
   WRITELN;
   WRITELN('     ROOT CAN NOT REACHED FOR  THE',
           '  GIVEN CONDITIONS');
   GOTO 300;
200 : WRITELN;
      WRITELN('         THE ROOT IS  ',X : 14);
300 : WRITELN
END.
```

```
PROGRAM SECANT;
(*....PROGRAM FOR COMPUTING ROOT OF NONLINEAR ..*)
(*....EQUATION USING THE SECANT METHOD        ..*)
(*....DEFINE THE FOLLOWING GIVEN VALUES :     ..*)
(*....    X0 = FIRST VALUE OF INITIAL GUESS OF X *)
(*....    X1 = SECOND VALUE OF INITIAL GUESS OF X*)
(*....    ES = STOPPING CRITERION TOLERANCE (%) .*)
LABEL 200,300;
VAR X0,X1,DF,DX,ES,F0,F1,TOL : real;
    ITER : integer;
(*----------------------------------------------*)
FUNCTION FUNC(X : real) : real;
BEGIN
   FUNC := EXP(-X/4)*(2.-X) - 1.
END;
(*----------------------------------------------*)
BEGIN
   X0 := 3.;
   X1 := 2.;
   ES := 0.001;
   WRITELN;
   WRITELN('          ITERATION NO.           X');
   WRITELN;
   FOR ITER := 1 TO 500 DO
      BEGIN
         F0 := FUNC(X0);
         F1 := FUNC(X1);
         DF := (F0-F1)/(X0-X1);
         DX := -F1/DF;
         X0 := X1;
         X1 := X1 + DX;
         WRITELN(' ',ITER : 8,'              ',X1 :13);
         TOL := ABS(DX*100./X1);
         IF TOL < ES THEN GOTO 200
      END;
   WRITELN;
   WRITELN('     ROOT CAN NOT BE REACHED FOR THE',
           '  GIVEN CONDITION');
   GOTO 300;
200 : WRITELN;
      WRITELN('         THE ROOT IS  ',X1 : 13);
300 : WRITELN
END.
```

```
PROGRAM NGELIM;
(*..PROGRAM FOR COMPUTING A SET OF SIMULTANEOUS *)
(*..LINEAR EQUATIONS  [A]{X} = {B}  USING    ...*)
(*..NAIVE GAUSS ELIMINATION                  ...*)
TYPE table = ARRAY [1..50] OF real;
VAR B,X : table;
    A : ARRAY [1..50,1..50] OF real;
    I,ICOL,IROW,N : integer;
    INPUTFILE,OUTPUTFILE : text;
(*-----------------------------------------------*)
PROCEDURE GAUSS(VAR X : table);
VAR IP,IE,IC : integer;
    SUM,RATIO : real;
(*..FORWARD ELIMINATION: PERFORM ACCORDING TO *)
(*..THE ORDER OF 'PRIME' FROM 1 TO N-1:     ...*)
BEGIN
   FOR IP := 1 TO N-1 DO
(*..LOOP OVER EACH EQUATION STARTING FROM THE *)
(*..ONE THAT COORESPONDS WITH THE ORDER OF    *)
(*..'PRIME' PLUS ONE:                         *)
      BEGIN
         FOR IE := IP+1 TO N DO
            BEGIN
               RATIO := A[IE,IP]/A[IP,IP];
(*..COMPUTE NEW COEFF. OF THE EQ. CONSIDERED: *)
               FOR IC := IP+1 TO N DO
                  A[IE,IC] := A[IE,IC] - RATIO*A[IP,IC];
               B[IE] := B[IE] - RATIO*B[IP]
            END;
(*..SET COEFF. ON LOWER LEFT PORTION TO ZERO: *)
            FOR IE := IP+1 TO N DO
               A[IE,IP] := 0.;
```

```pascal
PROGRAM TRISYS;
(*...PROGRAM FOR COMPUTING A SET OF SIMULTANEOUS *)
(*...LINEAR EQUATIONS   [A][X] = {B}  USING      *)
(*...GAUSS ELIMINATION FOR TRIDIAGONAL SYSTEM.   *)
TYPE table = ARRAY [1..50] OF real;
VAR A,B,C,D,X : table;
    I,IROW,N  : integer;
```

```pascal
PROCEDURE SCALE;
VAR IC,IE    : integer;
    AMAX,BIG : real;
BEGIN
(*...PERFORM SCALING:                       ...*)
   FOR IE := 1 TO N DO
      BEGIN
         BIG := ABS(A[IE,1]);
         FOR IC := 2 TO N DO
            BEGIN
               AMAX := ABS(A[IE,IC]);
               IF AMAX > BIG THEN BIG := AMAX
            END;
         FOR IC := 1 TO N DO
            A[IE,IC] := A[IE,IC]/BIG;
         B[IE] := B[IE]/BIG
      END;
```

```pascal
END;
(*................................................................*)
PROCEDURE PIVOT;
VAR I,J,JP    : integer;
    AMAX,BIG,DUMY : real;
BEGIN
(*...PERFORM PARTIAL PIVOTING:              ....*)
   JP := IP;
   BIG := ABS(A[IP,IP]);
   FOR I := IP+1 TO N DO
      BEGIN
         AMAX := ABS(A[I,IP]);
         IF AMAX > BIG THEN
            BEGIN
               BIG := AMAX;
               JP  := I
            END;
      END;
   IF JP <> IP THEN
      BEGIN
         FOR J := IP TO N DO
            BEGIN
               DUMY    := A[JP,J];
               A[JP,J] := A[IP,J];
               A[IP,J] := DUMY
            END;
         DUMY  := B[JP];
         B[JP] := B[IP];
         B[IP] := DUMY
      END;
END;
(*................................................................*)
PROCEDURE GAUSS;
VAR IP,IE,IC : integer;
    SUM,RATIO : real;
BEGIN
(*...PERFORM SCALING :                      ...*)
   SCALE;
(*...FORWARD ELIMINATION: PERFORM ACCORDING TO *)
(*...THE ORDER OF 'PRIME' FROM 1 TO N-1 :      *)
   FOR IP := 1 TO N-1 DO
      BEGIN
(*...PERFORM PARTIAL PIVOTING:                 *)
         PIVOT;
(*...LOOP OVER EACH EQUATION STARTING FROM THE *)
(*...ONE THAT COORESONDS WITH THE ORDER OF     *)
(*...'PRIME' PLUS ONE:                      ...*)
         FOR IE := IP+1 TO N DO
            :
            :
```

```pascal
END;
(*...BACK SUBSTITUTION:                      ...*)
(*...COMPUTE SOLUTION OF THE LAST EQUATION:  ...*)
   X[N] := B[N]/A[N,N];
(*...COMPUTE SOLUTIONS FROM EQUATION N-1 TO 1: *)
   FOR IE := N-1 DOWNTO 1 DO
      BEGIN
         SUM := 0.;
         FOR IC := IE+1 TO N DO
            SUM := SUM + A[IE,IC]*X[IC];
         X[IE] := (B[IE] - SUM)/A[IE,IE];
      END;
END;
(*................................................................*)
BEGIN
   ASSIGN(INPUTFILE,'INPUT.DAT');
   RESET(INPUTFILE);
   ASSIGN(OUTPUTFILE,'SOL.OUT');
   REWRITE(OUTPUTFILE);
(*...READ TOTAL NUMBER OF EQUATIONS TO BE SOLVED:*)
   READ(INPUTFILE,N);
(*...READ MATRIX [A] AND VECTOR {B}         ...*)
   FOR IROW := 1 TO N DO
      BEGIN
         FOR ICOL := 1 TO N DO
            READ(INPUTFILE,A[IROW,ICOL]);
         READLN(INPUTFILE,B[IROW])
      END;
   CLOSE(INPUTFILE);
   GAUSS(X);
   WRITELN(OUTPUTFILE);
   WRITELN(OUTPUTFILE,'          EQUATION NO.         ',
                      '          SOLUTION X');
   WRITELN(OUTPUTFILE);
   FOR I := 1 TO N DO
      WRITELN(OUTPUTFILE,I :12,'           ',
                         X[I] : 13);
   CLOSE(OUTPUTFILE)
END.
```

```
            INPUTFILE,OUTPUTFILE : text;
(*-------------------------------------------------*)
PROCEDURE TRIDG(VAR X : table);
(*..SOLVE TRIDIAGONAL SYSTEM FOR N EQUATIONS:    *)
VAR I : integer;
(*..PERFORM FORWARD ELIMINATION:           ....*)
BEGIN
   FOR I := 2 TO N DO
      BEGIN
         A[I] := A[I]/B[I-1];
         B[I] := B[I] - A[I]*C[I-1];
         D[I] := D[I] - A[I]*D[I-1];
      END;
(*..PERFORM BACKWARD SUBSTITUTION:          ....*)
   X[N] := D[N]/B[N];
   FOR I := N-1 DOWNTO 1 DO
      X[I] := (D[I] - C[I]*X[I+1])/B[I];
END;
(*-------------------------------------------------*)
BEGIN
   ASSIGN(INPUTFILE,'INPUT.DAT');
   RESET(INPUTFILE);
   ASSIGN(OUTPUTFILE,'SOL.OUT');
   REWRITE(OUTPUTFILE);
(*..READ TOTAL NUMBER OF EQUATIONS TO BE SOLVED:*)
   READ(INPUTFILE,N);
(*..READ VECTOR (A), (B), (C) AND (D) :    ....*)
   FOR IROW := 1 TO N DO
      BEGIN
         READ(INPUTFILE,A[IROW]);
         READ(INPUTFILE,B[IROW]);
         READ(INPUTFILE,C[IROW]);
         READLN(INPUTFILE,D[IROW]);
      END;
   CLOSE(INPUTFILE);
   TRIDG(X);
   WRITELN(OUTPUTFILE,'              EQUATION NO.',
   WRITELN(OUTPUTFILE,'                   SOLUTION X');
   FOR I := 1 TO N DO
      WRITELN(OUTPUTFILE,I :12,'
                         X[I] : 13);
   CLOSE(OUTPUTFILE)
END.

PROGRAM LUDCOM;
(*..PROGRAM FOR SOLVING A SET OF SIMULTANEOUS *)
(*..LINEAR EQUATIONS    [A]{X} = {B}          *)
(*..USING NAIVE LU DECOMPOSITION        ....*)
TYPE table = ARRAY [1..50] OF real;
     table = ARRAY [1..50,1..50] OF real;
VAR A,AL,AU             : table;
    B,X,Y               : table;
    I,ICOL,IROW,N       : integer;
    INPUTFILE,OUTPUTFILE : text;
(*-------------------------------------------------*)
PROCEDURE LU(VAR X : table);
LABEL 350;
VAR I,J,K : integer;
    SUM      : real;
BEGIN
(*..PERFORM DECOMPOSITION    [A] = [L][U]:  ....*)
   FOR J := 1 TO 50 DO
      FOR I := 1 TO 50 DO
         BEGIN
            AL[I,J] := 0.0;
            AU[I,J] := 0.0
         END;
   FOR I := 1 TO 50 DO AL[I,1] := A[I,1];
   FOR J := 2 TO N DO AU[1,J] := A[1,J]/AL[1,1];
   FOR J := 2 TO N-1 DO
      BEGIN
         FOR I := J TO N DO
            BEGIN
               SUM := 0.0;
               FOR K := 1 TO J-1 DO
                  SUM := SUM + AL[I,K]*AU[K,J];
               AL[I,J] := A[I,J]-SUM
            END;
         FOR K := J+1 TO N DO
            BEGIN
               SUM := 0.0;
               FOR I := 1 TO J-1 DO
                  SUM := SUM + AL[J,I]*AU[I,K];
               AU[J,K] := (A[J,K]-SUM)/AL[J,J]
            END;
      END;
   SUM := 0.0;
   FOR K := 1 TO N-1 DO
      SUM := SUM + AL[N,K]*AU[K,N];
   AL[N,N] := A[N,N] - SUM;
(*..PERFORM FORWARD PASS TO SOLVE [L]{Y} = {B} :*)
   Y[1] := B[1]/AL[1,1];
   FOR I := 2 TO N DO
      BEGIN
         SUM := 0.0;
         FOR J := 1 TO I-1 DO
            SUM := SUM + AL[I,J]*Y[J];
         Y[I] := (B[I]-SUM)/AL[I,I]
      END;
(*..PERFORM BACKWARD PASS TO SOLVE [U]{X} = {Y}:*)
   X[N] := Y[N];
   FOR I := N-1 DOWNTO 1 DO
      BEGIN
         SUM := 0.0;
         FOR J := I+1 TO N DO
            SUM := SUM + AU[I,J]*X[J];
         X[I] := Y[I] - SUM
      END
END;
(*-------------------------------------------------*)
BEGIN
   ASSIGN(INPUTFILE,'INPUT.DAT');
   RESET(INPUTFILE);
   ASSIGN(OUTPUTFILE,'SOL1.OUT');
   REWRITE(OUTPUTFILE);
(*..READ TOTAL NUMBER OF EQUATIONS TO BE SOLVED:*)
   READLN(INPUTFILE,N);
(*..READ MATRIX [A] AND VECTOR (B) :        ..*)
   FOR IROW := 1 TO N DO
      BEGIN
```

```
        FOR ICOL := 1 TO N DO
          READ(INPUTFILE,A[IROW,ICOL]);
        READLN(INPUTFILE,B[IROW])
      END;
    CLOSE(INPUTFILE);
    LU(X);
    WRITELN(OUTPUTFILE,'               EQUATION NO.           ',
    WRITELN(OUTPUTFILE,'                 SOLUTION X');
    WRITELN(OUTPUTFILE);
    FOR I := 1 TO N DO
        WRITELN(OUTPUTFILE,I :12,'
        X[I] : 13);
    CLOSE(OUTPUTFILE)
END.


PROGRAM CHOLESKY;
(*....PROGRAM FOR SOLVING A SET OF SIMULTANEOUS.*)
(*....LINEAR EQUATIONS  [A]{X} = [B]  USING ....*)
(*....CHOLESKY DECOMPOSITION IF [A] IS SYMMETRIC*)
TYPE table = ARRAY [1..50] OF real;
     table1 = ARRAY [1..50,1..50] OF real;
VAR A,AL : table1;
    B,X,Y : table;
    I,ICOL,IROW,N : integer;
    INPUTFILE,OUTPUTFILE : text;
(*----------------------------------------------*)
PROCEDURE CHOLES(VAR X : table);
LABEL 350;
VAR I,J,K : integer;
    SUM : real;
(*....PERFORM DECOMPOSITION  [A] = [L][LT] :    *)
BEGIN
    FOR I := 1 TO 50 DO
      FOR J := 1 TO 50 DO
        AL[I,J] := 0.0;
    AL[1,1] := SQRT(A[1,1]);
    FOR K := 2 TO N DO
      BEGIN
        FOR I := 1 TO K-1 DO
          BEGIN
            SUM := 0.0;
            IF I = 1 THEN GOTO 350;
            FOR J := 1 TO I-1 DO
              SUM := SUM + AL[I,J]*AL[K,J];
350 :       WRITELN;
            AL[K,I] := (A[K,I]-SUM)/AL[I,I]
          END;
        SUM := 0.0;
        FOR J := 1 TO K-1 DO
          SUM := SUM + AL[K,J]*AL[K,J];
        AL[K,K] := SQRT(A[K,K]-SUM)
      END;
(*....PERFORM FORWARD PASS :              ....*)
    Y[1] := B[1]/AL[1,1];
    FOR I := 2 TO N DO
      BEGIN
        SUM := 0.0;
        FOR J := 1 TO I-1 DO
          SUM := SUM + AL[I,J]*Y[J];
        Y[I] := (B[I]-SUM)/AL[I,I]
      END;
    X[N] := Y[N]/AL[N,N];
    FOR I := N-1 DOWNTO 1 DO
      BEGIN
        SUM := 0.0;
        FOR J := I+1 TO N DO
          SUM := SUM + AL[J,I]*X[J];
        X[I] := (Y[I] - SUM)/AL[I,I]
      END
END;
(*----------------------------------------------*)
BEGIN
    ASSIGN(INPUTFILE,'INPUT.DAT');
    RESET(INPUTFILE);
    ASSIGN(OUTPUTFILE,'SOL.OUT');
    REWRITE(OUTPUTFILE);
(*..READ TOTAL NUMBER OF EQUATIONS TO BE SOLVED:*)
    READLN(INPUTFILE,N);
(*..READ MATRIX [A] AND VECTOR [B]           ..*)
    FOR IROW := 1 TO N DO
      BEGIN
        FOR ICOL := 1 TO N DO
          READ(INPUTFILE,A[IROW,ICOL]);
        READLN(INPUTFILE,B[IROW])
      END;
    CLOSE(INPUTFILE);
    CHOLES(X);
    WRITELN(OUTPUTFILE);
    WRITELN(OUTPUTFILE,'               EQUATION NO.  ',
                       '         SOLUTION X');
    WRITELN(OUTPUTFILE);
    FOR I := 1 TO N DO
        WRITELN(OUTPUTFILE,I :12,'
        X[I] : 13);
    CLOSE(OUTPUTFILE)
END.


PROGRAM GSEIDEL;
(*...JACOBI ITERATION METHOD FOR EXAMPLE 3.13  *)
LABEL 400;
VAR XOLD,XNEW  : ARRAY [1..3] OF real;
    TOL,EPS    : real;
    I,ITER,IFLAG : integer;
BEGIN
    TOL := 0.05;
    FOR I := 1 TO 3 DO
        XOLD[I] := 100.0;
    WRITELN;
    WRITELN(' ITERATION NO.        X2           X1
                        X3');
    WRITELN;
    FOR ITER := 1 TO 500 DO
      BEGIN
        XNEW[1] := 100.0 + XOLD[2];
        XNEW[2] := 100.0 + 0.25*XOLD[1]+0.5*XOLD[3];
        XNEW[3] := 100.0 + 0.50*XOLD[2];
        WRITE(ITER : 6,'          ,XOLD[1] : 10:0);
        WRITELN('      ,XOLD[2] : 10:0,'
                XOLD[3] : 10:0);
```

```
PROGRAM GSEIDEL;
(*..GAUSS-SEIDEL ITERATION METHOD FOR EX 3.14*)
LABEL 400;
VAR XOLD,XNEW  : ARRAY [1..3] OF real;
    TOL,EPS    : real;
    I,ITER,IFLAG : integer;
BEGIN
  TOL := 0.05;
  FOR I := 1 TO 3 DO
    XOLD[I] := 100.0;
  WRITELN;
  WRITELN(' ITERATION NO.    X1      X2      X3');
  WRITELN;
  FOR ITER := 1 TO 500 DO
    BEGIN
      XNEW[1] := 100.0 + XOLD[2];
      XNEW[2] := 100.0+0.25*XNEW[1]+0.5*XOLD[3];
      XNEW[3] := 100.0 + 0.50*XNEW[2];
      WRITE(ITER : 6,'     ',XOLD[1] : 10:0);
      WRITELN('     ',XOLD[2] : 10:0,'     ',
              XOLD[3] : 10:0);
      IFLAG := 0;
      FOR I := 1 TO 3 DO
        BEGIN
          EPS:=ABS((XNEW[I]-XOLD[I])*100./XNEW[I]);
          IF EPS >= TOL THEN IFLAG := 1
        END;
      IF IFLAG = 0 THEN GOTO 400;
      FOR I := 1 TO 3 DO
        XOLD[I] := XNEW[I]
    END;
400 : WRITELN
END.
```

```
PROGRAM SOR;
(*..SUCCESSIVE OVER RELAXATION FOR EXAMPLE 3.15 *)
LABEL 400;
VAR XOLD,XNEW  : ARRAY [1..3] OF real;
    TOL,EPS,W  : real;
    I,ITER,IFLAG : integer;
BEGIN
  W   := 1.2;
  TOL := 0.05;
  FOR I := 1 TO 3 DO
    XOLD[I] := 100.0;
  WRITELN;
  WRITELN(' ITERATION NO.    X1      X2      X3');
  WRITELN;
  FOR ITER := 1 TO 500 DO
    BEGIN
      XNEW[1] := 100.0 + XOLD[2];
      XNEW[1] := W*XNEW[1] + (1.0-W)*XOLD[1];
      XNEW[2] := 100.0+0.25*XNEW[1]+0.5*XOLD[3];
      XNEW[2] := W*XNEW[2] + (1.0-W)*XOLD[2];
      XNEW[3] := 100.0 + 0.50*XNEW[2];
      XNEW[3] := W*XNEW[3] + (1.0-W)*XOLD[3];
      WRITE(ITER : 6,'     ',XOLD[1] : 10:0);
      WRITELN('     ',XOLD[2] : 10:0,'     ',
              XOLD[3] : 10:0);
      IFLAG := 0;
      FOR I := 1 TO 3 DO
        BEGIN
          EPS:=ABS((XNEW[I]-XOLD[I])*100./XNEW[I]);
          IF EPS >= TOL THEN IFLAG := 1
        END;
      IF IFLAG = 0 THEN GOTO 400;
      FOR I := 1 TO 3 DO
        XOLD[I] := XNEW[I]
    END;
400 : WRITELN
END.
```

```
PROGRAM ORGCG;
(*..PROGRAM FOR COMPUTING A SET OF SIMULTANEOUS *)
(*..LINEAR EQUATIONS  [A](X) = (B)   USING      *)
(*..CONJUGATE GRADIENT                        ..*)
TYPE table = ARRAY [1..50] OF real;
     tablel = ARRAY [1..50,1..50] OF real;
VAR A   : tablel;
    B,X : table;
    I,ICOL,IROW,N : integer;
    INPUTFILE,OUTPUTFILE : text;
(*...................................................*)
PROCEDURE CG(VAR X : table);
LABEL 400;
VAR D,R : ARRAY [1..50] OF real;
    ALAM,ALPHA,DOWN,RES,SUM,TOL,UP,XZERO : real;
    I,J,K : integer;
BEGIN
(*...ASSIGN INITIAL VALUES IN VECTOR X      ...*)
  XZERO := 0.;
  FOR I := 1 TO N DO
    X[I] := XZERO;
(*...ASSIGN TOLERANCE FOR STOPPING CRITERION:  *)
  TOL := 0.0001;
(*...COMPUTE INITIAL RESIDUAL & SEARCH         *)
(*...DIRECTION:                              ..*)
  FOR I := 1 TO N DO
    BEGIN
      SUM := 0.;
      FOR J := 1 TO N DO
        SUM := SUM + A[I,J]*X[J];
      R[I] := SUM - B[I];
```

```
    D[I] := -R[I];
    END;
(*....ENTER THE ITERATION LOOP:             ....*)
FOR K := 1 TO N+1 DO
    BEGIN
    UP := 0.;
    FOR I := 1 TO N DO
        UP := UP + D[I]*R[I];
    DOWN := 0.;
    FOR I := 1 TO N DO
        BEGIN
        SUM := 0.;
        FOR J := 1 TO N DO
            SUM := SUM + A[I,J]*D[J];
        DOWN := DOWN + D[I]*SUM;
        END;
    ALAM := -UP/DOWN;
    FOR I := 1 TO N DO
        X[I] := X[I] + ALAM*D[I];
    FOR I := 1 TO N DO
        BEGIN
        SUM := 0.;
        FOR J := 1 TO N DO
            SUM := SUM + A[I,J]*X[J];
        R[I] := SUM - B[I];
        END;
    RES := 0.;
    FOR I := 1 TO N DO
        RES := RES + R[I]*R[I];
    RES := SQRT(RES);
    IF RES < TOL THEN GOTO 400;
    UP := 0.;
    DOWN := 0.;
    FOR I := 1 TO N DO
        BEGIN
        SUM := 0.;
        FOR J := 1 TO N DO
            SUM := SUM + A[I,J]*D[J];
        UP := UP + R[I]*SUM;
        DOWN := DOWN + D[I]*SUM;
        END;
    ALPHA := UP/DOWN;
    FOR I := 1 TO N DO
        D[I] := -R[I] + ALPHA*D[I];
    END;
400 :
END;
(*-----------------------------------------------*)
BEGIN
ASSIGN(INPUTFILE,'INPUT.DAT');
RESET(INPUTFILE);
ASSIGN(OUTPUTFILE,'SOL.OUT');
REWRITE(OUTPUTFILE);
(*..READ TOTAL NUMBER OF EQUATIONS TO BE SOLVED:*)
READ(INPUTFILE,N);
(*..READ MATRIX [A] AND VECTOR {B} :        ....*)
FOR IROW := 1 TO N DO
    BEGIN
    FOR ICOL := 1 TO N DO
        READ(INPUTFILE,A[IROW,ICOL]);
    READLN(INPUTFILE,B[IROW]);
    END;
CLOSE(INPUTFILE);
CG(X);
WRITELN(OUTPUTFILE);
WRITELN(OUTPUTFILE,'                    EQUATION NO.',
                   '           SOLUTION X');
WRITELN(OUTPUTFILE);
FOR I := 1 TO N DO
    WRITELN(OUTPUTFILE,I :12,'          ',X[I] : 13);
CLOSE(OUTPUTFILE)
END.
```

```
PROGRAM IMPCG;
(*..PROGRAM FOR COMPUTING A SET OF SIMULTANEOUS *)
(*..LINEAR EQUATIONS  [A]{X} = {B}  USING    ....*)
(*..IMPROVED CONJUGATE GRADIENT             ....*)
TYPE table = ARRAY [1..50] OF real;
     table1 = ARRAY [1..50,1..50] OF real;
VAR A  : table1;
    B,X : table;
    I,ICOL,IROW,N : integer;
    INPUTFILE,OUTPUTFILE : text;
(*.................................................*)
PROCEDURE CGNEW(VAR X : table);
LABEL 400;
VAR D,R,U : ARRAY [1..50] OF real;
    ALAM,ALPHA,DEL,DEL1,DOWN,SUM,TOL,XZERO : real;
    I,J,K : integer;
BEGIN
(*....ASSIGN INITIAL VALUES IN VECTOR X      ....*)
XZERO := 0.;
FOR I := 1 TO N DO
    X[I] := XZERO;
(*....ASSIGN TOLERANCE FOR STOPPING CRITERION  *)
TOL := 0.0001;
(*....COMPUTE INITIAL RESIDUAL & SEARCH      ....*)
(*....DIRECTION:                             ....*)
FOR I := 1 TO N DO
    BEGIN
    SUM := 0.;
    FOR J := 1 TO N DO
        SUM := SUM + A[I,J]*X[J];
    R[I] := SUM - B[I];
    D[I] := -R[I];
    END;
DEL := 0.;
FOR I := 1 TO N DO
    DEL := DEL + R[I]*R[I];
(*....ENTER THE ITERATION LOOP:             ....*)
FOR K := 1 TO N+1 DO
    BEGIN
    FOR I := 1 TO N DO
        BEGIN
        U[I] := 0.;
        FOR J := 1 TO N DO
            U[I] := U[I] + A[I,J]*D[J];
        END;
    DOWN := 0.;
    FOR I := 1 TO N DO
        DOWN := DOWN + D[I]*U[I];
```

```pascal
    ALAM := DEL./DOWN;
    FOR I := 1 TO N DO
      BEGIN
        X[I] := X[I] + ALAM*D[I];
        R[I] := R[I] + ALAM*U[I];
      END;
    DEL1 := 0.;
    FOR I := 1 TO N DO
      DEL1 := DEL1 + R[I]*R[I];
    IF DEL1 < TOL THEN GOTO 400;
    ALPHA := DEL1/DEL;
    FOR I := 1 TO N DO
      D[I] := -R[I] + ALPHA*D[I];
    DEL := DEL1;
  END;
400 :
END;
(*-------------------------------------------------*)
BEGIN
  ASSIGN(INPUTFILE, 'INPUT.DAT');
  RESET(INPUTFILE);
  ASSIGN(OUTPUTFILE,'SOL.OUT');
  REWRITE(OUTPUTFILE);
(*..READ TOTAL NUMBER OF EQUATIONS TO BE SOLVED:*)
  READ(INPUTFILE,N);
(*..READ MATRIX [A] AND VECTOR {B} :        ....*)
  FOR IROW := 1 TO N DO
    BEGIN
      FOR ICOL := 1 TO N DO
        READ(INPUTFILE,A[IROW,ICOL]);
      READLN(INPUTFILE,B[IROW]);
    END;
  CLOSE(INPUTFILE);
  CGNEW(X);
  WRITELN(OUTPUTFILE);
  WRITELN(OUTPUTFILE,'          EQUATION NO.',
                     '     SOLUTION X');
  WRITELN(OUTPUTFILE);
  FOR I := 1 TO N DO
    WRITELN(OUTPUTFILE, I :12,'          ',
            X[I] : 13);
  CLOSE(OUTPUTFILE)
END.


PROGRAM NEWDIV;
(*...PROGRAM FOR COMPUTING  F(X)  AT A GIVEN X *)
(*...USING NEWTON'S DIVIDED-DIFFERENCE        *)
(*...INTERPOLATING POLYNOMIALS             ...*)
VAR X : ARRAY [1..10] OF real;
    FX : ARRAY [1..10,1..10] OF real;
    FF,XX,FAC : real;
    I,ICOL,IROW,N,M : integer;
BEGIN
  READ(N);
  FOR IROW := 1 TO N DO
    READ( X[IROW], FX[IROW,1]);
(*..COMPUTE DIVIDED-DIFFERENCE COEFFICIENTS: *)
  M := N;
  FOR ICOL := 2 TO N DO
    BEGIN
      M := M - 1;
      FOR IROW := 1 TO M DO
        BEGIN
          FX[IROW,ICOL] := FX[IROW+1, ICOL-1]
                           - FX[IROW, ICOL-1];
          FX[IROW,ICOL] := FX[IROW,ICOL]/
                           (X[IROW+ICOL-1]-X[IROW])
        END;
    END;
(*..COMPUTE DESIRED F(X) AT THE GIVEN X VALUE:*)
  READ(XX);
  FF := FX[1,1];
  FAC := 1.;
  FOR I := 2 TO N DO
    BEGIN
      FAC := FAC*(XX - X[I-1]);
      FF  := FF + FX[1,I]*FAC
    END;
  WRITE('    VALUE OF F(X) AT X = ', XX :10,' IS ',
        FF : 16)
END.


PROGRAM LAGPOL;
(*...PROGRAM FOR COMPUTING  F(X)   AT A GIVEN X *)
(*...USING LAGRANGE INTERPOLATION           ...*)
VAR X,FX : ARRAY [1..10] OF real;
    YY,AL,XX : real;
    I,J,N : integer;
BEGIN
(*..READ NUMBER OF DATA SETS, DATA OF X AND FX: *)
  READLN(N);
  FOR I := 1 TO N DO
    READLN(X[I],FX[I]);
(*..COMPUTE DESIRED F(X) AT THE GIVEN X VALUE: *)
  READLN(XX);
  YY := 0.;
  FOR I := 1 TO N DO
    BEGIN
      AL := 1.;
      FOR J := 1 TO N DO
        IF J <> I THEN
          AL := AL*(XX-X[J])/(X[I]-X[J]);
      YY := YY + AL*FX[I]
    END;
  WRITELN('    VALUE OF F(X) AT X =',XX : 10,
          '  IS ',YY : 16)
END.


PROGRAM CUBSPLN;
(*   A CUBIC SPLINE INTERPOLATING PROGRAM      *)
VAR A,B,C,D,E,X,FX : ARRAY [1..10] OF real;
    D1,D2,DD,T1,T2,T3,T4,FAC,FF,XX : real;
    I,J,N : integer;
BEGIN
(*..READ NUMBER OF DATA SETS, DATA OF X AND FX. *)
(*..THEN THE VALUE OF X THAT F(X) IS NEEDED:  *)
  READLN(N);
  FOR I := 1 TO N DO
```

```pascal
      READLN(X[I], FX[I]);
READLN(XX);
(*...FORM UP TRIDIAGONAL SYSTEM OF N EQUATIONS: *)
FOR I := 2 TO N-1 DO
  BEGIN
    A[I] := X[I] - X[I-1];
    B[I] := 2.0*(X[I+1] - X[I-1]);
    C[I] := X[I+1] - X[I];
    D[I] := 6.0*(FX[I+1]-FX[I])/(X[I+1]-X[I])
              + 6.0*(FX[I-1]-FX[I])/(X[I-1]-X[I-1]);
  END;
B[1] := 1.0;
C[1] := 0.0;
D[1] := 0.0;
A[N] := 0.0;
B[N] := 1.0;
D[N] := 0.0;
(*...SOLVE TRIDIAGONAL SYSTEM OF N EQUATIONS FOR *)
(*...2ND DERIVATIVES, RETURN SOLUTION IN E( ):  *)
(*...[STANDARD TRIDIAGONAL SYSTEM SOLVER - N EQS]*)
(*...COMPUTE  F(X)  AT THE GIVEN  X:            *)
FOR I := 2 TO N DO
  BEGIN
    A[I] := A[I]/B[I-1];
    B[I] := B[I] - A[I]*C[I-1];
  END;
FOR I := 2 TO N DO
  D[I] := D[I] - A[I]*D[I-1];
E[N] := D[N]/B[N];
FOR I := N-1 DOWNTO 1 DO
  E[I] := (D[I] - C[I]*E[I+1])/B[I];
FOR I := 2 TO N DO
  IF (XX >= X[I-1]) AND (XX <= X[I]) THEN
    BEGIN
      D1 := X[I] - XX;
      D2 := XX - X[I-1];
      DD := X[I] - X[I-1];
      T1 := E[I-1]*D1*D1*D1/(6.0*DD);
      T2 := E[I]*D2*D2*D2/(6.0*DD);
      T3 := (FX[I-1]/DD-E[I-1]*DD/6.0)*D1;
      T4 := (FX[I]/DD-E[I]*DD/6.0)*D2;
      FF := T1 + T2 + T3 + T4;
    END;
WRITELN(' THE VALUE OF F(X) AT X =',XX : 10:4,
        ' IS ',FF : 16 : 7)
END.


PROGRAM LREGRES;
(*...A LINEAR REGRESSION PROGRAM                 ....*)
VAR X,Y : ARRAY [1..100] OF real;
    I,N : integer;
    A0,A1,DET,SUMX,SUMY,SUMX2,SUMXY : real;
(*....READ NUMBER OF DATA SETS, DATA OF X AND Y:*)
BEGIN
  READLN(N);
  FOR I := 1 TO N DO READ(X[I], Y[I]);
(*....COMPUTE SUMMATION TERMS:                   ....*)
  SUMX  := 0.0;
  SUMY  := 0.0;
  SUMX2 := 0.0;
```

```pascal
SUMXY := 0.0;
FOR I := 1 TO N DO
  BEGIN
    SUMX  := SUMX  + X[I];
    SUMY  := SUMY  + Y[I];
    SUMX2 := SUMX2 + X[I]*X[I];
    SUMXY := SUMXY + X[I]*Y[I];
  END;
(*....SOLVE FOR COEFFICIENTS:                    ....*)
  DET := N*SUMX2 - SUMX*SUMX;
  A0  := (SUMY*SUMX2 - SUMXY*SUMX)/DET;
  A1  := (N*SUMXY - SUMX*SUMY)/DET;
  WRITELN(' COEFFICIENT A0 =', A0 : 13);
  WRITELN(' COEFFICIENT A1 =', A1 : 13);
END.


PROGRAM PREGRES;
(*...A POLYNOMIAL REGRESSION PROGRAM            ....*)
VAR X,Y : ARRAY [1..100] OF real;
    A   : ARRAY [1..10,1..10] OF real;
    B,XX : ARRAY [1..10] OF real;
    XT  : real;
    I,IC,IP,IR,IMl,J,K,N,M,MP1 : integer;
    INPUTFILE : text;
(*....READ NUMBER OF DATA SETS, DATA OF X AND Y:*)
BEGIN
  ASSIGN(INPUTFILE, 'INPUT.DAT');
  RESET(INPUTFILE);
  READLN(INPUTFILE,N);
  FOR I := 1 TO N DO
    READLN(INPUTFILE,X[I],Y[I]);
(*....READ ORDER OF POLINOMIAL NEEDED:          *)
  READ(INPUTFILE,M);
  FOR IR := 1 TO N DO
    BEGIN
      B[IR] := 0.0;
      FOR IC := 1 TO 10 DO
        A[IR,IC] := 0.0;
    END;
(*....COMPUTE SQUARE MATRIX ON LHS AND          ....*)
(*....VECTOR ON RHS OF SYSTEM EQUATIONS:        ....*)
  FOR IR := 1 TO M + 1 DO
    BEGIN
      FOR IC := 1 TO M + 1 DO
        BEGIN
          K := IR + IC - 2;
          FOR I := 1 TO N DO
            BEGIN
              XT := 1;
              FOR J := 1 TO K DO
                XT := XT*X[I];
              A[IR,IC] := A[IR,IC] + XT;
            END;
        END;
      FOR I := 1 TO N DO
        BEGIN
          XT := 1;
          FOR J := 1 TO IR-1 DO
            XT := XT*X[I];
          B[IR] := B[IR] + Y[I]*XT;
```

```
    END;
(*...CALL SUBROUTINE FOR SOLVING SYSTEM EQS:      *)
    MP1 := M + 1;
    GAUSS;
(*...PRINT OUT POLINOMIAL COEFFICIENTS:          ....*)
    WRITELN;
    WRITELN(' COEFFICIENTS OF FITTED POLINOMIAL'
            ' ARE:');
    FOR I := 1 TO M + 1 DO
      BEGIN
        IM1 := I - 1;
        WRITELN(' A(', IM1 : 1, ') =',XX[I] : 13);
      END;
END.
```

```
PROGRAM MREGRES;
(*...A MULTIPLE LINEAR REGRESSION PROGRAM     ....*)
VAR X    : ARRAY [1..100,1..10] OF real;
    Y    : ARRAY [1..100] OF real;
    A    : ARRAY [1..10,1..10] OF real;
    B,XX : ARRAY [1..10] OF real;
    I,IC,IM1,IP,IR,J,K,KP,N : integer;
    FC,FR : real;
    INPUTFILE : text;
(*...READ NUMBER OF DATA SETS N,               ...*)
(*...NUMBER OF INDEPENDENT VARIABLES K,         ...*)
(*...AND DATA OF X(I,K) AND Y(I):               ...*)
BEGIN
  ASSIGN(INPUTFILE,'INPUT.DAT');
  RESET(INPUTFILE);
  READLN(INPUTFILE,N,K);
  FOR I := 1 TO N DO
    BEGIN
      FOR J := 1 TO K DO
        READ(INPUTFILE,X[I,J]);
      READLN(INPUTFILE,Y[I]);
    END;
  FOR IR := 1 TO 10 DO
    BEGIN
      B[IR] := 0.0;
      FOR IC := 1 TO 10 DO
        A[IR,IC] := 0.0;
    END;
(*...COMPUTE SQUARE MATRIX ON LHS AND           ....*)
(*...VECTOR ON RHS OF SYSTEM EQUATIONS:          ....*)
(*...CALL SUBROUTINE FOR SOLVING SYSTEM EQS:     ..*)
  FOR IR := 1 TO N DO
    FOR IR := 1 TO K + 1 DO
      BEGIN
        IF IR = 1 THEN FR := 1.0;
        IF IR > 1 THEN FR := X[I,IR-1];
        FOR IC := 1 TO K + 1 DO
          BEGIN
            IF IC = 1 THEN FC := 1.0;
            IF IC > 1 THEN FC := X[I,IC-1];
            A[IR,IC] := A[IR,IC] + FR*FC;
          END;
        B[IR] := B[IR] + FR*Y[I];
      END;
```

```
        KP := K + 1;
        GAUSS;
(*...PRINT OUT COEFFICIENTS:                      ....*)
        WRITELN;
        WRITELN(' COEFFICIENTS OF FITTED FUNCTION',
                ' ARE:');
        FOR I := 1 TO K + 1 DO
          BEGIN
            IM1 := I - 1;
            WRITELN(' A(', IM1 : 1, ') =',XX[I] : 12);
          END;
END.
```

```
PROGRAM TRAPEZ;
(*...A MULTIPLE-SEGMENT TRAPEZOIDAL PROGRAM....*)
(*...FOR ESTIMATING INTEGRAL OF F(X)            ....*)
VAR A,B,H,FX,FX0,FXN,SOL,SUM,X : real;
    I,N : integer;
FUNCTION FUNC( X: real) : real;
BEGIN
  FUNC := 2.0*X*X*X - 5.0*X*X + 3.0*X + 1.0;
END;
(*------------------------------------------------*)
BEGIN
  A := 0.0;
  B := 2.0;
(*...READ NUMBER OF SEGMENTS REQUIRED:           ....*)
  READLN(N);
  H := (B - A)/N;
  SUM := 0.0;
  X := A + H;
  FOR I := 1 TO N-1 DO
    BEGIN
      FX  := FUNC(X);
      SUM := SUM + FX;
      X   := X + H;
    END;
  FX0 := FUNC(A);
  FXN := FUNC(B);
  SOL := (FX0 + FXN + 2.0*SUM)*H/2.0;
  WRITELN(' INTEGRAL OF F(X) USING', N : 3,
          ' SEGMENTS IS', SOL : 10 :6);
END.
```

```
PROGRAM SIMPSON;
(*...A MULTIPLE-SEGMENT SIMPSON'S 1/3 PROGRAM.*)
(*...FOR ESTIMATING INTEGRAL OF F(X)            ..*)
LABEL 10,20,50,100;
VAR A,B,FX,FX0,FXN,H,SOL,SUM1,SUM2,X : real;
    I,M,N : integer;
(*------------------------------------------------*)
FUNCTION FUNC(X : real) : real;
BEGIN
  FUNC := 2.*X*X*X - 5.*X*X + 3.*X + 1.;
END;
(*------------------------------------------------*)
BEGIN
```

```pascal
A := 0.;
B := 2.;
(*...READ NUMBER OF SEGMENTS REQUIRED:   ...*)
READLN(N);
M := N - trunc(N/2)*2;
IF M = 0  THEN GOTO 50;
WRITELN(' NUMBER OF SEGMENTS MUST BE EVEN');
GOTO 100;
50 : H := (B - A)/N;
SUM1 := 0.;
X := A + H;
I := 1;
WHILE I <= N-1 DO
  BEGIN
  FX := FUNC(X);
  SUM1 := SUM1 + FX;
  X := X + 2.*H;
  I := I + 2;
  END;
SUM2 := 0.;
X := A + 2.*H;
I := 2;
WHILE I <= N-2 DO
  BEGIN
  FX := FUNC(X);
  SUM2 := SUM2 + FX;
  X := X + 2.*H;
  I := I + 2;
  END;
FX0 := FUNC(A);
FXN := FUNC(B);
SOL := (FX0 + FXN + 4.*SUM1 + 2.*SUM2)*H/3.;
WRITELN(' INTEGRAL OF F(X) USING', N : 3 ,
'  SEGMENTS IS ', SOL : 10 : 6);
100 : WRITELN
END.
```

```pascal
PROGRAM ROMBERG;
(*A ROMBERG INTEGRATING PROGRAM FOR ESTIMATING*)
(* INTEGRAL OF F(X) WITHIN SPECIFIED % ERROR  *)
LABEL 100;
VAR R : ARRAY[1..10,1..10] OF real;
A,AREA,B,ERR,EPS,FX0,FXN,PI : real;
I,IC,IR,J,K,K1,L,N : integer;
(*--------------------------------------------*)
FUNCTION  FUNC( X : real) : real;
BEGIN
FUNC := SIN(X);
END;
(*--------------------------------------------*)
PROCEDURE TRAP;
(*....MULTIPLE-SEGMENT TRAPEZOIDAL RULE......*)
VAR FX,H,X,SUM : real;
    M : integer;
BEGIN
H := (B - A)/N.;
SUM := 0.;
X := A + H;
FOR M := 1 TO N-1 DO
  BEGIN
  FX := FUNC(X);
  SUM := SUM + FX;
  X := X + H;
  END;
FX0 := FUNC(A);
FXN := FUNC(B);
AREA := (FX0 + FXN + 2.*SUM)*H/2.;
END;
(*--------------------------------------------*)
BEGIN
PI := 4.0*ARCTAN(1.0);
A := 0.0;
B := PI/2.0;
EPS := 0.0001;
(*....COMPUTE R(1,1):  *)
FX0 := FUNC(A);
FXN := FUNC(B);
R[1,1] := (FX0+FXN)*(B-A)/2.0;
(*..LOOP OVER NUMBER OF ROMBERG APPLICATIONS:..*)
FOR I := 1 TO 9 DO
  BEGIN
  N := 1;
  FOR J := 1 TO I DO
    BEGIN
    N := N*2;
    END;
  TRAP;
  R[I+1,1] := AREA;
  FOR IC := 2 TO I+1 DO
    BEGIN
    K := IC - 1;
    IR := 2 + I - IC;
    K1 := 1;
    FOR L := 1 TO K DO
      BEGIN
      K1 := K1*4;
      END;
    R[IR,IC] := (K1*R[IR+1,K] - R[IR,K])
                     /(K1 - 1.0);
    END;
  ERR := 100.0*(R[1,K+1]-R[2,K])/R[1,K+1];
  ERR := ABS(ERR);
  IF ERR < EPS  THEN GOTO 100
  END;
100 : WRITELN;
WRITELN(' FINAL INTEGRAL VALUE =', R[1,K+1]:16);
WRITELN(' WITH RELATIVE ERROR  =',ERR:12, ' %');
END.
```

```pascal
PROGRAM  GAUSINT;
(*....A GAUSS-LEGENDRE INTEGRATION PROGRAM FOR*)
(*....ESTIMATING INTEGRATION OF FUNCTION F(X) *)
(*....USING 1 THROUGH 6 GAUSS POINTS          *)
VAR  XI,W : ARRAY[1..21] OF real;
    A,A0,A1,B,SUM,X : real;
    IC,ITERMS,NG : integer;
(*--------------------------------------------*)
FUNCTION FUNC(X : real) : real;
BEGIN
FUNC := 2.*X*X*X - 5.*X*X + 3.*X + 1.;
```

```pascal
END;
(*.....COMPUTE FUNCTION VALUES AT POINTS.......*)
BEGIN
  FOR I := 1 TO N DO
    BEGIN
      FX[I] := FUNC(X);
      X := X + H;
    END;
(*.....COMPUTE DERIVATIVES AT POINTS..........*)
  DIFF[1] := (FX[2] - FX[1])/H;
  FOR I := 2 TO N-1 DO
    BEGIN
      DIFF[I] := (FX[I+1] - FX[I-1])/(2.*H);
    END;
  DIFF[N] := (FX[N] - FX[N-1])/H;
  WRITELN('        X',',        FX',
          '        DERIVATIVE');
  WRITELN;
  X := A;
  FOR I := 1 TO N DO
    BEGIN
      WRITELN('    ',X :5:2,'    ',FX[I] :10:3,
              '    ',DIFF[I] :10:3);
      X := X + H;
    END;
END.
```

```pascal
PROGRAM EULER;
(*.A PROGRAM FOR SOLVING ORDINARY DIFFERENTIAL.*)
(*.EQUATION USING THE EULER'S METHOD           *)
(*.READ INITIAL CONDITIONS, NUMBER OF STEPS,   *)
(*.AND STEP SIZE:                              *)
VAR H,SLOPE,X,Y : real;
    I,N : integer;
(*------------------------------------------*)
FUNCTION FUNC(X,Y : real) : real;
BEGIN
  FUNC := Y*COS(X);
END;
(*------------------------------------------*)
BEGIN
  READLN(X, Y, N, H);
  WRITELN(' SOLUTION WITH STEP SIZE =',H:10:4,
          '     IS:');
  WRITELN('        X','        Y');
  WRITELN('    ',X : 13,'    ', Y : 13);
  FOR I := 1 TO N DO
    BEGIN
      SLOPE := FUNC(X,Y);
      Y := Y + SLOPE*H;
      X := X + H;
      WRITELN('    ',X : 13,'    ', Y : 13);
    END;
END.
```

```pascal
PROGRAM HEUN;
(*.A PROGRAM FOR SOLVING ORDINARY DIFFERENTIAL.*)
(*.EQUATION USING THE HEUN'S METHOD            *)
(*.READ INITIAL CONDITIONS, NUMBER OF STEPS,   *)
```

```pascal
END;
(*-------------------------------------------*)
BEGIN
  XI[1]  :=  0.0000000;   XI[2]  := -0.5773503;
  XI[3]  :=  0.5773503;   XI[4]  := -0.7745967;
  XI[5]  :=  0.0000000;   XI[6]  := -0.7745967;
  XI[7]  := -0.8611363;   XI[8]  := -0.3399810;
  XI[9]  :=  0.3399810;   XI[10] :=  0.8611363;
  XI[11] := -0.9061798;   XI[12] := -0.5384693;
  XI[13] :=  0.0000000;   XI[14] :=  0.5384693;
  XI[15] :=  0.9061798;   XI[16] := -0.9324695;
  XI[17] := -0.6612094;   XI[18] := -0.2386192;
  XI[19] := +0.2386192;   XI[20] := +0.6612094;
  XI[21] := +0.9324695;
  W[1] :=2.0000000;W[2] :=1.0000000;W[3] :=1.0000000;
  W[4] :=0.5555556;W[5] :=0.8888889;W[6] :=0.5555556;
  W[7] :=0.3478549;W[8] :=0.6521452;W[9] :=0.6521452;
  W[10]:=0.3478549;W[11]:=0.2369269;
  W[12]:=0.4786287;W[13]:=0.5688889;
  W[14]:=0.4786287;W[15]:=0.2369269;
  W[16]:=0.1713245;W[17]:=0.3607616;
  W[18]:=0.4679139;W[19]:=0.4679139;
  W[20]:=0.3607616;W[21]:=0.1713245;
(*.....READ LIMITS OF INTEGRATION            *)
  READLN(A,B);
  A0 := (A + B)/2.;
  A1 := (B - A)/2.;
(*.PERFORM 1 THRU 6 GUASS POINTS COMPUTATION:*)
  IC := 1;
  FOR NG := 1 TO 6 DO
    BEGIN
      SUM := 0.;
      FOR ITERMS := 1 TO NG DO
        BEGIN
          X := A0 + A1*XI[IC];
          AI := FUNC(X);
          SUM := SUM + W[IC]*AI;
          IC := IC + 1;
        END;
      SUM := A1*SUM;
      WRITE(' RESULT OF INTEGRATION WITH', NG : 2);
      WRITELN(' GAUSS POINT(S) IS', SUM :13);
    END;
END.
```

```pascal
PROGRAM NUMDIF;
(*....A NUMERICAL DIFFERENTIATION PROGRAM.....*)
VAR FX,DIFF : ARRAY[1..100] OF real;
    A,B,H,X : real;
    I,N : integer;
(*-------------------------------------------*)
FUNCTION FUNC(X : real) : real;
BEGIN
  FUNC := 2.*X*X*X - 5.*X*X + 3.*X + 1.;
END;
(*-------------------------------------------*)
BEGIN
(*.....READ END LOCATIONS AND NO. OF POINTS:...*)
  READ(A, B, N);
  H := (B - A)/(N - 1);
```

```
PROGRAM RK3;
(*.A PROGRAM FOR SOLVING ORDINARY DIFFERENTIAL.*)
(*.EQUATION BY THIRD-ORDER RUNGE-KUTTA METHOD  *)
(*.READ INITIAL CONDITIONS, NUMBER OF STEPS,   *)
(*.AND STEP SIZE:                              *)
VAR AK1,AK2,AK3,H,X,XX,Y,YY : real;
    I,N : integer;
(*------------------------------------------------*)
FUNCTION FUNC(X,Y : real) : real;
BEGIN
   FUNC := Y*COS(X);
END;
(*------------------------------------------------*)
BEGIN
   READ(X, Y, N, H);
   WRITELN(' SOLUTION WITH STEP SIZE =',H:10,
           ' IS:');
   WRITELN('         X',' ',
                          'Y');
   WRITELN('      ',X : 13,'     ', Y : 13);
   FOR I := 1 TO N DO
     BEGIN
       AK1 := FUNC(X,Y);
       XX  := X + H/2.;
       YY  := Y + H*AK1/2.;
       AK2 := FUNC(XX,YY);
       XX  := X + H;
       YY  := Y - H*AK1 + 2.*H*AK2;
       AK3 := FUNC(XX,YY);
       Y   := Y + (AK1 + 4.*AK2 + AK3)*H/6.;
       X   := X + H;
       WRITELN('      ',X : 13,'     ', Y : 13);
     END;
END.
```

```
PROGRAM RK4;
(*.A PROGRAM FOR SOLVING ORDINARY DIFFERENTIAL.*)
(*.EQUATION BY FOURTH-ORDER RUNGE-KUTTA METHOD *)
(*.READ INITIAL CONDITIONS, NUMBER OF STEPS,   *)
(*.AND STEP SIZE:                              *)
VAR AK1,AK2,AK3,AK4,H,X,XX,Y,YY : real;
    I,N : integer;
(*------------------------------------------------*)
FUNCTION FUNC(X,Y : real) : real;
BEGIN
   FUNC := Y*COS(X);
END;
(*------------------------------------------------*)
BEGIN
   READ(X, Y, N, H);
   WRITELN(' SOLUTION WITH STEP SIZE =',H:10,
           ' IS:');
   WRITELN('         X',' ',
                          'Y');
   WRITELN('      ',X : 13,'     ', Y : 13);
   FOR I := 1 TO N DO
     BEGIN
       AK1 := FUNC(X,Y);
       XX  := X + H/2.;
       YY  := Y + H*AK1/2.;
       AK2 := FUNC(XX,YY);
       YY  := Y + H*AK2/2.;
```

```
(*.AND STEP SIZE:                              *)
VAR H,S0,S1,SA,X,X1,Y,Y1 : real;
    I,N : integer;
(*------------------------------------------------*)
FUNCTION FUNC(X,Y : real) : real;
BEGIN
   FUNC := Y*COS(X);
END;
(*------------------------------------------------*)
BEGIN
   READ(X, Y, N, H);
   WRITELN(' SOLUTION WITH STEP SIZE =',H:10,
           ' IS:');
   WRITELN('         X',' ',
                          'Y');
   WRITELN('      ',X : 13,'     ', Y : 13);
   FOR I := 1 TO N DO
     BEGIN
       S0 := FUNC(X,Y);
       Y1 := Y + S0*H;
       X  := X + H;
       S1 := FUNC(X,Y1);
       SA := (S0 + S1)/2.;
       Y  := Y + SA*H;
       WRITELN('      ',X : 13,'     ', Y : 13);
     END;
END.
```

```
PROGRAM MEULER;
(*.A PROGRAM FOR SOLVING ORDINARY DIFFERENTIAL.*)
(*.EQUATION USING THE MODIFIED EULER'S METHOD  *)
(*.READ INITIAL CONDITIONS, NUMBER OF STEPS,   *)
(*.AND STEP SIZE:                              *)
VAR H,S0,SA,X,X1,Y,Y1 : real;
    I,N : integer;
(*------------------------------------------------*)
FUNCTION FUNC(X,Y : real) : real;
BEGIN
   FUNC := Y*COS(X);
END;
(*------------------------------------------------*)
BEGIN
   READ(X, Y, N, H);
   WRITELN(' SOLUTION WITH STEP SIZE =',H:10,
           ' IS:');
   WRITELN('         X',' ',
                          'Y');
   WRITELN('      ',X : 13,'     ', Y : 13);
   FOR I := 1 TO N DO
     BEGIN
       S0 := FUNC(X,Y);
       Y1 := Y + S0*H/2.;
       X1 := X + H/2.;
       SA := FUNC(X1,Y1);
       Y  := Y + SA*H;
       X  := X + H;
       WRITELN('      ',X : 13,'     ', Y : 13);
     END;
END.
```

```pascal
    AK3  := FUNC(XX,YY);
    XX   := X + H;
    YY   := Y + H*AK3;
    AK4  := FUNC(XX,YY);
    Y    := Y + (AK1 + 2.*AK2 + 2.*AK3 + AK4)*H/6.;
    X    := X + H;
    WRITELN('       ',X : 13,'       ', Y : 13);
  END;
END.

(*------------------------------------------------------*)

PROGRAM SYSEUL;
(*..A PROGRAM FOR SOLVING A SET OF TWO ORDINARY *)
(*..FIRST-ORDER DIFFERENTIAL EQUATIONS USING    *)
(*..THE EULER'S METHOD                          *)
(*..READ INITIAL CONDITIONS, NUMBER OF STEPS,   *)
(*..AND STEP SIZE:                              *)
VAR F1,F2,H,X,Y,Z : real;
    I,N : integer;
(*------------------------------------------------------*)
FUNCTION FUNC1(X,Y,Z : real) : real;
BEGIN
  FUNC1 := Z;
END;
(*------------------------------------------------------*)
FUNCTION FUNC2(X,Y,Z : real) : real;
BEGIN
  FUNC2 := -2.*Z - 4.*Y;
END;
(*------------------------------------------------------*)
BEGIN
  READLN(X, Y, Z, N, H);
  WRITELN(' SOLUTION WITH STEP SIZE =',H:10:4,
          ' IS:');
  WRITELN('       X','       Y',
          'Z');
  WRITELN('       ',X : 13,'       ', Y : 13,'  ',
          Z : 13);
  FOR I := 1 TO N DO
  BEGIN
    F1 := FUNC1(X,Y,Z);
    F2 := FUNC2(X,Y,Z);
    Y  := Y + F1*H;
    Z  := Z + F2*H;
    X  := X + H;
    WRITELN('       ',X : 13,'       ', Y : 13,'  ',
            Z : 13);
  END;
END.

(*------------------------------------------------------*)

PROGRAM SYSRK4;
(*..A PROGRAM FOR SOLVING A SET OF TWO ORDINARY *)
(*..FIRST-ORDER DIFFERENTIAL EQUATIONS USING    *)
(*..THE FOURTH-ORDER RUNGE-KUTTA METHOD         *)
(*..READ INITIAL CONDITIONS, NUMBER OF STEPS,   *)
(*..AND STEP SIZE:                              *)
VAR AK1Y,AK2Y,AK3Y,AK4Y,H,X,XX,Y,YY,Z,ZZ : real;
    AK1Z,AK2Z,AK3Z,AK4Z : real;
    I,N : integer;
```

```pascal
(*------------------------------------------------------*)
FUNCTION FUNC1(X,Y,Z : real) : real;
BEGIN
  FUNC1 := Z;
END;
(*------------------------------------------------------*)
FUNCTION FUNC2(X,Y,Z : real) : real;
BEGIN
  FUNC2 := -2.*Z - 4.*Y;
END;
(*------------------------------------------------------*)
BEGIN
  READ(X, Y, Z, N, H);
  WRITELN(' SOLUTION WITH STEP SIZE =',H:10,
          ' IS:');
  WRITELN('       X','       Y',
          'Z');
  WRITELN('       ',X : 13,'       ', Y : 13,'  ',
          Z : 13);
  FOR I := 1 TO N DO
  BEGIN
    AK1Y := FUNC1(X,Y,Z);
    AK1Z := FUNC2(X,Y,Z);
    XX   := X + H/2.;
    YY   := Y + H*AK1Y/2.;
    ZZ   := Z + H*AK1Z/2.;
    AK2Y := FUNC1(XX,YY,ZZ);
    AK2Z := FUNC2(XX,YY,ZZ);
    YY   := Y + H*AK2Y/2.;
    ZZ   := Z + H*AK2Z/2.;
    AK3Y := FUNC1(XX,YY,ZZ);
    AK3Z := FUNC2(XX,YY,ZZ);
    XX   := X + H;
    YY   := Y + H*AK3Y;
    ZZ   := Z + H*AK3Z;
    AK4Y := FUNC1(XX,YY,ZZ);
    AK4Z := FUNC2(XX,YY,ZZ);
    Y    := Y + (AK1Y+2.*AK2Y+2.*AK3Y+AK4Y)*H/6.;
    Z    := Z + (AK1Z+2.*AK2Z+2.*AK3Z+AK4Z)*H/6.;
    X    := X + H;
    WRITELN('       ',X : 13,'       ', Y : 13,'  ',
            Z : 13);
  END;
END.

(*------------------------------------------------------*)

PROGRAM ADAMBAS;
(*..A PROGRAM FOR SOLVING ORDINARY DIFFERENTIAL.*)
(*..EQ. BY FOURTH-ORDER ADAMS-BASHFORTH METHOD  *)
(*..READ INITIAL CONDITIONS OF X0, Y0, Y-1,     *)
(*..Y-2, Y-3, NUMBER OF STEPS AND STEP SIZE:    *)
VAR F0,F1,F2,F3,H,SLOPE,X,XM1,XM2,XM3 : real;
    Y,YM1,YM2,YM3 : real;
    I,N : integer;
(*------------------------------------------------------*)
FUNCTION FUNC(X,Y : real) : real;
BEGIN
  FUNC := Y*COS(X);
END;
(*------------------------------------------------------*)
BEGIN
```

```
READLN(X, Y, YM1, YM2, YM3, N, H);
WRITELN(' SOLUTION WITH STEP SIZE =',H:10:4,
        ' IS:');
WRITELN('            X',        X',           Y');
WRITELN('   ',X : 13,'    ',Y : 13);
FOR I := 1 TO N DO
  BEGIN
    F0   := FUNC(X,Y);
    XM1  := X - H;
    F1   := FUNC(XM1,YM1);
    XM2  := X - 2.*H;
    F2   := FUNC(XM2,YM2);
    XM3  := X - 3.*H;
    F3   := FUNC(XM3,YM3);
    YM3  := YM2;
    YM2  := YM1;
    YM1  := Y;
    Y    := Y + (55*F0 - 59*F1 + 37*F2 - 9*F3)
                *H/24.;
    X    := X + H;
    WRITELN('   ',X : 13,'    ',Y : 13);
  END;
END.
```

```
PROGRAM ELLIP;
(*...A FINITE DIFFERENCE PROGRAM FOR SOLVING...*)
(*...TEMPERATURE DISTRIBUTION IN A PLATE       *)
LABEL 300,1000;
VAR T : ARRAY[1..9,1..5] OF real;
    DIFF,DX,PI,TEMP,TOL,X : real;
    I,IFLAG,J,ITER,MXITER : integer;
(*...ASSIGN TOLERANCE AND MAX. NO. OF ITERATIONS:*)
BEGIN
  TOL := 0.00001;
  MXITER := 100;
(*...SET UP BOUNDARY CONDITIONS:               *)
  FOR I := 1 TO 9 DO
    BEGIN
      T[I,1] := 0.;
    END;
  FOR J := 1 TO 5 DO
    BEGIN
      T[1,J] := 0.;
      T[9,J] := 0.;
    END;
  X  := 0.25;
  DX := X;
  PI := 4.*ARCTAN(1.0);
  FOR I := 2 TO 8 DO
    BEGIN
      T[I,5] := SIN(PI*X/2.0);
      X := X + DX;
    END;
(*....SET UP INITIAL TEMPERATURE VALUES:        *)
  FOR I := 2 TO 8 DO
    BEGIN
      FOR J := 2 TO 4 DO
        T[I,J] := J*T[I,5]/5
    END;
(*....SOLVE UNKNOWN TEMPERATURES AT GRID POINTS*)
```

```
(*...USING GAUSS-SEIDEL ITERATION TECHNIQUE:   *)
  FOR ITER := 1 TO MXITER DO
    BEGIN
      IFLAG := 0;
      FOR I := 2 TO 8 DO
        BEGIN
          FOR J := 2 TO 4 DO
            BEGIN
              TEMP := ( T[I-1,J] + T[I+1,J] +
                        T[I,J+1] + T[I,J-1] )/4.;
              DIFF := T[I,J] - TEMP;
              IF ABS(DIFF) > TOL THEN
                IFLAG := 1;
              T[I,J] := TEMP;
            END;
        END;
      IF IFLAG = 0 THEN GOTO 300;
    END;
  WRITELN(' SOLUTION NOT CONVERGED WITHIN THE',
          ' SPECIFIED NO. OF ITERATIONS & TOLERANCE');
  GOTO 1000;
300 : WRITELN;
(*.PRINT OUT TEMPERATURES AT GRID POINTS IN THE.*)
(*.FORMAT CORRESPONDING TO THE PROBLEM FIGURE.:*)
  FOR J := 5 DOWNTO 1 DO
    BEGIN
      FOR I := 1 TO 9 DO
        WRITE(T[I,J] : 6 : 3);
      WRITELN;
    END;
1000 : WRITELN;
END.
```

```
PROGRAM PARAEXP;
(*...A FINITE DIFFERENCE PROGRAM FOR SOLVING    *)
(*...TRANSIENT TEMPERATURE DISTRIBUTION IN A    *)
(*...ROD USING EXPLICIT METHOD                  *)
VAR TOLD,TNEW : ARRAY[1..11] OF real;
    ALPHA,DTIME,DX,PI,TIME,X : real;
    I,IP,ISTEP,J,NSTEPS : integer;
(*...ASSIGN TIME STEP AND NO. OF TIME STEPS:    *)
BEGIN
  DTIME := 0.005;
  NSTEPS := 40;
(*...SET UP INITIAL AND BOUNDARY CONDITIONS:    *)
  X  := 0.;
  DX := 0.1;
  PI := 4.*ARCTAN(1.0);
  FOR I := 1 TO 11 DO
    BEGIN
      TOLD[I] := SIN(PI*X);
      TNEW[I] := TOLD[I];
      X := X + DX;
    END;
(*...SOLVE FOR TEMPERATURE RESPONSE:            *)
  ALPHA := DTIME/(DX*DX);
  TIME := DTIME;
  FOR ISTEP := 1 TO NSTEPS DO
    BEGIN
      FOR I := 2 TO 10 DO
        BEGIN
```

```
      TNEW[I] := TOLD[I] + ALPHA*
                 (TOLD[I+1] - 2.*TOLD[I] + TOLD[I-1]);
(*..PRINT OUT AT EVERY 4 STEPS:               *)
    IP := ISTEP - TRUNC(ISTEP/4)*4;
    IF IP = 0 THEN
      BEGIN
        WRITE(TIME : 6 : 2,' ');
        FOR J := 1 TO 11 DO
          WRITE(TNEW[J] : 6 : 4);
        WRITELN;
      END;
    FOR I := 2 TO 10 DO
      TOLD[I] := TNEW[I];
    TIME := TIME + DTIME;
  END.
END.


PROGRAM PARAIMP;
(*..A FINITE DIFFERENCE PROGRAM FOR SOLVING           *)
(*..TRANSIENT TEMPERATURE DISTRIBUTION IN A           *)
(*..ROD USING IMPLICIT METHOD                         *)
VAR TEMP : ARRAY[1..11] OF real;
    A,B,C,D,E : ARRAY[1..9] OF real;
    ALPHA,COEF,DTIME,DX,PI,TIME,X : real;
    I,IP,ISTEP,J,N,NSTEPS : integer;
(*..ASSIGN TIME STEP AND NO. OF TIME STEPS:           *)
BEGIN
  DTIME := 0.01;
  NSTEPS := 20;
(*..SET UP INITIAL AND BOUNDARY CONDITIONS:           *)
  X := 0.;
  DX := 0.1;
  PI := 4.*ARCTAN(1.0);
  FOR I := 1 TO 11 DO
    BEGIN
      TEMP[I] := SIN(PI*X);
      X := X + DX;
    END;
  ALPHA := DTIME/(DX*DX);
  COEF := 1. + 2.*ALPHA;
  TIME := DTIME;
  N := 9;
  FOR ISTEP := 1 TO NSTEPS DO
    BEGIN
(*..FORM UP TRIDIAGONAL SYSTEM OF N EQUATIONS         *)
(*..FOR INTERIOR GRIDS (HERE N=9):                    *)
      B[1] := COEF;
      C[1] := -ALPHA;
      D[1] := TEMP[2];
      FOR I := 2 TO N-1 DO
        BEGIN
          A[I] := -ALPHA;
          B[I] := COEF;
          C[I] := -ALPHA;
          D[I] := TEMP[I+1];
        END;
      A[N] := -ALPHA;
      B[N] := COEF;
      D[N] := TEMP[N+1];
(*..SOLVE SUCH TRIDIAGONAL SYSTEM OF N EQUATIONS      *)
(*..FOR TEMPERATURES AT INTERIOR GRIDS, RETURN        *)
(*..SOLUTION IN E( ).                                 *)
      FOR I := 2 TO N DO
        BEGIN
          A[I] := A[I]/B[I-1];
          B[I] := B[I] - A[I]*C[I-1];
        END;
      FOR I := 2 TO N DO
        D[I] := D[I] - A[I]*D[I-1];
      E[N] := D[N]/B[N];
      FOR I := N-1 DOWNTO 1 DO
        E[I] := (D[I] - C[I]*E[I+1])/B[I];
      FOR I := 2 TO 10 DO
        TEMP[I] := E[I-1];
(*..PRINT OUT AT EVERY 1 STEPS:                       *)
      IP := ISTEP - TRUNC(ISTEP/2)*2;
      IF IP = 0 THEN
        BEGIN
          WRITE(TIME : 6 : 2,' ');
          FOR J := 1 TO 11 DO
            WRITE(TEMP[J] : 6 : 4);
          WRITELN;
        END;
      TIME := TIME + DTIME;
END.


PROGRAM PARACN;
(*..A FINITE DIFFERENCE PROGRAM FOR SOLVING           *)
(*..TRANSIENT TEMPERATURE DISTRIBUTION IN A           *)
(*..ROD USING CRANK-NICOLSON METHOD                   *)
VAR TEMP : ARRAY[1..11] OF real;
    A,B,C,D,E : ARRAY[1..9] OF real;
    ALPHA,COEFP,COEFM,DTIME,DX,PI,TIME,X : real;
    I,IP,ISTEP,J,N,NSTEPS : integer;
(*..ASSIGN TIME STEP AND NO. OF TIME STEPS:           *)
BEGIN
  DTIME := 0.02;
  NSTEPS := 10;
(*..SET UP INITIAL AND BOUNDARY CONDITIONS:           *)
  X := 0.;
  DX := 0.1;
  PI := 4.*ARCTAN(1.0);
  FOR I := 1 TO 11 DO
    BEGIN
      TEMP[I] := SIN(PI*X);
      X := X + DX;
    END;
  ALPHA := DTIME/(DX*DX);
  COEFP := 2.*(1. + ALPHA);
  COEFM := 2.*(1. - ALPHA);
  TIME := DTIME;
  N := 9;
  FOR ISTEP := 1 TO NSTEPS DO
    BEGIN
(*..FORM UP TRIDIAGONAL SYSTEM OF N EQUATIONS         *)
(*..FOR INTERIOR GRIDS (HERE N=9):                    *)
      B[1] := COEFP;
      C[1] := -ALPHA;
      D[1] := COEFM*TEMP[2] + ALPHA*TEMP[3];
```

```
FOR I := 2 TO N-1 DO
    BEGIN
    A[I] := -ALPHA;
    B[I] := COEFP;
    C[I] := -ALPHA;
    D[I] := ALPHA*TEMP[I]+COEFM*TEMP[I+1]
            + ALPHA*TEMP[I+2];
    END;
A[N] := -ALPHA;
B[N] := COEFP;
D[N] := ALPHA*TEMP[N] + COEFM*TEMP[N+1];
(*.SOLVE SUCH TRIDIAGONAL SYSTEM OF N EQUATIONS *)
(*.FOR TEMPERATURES AT INTERIOR GRIDS, RETURN   *)
(*.SOLUTION IN E().                             *)
FOR I := 2 TO N DO
    BEGIN
    A[I] := A[I]/B[I-1];
    B[I] := B[I] - A[I]*C[I-1];
    END;
FOR I := 2 TO N DO
    D[I] := D[I] - A[I]*D[I-1];
E[N] := D[N]/B[N];
FOR I := N-1 DOWNTO 1 DO
    E[I] := (D[I] - C[I+1]*E[I+1])/B[I];
FOR I := 2 TO 10 DO
    TEMP[I] := E[I-1];
(*.PRINT OUT AT EVERY 1 STEPS:                  *)
IP := ISTEP - TRUNC(ISTEP/1)*1;
IF IP = 0 THEN
    BEGIN
    WRITE(TIME : 3 : 2,' ');
    FOR J := 1 TO 11 DO
        WRITE(TEMP[J] : 6 : 4);
    WRITELN;
    END;
TIME := TIME + DTIME;
END.


PROGRAM HYPER;
(*.A FINITE DIFFERENCE PROGRAM FOR SOLVING      *)
(*.VIBRATION IN STRING                          *)
VAR UN,UNM1,UNP1 : ARRAY[1..7] OF real;
    DTIME,DX,TIME,X : real;
    I,ISTEP,J,NSTEPS : integer;
(*.ASSIGN TIME STEP AND NO. OF TIME STEPS:      *)
BEGIN
DTIME := 0.0025;
NSTEPS := 12;
(*.SET UP INITIAL AND BOUNDARY CONDITIONS:      *)
X := 0.0;
DX := 0.25;
TIME := 0.0;
FOR I := 1 TO 5 DO
    BEGIN
    UN[I] := 0.07*X;
    X := X + DX;
    END;
X := 5.0*DX;
FOR I := 6 TO 7 DO
    BEGIN
    UN[I] := 0.21 - 0.14**x;
    X := X + DX;
    END;
WRITE(TIME : 6 : 4,' ');
FOR J := 1 TO 7 DO
    WRITE(UN[J] : 8 : 5,' ');
WRITELN;
TIME := TIME + DTIME;
(*.COMPUTE DISPLACEMENTS AT FIRST TIME STEP:   *)
UNP1[1] := UN[1];
FOR I := 2 TO 6 DO
    UNP1[I] := 0.5*(UN[I+1] + UN[I-1]);
UNP1[7] := UN[7];
WRITE(TIME : 6 : 4,' ');
FOR J := 1 TO 7 DO
    WRITE(UNP1[J] : 8 : 5,' ');
WRITELN;
(*.COMPUTE DISPLACEMENTS AFTER FIRST TIME STEP:*)
TIME := TIME + DTIME;
FOR ISTEP := 2 TO NSTEPS DO
    BEGIN
    FOR I := 1 TO 7 DO
        BEGIN
        UNM1[I] := UN[I];
        UN[I]   := UNP1[I];
        END;
    FOR I := 2 TO 6 DO
        UNP1[I] := -UNM1[I] + UN[I+1] + UN[I-1];
    WRITE(TIME : 6 : 4,' ');
    FOR J := 1 TO 7 DO
        WRITE(UNP1[J] : 8 : 5,' ');
    WRITELN;
    TIME := TIME + DTIME;
    END;
END.


PROGRAM FINITE;
(*                                                            *)
(*   A FINITE ELEMENT COMPUTER PROGRAM FOR SOLVING PARTIAL    *)
(*   DIFFERENTIAL EQUATION IN THE FORM OF POISSON'S EQUATION  *)
(*   FOR TWO-DIMENSIONAL STEADY-STATE HEAT CONDUCTION WITH    *)
(*   INTERNAL HEAT GENERATION.          PROF. DR. PRAMOTE DECHAUMPHAI *)
(*                                      FACULTY OF ENGINEERING *)
(*                                      CHULALONGKORN UNIVERSITY *)
(*                                                            *)
(*   THE VALUES DECLARED IN THE PARAMETER STATEMENT BELOW SHOULD *)
(*   BE ADJUSTED ACCORDING TO THE SIZE OF THE PROBLEMS AND TYPES *)
(*   OF COMPUTERS:                                            *)
(*       MXPOI = MAXIMUM NUMBER OF NODES IN THE MODEL.        *)
(*       MXELE = MAXIMUM NUMBER OF ELEMENTS IN THE MODEL.     *)
(*                                                            *)
LABEL 1000;
CONST MXPOI=100; MXELE=100;
TYPE TABLE  = ARRAY[1..MXPOI] OF real;
     TABLE1 = ARRAY[1..MXPOI,1..MXPOI] OF real;
     TABLE2 = ARRAY[1..3] OF real;
     TABLE3 = ARRAY[1..3,1..3] OF real;
     TABLE4 = ARRAY[1..MXELE,1..3] OF integer;
```

```pascal
   TABLE5 = ARRAY[1..MXPOI] OF integer;
VAR COORD : ARRAY[1..MXPOI,1..2] OF real;
   TEMP,SYSQ : TABLE;
   QELE : ARRAY[1..MXELE] OF real;
   SYSK : TABLE1;
   IBC : TABLE5;
   INTMAT : TABLE4;
   INPUTFILE,OUTPUTFILE : text;
   NAME1, NAME2, TEXT1 : string[80];
   THICK,TK : real;
   I,IE,IP,IQ,ILINE,J,N,NELEM,NEQ,NLINES,NPOIN : integer;            *)
                                                                      *)
(*-------------------------------------------------------------------- *)
PROCEDURE ASSEMBLE(VAR IE : integer;VAR AKC : TABLE3; VAR QQ : TABLE2;
                   VAR SYSK : TABLE1; VAR SYSQ : TABLE);
(*                                                                    *)
(*     ASSEMBLE ELEMENT EQUATIONS INTO SYSTEM EQUATIONS               *)
(*                                                                    *)
VAR IC,ICOL,IR,IROW,NNODE : integer;                                 *)
BEGIN
   NNODE := 3;
   FOR IR := 1 TO NNODE DO
      BEGIN
         FOR IC := 1 TO NNODE DO
            BEGIN
               IROW := INTMAT[IE,IR];
               ICOL := INTMAT[IE,IC];
               SYSK[IROW,ICOL] := SYSK[IROW,ICOL] + AKC[IR,IC];
            END;
         SYSQ[IROW] := SYSQ[IROW] + QQ[IR];
      END;
END;
(*                                                                    *)
(*------------------------------------------------------------------- *)
(*                                                                    *)
PROCEDURE TRI;
(*                                                                    *)
(*     ESTABLISH ALL ELEMENT MATRICES AND ASSEMBLE THEM TO FORM       *)
(*     UP SYSTEM EQUATIONS                                            *)
(*                                                                    *)
LABEL 600;
VAR QQ : TABLE2;
   B : ARRAY[1..2,1..3] OF real;
   BT : ARRAY[1..3,1..2] OF real;
   AKC : TABLE3;
   II,JJ,KK,IE,K : integer;
   AREA,B1,B2,B3,C1,C2,C3,FAC,XG1,XG2,XG3,YG1,YG2,YG3 : real;
(*     LOOP OVER THE NUMBER OF ELEMENTS:                              *)
(*                                                                    *)
BEGIN
   FOR IE := 1 TO NELEM DO
(*     FIND ELEMENT LOCAL COORDINATES:                                *)
(*                                                                    *)
      BEGIN
         II := INTMAT[IE,1];
         JJ := INTMAT[IE,2];
         KK := INTMAT[IE,3];
         XG1 := COORD[II,1];
         XG2 := COORD[JJ,1];
         XG3 := COORD[KK,1];
         YG1 := COORD[II,2];
         YG2 := COORD[JJ,2];
         YG3 := COORD[KK,2];
         AREA := 0.5*(XG2*(YG3-YG1) + XG1*(YG2-YG3) + XG3*(YG1-YG2));
         IF AREA <= 0. THEN
            BEGIN
               WRITELN;
               WRITELN(' !!! ERROR !!! ELEMENT NO.', IE : 5,
               WRITELN('  HAS NEGATIVE OR ZERO AREA ');
               WRITELN('  --- CHECK F.E. MODEL FOR NODAL COORDINATES',
                       '  AND ELEMENT NODAL CONNECTIONS ---');
            END;                                                      *)

         IF AREA <= 0. THEN GOTO 600;                                (*

         B1 := YG2 - YG3;                                            (*
         B2 := YG3 - YG1;
         B3 := YG1 - YG2;
         C1 := XG3 - XG2;
         C2 := XG1 - XG3;
         C3 := XG2 - XG1;

         FOR I := 1 TO 2 DO                                          (*
            FOR J :=1 TO 3 DO
               B[I,J] := 0.;

         B[1,1] := B1;                                               (*
         B[1,2] := B2;
         B[1,3] := B3;
         B[2,1] := C1;
         B[2,2] := C2;
         B[2,3] := C3;

         FOR I := 1 TO 2 DO                                          (*
            FOR J := 1 TO 3 DO
               BEGIN
                  B[I,J] := B[I,J]/(2.*AREA);
                  BT[J,I] := B[I,J];
               END;

(*     ELEMENT CONDUCTION MATRIX:                                    *)

         FOR I := 1 TO 3 DO                                          (*
            FOR J := 1 TO 3 DO
               BEGIN
                  AKC[I,J] := 0.;
                  FOR K := 1 TO 2 DO
                     AKC[I,J] := AKC[I,J] + BT[I,K]*B[K,J];
                  AKC[I,J] := TK*AREA*THICK*AKC[I,J];
               END;

(*     ELEMENT LOAD VECTOR DUE TO INTERNAL HEAT GENERATION:          *)

         FAC := QELE[IE]*AREA*THICK/3.;                              (*
         FOR I := 1 TO 3 DO
            QQ[I] := FAC;

(*     ASSEMBLE THESE ELEMENT MATRICES TO FORM SYSTEM EQUATIONS:     *)

         ASSMBLE(IE,AKC,QQ,SYSK,SYSQ);                               (*

      END;
```

```pascal
      (*                                                            *)
      END;
(*------------------------------------------------------------------*)
(*                                                                  *)
PROCEDURE APPLYBC;
(*                                                                  *)
(*    APPLY TEMPERATURE BOUNDARY CONDITIONS WITH CONDITION CODES OF:*)
(*          0 = FREE TO CHANGE (TO BE COMPUTED)                     *)
(*          1 = FIXED AS SPECIFIED                                  *)
(*                                                                  *)
LABEL 100,200;
VAR IC,IEQ,IR : integer;
(*                                                                  *)
BEGIN
FOR IEQ := 1 TO NPOIN DO
    BEGIN
    IF IBC[IEQ] = 0 THEN GOTO 100;

    (*                                                              *)
    FOR IR := 1 TO NPOIN DO
        BEGIN
        IF IR = IEQ THEN GOTO 200;
        SYSQ[IR] := SYSQ[IR] - SYSK[IR,IEQ]*TEMP[IEQ];
        SYSK[IR,IEQ] := 0.0;
        200 : ;
        END;

    FOR IC := 1 TO NPOIN DO
        SYSK[IEQ,IC] := 0.0;
    SYSK[IEQ,IEQ] := 1.0;
    SYSQ[IEQ] := TEMP[IEQ];

    100 : ;
    END;
(*                                                                  *)
END;
(*                                                                  *)
(*------------------------------------------------------------------*)
PROCEDURE SCALE (N : integer; VAR A : TABLE1; VAR B : TABLE) ;
VAR IC,IE   : integer;
    AMAX,BIG : real;
BEGIN
(*...PERFORM SCALING:                                      ....*)
(*                                                                  *)
FOR IE := 1 TO N DO
    BEGIN
    BIG := ABS(A[IE,1]);
    FOR IC := 2 TO N DO
        BEGIN
        AMAX := ABS(A[IE,IC]);
        IF AMAX > BIG THEN BIG := AMAX
        END;
    FOR IC := 1 TO N DO
        A[IE,IC] := A[IE,IC]/BIG;
    B[IE] := B[IE]/BIG
    END;
(*                                                                  *)
END;
(*                                                                  *)
(*------------------------------------------------------------------*)
PROCEDURE PIVOT (VAR IP,N : integer; VAR A : TABLE1; VAR B : TABLE) ;
VAR I,J,JP        : integer;
    AMAX,BIG,DUMY : real;

BEGIN
(*....PERFORM PARTIAL PIVOTING:                          ....*)
JP := IP;
BIG := ABS(A[IP,IP]);
FOR I := IP+1 TO N DO
    BEGIN
    AMAX := ABS(A[I,IP]);
    IF AMAX > BIG THEN
        BEGIN
        BIG := AMAX;
        JP  := I
        END;
    END;

IF JP <> IP THEN
    BEGIN
    FOR J := IP TO N DO
        BEGIN
        DUMY    := A[JP,J];
        A[JP,J] := A[IP,J];
        A[IP,J] := DUMY
        END;

    DUMY   := B[JP];
    B[JP]  := B[IP];
    B[IP]  := DUMY
    END;

END;
(*                                                                  *)
(*------------------------------------------------------------------*)
(*                                                                  *)
PROCEDURE GAUSS( VAR N : integer; VAR A : TABLE1;VAR B,X : TABLE);
VAR IP,IE,IC : integer;
    SUM,RATIO : real;
BEGIN
(*....PERFORM SCALING :                                    ....*)
SCALE(N,A,B);
(*....FORWARD ELIMINATION:  PERFORM ACCORDING TO *)
(*....THE ORDER OF 'PRIME' FROM 1 TO N-1 :              *)
FOR IP := 1 TO N-1 DO
    BEGIN
    (*....PERFORM PARTIAL PIVOTING:                    ....*)
    PIVOT(IP,N,A,B);
    (*....LOOP OVER EACH EQUATION STARTING FROM THE  *)
    (*....ONE THAT COORESONDS WITH THE ORDER OF      *)
    (*....'PRIME' PLUS ONE:                          ....*)
    FOR IE := IP+1 TO N DO
        BEGIN
        RATIO := A[IE,IP]/A[IP,IP];
        FOR IC := IP+1 TO N DO
            A[IE,IC] := A[IE,IC] - RATIO*A[IP,IC];
        B[IE] := B[IE] - RATIO*B[IP]
        END;
    FOR IE := IP+1 TO N DO
        A[IE,IP] := 0.;
    END;
X[N] := B[N]/A[N,N];
FOR IE := N-1 DOWNTO 1 DO
    BEGIN
    SUM := 0.;
    FOR IC := IE+1 TO N DO
        SUM := SUM + A[IE,IC]*X[IC];
    X[IE] := (B[IE] - SUM)/A[IE,IE];
    END;
```

```pascal
  WRITELN;
  WRITE(' *** THE FINITE ELEMENT MODEL CONSISTS OF', NPOIN : 5,
        '  NODES AND', NELEM : 5,' ELEMENTS ***');             *)

(*     ESTABLISH ALL ELEMENT MATRICES AND ASSEMBLE THEM TO
(*     FORM UP SYSTEM EQUATIONS
(*

  WRITELN;
  WRITELN(' *** ESTABLISHING ELEMENT MATRICES AND',
          '  ASSEMBLING ELEMENT EQUATIONS ***');

(*   TRI;

  WRITELN;
  WRITELN(' *** APPLYING BOUNDARY CONDITIONS OF NODAL',
          '  TEMPERATURES ***');

  APPLYBC;

(*
  WRITELN;
  WRITELN(' *** SOLVING A SET OF SIMULTANEOUS EQUATIONS',
          '  FOR TEMPERATURE SOLUTIONS ***'
  WRITELN('      ( TOTAL OF', NEQ : 5,' EQUATIONS TO BE S
  WRITELN;
  GAUSS(NEQ,SYSK,SYSQ,TEMP);

(*
(*     PRINT OUT NODAL TEMPERATURE SOLUTIONS:

  WRITELN(' PLEASE ENTER FILE NAME FOR TEMPERATURE SOLUT
  READLN(NAME2);
  ASSIGN(OUTPUTFILE,NAME2);
  REWRITE(OUTPUTFILE);
  WRITELN(OUTPUTFILE,'   NODAL TEMPERATURE SOLUTIONS [', NPOIN:5,']:');
  WRITELN(OUTPUTFILE); WRITELN(OUTPUTFILE,'   NODE',' TEMPERATURE');
  WRITELN(OUTPUTFILE);
  FOR IP := 1 TO NPOIN DO
     WRITELN(OUTPUTFILE, IP : 6,'  ', TEMP[IP] : 13);
(*                                                          *)
1000 : CLOSE(OUTPUTFILE);

END
```

```pascal
END;
(*                                                          *)
(*--------------------------------------------------------*)
(*                                                          *)
BEGIN
  WRITELN;
  WRITELN(' PLEASE ENTER THE INPUT FILE NAME:');
  READLN(NAME1);
  ASSIGN(INPUTFILE,NAME1);
  RESET(INPUTFILE);

(*     READ TITLE OF COMPUTATION:                          *)
(*                                                          *)

  READLN(INPUTFILE, NLINES);
  FOR ILINE := 1 TO NLINES DO
     READLN(INPUTFILE,TEXT1);

(*     READ INPUT DATA:                                    *)
(*                                                          *)

  READLN(INPUTFILE, TEXT1);
  READLN(INPUTFILE, NPOIN, NELEM);
  IF NPOIN > MXPOI THEN
     WRITE(' PLEASE INCREASE THE PARAMETER MXPOI TO ', NPOIN : 5);
  IF NPOIN > MXPOI THEN GOTO 1000;
  IF NELEM > MXELE THEN
     WRITE(' PLEASE INCREASE THE PARAMETER MXELE TO ', NELEM : 5);
  IF NELEM > MXELE THEN GOTO 1000;
  READLN(INPUTFILE, TEXT1);
  READLN(INPUTFILE, TK,THICK);
  READLN(INPUTFILE, TEXT1);
  FOR IP := 1 TO NPOIN DO
  BEGIN
     READLN(INPUTFILE, I, IBC[I], COORD[I,1], COORD[I,2], TEMP[I]);
     IF I <> IP THEN
        WRITE(' NODE NO.', IP : 5,'  IN DATA FILE IS MISSING');
     IF I <> IP THEN GOTO 1000;
  END;
  IQ := 0;
  READLN(INPUTFILE, TEXT1);
  FOR IE := 1 TO NELEM DO
  BEGIN
     READLN(INPUTFILE, I,INTMAT[I,1],INTMAT[I,2],INTMAT[I,3], QELE[I]);
     IF I <> IE THEN
        WRITE(' ELEMENT NO.', IE : 5,' IN DATA FILE IS MISSING');
     IF I <> IE THEN GOTO 1000;
     IF QELE[I] <> 0.0 THEN IQ := 1;
  END;
  CLOSE(INPUTFILE);
  WRITELN;
  WRITELN(' THE F.E. MODEL INCLUDES THE FOLLOWING',
          ' HEAT TRANSFER MODE(S):');
  IF IQ = 1 THEN
     WRITELN('     -- HEAT CONDUCTION                     ');
     WRITELN('     -- INTERNAL HEAT GENERATION            ');
(*                                                          *)

  NEQ := NPOIN;
  FOR I := 1 TO NEQ DO
     SYSQ[I] := 0.0;
  FOR I :=1 TO NEQ DO
     FOR J := 1 TO NEQ DO
        SYSK[I,J] := 0.0;
(*                                                          *)
```

Index